D1800009

Progress in Mathematics

Volume 339

Series Editors
Antoine Chambert-Loir ⓘ, *Université Paris-Diderot, Paris, France*
Jiang-Hua Lu, *The University of Hong Kong, Hong Kong SAR, China*
Michael Ruzhansky, *Ghent University, Belgium and Queen Mary University of London, UK*
Yuri Tschinkel, *Courant Institute of Mathematical Sciences, New York, USA*

More information about this series at http://www.springer.com/series/4848

Vladimir Rovenski • Paweł Walczak

Extrinsic Geometry
of Foliations

 Birkhäuser

Vladimir Rovenski ⓘD
Department of Mathematics
University of Haifa
Haifa, Israel

Paweł Walczak ⓘD
Faculty of Mathematics
and Computer Science
University of Lodz
Łódź, Poland

ISSN 0743-1643   ISSN 2296-505X   (electronic)
Progress in Mathematics
ISBN 978-3-030-70069-0   ISBN 978-3-030-70067-6   (eBook)
https://doi.org/10.1007/978-3-030-70067-6

© Springer Nature Switzerland AG 2021
This work is subject to copyright. All rights are reserved by the Publisher, whether the whole or part of the material is concerned, specifically the rights of translation, reprinting, reuse of illustrations, recitation, broadcasting, reproduction on microfilms or in any other physical way, and transmission or information storage and retrieval, electronic adaptation, computer software, or by similar or dissimilar methodology now known or hereafter developed.
The use of general descriptive names, registered names, trademarks, service marks, etc. in this publication does not imply, even in the absence of a specific statement, that such names are exempt from the relevant protective laws and regulations and therefore free for general use.
The publisher, the authors, and the editors are safe to assume that the advice and information in this book are believed to be true and accurate at the date of publication. Neither the publisher nor the authors or the editors give a warranty, expressed or implied, with respect to the material contained herein or for any errors or omissions that may have been made. The publisher remains neutral with regard to jurisdictional claims in published maps and institutional affiliations.

This book is published under the imprint Birkhäuser, www.birkhauser-science.com, by the registered company Springer Nature Switzerland AG
The registered company address is: Gewerbestrasse 11, 6330 Cham, Switzerland

Preface

In ancient times, Euclidean geometry was the basis of a philosophical worldview. Modern geometry has many branches such as differential geometry, among others, and underlies many fields in physics. The key concept of differential geometry is curvature and its relation with the topology of a manifold endowed with such structures as Riemannian, metric-affine, and Finsler. The foundations for such structures were laid by K. Gauss, B. Riemann, P. Finsler, and E. Cartan. The Einstein theory of relativity continues to have a great influence on the development of differential geometry.

Recently, interest in geometric flows on manifolds has increased, thanks to the works of R. Hamilton, G. Perelman, and many others. Poincaré's work on the curves defined by a dynamical system on the torus serves as the starting point of topology of foliations. Just as dynamical systems constitute the geometric theory of ordinary differential equations, foliations can be considered the geometric theory of certain partial differential equations. A systematic study of foliations began in the 1940s in a series of papers by G. Reeb and Ch. Ehresmann, culminating in the article [Re2]. Since then, the subject has enjoyed rapid development. Foliations are related to such topics in differential geometry as distributions, almost-product structures, submersions, fiber bundles, pseudogroups, characteristic classes; see [Bo, CC, Mo2]. Explicit constructions (suspensions, pullbacks, tangential and transverse gluings, turbulizations, etc.) provide, in particular, Hopf and Reeb foliations. The theory of foliations has at least three aspects: topological, dynamical, and geometric. The topology of foliations has been described in a number of books such as [CN, CC, HH, Tam]. The second author published a book [W1] on dynamics of foliations; also, several aspects of geometry of foliations (on Riemannian manifolds) have been described in several books such as [BF, GW, Mol, Rei, To] and in the book [R1] by the first author.

The present book is devoted to particular geometric problems of foliation theory, namely, those related to *extrinsic geometry*, which, roughly speaking, describes how the leaves (or single submanifolds) are located in the ambient Riemannian manifold. The properties of a foliation (totally geodesic, totally umbilical and harmonic, or minimal) that can be expressed in terms of the second fundamental tensor of the leaves and its invariants belong to extrinsic geometry.

One of the principal problems of extrinsic geometry of foliations is as follows:

Given a foliation $\mathscr{F}$ on a manifold M and an extrinsic geometric property (P), does there exist a Riemannian metric g on M such that $\mathscr{F}$ enjoys (P) with respect to g?

This problem (first posed by H. Gluck for geodesic foliations) was studied in the 1970s, when D. Sullivan [Su1, Su2] provided a topological condition (called *topological tautness*) for a foliation, equivalent to geometrical tautness, that is existence of a Riemannian metric making all the leaves minimal. It follows from this result and the classical theorem of S. Novikov [No] that the three-dimensional sphere S^3 does not admit a two-dimensional foliation that is minimal with respect to some Riemannian metric. Another example of this type belongs to R. Langevin and P. Walczak [LaW]: closed Riemannian manifolds of negative Ricci curvature admit no codimension-one totally umbilical foliations. In recent decades, several tools providing the results of this sort have been developed, e.g., foliated cycles, integral formulae, variation formulae, and extrinsic geometric flows on foliations; see the surveys [R13, RW2] and articles [ARW, LuW, R4, R5, R8, R10, RW5, RW7, RWo1, RZ1, RZ2, RZ3, RZe2, RZe3].

Although Riemannian curvature belongs to intrinsic geometry, a special part called the mixed curvature is also part of the extrinsic geometry of foliations [R1, R2, R3, R6]. The *mixed sectional curvature* K_{mix} (of a plane spanned by a tangent vector and a vector orthogonal to $\mathscr{F}$) relates through the Jacobi equation to the deviation of leaves. For example, foliations produced by Reeb vector fields on Sasakian manifolds have $K_{\mathrm{mix}} \equiv \mathrm{const}$, and totally geodesic foliations with $K_{\mathrm{mix}} \equiv 0$ split. The theory of foliations with nonnegative mixed curvature is less systematic than the theory of Riemannian manifolds of nonnegative curvature.

This book focuses on the concept of mixed curvature, associated with Toponogov's problem on foliations with $K_{\mathrm{mix}} > 0$, the partial Ricci flow, and the modified theory of gravity based on the mixed scalar curvature S_{mix}. Thus, the book examines a version of the deep problem of understanding the Ricci curvature (described in the books by A. L. Besse "Einstein Manifolds" and M. Gromov "Metric Structures for Riemannian and Non-Riemannian Spaces") for the case of mixed curvature of foliations. The mean value of sectional curvatures over all mixed planes containing a given unit vector of a distribution is called the *partial Ricci curvature*. The trace of the partial Ricci curvature tensor $\mathrm{Ric}^\perp$ is S_{mix}, whose investigation led to multiple results regarding the existence of foliations with interesting geometry [BW, W4]. One of the goals of the book is to argue that the study of $K_{\mathrm{mix}}, \mathrm{Ric}^\perp$, and S_{mix} is a fundamental problem of extrinsic geometry of foliations.

The authors have long been interested in the extrinsic geometry of foliations, and this work, in a sense, is an expression of this interest, as well as a continuation of their earlier book [RW2], devoted to some of these problems for codimension-one foliations. It contains many original results and new investigations of the authors. The topics discussed in the book, divided into five chapters, center around extrinsic geometry as a modern branch of Riemannian geometry, dealing with integral and variation formulas, curvature, and dynamics of foliations. Extrinsic differential geometry is considered by the authors to be a bridge between the geometries of submanifolds and of foliations. The readers can find several classical and recent books devoted to these two subjects and to global Riemannian geometry. The central problem discussed here is that of prescribing the extrinsic geometry and (mean and mixed) curvature of foliated and almost-product manifolds. Different approaches and methods in solving the problem are described: using integral formulas, variation formulas, and extrinsic geometric flows, generalizations of Ricci and scalar curvatures, pseudo-Riemannian and metric-affine geometries, and "computable" Finsler metrics; see [R9, R10, RW6, RW7, SS].

The book is intended for both new and experienced mathematicians working in Riemannian geometry, theory of foliations, differential topology, and related fields as well as for students interested in these topics. We hope that the book will inspire scientists in searching for interesting topics to study. In the Bibliography, we list only references most related to the material treated in the book and apologize if some important related works have been missed.

Acknowledgments

The book is a recapitulation of several years (even, decades) of our research on the subject of extrinsic geometry of foliations and arbitrary distributions. Thus, here we list the names and express our gratitude to all our collaborators and students in this field: K. Andrzejewski, F. Brito, V.P. Golubyatnikov, S. Dragomir, B. Hajduk, R. Langevin, M. Luźyńczyk (now, Wojciechowska), D.S. Patra, P. Popescu, V. Sharafutdinov, P. Schweitzer, S.E. Stepanov, R. Wolak, T. Zawadzki, L. Zelenko, and D. Blanc. Best thanks to all of you for successful cooperation in the course of previous studies.

Our research has received ongoing and substantial support from the administration of our universities in Lodz and Haifa—we thank them for this support.

The authors warmly thank Elizabeth Loew and Ann Kostant and the whole Springer team for their support in the publishing process.

In addition, our best thanks are due to our wives: Irina Albinsky for her logistic help in organization of conferences and our mutual trips to our universities and Zofia Walczak for her help in the use of TeX.

Haifa, Israel Vladimir Rovenski
Łódź, Poland Paweł Walczak
November 2020

Contents

1 Foliations and the Mixed Curvature 1
  1.1 Foliations and Holonomy 1
      1.1.1 Basic Notions 2
      1.1.2 Holonomy 4
      1.1.3 Saturated and Minimal Sets, Generic Leaves 9
  1.2 Metric Structures on Manifolds 12
      1.2.1 Riemannian Structure.. 13
      1.2.2 Finsler Structure 20
  1.3 Basic of the Extrinsic Geometry of Foliations 23
      1.3.1 Fundamental Tensors 24
      1.3.2 The Mixed Curvature 27
  1.4 The Partial Ricci Curvature 32
      1.4.1 Structures with Constant Partial Ricci Curvature 32
      1.4.2 Prescribing the Partial Ricci Curvature 36
      1.4.3 The Weighted Mixed Curvature 40
      1.4.4 Toponogov Conjecture.. 43
  1.5 Appendix 48
      1.5.1 Tensors and Differential Forms.. 48
      1.5.2 Frobenius Theorem 54
      1.5.3 The Elementary Symmetric Functions 57

2 Integral Formulas 61
  2.1 Codimension One Foliations of Riemannian Manifolds 61
      2.1.1 Using a Family of Diffeomorphisms 63
      2.1.2 Applications 65
      2.1.3 Using the Divergence Theorem.. 67
  2.2 Foliations and Singularities 71
      2.2.1 Adapted Singular Foliations 72
      2.2.2 Improper Integrals 74
      2.2.3 Civilized Foliations 76
  2.3 Foliations of Arbitrary Codimension.. 79

        2.3.1   Using a Family of Diffeomorphisms 79
        2.3.2   Using the Divergence Theorem........................ 84
        2.3.3   Splitting of Weighted Generalized Products 90
        2.3.4   Multi-Product Structures 91
    2.4   Foliations of Metric-Affine Manifolds 96
        2.4.1   Integral Formulas with the Mixed Scalar Curvature 96
        2.4.2   Integral Formula with the Ricci Curvature 99
        2.4.3   Splitting Results 101
    2.5   Codimension One Foliations of Finsler Spaces 103
        2.5.1   The Generalization of (α, β)-Norm 104
        2.5.2   The Modified Scalar Product.......................... 108
        2.5.3   The Shape Operator 114
        2.5.4   Around the Reeb Integral Formula and Its Counterpart 118

3   Prescribing the Mean Curvature 123
    3.1   Minimal Submanifolds 123
    3.2   Tautness of Foliations 124
        3.2.1   Rummler Formula 124
        3.2.2   Foliation Currents................................. 126
        3.2.3   Tautness... 128
        3.2.4   Tautness and Holonomy............................. 130
    3.3   Prescribing Mean Curvature in Codimension One 133
        3.3.1   Novikov Components 134
        3.3.2   Consequences of Rummler Formula 136
        3.3.3   Characterization of Mean Curvature Functions........... 136
    3.4   Prescribing Mean Curvature in Higher Codimension 140
        3.4.1   Notation... 141
        3.4.2   Away from Singularities 143
        3.4.3   At Singular Sets 148

4   Variational Formulae 153
    4.1   "Optimally Placed" Distributions 153
    4.2   Adapted Variations of Metric 156
        4.2.1   Variational Formulae 157
        4.2.2   Euler–Lagrange Equations for the Total S_{mix} 161
        4.2.3   Particular Cases 167
    4.3   General Variations of Metric 175
        4.3.1   Variational Formulae 176
        4.3.2   Euler–Lagrange Equations for the Total S_{mix} 178
        4.3.3   Particular Cases 181
    4.4   Einstein–Hilbert Type Action 188
        4.4.1   Variable Metric 189
        4.4.2   The Mixed Field Equations for Space-Times 192
        4.4.3   Variable Connection................................ 194
    4.5   The Godbillon-Vey Type Invariant 198

	4.5.1	Construction	200
	4.5.2	Variations of (ω, T) and the Index Form	202
	4.5.3	Integrability in Average	206
	4.5.4	Concordance and Homotopy	209
	4.5.5	Critical Foliations	211
	4.5.6	Around the Reinhart–Wood Formula	215
	4.5.7	The Bott Invariant.....................................	219
	4.5.8	Higher Dimensional Cases	220

5 Extrinsic Geometric Flows 223
	5.1	Prescribing the Mean Curvature Vector	223
		5.1.1	$\widetilde{\mathscr{D}}$- and $\mathscr{D}$-Related Geometric Quantities	225
		5.1.2	Existence and Uniqueness	228
		5.1.3	The Codimension-One Case	233
		5.1.4	The Doubly Twisted Products	235
	5.2	Flows of Metrics on Codimension-One Foliations	235
		5.2.1	$g^{\top}$-Variations of Mean Curvatures	237
		5.2.2	The Extrinsic Geometric Flow Depending on $\{f_m\}$	238
		5.2.3	The Generalized Companion Matrix	240
		5.2.4	Searching for Power Sums	242
		5.2.5	Existence and Uniqueness	244
		5.2.6	Extrinsic Geometric Flow on a Foliated Surface..........	249
	5.3	The Partial Ricci Flow	251
		5.3.1	Preliminaries	251
		5.3.2	Time-Dependent Adapted Metrics	253
		5.3.3	The Leafwise Laplacian of the Curvature Tensor	255
		5.3.4	Toward the Linearization of the Partial Ricci Flow	258
		5.3.5	Evolution of the Curvature Tensor	260
		5.3.6	Evolution of the Extrinsic Geometry	263
		5.3.7	Examples with (Co)Dimension One Foliations	264
		5.3.8	Around an Almost Contact Structure	268
	5.4	Prescribing the Mixed Scalar Curvature	272
		5.4.1	Leafwise Constant Mixed Scalar Curvature	273
		5.4.2	$\mathscr{D}$-Conformal Change of Metric	275
		5.4.3	$\mathscr{D}$-Conformal Flows of Metrics	277
		5.4.4	Prescribing S_{mix} on Warped Products	279
		5.4.5	Prescribing S_{mix} by a $\mathscr{D}$-Conformal Change of Metric	285
	5.5	The Nonlinear Heat Equation................................	288
		5.5.1	Parabolic PDE's	288
		5.5.2	Stabilization of Solutions of the Nonlinear Heat Equation ..	292

References 303

Index 313

About the Authors 319

List of Figures

1.1   Tangential and transverse boundaries 4
1.2   A chain of plaques... 4
1.3   A Reeb component 10
1.4   Six surfaces: (**a**) plane, (**b**) Loch Ness monster, (**c**) cylinder, (**d**) Jacob ladder, (**e**) Cantor tree, (**f**) blooming Cantor tree.......... 11
1.5   (**a**) A resilient leaf. (**b**) Pair of pants 12
1.6   Indicatrix of Minkowski norm F 20
1.7   Twisted product 31
1.8   The $\mathscr{F}$-parallel Jacobi field along the "extremal" geodesic 47
1.9   A contact structure $\ker(dz + xdy)$ in $\mathbb{R}^3$ 56

2.1   Two directions $\mathbf{n}$ and $\mathbf{n}'$ normal to W 108
2.2   Dependence of $s = \beta(n)$ on $\beta(N)$ for the norms: (**a**) Kropina, (**b**) Matsumoto, (**c**) quadratic, and (**d**) exponential 113

3.1   A submanifold spanned by X and $\mathscr{F}$ 145
3.2   A vector field, which cannot be mean curvature 146
3.3   Some non-gradient phenomena 150

4.1   Twisting of the leaves (the helical wobble of a foliation) 199
4.2   Construction of η 201
4.3   Reeb foliation: strip, annulus and torus........................ 212
4.4   Family of solutions $f(r)$ to (4.129) with $A_0 = 1, A_2 = 0$ and $A_1 = i/8$ ($i = 1 \ldots 5$), producing singular Reeb foliations by rotation about f-axis 214

5.1   Graphs of functions $f_\pm(y) = P_{\phi\pm}(y^2)/y^3$ and f'. (**a**) $S_{\mathfrak{F}^\top} > 0$, $\varPsi_1 > 0$ and $\varPsi_2 > 0$;   (**b**) $S_{\mathfrak{F}^\top} < 0$, $\varPsi_1 < 0$ and $\varPsi_2 > 0$ 287
5.2 $f_\pm(y) = P_{\phi\pm}(y^2)/y^3$, with $\beta < 0$ and $4|\beta|\varPsi_2 < \varPsi_1^2$. (**a**) Graphs of $f_\pm, f'_\pm$ for $S_{\mathfrak{F}^\top} \equiv 0$, $\varPsi_1 > 0$, $\varPsi_2 > 0$. (**b**) y_1—unstable, y_2—stable... 287
5.3   Example 5.113: (**a**) $\beta < 0$, (**b**) $\beta > 0$........................ 296

Chapter 1
Foliations and the Mixed Curvature

In this introductory chapter several notions from differential geometry (vector fields, differential forms and tensor fields etc.) are discussed in brief. The reader is assumed to have the necessary background in topology, manifolds and differential geometry. Some knowledge of differential equations would also be helpful. For the reader's convenience, we provide also the basics of foliation theory. This chapter contains a concise survey of notions and facts concerning these topics. We present structures with constant partial Ricci curvature (playing the key role in the book) and define several functions which interpolate between the weighted sectional and Ricci curvature. Toponogov's problem on totally geodesic foliations with positive mixed sectional curvature is discussed with relation to the weighted curvature.

1.1 Foliations and Holonomy

Foliations, which are defined as partitions of a manifold M into collections of submanifolds of the same dimension, called leaves, appeared in 1940s in the works of G. Reeb and Ch. Ehresmann. Since then, the subject has enjoyed a rapid development, see e.g. [CC]. The leaves of a foliation $\mathscr{F}$ are tangent to involutive (or integrable) distribution $T\mathscr{F} = \widetilde{\mathscr{D}}$ — subbundle of the tangent bundle TM. Foliations relate to such topics as vector fields, submersions, fiber bundles, pseudogroups, Lie groups actions; many models in physics are foliated.

Here, we offer a brief overview of the basics of the foliations theory: we list several definitions (e.g., these of a foliation and its holonomy), mention some examples and a few results, some of them together with short proofs that are supposed to show how complicated the topology of foliations can be. This should convince the reader that also extrinsic geometry of foliations is non-trivial: in general, topologically complicated foliations cannot be built of leaves carrying "nice" geometry.

© Springer Nature Switzerland AG 2021
V. Rovenski, P. Walczak, *Extrinsic Geometry of Foliations*, Progress in Mathematics 339, https://doi.org/10.1007/978-3-030-70067-6_1

1.1.1 Basic Notions

Roughly speaking, a p-dimensional *foliation* of a manifold M, $0 < p < n = \dim M$, is a decomposition of M into a family $\mathscr{F}$ of pairwise disjoint, connected, p-dimensional submanifolds, called *leaves*, of M, located in M "regularly". The simplest example of a foliation is provided by a single submersion $F : M \to N$, where M and N are manifolds. Such a foliation consists of the connected components of the fibres $F^{-1}(x), x \in N$, which are $(n - q)$-dimensional ($q = \dim N$) submanifolds of M.

General foliations are defined locally by submersions. Their fibres should fit nicely when intersect to produce differentiable submanifolds, leaves of a foliation. This property can be expressed by the existence of local homeomorphisms of the base space which should connect one of such submersions to another. That is, if the domains of two such submersions F_1 and F_2 overlap, then any point of the intersection of these domains admits a neighborhood in which $F_1 = \phi \circ F_2$ for some local homeomorphism ϕ of the base space. All such maps ϕ generate a pseudogroup, that is, a family of transformations satisfying the following.

Definition 1.1. A family $\mathscr{G}$ of homeomorphisms between open subsets of a given topological space X is said to be a *pseudogroup* if it is closed under composition, inversion, restriction to open subdomains and unions. More precisely, $\mathscr{G}$ should satisfy the following conditions:

(i) $g \circ h \in \mathscr{G}$, whenever g and $h \in \mathscr{G}$,
(ii) $g^{-1} \in \mathscr{G}$, whenever $g \in \mathscr{G}$,
(iii) $g|_U \in \mathscr{G}$, whenever $g \in \mathscr{G}$ and $U \subset D_g$ is open,
(iv) if $g : D_g \to R_g$ is a homeomorphism between open subsets D_g and R_g of X, $\mathscr{U}$ is an open cover of D_g and $g|U \in \mathscr{G}$ for any $U \in \mathscr{U}$, then $g \in \mathscr{G}$.

Moreover, we shall always assume that

(v) $\mathrm{id}_X \in \mathscr{G}$ (or, equivalently, $\bigcup\{D_g : g \in \mathscr{G}\} = X$).

For any set Γ of local homeomorphisms of X satisfying $\bigcup\{D_g \cup R_g : g \in \Gamma\} = X$ there exists a unique smallest pseudogroup $\mathscr{G}(\Gamma)$ containing Γ: $g \in \mathscr{G}(\Gamma)$ if and only if $g : D_g \to R_g$ is a homeomorphism and for any $x \in D_g$ there exist maps $g_1 \ldots, g_n \in \Gamma$, exponents $e_1 \ldots, e_n \in \{\pm 1\}$ and an open neighborhood U of x, such that $U \subset D_g$ and $g|_U = g_1^{e_1} \circ \cdots \circ g_n^{e_n}|_U$.

Definition 1.2. $\mathscr{G}(\Gamma)$ is said to be *generated* by Γ. If $\mathscr{G} = \mathscr{G}(\Gamma)$ for a finite set Γ, then $\mathscr{G}$ is said to be *finitely generated*.

A pseudogroup generated by the transition maps between local submersions defining a foliation $\mathscr{F}$ is called the *holonomy pseudogroup* of the foliation. In the case of a foliation defined by a single submersion, its holonomy is trivial and is generated just by the identity. If M is compact, then holonomy pseudogroups of all foliations of M are finitely generated. The property discussed above can also be expressed in terms of good (called distinguished or foliated) charts on a foliated manifold. To make the above rough description accurate, we shall formulate a number of definitions. So, let M be an n-dimensional manifold, possibly with non-empty boundary ∂M. Recall that all the manifolds here are supposed to be Hausdorff, paracompact and smooth, i.e., C^∞-differentiable.

Definition 1.3. A p-dimensional C^r-*foliation* $\mathscr{F}$ $(r = 0, 1, \ldots, \infty)$ of M is a decomposition of M into connected submanifolds (called *leaves*) such that for any $x \in M$ there exists a C^r-differentiable chart $\phi = (\phi', \phi'') : U \to \mathbb{R}^n = \mathbb{R}^p \times \mathbb{R}^q$ defined on a neighborhood U of x and satisfying the condition

(i) for any leaf L of $\mathscr{F}$ the connected components (called *plaques*) of $L \cap U$ are given
by the equation $\phi'' = \text{const}$.

Charts satisfying (i) are said to be *distinguished* (or, *foliated*) by $\mathscr{F}$. An atlas built of foliated charts is also said to be *foliated*. The number $q = \text{codim}\, \mathscr{F} = n - p$ is the *codimension* of $\mathscr{F}$.

Generally speaking, the manifold topology of leaves is stronger than that induced from M. Leaves for which these topologies coincide are called *proper*. Certainly, compact leaves are proper. Simple examples show that non-compact proper leaves may exist. In terms of distinguished charts U this property can be expressed as follows: A leaf L is proper if and only if any plaque $P \subset U \cap L$ has an open neighborhood $V \subset U$ such that $V \cap L = P$. If the boundary ∂M of M is non-empty, then any point $x \in \partial M$ should be contained in the domain of a distinguished chart $\phi : U \to \mathbb{R}^n$ which maps

(1) U either into $\mathbb{R}^p \times \mathbb{R}^q_+$ or into $\mathbb{R}^p_+ \times \mathbb{R}^q$ (where $\mathbb{R}^k_+ = \{(u_1, \ldots, u_k); u_k \geq 0\}$),
(2) $U \cap \partial M$ either into $\mathbb{R}^p \times \mathbb{R}^q_0$ or into $\mathbb{R}^p_0 \times \mathbb{R}^q$ (where $\mathbb{R}^k_0 = \{(u_1, \ldots, u_k); u_k = 0\}$).

The plaques containing x coincide either with $\phi^{-1}(\mathbb{R}^p \times \{u_0\})$ or with $\phi^{-1}(\mathbb{R}^p_+ \times \{v\})$, where $u_0 \in \mathbb{R}^q_0$ and $v \in \mathbb{R}^q$. Therefore, the boundary ∂M decomposes as $\partial^\top M \cup \partial^\pitchfork M$, where $\partial^\top M$ is a union of leaves, while $\mathscr{F}$ is transverse to ∂M at any point $x \in \partial^\pitchfork M$. $\partial^\top M$ is called the *tangential boundary*, while $\partial^\pitchfork M$ the *transverse boundary* of $(M, \mathscr{F})$ (Figure 1.1). Clearly, both $\partial^\top M$ and $\partial^\pitchfork M$ are unions of connected components of ∂M. The situation can be different if we allow M to have some "corners", see [CC]. If $\mathscr{A}$ is an atlas on M such that

(ii) for any $\phi, \psi \in \mathscr{A}$ the mapping $\phi \circ \psi^{-1} : V \to \mathbb{R}^p \times \mathbb{R}^q$ $(V \subset \mathbb{R}^p \times \mathbb{R}^q)$ is of the
form

$$\phi \circ \psi^{-1}(x, y) = (\alpha(x, y), \gamma(y)), \quad x \in \mathbb{R}^p, \ y \in \mathbb{R}^q, \tag{1.1}$$

then there exists a unique foliation $\mathscr{F}$ on M such that all the charts of $\mathscr{A}$ are distinguished by $\mathscr{F}$. Any foliated atlas is contained in a maximal (with respect to the relation "$\subset$") one. Such a maximal atlas $\mathscr{A}$ determines a unique foliation $\mathscr{F}$: the leaves of $\mathscr{F}$ are unions of plaques of charts of $\mathscr{A}$, Figure 1.2. Two plaques P and P' belong to the same leaf L, whenever they can be connected by a chain $P_0, \ldots, P_k$ of plaques in such a way that $P = P_0$, $P' = P_k$ and $P_i \cap P_{i+1} \neq \emptyset$ for $i = 0, \ldots, k-1$.

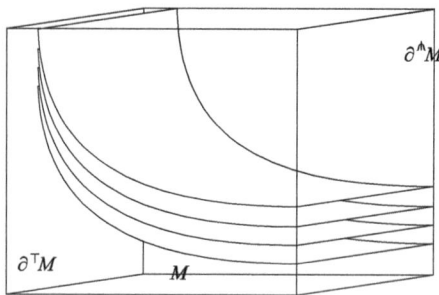

Fig. 1.1 Tangential and transverse boundaries

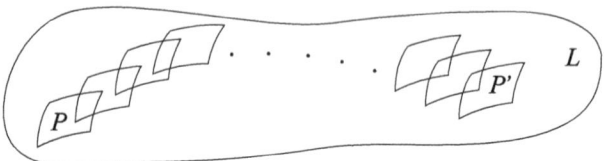

Fig. 1.2 A chain of plaques

1.1.2 Holonomy

Holonomy is one of the most important tools for studying foliations, especially for studying their topology. However, as we shall see later on in this book, for instance in Chapter 3, it has some influence also on the foliation geometry.

Definition 1.4. A foliated atlas $\mathscr{A}$ is said to be *nice* or *regular* (also, *nice* or *regular* is the covering of M by the domains D_ϕ of the charts $\phi \in \mathscr{A}$) if

(i) the covering $\{D_\phi : \phi \in \mathscr{A}\}$ is locally finite,
(ii) for any $\phi \in \mathscr{A}$, $R_\phi = \phi(D_\phi) \subset \mathbb{R}^n$ is an open cube,
(iii) if $\phi, \psi \in \mathscr{A}$ and $D_\phi \cap D_\psi \neq \emptyset$, then there is a chart χ distinguished by $\mathscr{F}$ and such that R_χ is an open cube, D_χ contains the closure of $D_\phi \cup D_\psi$ and $\phi = \chi|_{D_\phi}$.

Since manifolds are assumed to be paracompact here, they are separable, hence nice coverings are denumerable. Nice coverings on compact manifolds are finite.

Lemma 1.5. *For any foliation $\mathscr{F}$ on any manifold M nice coverings of M exist.*

Proof. Let $\mathscr{A}$ be a foliated atlas on M such that R_ϕ is a cube for any $\phi \in \mathscr{A}$. Denote by $\mathscr{U}$ the open covering of M by the domains of all the charts of $\mathscr{A}$. Since M is paracompact, we can find a locally finite open covering $\mathscr{V}$ of M subordinated to $\mathscr{U}$ and such that for any $V \in \mathscr{V}$ there exists $U \in \mathscr{U}$ for which

$$\bigcup \{W \in \mathscr{V} : W \cap V \neq \emptyset\} \subset U. \tag{1.2}$$

For any $V \in \mathcal{V}$, choose $U \in \mathcal{U}$ satisfying (1.2) and let $\phi' = \phi|V$, where ϕ is a chart of $\mathscr{A}$ defined on U. Denote by $\mathscr{A}'$ the foliated atlas built of all the charts ϕ' obtained in this way. Now, for any $\phi' \in \mathscr{A}'$ find a locally finite covering of $R_{\phi'}$ by open cubes Q contained in $R_{\phi'}$ together with the closures and construct a new atlas $\mathscr{A}''$ consisting of all the restrictions of the charts $\phi' \in \mathscr{A}'$ to the sets of the form $\phi'^{-1}(Q)$. $\mathscr{A}''$ and the covering $\mathcal{U}'' = \{D_{\phi''}; \phi'' \in \mathscr{A}''\}$ are nice. $\square$

Now, fix a nice covering $\mathcal{U}$ of a foliated manifold $(M, \mathscr{F})$. For any $U \in \mathcal{U}$, let T_U be the space of the plaques of $\mathscr{F}$ contained in U. Equip $T_U = U/(\mathscr{F}|_U)$ with the quotient topology: two points of U are equivalent iff they belong to the same plaque. T_U is homeomorphic (C^r-diffeomorphic when $\mathscr{F}$ is C^r-differentiable and $r \geq 1$) to an open cube $Q \subset \mathbb{R}^q$ ($q = \mathrm{codim}\,\mathscr{F}$) via the mapping ϕ'', where $\phi = (\phi', \phi'') : U \to \mathbb{R}^p \times \mathbb{R}^q$ is a distinguished chart on U.

Definition 1.6. The disjoint union $T = \bigsqcup_{U \in \mathcal{U}} T_U$ is called a *complete transversal* for $\mathscr{F}$.

Transversality refers to the fact that, if $\mathscr{F}$ is differentiable, each of the spaces T_U can be mapped homeomorphically onto a C^r-submanifold $T'_U \subset U$ transverse to U: if $x \in T'_U$ and L is the leaf of $\mathscr{F}$ passing through x, then $T_x M = T_x T'_U \oplus T_x L$. Completeness of T means that every leaf of $\mathscr{F}$ intersects at least one of the submanifolds T'_U.

Definition 1.7. Given two sets U and $V \in \mathcal{U}$ such that $U \cap V \neq \emptyset$ the *holonomy map* $h_{VU} : D_{VU} \to T_V$, D_{VU} being the open subset of T_U which consists of all the plaques P of U such that $P \cap V \neq \emptyset$, is defined in the following way:

$$h_{VU}(P) = P' \iff \text{the plaques } P \subset U \text{ and } P' \subset V \text{ intersect.} \qquad (1.3)$$

From (v), it follows that the mapping h_{VU} is well defined on an open subset D_{VU} of T_U. In fact, any plaque of U intersects at most one plaque of V. From (1.1), it follows that h_{VU} maps D_{VU} homeomorphically (C^r-diffeomorphically when $\mathscr{F}$ is C^r-differentiable and $r \geq 1$) onto an open subset D_{UV} of T_V. Clearly, $h_{UV} = h_{VU}^{-1}$ and $h_{WU} = h_{WV} \circ h_{VU}$, whenever all the holonomy maps involved are defined. All the maps h_{UV} ($U, V \in \mathcal{U}$) generate a pseudogroup $\mathscr{H}$ on T. $\mathscr{H}$ is called the *holonomy pseudogroup* of $\mathscr{F}$.

To continue, let us consider two pseudogroups $\mathscr{G}$ and $\mathscr{H}$ of local homeomorphisms on X and Y, respectively.

Definition 1.8. A family Φ of homeomorphisms $\phi : D_\phi \to R_\phi$ between open sets $D_\phi \subset X$ and $R_\phi \subset Y$, is a *morphism* of $\mathscr{G}$ into $\mathscr{H}$, $\Phi : \mathscr{G} \to \mathscr{H}$, if the domains D_ϕ, $\phi \in \Phi$, cover X and $\phi \circ g \circ \psi^{-1} \in \mathscr{H}$ for all $g \in \mathscr{G}$ and $\phi, \psi \in \Phi$. Here, Φ is an *isomorphism* if $\Phi^{-1} = \{\phi^{-1} : \phi \in \Phi\}$ is a morphism of $\mathscr{H}$ into $\mathscr{G}$. In this case,

$$\bigcup_{\phi \in \Phi} D_\phi = X, \quad \bigcup \phi \in \Phi R_\phi = Y,$$
$$\phi \circ g \circ \psi^{-1} \in \mathscr{H} \quad (\phi \in \Phi, \psi \in \Psi) \quad \Longleftrightarrow \quad g \in \mathscr{G}.$$

If spaces X and Y are compact and pseudogroups $\mathscr{G}$ on X and $\mathscr{H}$ on Y are isomorphic, then there exists a finite family Φ_0 establishing an isomorphism between $\mathscr{G}$ and $\mathscr{H}$. In fact, if $\Phi : \mathscr{G} \to \mathscr{H}$ is any isomorphism, $\phi_1, \ldots, \phi_{m+n} \in \Phi$, $\bigcup_{i=1}^{m} D_{\phi_i} = X$ and $\bigcup_{j=m+1}^{m+n} R_{\phi_j} = Y$, then one could take $\Phi_0 = \{\phi_1, \ldots, \phi_{m+n}\}$. Also, any isomorphism $\Phi : \mathscr{G} \to \mathscr{H}$ can be enlarged to the family $\tilde{\Phi} = \{h \circ \phi \circ g : g \in \mathscr{G}, h \in \mathscr{H} \text{ and } \phi \in \Phi\}$ which maps isomorphically $\mathscr{G}$ onto $\mathscr{H}$ again. $\tilde{\Phi}$ satisfies the condition

$$g \in \mathscr{G}, \ h \in \mathscr{H} \text{ and } \phi \in \Phi \Rightarrow h \circ \phi \circ g \in \tilde{\Phi},$$

so it becomes an isomorphism of pseudogroups in the sense of [Ha].

The following result is essentially due to Haefliger.

Proposition 1.9. *The holonomy pseudogroups $\mathscr{H}$ and $\mathscr{H}'$ corresponding to two nice coverings $\mathscr{U}$ and $\mathscr{U}'$ of a foliated manifold $(M, \mathscr{F})$ are isomorphic.*

Proof. First, assume that $\mathscr{U}'$ is subordinated to $\mathscr{U}$. Then, the covering $\mathscr{U}'' = \mathscr{U} \cup \mathscr{U}'$ is nice. Let Φ be the family of all the maps of the form $h \circ h''_{UU'} \circ h'$, where $h \in \mathscr{H}$, $h' \in \mathscr{H}'$, $U \in \mathscr{U}$, $U' \in \mathscr{U}'$, $U' \subset U$ and $h''_{UU'} \in \mathscr{H}''$, the holonomy pseudogroup $\mathscr{H}''$ corresponding to $\mathscr{U}''$. Since each set U' of $\mathscr{U}'$ lies in some U of $\mathscr{U}$, $T = \bigsqcup T_U$ meets all leaves, and for any two points x and $y \in T$ of one leaf L there exists a holonomy map $h \in \mathscr{H}$ mapping x to y, the domains (respectively, ranges) of all the mappings of Φ cover $T' = \bigsqcup T'_{U'}$ (respectively, T), therefore, Φ becomes a pseudogroup isomorphism: $\Phi : \mathscr{H}' \to \mathscr{H}$. For any $\mathscr{U}$ and $\mathscr{U}'$ one can find a nice covering $\mathscr{U}''$ subordinated to both coverings. The holonomy pseudogroup $\mathscr{H}''$ corresponding to $\mathscr{U}''$ is isomorphic to both holonomy pseudogroups, $\mathscr{H}$ and $\mathscr{H}'$. $\qquad\square$

Remark 1.10. In a similar way, the holonomy of $\mathscr{F}$ can be defined on any submanifold T of M ($\dim T = \operatorname{codim} \mathscr{F}$) transverse to $\mathscr{F}$ and intersecting all the leaves. T like that is also said to be a complete transversal for $\mathscr{F}$. The holonomy pseudogroup on T is again isomorphic to those arising from nice coverings.

For any leaf L of $\mathscr{F}$ and a point $x \in L \cap U$, (U, ϕ) being a distinguished chart of a nice covering $\mathscr{U}$, consider the germ $[h]_x$ at $\phi(x)$ of holonomy maps h of the form

$$h = h_{U_0 U_1} \circ \cdots \circ h_{U_{k-1} U_k},$$

where $U_0 = U_k = U$. If $\gamma : [0,1] \to L$ is a leaf curve covered by plaques $P_0, P_1, \ldots, P_k$, $P_i \subset U_i$, in such a way that $\gamma([t_i, t_{i+1}]) \subset P_i$ for some t_i's such that $0 = t_0 < t_1 < \cdots < t_k = 1$, then the map h defined above is called a *holonomy along γ* and is denoted sometimes by h_γ. This notation can be accepted because of the following.

If two chains $U_0, U_1, \ldots, U_k$ and $U'_0, U'_1, \ldots, U'_l$ cover homotopic leaf curves $\gamma, \gamma' : [0,1] \to L$, $\gamma(0) = \gamma'(0) = x$ and $\gamma(1) = \gamma'(1) = y$, (in particular, the same curve γ), then the corresponding germs $[h]_x$ and $[h']_x$ are equal. Therefore, we have a properly defined holonomy homomorphism $\Phi_L : \pi_1(L, x) \ni [\gamma] \mapsto [h_\gamma]_x = [h]_x$, where h is any holonomy map corresponding to a chain of charts covering γ. Then, $H_L = \operatorname{im} \Phi_L$ is called the *holonomy group* of L.

Holonomy groups of L corresponding to different points x, different charts U and different nice coverings $\mathscr{U}$ are isomorphic. In general, they describe the behaviour of $\mathscr{F}$ in a neighborhood of L while the whole holonomy pseudogroup $\mathscr{H}$ reflects some global properties of the foliation. For example, one has the following Reeb Stability Theorem: *If a compact leaf L has trivial holonomy group, then there exists an open neighborhood of L that is a union of leaves homeomorphic to L.* Moreover, *if the holonomy group of a compact leaf L is finite, then all the leaves close to L are finite covering spaces of L, therefore are also compact.* Sometimes the holonomy group of a single leaf can influence the global structure of a foliation. For example, one has the following Global Reeb Stability Theorem: *If M is closed (i.e., a compact manifold without boundary) and connected,* codim $\mathscr{F} = 1$, $\mathscr{F}$ *is transversely orientable and contains a compact leaf with finite holonomy group, then all leaves of $\mathscr{F}$ are homeomorphic to L and form the fibers of a bundle M over S^1.*

To explain better the notion of holonomy, it would be a good idea to start by constructing the *suspension* of a single map. Let f be a diffeomorphism of a compact manifold T. Consider $\tilde{M} = T \times \mathbb{R}$ and identify two points (x,t) and (x',t') of $\tilde{M}$, whenever there exists $n \in \mathbb{Z}$ such that $t' - t = n$ and $x' = f^n(x)$. This identification is an equivalence relation, and one can consider the corresponding quotient space M. This M is a manifold, which has a natural structure of a fibre bundle over S^1 (with the typical fibre T). The vector field d/dt projects to M. Trajectories of this projection constitute an one-dimensional foliation transverse to the fibres of M. Any fibre T of this bundle can serve as a complete transversal of our foliation. The holonomy pseudogroup on T is isomorphic to $\mathscr{G}(f)$, the pseudogroup generated by the map f. The foliation obtained in this way is called the *suspension* of f (or, of the group $\mathbb{Z}$ generated by f and acting on $\tilde{M}$). The next example shows how to generalize the suspension construction to arbitrary (finitely presented) groups.

Example 1.11. Let Γ be a finitely generated group isomorphic to the fundamental group $\pi_1(B)$ of a compact manifold B. Recall that any finitely presented group is isomorphic to the fundamental group of a certain compact manifold. Take another manifold F and a homomorphism h of Γ into the group of C^r-diffeomorphisms of F. Γ acts on the universal covering $\tilde{B}$ via the covering maps and on $\tilde{B} \times F$ as

$$\gamma(x,z) = (\gamma(x), h(\gamma)(z)), \quad x \in \tilde{B}, \, z \in M.$$

The foliation $\tilde{\mathscr{F}} = \{\tilde{B} \times \{z\} : z \in F\}$ projects (*via* the canonical projection) onto a C^r-foliation $\mathscr{F}$ of $M = (\tilde{B} \times F)/\Gamma$. The holonomy pseudogroup $\mathscr{H}$ of $\mathscr{F}$ is isomorphic to the pseudogroup $\mathscr{G}(\Gamma_0)$ generated by $\Gamma_0 = \mathrm{im}(h)$. Hence, any pseudogroup of the form $\mathscr{G}(\Gamma)$, Γ being a finitely presented group of diffeomorphisms of a manifold, is isomorphic to the holonomy pseudogroup of a foliation. This foliation $\mathscr{F}$ is called the *suspension* of the homomorphism h. It is easy to see that the manifold M is a fibre bundle over B and F coincides with its fibre. Finally, $\mathscr{F}$ is transverse to the fibres of this bundle and dim $\mathscr{F} = \dim B$. This is why $(M, \mathscr{F})$ is said to be a *foliated bundle*. Note that any compact foliated bundle (i.e., a fibre bundle with a foliation transverse to the fibres and of dimension equal to the dimension of the base) can be obtained as a suspension of a homomorphism h as above.

Example 1.12. The fundamental group Γ_g of the compact oriented surface Σ_g of genus $g \geq 1$ is generated by $2g$ elements $a_1, \ldots, a_g, b_1, \ldots, b_g$ subject to one relation $\prod_{k=1}^{g} a_k b_k a_k^{-1} b_k^{-1} = 1$. (So, it is abelian if and only if $g = 1$, Σ_g being the torus.) Taking any closed manifold F and arbitrary diffeomorphisms $f_1, \ldots, f_g$ of F, and defining a homomorphism $h : \Gamma_g \to \text{Diff}(\Sigma_g)$ by $h(a_k) = f_k$ and $h(b_k) = \text{id}_F$, one can obtain a foliated bundle with base space Σ_g and fibre F, leaves being surfaces, covering spaces of Σ_g.

Now, let $f : N \to M$ be a differentiable map transverse to a C^1-foliation $\mathscr{F}$ of M. It means that

$$T_{f(x)}M = f_*(T_xN) + T_{f(x)}\mathscr{F} \quad (x \in N), \tag{1.4}$$

where $T_y\mathscr{F}$ is the fibre of $T\mathscr{F}$ over y (i.e., $T_y\mathscr{F} = T_yL_y$, L_y being the leaf of $\mathscr{F}$ through y). For any $z \in f(N)$, $f^{-1}(L_z)$ is a submanifold of N. The connected components of all such submanifolds provide a foliation $f^*\mathscr{F}$ of N, codim $f^*\mathscr{F} = $ codim $\mathscr{F}$. The foliation $f^*\mathscr{F}$ is called the *pullback* of $\mathscr{F}$ via f. In particular, if $(M, \mathscr{F})$ has non-empty transverse boundary, then the inclusion map $\partial^{\pitchfork}M \hookrightarrow M$ is transverse to $\mathscr{F}$ and $\mathscr{F}$ induces the foliation $\mathscr{F}^{\pitchfork}$ of $\partial^{\pitchfork}M$.

Proposition 1.13. *If $f(N)$ intersects all the leaves, then the holonomy pseudogroup $\mathscr{H}'$ of $\mathscr{F}' = f^*\mathscr{F}$ is isomorphic to a subpseudogroup of $\mathscr{H}$, the holonomy pseudogroup of $\mathscr{F}$.*

Proof. Let $T' \subset N$ be a complete transversal for $\mathscr{F}'$, $T = f(T')$. Then T is a complete transversal for $\mathscr{F}$ and $\tilde{f} = f|_{T'} : T' \to T$ is a local diffeomorphism. In fact, if $x \in N$ then

$$T_xN = T_x\mathscr{F}' \oplus T_xT' \quad \text{and} \quad f_*(T_x\mathscr{F}') \subset T_{f(x)}\mathscr{F}. \tag{1.5}$$

From (1.4) and (1.5) we get $T_{f(x)}M = \tilde{f}_*(T_xT') + T_{f(x)}\mathscr{F}$. Thus, $\dim \tilde{f}_*(T_xT') \geq$ codim $\mathscr{F}$. On the other hand, $\dim \tilde{f}_*(T_xT') \leq \dim T_xT' = \text{codim }\mathscr{F}$. Consequently, $\dim \tilde{f}_*(T_xT') = \dim T_xT'$ and the differential $\tilde{f}_*$ occurs to be an isomorphism of tangent spaces. Let Φ be a family of all the restrictions $\tilde{f}|_W$, where $W \subset T'$ is any open set which is mapped by $\tilde{f}$ diffeomorphically onto $\tilde{f}(W)$. Φ generates an isomorphism of $\mathscr{H}'$, the holonomy pseudogroup of $\mathscr{F}'$ acting on T', onto the subpseudogroup of $\mathscr{H}$, the holonomy pseudogroup of $\mathscr{F}$ acting on T, generated by all the maps of the form $(\tilde{f}|_{W'}) \circ h' \circ (\tilde{f}|_W)^{-1}$, where $h' \in \mathscr{H}'$, W and W' are sufficiently small open subsets of T'. $\square$

Remark 1.14. Holonomy pseudogroups can be defined in the same way for laminations (called foliated spaces in [CC]). Laminations are "generalized foliations". The simplest example of a lamination is that of a "bunch of leaves" $\mathscr{L}$ (of a regular foliation $\mathscr{F}$) with the compact union $X = \bigcup \mathscr{L}$. The holonomy pseudogroup of such a lamination is isomorphic to the restriction of the holonomy pseudogroup of $\mathscr{F}$ to the intersection of a complete transversal T (for $\mathscr{F}$) with X. The generalization consists in replacing a transverse manifold T by a "transverse topological space". In other words, foliations are built of plaques parameterized by points of a manifold while laminations consist of plaques parameterized by points of a topological space.

More precisely, a *p-dimensional lamination* $\mathscr{L}$ is a separable locally compact space X equipped with an open covering $\mathscr{U}$ and homeomorphisms (*charts distinguished by* $\mathscr{L}$) $\phi : U \to D_U \times T_U$ ($U \in \mathscr{U}$), where D_U is an open subset of $\mathbb{R}^p$ and T_U is a topological space (a *transversal*). For any other distinguished chart $\psi : V \to D_V \times T_V$, the transition map $\phi \circ \psi^{-1}$ is of the form (1.1) with $x \in D_V$ and $y \in T_V$. As for foliations, the sets $\phi^{-1}(D_u \times \{y\})$, $y \in T_U$, are called *plaques* and the plaques glue together to form maximal connected sets $L \subset X$ (*leaves of* $\mathscr{L}$). Usually, some smoothness conditions are required. We assume that the maps α in (1.1) are C^r-differentiable in the first variable and that all their partial derivatives with respect to the first variable are continuous in all the variables.

As far as foliations are concerned, if $\mathscr{U}$ is a nice covering of X (i.e. a covering by distinguished charts such that any plaque of one chart intersects at most one plaque of another), then we can define the holonomy maps h_{VU} by (1.3) and generate a pseudogroup $\mathscr{H}$ (the *holonomy pseudogroup* of $\mathscr{L}$) acting on the space $T = \bigsqcup T_U$.

Example 1.15. If $\mathscr{F}$ is a foliation of a manifold M and X is a closed *saturated* subset of M (i.e., X is the union of a family of leaves of $\mathscr{F}$), then the leaves contained in X form a lamination of X. If $\mathscr{A}$ is a foliated atlas on $(M, \mathscr{F})$, then the maps $\phi|_{X \cap D_\phi}$, $\phi \in \mathscr{A}$, become charts distinguished by $\mathscr{L}$. If T is a complete transversal for $\mathscr{F}$, then $T \cap X$ becomes a complete transversal for $\mathscr{L}$ and the holonomy pseudogroup of $\mathscr{L}$ coincides with $\mathscr{H}|_{T \cap X}$, $\mathscr{H}$ being the holonomy pseudogroup of $\mathscr{F}$ on T.

Remark 1.16. Note that there exist foliated spaces (laminations, see [CC]) essentially different from those of Example 1.15.

1.1.3 Saturated and Minimal Sets, Generic Leaves

Definition 1.17. A subset A of M is $\mathscr{F}$-*saturated* when any leaf of a foliation $\mathscr{F}$, which intersects A, is contained in A. The union $\tilde{A}$ of all the leaves intersecting an arbitrary set $A \subset M$ is obviously saturated. $\tilde{A}$ is called the *saturation* of A. A nonempty closed saturated set is said to be *minimal*, if it does not contain proper subsets, non-empty, closed and saturated.

Therefore, A is saturated if and only if $\tilde{A} = A$. Obviously, single closed leaves are always minimal. If all leaves of $\mathscr{F}$ are dense, then M itself is minimal and we also call the foliation $\mathscr{F}$ *minimal*.

Definition 1.18. Minimal subsets of $(M, \mathscr{F})$ different from M and single leaves are called *exceptional*.

Existence of exceptional minimal sets implies complexity of geometry and topology of foliations. Zorn Lemma implies directly the following.

Proposition 1.19. *The closure of any leaf of a foliation on a compact manifold contains a minimal set.*

Example 1.20. One of the oldest non-trivial examples in foliation theory is the *Reeb foliation* $\mathscr{F}$ of S^3. It is built of two *Reeb components* $\mathscr{R}$ (Figure 1.3), foliations of $D^2 \times S^1$ obtained from a graph Γ of a function $f : (-1,1) \to \mathbb{R}$, which is smooth, symmetric, and satisfies $f(0) = 0$, $f(-t) = f(t)$ and $\lim_{t \to \pm 1} f(t) = \infty$. To obtain $\mathscr{R}$, one has to shift Γ up and down to fill the strip $[-1,1] \times \mathbb{R}$, to rotate this strip around the axis of symmetry $\{0\} \times \mathbb{R}$, and to project the whole picture to $D^2 \times S^1 = D^2 \times (\mathbb{R}/\mathbb{Z})$. To get $\mathscr{F}$, one has to take two such Reeb components $A = \{(w,z) \in S^3 \subset \mathbb{C}^2 : |w| \le \frac{1}{\sqrt{2}}\}$ and $B = \{(w,z) \in S^3 \subset \mathbb{C}^2 : |w| \ge \frac{1}{\sqrt{2}}\}$ and glue them together along the common boundary $T^2 = S^1 \times S^1 = \{(w,z) : |w| = |z| = \frac{1}{\sqrt{2}}\}$.

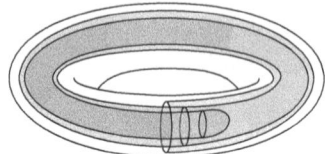

Fig. 1.3 A Reeb component

Let us show that all the non-compact leaves of the Reeb foliations have trivial holonomy groups, while its only compact leaf has holonomy group isomorphic to $\mathbb{Z}^2$. Since interiors of Reeb components are equivalent to products $\mathbb{R}^2 \times (a,b)$ equipped with trivial foliations by planes, the planes foliating these interiors have trivial holonomies. Identify the segment $(-1,1)$ with a transversal of the Reeb foliation in such a way that the origin 0 lies on the toral leaf. The holonomy group of the toral leaf can be generated by two maps f_1 and f_2 such that: $f_1(t) = t$ for $t \le 0$, $f_1(t) < t$ for $t > 0$, $f_2(t) = t$ for $t \ge 0$ and $f_2(t) > t$ for $t < 0$. These maps commute, so they generate a group isomorphic to $\mathbb{Z}^2$. These two maps together with their inverses and the identity on $(-1,1)$ form a finite symmetric set generating the holonomy pseudogroup of the Reeb foliation.

Remark 1.21. Choose $F = S^1$ in Example 1.12 and take as h_k's Morse–Smale diffeomorphisms with one source x_k and one sink $y_k \in S^1$. Let us discuss the structure (types of leaves, minimal sets etc.) of the foliated bundle obtained from the suspension construction subject to $g, x_1, \ldots, x_g, y_1, \ldots, y_g$. All the leaves are covering spaces of Σ_g. (Classification of such coverings can be found in many books on algebraic topology.) The types of leaves strongly depend on the relative position of the points x_k and y_k. For example, (1) if all these points belong to the same orbit of the group generated by h_k's, then the corresponding leaf is closed; (2) the leaves corresponding to the points outside the union of the orbits of x_k's and y_k's are universal coverings of Σ_g, so they are diffeomorphic to the plane $\mathbb{R}^2$; (3) the leaves of (2) accumulate on those corresponding to the orbits of x_k's and y_k's, etc.

Recall that a subset of a locally compact Hausdorff space X is *residual* if it contains a countable intersection of open dense sets. The leaves contained in a saturated residual subset of a foliated manifold are called *generic leaves*.

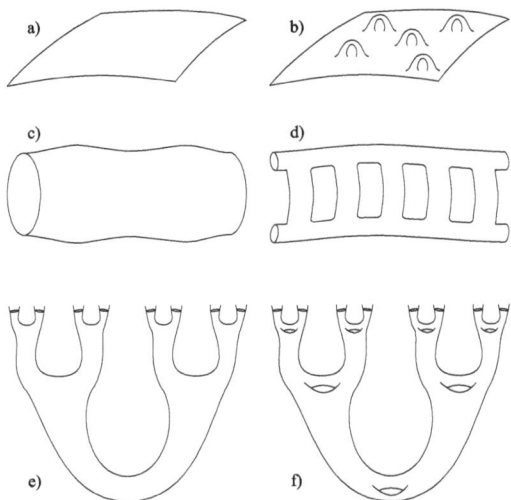

Fig. 1.4 Six surfaces: (**a**) plane, (**b**) Loch Ness monster, (**c**) cylinder, (**d**) Jacob ladder, (**e**) Cantor tree, (**f**) blooming Cantor tree

Proposition 1.22. *The leaves with trivial holonomy groups are generic.*

Proof. The foliated manifold $(M, \mathscr{F})$ under consideration admits a regular foliated atlas $\mathscr{A}$ which — by its definition — has to be countable. Therefore, the corresponding holonomy pseudogroup (acting on the space T of all the plaques of charts of $\mathscr{A}$) is generated by a countable set h_i ($i \in \mathbb{N}$) of holonomy transformations. For each $i \in \mathbb{N}$, the set F_i of all fixed points of h_i is closed in the domain D_i of h_i. The set-theoretic boundary ∂F_i is closed in D_i and is nowhere dense. By local compactness of M, he set ∂F_i is a countable union of compact, nowhere dense sets. For each i denote by Z_i the union of the closures of the plaques $P \subset \partial F_i$. Again, Z_i is a countable union of compact, nowhere-dense sets. The same holds for the union $Z = \bigcup_i Z_i$. Both sets, Z and the complement $M \setminus Z$, are saturated, and the second is the union of all leaves with nontrivial holonomy groups. Of course, the set $M \setminus Z$ is residual. $\square$

As is known, the generic leaves of two-dimensional oriented foliations of closed manifolds are either closed manifolds or homeomorphic to one of the six noncompact surfaces shown in Figure 1.4. Recall that a leaf L of a foliation $\mathscr{F}$ is called *resilient* (compare Figure 1.5), if it contains a loop Γ at $x \in L$ and a point y, $y \neq x$, in the domain of the corresponding holonomy map $h = h_\Gamma$ for which the sequence (y_n), $y_n = h^n(y)$, converges to x as $n \to \infty$.

Example 1.23. Let $N = D^2 \times S^1$ be the solid torus, $\phi : N \to N$ be given by $\phi(z, w) = (\frac{1}{2}w + \frac{1}{4}z, w^2)$, where $|z| \le 1$, $|w| = 1$, and N_0 be the complement of the interior of $\phi(N)$ in N. Then N_0 is a compact three-manifold with the boundary $\partial N_0 = \partial_0 \cup \partial_1$, ∂_i being diffeomorphic to the torus T^2. Such N_0 carries the natural foliation $\mathscr{F}_0$ by the surfaces $w = \text{const}$. The foliation $\mathscr{F}_0$ is transverse to the boundary of N_0, and its

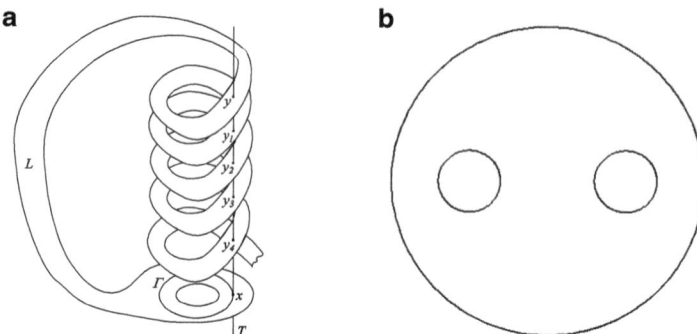

Fig. 1.5 (a) A resilient leaf. (b) Pair of pants

leaves look like *pairs of pants*, surfaces with boundaries homeomorphic to closed discs with two disjoint smaller open discs removed (Figure 1.5).

Let M be a closed (that is, compact without boundary) manifold obtained from N_0 by identifying ∂_0 and ∂_1 *via* ϕ and $\mathscr{F}$ — the foliation induced on M from $\mathscr{F}_0$. This foliation $\mathscr{F}$ carries the name of *Hirsch foliation*.

Since, by construction, the leaves of the Hirsch foliation are obtained by gluing together infinite families of pairs of pants, they are of the type of *Cantor tree* or *blooming Cantor tree*. Since the roots of the unity (of all degrees) are dense in the unit circle, all leaves are dense in M, the manifold carrying the Hirsch foliation. Hence, none of them is compact. Since the holonomy pseudogroup of the Hirsch foliation is generated by the local restrictions of the map $S^1 \ni z \to z^2$ and their local inverses contract, resilient leaves do exist.

1.2 Metric Structures on Manifolds

Here, a brief introduction to Riemannian, metric-affine and Finsler geometries (which are used later in the book) is given.

1.2.1 Riemannian Structure

A *Riemannian structure* (or, *Riemannian tensor*, or *Riemannian metric*) on a manifold M is a tensor field g of type $(0,2)$ such that g_x $(x \in M)$ is a positive definite inner product on the tangent space T_xM. In other words, $g : \mathfrak{X}_M \times \mathfrak{X}_M \to C^\infty(M)$ is bilinear (over $C^\infty(M)$), symmetric ($g(Y,X) = g(X,Y)$ for all X and Y) and positive definite ($g(X(x),X(x)) > 0$, whenever $X(x) \neq 0$). A manifold equipped with a Riemannian structure is called briefly a *Riemannian manifold*, see e.g. [Ber, GKM, Jo, Kl, Pe].

We write often $\langle X,Y \rangle$ instead of $g(X,Y)$. The *norm* $\|v\|$ of a vector $v \in TM$ (on a Riemannian manifold) is defined as the square root of $\langle v,v \rangle$. The *Schwarz's inequality*, $|\langle v,w \rangle| \leq \|v\| \cdot \|w\|$, is valid for all v and w tangent to M (at the same point) and allows to define the angle $\angle(v,w)$ between non-zero vectors: $\cos \angle(v,w) = \langle v,w \rangle / (\|v\| \cdot \|w\|)$.

The "musical" isomorphisms $\sharp : T^*M \to TM$ and $\flat : TM \to T^*M$ are used for rank one tensors, e.g. if $\omega \in T_0^1(M)$ is a 1-form and $X \in \mathfrak{X}_M$ then $\omega(X) = \langle \omega^\sharp, X \rangle = X^\flat(\omega^\sharp)$. For $(0,2)$-tensors A and B we have $\langle A,B \rangle = \mathrm{trace}(A^\sharp B^\sharp) = \langle A^\sharp, B^\sharp \rangle$.

A Riemannian structure allows to calculate the *length* $l(c)$ of any smooth curve c, $c : [a,b] \to M$, on M: $l(c) = \int_a^b \|\dot{c}(t)\| \, dt$. Note that $l(c)$ does not depend on the parametrization of c: $l(c) = l(\tilde{c})$ when $\tilde{c} = c \circ h$ and $h : [\tilde{a}, \tilde{b}] \to [a,b]$ is a diffeomorphism. If $c : (a,b) \to M$ has an *arc length parametrization*, that is when $\|\dot{c}(t)\| = 1$ for any t, then obviously $l(c) = b - a$. Any *regular* ($\dot{c}(t) \neq 0$ for all t) curve c admits a regular reparametrization. If M is connected, then the formula

$$d(x,y) = \inf\{l(c) : \ c \in \Omega_{xy}\}, \tag{1.6}$$

where Ω_{xy} is the space of all the smooth curves $c : [0,1] \to M$ for which $c(0) = x$ and $c(1) = y$, defines a metric on M. The topology of the metric space (M,d) coincides with the original topology of our manifold M. An *isometry* between Riemannian manifolds (M, g_M) and (N, g_N) is a diffeomorphism $F : M \to N$ such that $F^* g_N = g_M$, i.e., $g_N(dF(v), dF(w)) = g_M(v,w)$ for all tangent vectors $v,w \in T_xM$ and all $x \in M$. A vector field X on a Riemannian manifold (M,g) is called a *Killing field* if the local flows generated by X act by isometries. This translates into the following simple characterization: $\mathscr{L}_X g = 0$, or equivalently,

$$X g(Y,Z) = \langle [X,Y], Z \rangle + \langle [X,Z], Y \rangle,$$

for all vector fields Y,Z on M. A vector field X is called *geodesic* if the orbits are geodesics ($\nabla_X X = 0$, see definition of ∇ below).

Remark 1.24. Using the identities

$$\mathscr{L}_X(X,Y) = X\langle X,Y \rangle - \langle X, [X,Y] \rangle = \langle \nabla_X X, Y \rangle + \frac{1}{2} Y\langle X,X \rangle,$$

one may show that for a Killing field X the following properties are equivalent: (a) X has a constant length, (b) X is geodesic.

Paracompactness of manifolds implies that *Riemannian tensors exist on all smooth manifolds*: they can be defined locally as in Example 1.25(ii) and then "glued" together with the use of a suitable smooth partition of unity. Consequently, *any connected paracompact manifold is metrizable and has a countable basis of open sets.*

A (linear) *connection* on M is an operator $\nabla : \mathfrak{X}_M \times \mathfrak{X}_M \to \mathfrak{X}_M$ such that

$$\nabla_{fX_1+X_2}Y = f\nabla_{X_1}Y + \nabla_{X_2}Y, \quad \nabla_X(fY_1+Y_2) = f\nabla_X Y_1 + X(f)\cdot Y_1 + \nabla_X Y_2$$

for any vector fields X, Y, X_1, X_2, Y_1, Y_2 and a smooth function f on M. Given ∇ and using the projection $\pi : TM \to M$, split TTM (the tangent bundle of TM) into the sum $TTM = V \oplus H$ of vertical and horizontal subbundles. For any $\xi \in T_xM$ the subspaces $V_{(x,\xi)}, H_{(x,\xi)}$ can be identified with T_xM *via* $\pi_* : TTM \to TM$ and the connection map $K : TTM \to TM$. The linear map K is determined by the condition

$$K(Z_*\xi) = \nabla_\xi Z \quad \text{(for all } \xi \in TM \text{ and any vector field } Z \text{ on } M). \tag{1.7}$$

For any linear connection ∇, one can define the tensors T of *torsion* and R of *curvature*: T is of type $(1,2)$, while R is of type $(1,3)$ and

$$\mathrm{T}(X,Y) = \nabla_X Y - \nabla_Y X - [X,Y],$$
$$R(X,Y)Z = \nabla_X \nabla_Y Z - \nabla_Y \nabla_X Z - \nabla_{[X,Y]}Z$$

for any vector fields X, Y and Z. A connection is called *symmetric* when $\mathrm{T} = 0$ and *flat* when $R = 0$. A ∇ is *metric compatible* when $\nabla g = 0$, equivalently when

$$X\langle Y,Z\rangle = \langle \nabla_X Y, Z\rangle + \langle Y, \nabla_X Z\rangle \tag{1.8}$$

for all X, Y and Z. *On any Riemannian manifold (M,g) there exists a unique symmetric metric compatible connection ∇ (called the Levi-Civita connection of g). This is a unique torsion free connection on (M,g) preserving g*; and it is given by

$$2\langle \nabla_X Y, Z\rangle = X\langle Y,Z\rangle + Y\langle X,Z\rangle - Z\langle X,Y\rangle$$
$$+ \langle [X,Y], Z\rangle - \langle [X,Z], Y\rangle - \langle [Y,Z], X\rangle. \tag{1.9}$$

Locally, in a chart $\phi = (\phi_1, \ldots, \phi_n)$, the Levi-Civita connection ∇ is determined by

$$\nabla_{X_i}X_j = \sum_{k=1}^n \Gamma_{ij}^k \cdot X_k = \Gamma_{ij}^k X_k, \tag{1.10}$$

where we use Einstein summation convention, $X_i = \partial/\partial\phi_i$ and $\Gamma_{ij}^k = g^{kl}\frac{1}{2}\left(\frac{\partial g_{jl}}{\partial\phi_i} + \frac{\partial g_{il}}{\partial\phi_j} - \frac{\partial g_{ij}}{\partial\phi_l}\right)$ are smooth functions, which are called the *Christoffel symbols*, g^{ij} being the elements of the matrix inverse to $[g_{ij}]$ of Example 1.25(ii).

Throughout the book, ∇ denotes always (if not stated different) the Levi-Civita connection of the Riemannian structure under consideration.

Example 1.25.

(i) The standard inner product on $\mathbb{R}^n$, $\langle (x_i), (y_j) \rangle = \Sigma_i x_i y_i$, defines — via the canonical isomorphism $\iota : \mathbb{R}^n \to T_x \mathbb{R}^n$ — the Riemannian metric on $\mathbb{R}^n$ (and on any open subset U of $\mathbb{R}^n$).

(ii) Locally, in a chart $\phi = (\phi_1, \dots, \phi_n)$, any Riemannian metric has the form

$$\langle X, Y \rangle = \sum_{i,j=1}^{n} g_{ij} X^i Y^j, \quad X = \sum_i X^i \partial/\partial \phi_i, \quad Y = \sum_j Y^j \partial/\partial \phi_j,$$

where g_{ij} are smooth functions on the domain D_ϕ, $g_{ij} = g_{ji}$ and the matrix $[g_{ij}(x)]$ is positive definite for any $x \in D_\phi$.

(iii) If M and N are manifolds, $x \in M$ and $y \in N$, then the tangent space $T_{(x,y)} M \times N$ is canonically isomorphic to the direct sum $T_x M \oplus T_y N$. Therefore, the Riemannian metrics g_M on M and g_N on N determine the *Riemannian product* $(M \times N, g_M \times g_N)$ of Riemannian manifolds (M, g_M) and (N, g_N).

(iv) If g is a Riemannain metric on M and $f \in C^\infty(M)$, then the product $\tilde{g} = e^{2f} \cdot g$ is a new Riemannian metric on M. In this situation, we say that Riemannian structures g and $\tilde{g}$ are *conformally equivalent*: g and $\tilde{g}$ determine the same angles between arbitrary vectors.

(v) If $F : M \to N$ is an immersion and g_N is a Riemannian tensor on N, then the formula $g_M(v, w) = g_N(F_{*x}(v), F_{*x}(w))$ for $v, w \in T_x M$ and $x \in M$, defines the Riemannian tensor g_M on M. This Riemannian structure on M is said to be *induced* from N.

(vi) A *Sasaki metric* $\bar{g}$ on TM is the unique Riemannian metric, for which V and H are orthogonal and all maps $\pi_{*|H} : H_{(x,\xi)} \to T_x M$ and $K_{|V} : V_{(x,\xi)} \to T_x M$, see (1.7), are isometries. The *unit tangent sphere bundle* over (M, g) is $T_1 M = \{(x, \xi) \in TM : |\xi| = 1\}$ (the hypersurface of TM). The natural projection $\pi : TM \to M$ is a *Riemannian submersion* with totally geodesic fibers $\{T_x M\}_{x \in M}$, see [GW].

(vii) An *almost complex structure* is a $(1,1)$-tensor J such that $J^2 = -\mathrm{id}$. This is a *complex structure* if $N_J = 0$ (the theorem of Newlander and Nirenberg), where

$$N_J(X, Y) := [JX, JY] - J[JX, Y] - J[X, JY] - [X, Y]$$

is the *Nijenhuis torsion* of J. A *Hermitian structure* on a Riemannian manifold (M, g) is an almost complex structure J such that $g(JX, JY) = g(X, Y)$. The *Kähler form* of a Hermitian structure is $\omega(X, Y) = g(JX, Y)$. If $d\omega = 0$ and J is a complex structure, then $\nabla J = 0$, and in this case we have a *Kähler structure*.

Geodesic lines (briefly, *geodesics*) play an important role in Riemannian geometry. To define them, let us say first that a *vector field along* a curve $c : (a, b) \to M$ is a smooth map $X : (a, b) \to TM$ such that $X(t) \in T_{c(t)} M$ for any t. Local description of a connection (1.10) shows that for any such X the covariant derivative $\nabla_{\dot{c}(t)} X$ in the direction of $\dot{c}$ can be defined as $\nabla_Y Z(c(t))$, where Y and Z are vector fields on M such that $Y(c(t)) = \dot{c}(t)$ and $Z(c(t)) = X(t)$ for any t. A vector field X along c is said to be *parallel*, whenever $\nabla_{\dot{c}} X = 0$. The theory of ordinary differential equations implies the existence of parallel fields: *if $c : (a,b) \to M$, $t_0 \in (a,b)$ and $v \in T_{t_0}M$, then there exists a unique parallel field X_v along c such that $X(t_0) = v$.* If also $t_1 \in (a,b)$, then the map $\tau_c : T_{c(t_0)}M \to T_{c(t_1)}M$ given by $\tau_c(v) = X_v(t_1)$ is a linear isomorphism called the *parallel transport* along c. Condition (1.8) says that *parallel transports* defined by Riemannian connections are isometric. Now, a curve c is called a *geodesic line*, whenever $\nabla_{\dot{c}} \dot{c} = 0$ everywhere along c. Again the theory of ODE's says that *for any $x \in M$ and $v \in T_xM$ there exists a unique (up to the domain) geodesic line $c : I \to M$ such that $0 \in I$, $c(0) = x$ and $\dot{c}(0) = v$.* Geodesic lines are known to be the critical points of the length: if $c : [a,b] \to M$ is a geodesic and c_s, $s \in (-\varepsilon, \varepsilon)$ is a smooth variation of c preserving the end points (that is, $c_0 = c$, $c_s(a) = c(a)$ and $c_s(b) = c(b)$ for any s), then $L'(0) = 0$, where $L(s) = l(c_s)$ is the length of c_s.

A Riemannian manifold (M,g) is said to be *geodesically complete*, whenever all the geodesic lines can be extended to the whole real line $\mathbb{R}$. On connected manifolds, the classical Hopf-Rinow Theorem provides the equivalence of the following conditions: (i) *M is geodesically complete*, (ii) *the metric space (M,d), d being defined by (1.6), is complete*, (iii) *all the bounded closed subsets of (M,d) are compact*. Consequently, *any connected compact Riemannian manifold is complete and geodesically complete*. Moreover, each of conditions (i)–(iii) implies that (iv) *for any two points $x,y \in M$ there exists a geodesic c joining x and y such that $l(c) = d(x,y)$.*

A connection ∇ allows to differentiate arbitrary tensors. For tensors S of types $(0,j)$ and $(1,j)$ the formulae for $\nabla_X S$ are similar to those for the Lie derivative:

$$(\nabla_X S)(Y_1,\ldots,Y_j) = \nabla_X S(Y_1,\ldots,Y_j)$$
$$- \sum_{i \le j} S(Y_1,\ldots,\nabla_X Y_i,\ldots,Y_j), \quad X,Y_i \in \mathfrak{X}_M.$$

In the book, we call often ∇ (and ∇S) the *covariant derivative* (of S).

A frame $(E_1,\ldots,E_n)$ on TM built either of single vectors or of (local) vector fields is *orthonormal*, whenever $\langle E_i, E_j \rangle = \delta_{ij}$ for all i and j. For a $(r, j+1)$-tensor field S, where $r = 0,1$, we also define a (r,j)-tensor $\nabla^* S$ by

$$(\nabla^* S)(Y_1,\ldots,Y_j) = -\sum_i (\nabla_i S)(E_i, Y_1 \ldots Y_j), \qquad (1.11a)$$

called the *adjoint of the covariant derivative on tensors*. The *divergence* of a $(0,k+1)$-tensor S is a $(0,k)$-tensor $\operatorname{div} S = -\nabla^* S$. Using a metric g on a manifold, they identify $(1,k)$-tensors with $(0,k+1)$-tensors: $S^j{}_{i_1,\ldots i_k} = g^{mj} S_{m,i_1,\ldots i_k}$. Thus, for a $(1,k)$-tensor S we can equivalently write $\operatorname{div} S = \operatorname{trace}(Y \to \nabla_Y S)$, that is

$$(\operatorname{div} S)(X_1,\ldots,X_k) = \sum_i \langle (\nabla_i S)(X_1,\ldots,X_k), E_i \rangle, \qquad (1.11b)$$

or, $(\operatorname{div} S)_{i_1,\ldots i_k} = \nabla_j S^j{}_{\cdot i_1,\ldots i_k}$ in coordinates. For example, if S is a $(1,1)$-tensor (a linear operator) then $\nabla^* S$ is a vector field and $\operatorname{div} S$ is a 1-form.

Remark 1.26. Let X be a vector field on (M,g). Using the identities

$$\nabla^* X^\flat = -\sum_i (\nabla_i X^\flat)(E_i) = -\sum_i (E_i \langle X, E_i \rangle - \langle X, \nabla_i E_i \rangle) = -\sum_i \langle \nabla_i X, E_i \rangle,$$

one may show that $\operatorname{div} X = -\nabla^*(X^\flat)$, where $X^\flat = \langle X, \cdot \rangle$ is the 1-form dual to X, and $\operatorname{div} X$ is also given by $\operatorname{div} X = \sum_i \langle \nabla_i X, E_i \rangle$.

The *(rough) Laplacian* of S is the divergence of the covariant derivative of the tensor: $\Delta S = \operatorname{div}(\nabla S)$. When applied to functions, we get the *Laplacian* $\Delta u = \operatorname{div}(\nabla u) = \operatorname{trace} \operatorname{Hess}_u$. The *Hessian of a function* u is the symmetric $(0,2)$-tensor $h_u(X,Y) := \langle \operatorname{Hess}_u X, Y \rangle$, where $\operatorname{Hess}_u X := \nabla_X \nabla u$ is a self-adjoint $(1,1)$-tensor.

If S is a vector, the gradient is a covariant derivative which results in a $(1,1)$-tensor, and the divergence of this is again a vector. The *divergence operator* δ is adjoint to d with respect to L^2-product on TM and is defined as follows: for $\alpha \in \Lambda^{p-1}(M)$ and $\beta \in \Lambda^p(M)$ with compact supports we have, see [Jo],

$$\int_M d\alpha \wedge \star \beta = \int_M \alpha \wedge \star \delta \beta,$$

and $\star : \Lambda^p(T^*TM) \to \Lambda^{n-p}(T^*TM)$ is the Hodge *star operator*, defined by

$$\langle \gamma, \eta \rangle \, d\operatorname{vol}_g = \gamma \wedge \star \eta$$

for any $\gamma, \eta \in \Lambda^p(M)$. The *Hodge Laplacian* acting on differential forms is defined as $\Delta_H = -(d\delta + \delta d)$, it is selfadjoint and non-positive definite. The rough Laplacian and the Hodge Laplacian on scalar functions are the same as the *Laplace-Beltrami operator*. On any oriented Riemannian manifold (M,g) one has its *Riemannian volume* form $d\operatorname{vol}_g$, the unique n-form ($n = \dim M$), which satisfies the condition $d\operatorname{vol}_g(E_1, \ldots, E_n) = 1$ for any positively oriented orthonormal frame $(E_1, \ldots, E_n)$. In other words,

$$d\operatorname{vol}_g(X_1, \ldots X_N) = \det[\langle X_i, E_j \rangle : i, j \leq n]$$

for all $X_1, \ldots, X_n$, (E_i) being a frame as before. With this volume form in mind, one has the formula $\operatorname{div} X = \operatorname{trace}(Z \to \nabla_Z X)$. Since the Levi-Civita connection is Riemannian ($\nabla g = 0$), one has also $\nabla d\operatorname{vol}_g = 0$. Certainly, if $\tilde{g} = e^{2f} \cdot g$, i.e., $\tilde{g}$ is conformally equivalent to g, then $d\operatorname{vol}_{\tilde{g}} = e^{nf} \cdot d\operatorname{vol}_g$.

For a compact manifold M with boundary and inner normal n, the *Divergence Theorem* reads as

$$\int_M (\operatorname{div} X) \, d\operatorname{vol}_g = \int_{\partial M} \langle X, n \rangle \, d\omega. \tag{1.12}$$

Thus, if either X has compact support or M is closed, then

$$\int_M (\operatorname{div} X) \, d\operatorname{vol}_g = 0. \tag{1.13}$$

Tracing the identity $\nabla(fX) = f \cdot \nabla X + df \otimes X$, yields the equality for a function $f \in C^1(M)$ and a vector field X,

$$\operatorname{div}(f \cdot X) = f \cdot \operatorname{div} X + X(f). \tag{1.14}$$

Example 1.27. Let θ be a one-form in some cohomology class of a closed oriented differentiable manifold M. Recall that the set of equivalence classes of closed 1-

forms is a finite-dimensional vector space over $\mathbb{R}$, called the *first de Rham cohomology group* and denoted by $H^1(M,\mathbb{R})$. A 1-form θ is said to be closed if $d\theta = 0$, and θ is exact, if there is a function f with $df = \theta$. Because of $d \circ d = 0$, exact forms are always closed. The closed 1-forms θ_0, θ_1 are cohomologous if $\theta_0 - \theta_1$ is exact.

Notice that $\delta\theta = -\operatorname{div}\theta$ for 1-forms. A 1-form θ is called harmonic if $\Delta_H\,\theta = 0$; on a closed manifold we have $d\,\theta = 0$ and $\delta\,\theta = 0$. In a local chart, we have

$$d\theta = \sum_k \frac{\partial\theta_i}{\partial x^k}\,dx^k \wedge dx^i, \quad \delta\theta = -\sum_{k,l} g^{kl}\Big(\frac{\partial\theta_k}{\partial x^l} - \Gamma^j_{kl}\theta_j\Big), \quad \Delta_H\,\theta = \sum_k \frac{\partial^2\theta_i}{(\partial x^k)^2}\,dx^i$$

for a 1-form $\theta = \theta_i dx^i$. By Hodge Theorem, *every cohomology class in $H^1(M,\mathbb{R})$ contains precisely one harmonic 1-form.* A 1-form $X^\flat$ (dual to the vector field X) is harmonic if and only if $\operatorname{div} X = 0$ and the $(1,1)$-tensor ∇X is symmetric.

The *sectional curvature K* of a Riemannian manifold (M,g) is defined for any plane $P \subset T_x M$ $(x \in M)$ and any pair of vectors X, Y spanning P:

$$K(P) = \frac{\langle R(X,Y)Y, X\rangle}{\|X\|^2\|Y\|^2 - \langle X, Y\rangle^2}.$$

A Riemannan manifold (M,g) is of *constant curvature c*, whenever $K(P) = c$ for any P. Pointwise and local orthonormal frames always exist and can be produced from arbitrary frames by the procedure of Schmidt orthogonalization. The *Ricci tensor* Ric on (M,g) is defined as the suitable trace of the curvature tensor R:

$$\operatorname{Ric}_{X,Y} = \operatorname{trace}(Z \to R(Z,X)Y) = \sum_{i=1}^n \langle R(E_i,X)Y, E_i\rangle$$

when (E_i) is an orthonormal frame. The Ricci tensor is, obviously, of type $(0,2)$ and is symmetric: $\operatorname{Ric}_{X,Y} = \operatorname{Ric}_{Y,X}$ for all X and Y. The *Ricci curvature* of (M,g) in the direction of a unit X is defined as $\operatorname{Ric}_{X,X}$. If $\operatorname{Ric} = \lambda \cdot g$ for a smooth function λ, then (M,g) is called *Einstein* [Bes]. In this case, $\lambda(x) = \operatorname{Ric}_{X,X}(x)$ for any unit vector X. Finally, the *scalar curvature S* of (M,g) is defined as the trace of Ric:

$$S = \sum_{i=1}^n \operatorname{Ric}_{E_i,E_i}$$

for any orthonormal E_i's. *Riemannian manifolds of constant sectional curvature are Einstein and have constant scalar curvature.*

Example 1.28. The Euclidean space $\mathbb{R}^n$ (equipped with the Riemannnian tensor of Example 1.25 (i)) is flat. Also, all the tori T^n equipped with the Riemannian product structure arising from that of S^1 are flat. The unit sphere S^n equipped with the Riemannian structure induced from the Euclidean space $\mathbb{R}^{n+1}$ have constant sectional curvature 1, constant Ricci curvature $n-1$ and constant scalar curvature $n(n-1)$. The unit ball $B = B(0,1) \subset \mathbb{R}^n$ equipped with the metric $g = f \cdot g_0$, g_0 being the standard inner product of Example 1.25 (i) and $f : B \to \mathbb{R}$—a smooth function given by $f(x) = 4/(1 - |x|^2)^2$, has constant sectional curvature -1, constant Ricci curvature $1 - n$ and constant scalar curvature $n(1 - n)$. All these manifolds are complete and geodesically complete.

The metric-affine geometry (founded by E. Cartan) generalizes Riemannian geometry: it uses a linear connection with torsion, $\overline{\nabla}$, instead of the Levi-Civita connection ∇ of g, and appears in such context as homogeneous and almost Hermitian manifolds, Finsler and generalized geometries, e.g. [Mik]. The important distinguished cases are: *Riemann–Cartan manifolds*, where metric connections ($\overline{\nabla}g = 0$) are used, e.g. [GPS], and *statistical manifolds*, e.g. [Op], where $\overline{\nabla}$ is torsionless and the tensor $\overline{\nabla}g$ is symmetric in all its entries. The main notion of information geometry is that of statistical manifold; affine hypersurfaces in $\mathbb{R}^{n+1}$ are a natural source of such manifolds. Riemann–Cartan spaces are central in gauge theory of gravity, where the torsion is involved in the Cartan spin-connection equation. The ∇ is a distinguished point in affine space of all connections on M, and the difference $\mathfrak{T} := \overline{\nabla} - \nabla$ is called the *contorsion tensor*. Define (1,2)-tensors $\mathfrak{T}^*$ and $\widehat{\mathfrak{T}}$,

$$\langle \mathfrak{T}_X^* Y, Z \rangle = \langle \mathfrak{T}_X Z, Y \rangle, \quad \widehat{\mathfrak{T}}_X Y = \mathfrak{T}_Y X, \quad X, Y, Z \in \mathfrak{X}_M. \tag{1.15}$$

Note that generally $(\widehat{\mathfrak{T}})^* \neq \widehat{\mathfrak{T}^*}$. Indeed,

$$\langle (\widehat{\mathfrak{T}})_X^* Y, Z \rangle = \langle \widehat{\mathfrak{T}}_X Z, Y \rangle = \langle \mathfrak{T}_Z X, Y \rangle, \quad \langle (\widehat{\mathfrak{T}^*})_X Y, Z \rangle = \langle \mathfrak{T}_Y^* X, Z \rangle = \langle \mathfrak{T}_Y Z, X \rangle.$$

A connection $\overline{\nabla} = \nabla + \mathfrak{T}$ is *metric compatible* if $\overline{\nabla}g = 0$; in this case, $\mathfrak{T}^* = -\mathfrak{T}$,

$$(\overline{\nabla}_X g)(Y, Z) = \langle \mathfrak{T}_X Y, Z \rangle + \langle \mathfrak{T}_X Z, Y \rangle = 0 \quad (X, Y, Z \in TM). \tag{1.16}$$

If $\overline{\nabla}$ is torsionless and tensor $\overline{\nabla}g$ is symmetric in all its entries then $\overline{\nabla}$ is called a *statistical connection* in literature; in this case, $\widehat{\mathfrak{T}} = \mathfrak{T}$ and $\mathfrak{T}^* = \mathfrak{T}$.

Comparing the curvature tensor $\overline{R}_{X,Y} = [\overline{\nabla}_Y, \overline{\nabla}_X] + \overline{\nabla}_{[X,Y]}$ of $\overline{\nabla}$ with R, we find

$$\overline{R}(X, Y) - R(X, Y) = (\nabla_Y \mathfrak{T})_X - (\nabla_X \mathfrak{T})_Y + [\mathfrak{T}_Y, \mathfrak{T}_X], \quad X, Y \in \mathfrak{X}_M. \tag{1.17}$$

The tensor $\overline{R}$ has some symmetry properties, e.g. $\langle \overline{R}(X, Y)Z, U \rangle = -\langle \overline{R}(Y, X)Z, U \rangle$.

Example 1.29.

(a) Connections $\overline{\nabla}$ and ∇ are *projectively equivalent*, i.e., their systems of geodesics coincide, if $\mathfrak{T}_Y X = \omega(X)Y + \omega(Y)X$ for some closed one-form ω and any $X, Y \in \mathfrak{X}_M$. Of course, since ω is closed, any point x of M has an open neighbourhood U such that $\omega = df$ for some smooth function f on U. If $\nabla_X X = 0$ and $\phi = e^{2f}$, then $\overline{\nabla}_{\phi X}(\phi X) = 2\phi \cdot (\omega(X) + df(X)) \cdot X = 0$. Therefore, $\overline{\nabla}$ and ∇ have the same geodesics (up to a parametrization) on U. Since x was arbitrary, all the geodesics of $\overline{\nabla}$ and ∇ coincide up to a parametrization.

(b) The *semi-symmetric connections*, i.e., $\mathfrak{T}_X Y = \langle U, Y \rangle X - \langle X, Y \rangle U$, where U is a given vector field, are metric connections. Of course, since ∇ is the Levi-Civita connection, $\nabla g = 0$ and

$$(\overline{\nabla}_X g)(Y, Z) = -\langle \mathfrak{T}_X Y, Z \rangle - \langle Y, \mathfrak{T}_X Z \rangle$$
$$= -\langle U, Y \rangle \cdot \langle X, Z \rangle + \langle X, Y \rangle \cdot \langle U, Z \rangle - \langle U, Z \rangle \cdot \langle X, Y \rangle + \langle X, Z \rangle \cdot \langle U, Y \rangle = 0$$

for all vector fields X, Y and Z. Therefore, $\overline{\nabla}g = 0$ and $\overline{\nabla}$ is metric.

1.2.2 Finsler Structure

A *Finsler metric* on a smooth manifold M is a Minkowski norm F in tangent spaces T_xM, which smoothly depends on a point $x \in M$, e.g., [SS]. A *Minkowski norm* on a vector space $\mathbb{R}^n$ ($n \geq 2$) is a function $F : \mathbb{R}^n \to [0, \infty)$ with the following properties of regularity, positive 1-homogeneity and strong convexity:

$\mathrm{M}_1 : F \in C^\infty(\mathbb{R}^n \setminus 0)$, $\mathrm{M}_2 : F(\lambda y) = \lambda F(y)$ for $\lambda > 0$ and $y \in \mathbb{R}^n$,

$\mathrm{M}_3 :$ For any $y \in \mathbb{R}^n \setminus 0$, the following bilinear form is *positive definite*:

$$g_y(u, v) = \frac{1}{2} \frac{\partial^2}{\partial s \, \partial t} \left[F^2(y + t_1 u + t_2 v) \right]_{|t_1=t_2=0}. \tag{1.18}$$

By (1.18) and M_2, $g_{\lambda y} = g_y$ ($\lambda > 0$) and $g_y(y, y) = F^2(y)$. As a result of M_3, the *indicatrix* $S := \{y \in \mathbb{R}^n : F(y) = 1\}$ is a closed, convex smooth hypersurface (generally, not central symmetric) that surrounds the origin, see Figure 1.6.

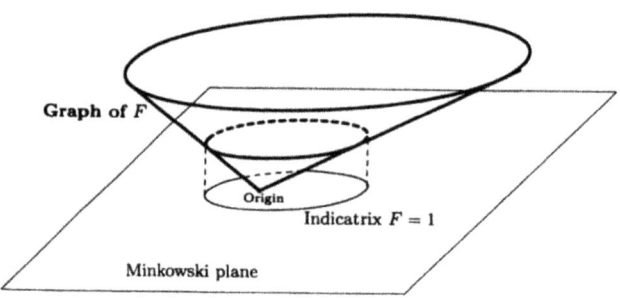

Fig. 1.6 Indicatrix of Minkowski norm F

Remark 1.30. In some cases (when F is not positive definite or smoothly inextendible in some directions, so that these directions must be removed from the domain of F) Minkowski norms are defined only in conic domains $A \subset \mathbb{R}^n$, e.g., [JS]. Such cases called *conic Minkowski norms*, are considered below only in examples (e.g., Kropina norm or slope Matsumoto norm).

The following symmetric trilinear form on $\mathbb{R}^n$:

$$C_y(u, v, w) = \frac{1}{4} \frac{\partial^3}{\partial r \, \partial s \, \partial t} \left[F^2(y + ru + sv + tw) \right]_{|r=s=t=0} \tag{1.19}$$

is called the *Cartan torsion* for F. Note that $C_y(u, v, y) = 0$ and $C_{\lambda y} = \lambda^{-1} C_y$ for $\lambda > 0$. Vanishing of a 1-form $I_y(u) = \mathrm{trace}_{g_y} C_y(u, \cdot, \cdot)$ (called the *mean Cartan torsion*) characterizes Euclidean norms among all Minkowski norms, see e.g. [SS]. In coordinates, we have $C_{ijk} = \frac{1}{4} [F^2]_{y^i y^j y^k} = \frac{1}{2} \frac{\partial g_{ij}}{\partial y^k}$ and $g_{ij} = \frac{1}{2} [F^2]_{y^i y^j}$; hence,

$C_{ijk}y^k = 0$ and $C_k = C_{ijk}g^{ij}$. Given a unit vector field N, one may define a particular Riemannian metric $g := g_N$ on M, The *Chern connection* D^N is determined by

$$g(D_u^N v, w) - g(\nabla_u v, w) = C_N(D_w^N N, u, v) - C_N(D_u^N N, v, w) - C_N(D_v^N N, u, w),$$
(1.20)

where $u, v, w \in \mathfrak{X}_M$ and ∇ is the Levi-Civita connection of g. It is torsion free, i.e., $D_u^N v - D_v^N u = [u, v]$, and 'almost metric':

$$u(g(v, w)) = g(D_u^N v, w) + g(D_v^N u, w) + 2C_N(D_u^N N, v, w).$$

By (1.20), $D_u^N v|_x$ depends only on N_x, u_x and v near $x \in M$. The *Chern curvature tensor* is defined by

$$R^N(u, v)w = D_u^N D_v^N w - D_v^N D_u^N w - D_{[u,v]}^N w.$$

It obeys the first Bianchi identity $R^N(u, v)w + R^N(v, w)u + R^N(w, u)v = 0$.

One can define the following notions for (M, F): covariant derivative, parallel translation (and parallel vectors) along a curve, geodesics and curvature, see [SS].

The next lemma is used to compute the volume forms of (α, β)-norm, see Example 1.32 below. It extends the Silvester's identity.[1] $\det(\mathrm{id}_n + C_1 P_1^t) = 1 + C_1^t P_1$, where C_1, P_1 are n-vectors (columns), and id_n is the identity n-matrix.

Lemma 1.31. *Given real c_1, c_2, vectors b^1, b^2 in $\mathbb{R}^n$, and reversible symmetric $n \times n$ matrix a_{ij}, define matrices $A_{ij} = a_{ij} + c_1 b_i^1 b_j^1$ and $g_{ij} = a_{ij} + c_1 b_i^1 b_j^1 + c_2 b_i^2 b_j^2$. Then*

$$\det[A_{ij}] = \det[a_{ij}](1 + c_1 |b^1|_a^2),$$
$$\det[g_{ij}] = \det[a_{ij}] \cdot [(1 + c_1 |b^1|_a^2)(1 + c_2 |b^2|_a^2) - c_1 c_2 \langle b^1, b^2 \rangle_a^2].$$

Proof. The first claim is straightforward. If $1 + c_1 |b^1|_a^2 \neq 0$ then the inverse matrix $A^{kl} = a^{kl} - \frac{c_1}{1 + c_1 |b^1|_a^2} b_k^1 b_l^1$ exists. For any vectors u, v we get

$$\langle u, v \rangle_A = A^{kl} u_k v_l = \left(a^{kl} - \frac{c_1 b^{1k} b^{1l}}{1 + c_1 |b^1|_a^2} \right) u_k v_l = \langle u, v \rangle_a - \frac{c_1}{1 + c_1 |b^1|_a^2} \langle b^1, u \rangle_a \langle b^1, v \rangle_a.$$

Hence, $|b^2|_A^2 = A^{kl} b_k^2 b_l^2 = |b^2|_a^2 - \frac{c_1}{1 + c_1 |b^1|_a^2} \langle b^1, b^2 \rangle_a^2$. Using the first claim, we have $\det[g_{ij}] = \det[A_{ij}](1 + c_2 |b^2|_A^2)$. From the above the second claim follows. $\qquad\square$

Example 1.32. Let $a = \langle \cdot, \cdot \rangle$ be a scalar product on $\mathbb{R}^n$ with a Euclidean norm α, and β a 1-form with the norm $b := \alpha(\beta) < \delta_0$. The function $F = \alpha\phi(\beta/\alpha)$ is a Minkowski norm on $\mathbb{R}^n$ (called the (α, β)-*norm*) for any such α and β if and only if $\phi : (-\delta_0, \delta_0) \to (0, \infty)$ satisfies

$$\phi - s\dot{\phi} + (b^2 - s^2)\ddot{\phi} > 0,$$
(1.21)

---

[1] J.J. Silvester, On the relation between the minor determinants of linearly equivalent quadratic functions. Philosophical Magazine, 1, (1851), 295–305.

where $|s| \leq b$. Taking $s = b$ in (1.21), we get $\phi - s\dot{\phi} > 0$. A better criterion than (1.21) can be found in [JS, Section 4.3]. Next, by (1.18), we find

$$g_y(u,v) = \rho\langle u,v\rangle + \rho_0\beta(u)\beta(v) + \rho_1(\beta(u)\langle y,v\rangle + \beta(v)\langle y,u\rangle)/\alpha(y)$$
$$- \rho_1\beta(y)\langle y,u\rangle\langle y,v\rangle/\alpha^3(y), \tag{1.22}$$

where $\rho = \phi(\phi - s\dot{\phi}) > 0$, $\rho_0 = \phi\ddot{\phi} + \dot{\phi}^2$ and $\rho_1 = \phi\dot{\phi} - s(\phi\ddot{\phi} + \dot{\phi}^2)$ are functions of $s = \beta/\alpha$. The following relations hold: $\dot{\rho} = \rho_1$, $\ddot{\rho} = \dot{\rho}_1 = -s\dot{\rho}_0$.

To show (1.21), set $\tilde{Y} = s^{-1}\beta^\sharp - y/\alpha(y)$ and $\varepsilon = s\rho_1$. Then (1.22) takes the form

$$g_y(u,v) = \rho\langle u,v\rangle + (\rho_0 + \rho_1^2/\varepsilon)\beta(u)\beta(v) - \varepsilon\langle\tilde{Y},u\rangle\langle\tilde{Y},v\rangle. \tag{1.23}$$

From (1.23) and Lemma 1.31 with $c_1 = (\rho_0 + \rho_1^2/\varepsilon)\rho^{-1}$, $c_2 = -\varepsilon$ and $b^1 = \beta^\sharp$, $b^2 = \tilde{Y}^\sharp$, for the volume form $d\,\mathrm{vol}_{g_y} = \mu_{g_y}(y)\,d\,\mathrm{vol}_a$ we obtain

$$\mu_{g_y}(y) = \rho^{n-2}(\rho_0\rho_1 s^3 + \rho_1^2 s^2 + (\rho - \rho_0 b^2)\rho_1 s + (\rho\rho_0 - \rho_1^2)b^2 + \rho^2)$$
$$= \phi^{n+1}(\phi - s\dot{\phi})^{n-2}[\phi - s\dot{\phi} + (b^2 - s^2)\ddot{\phi}]. \tag{1.24}$$

From $\mu_{g_y}(y) > 0$ the inequality (1.21) follows.

Set $p_y = s\tilde{Y} = \beta^\sharp - sy/\alpha(y)$. The Cartan torsion of (α,β)-norm has a special form

$$2C_y(u,v,w) = (\rho_1/\alpha(y))(K_y(u,v)\langle p_y,w\rangle + K_y(v,w)\langle p_y,u\rangle + K_y(w,u)\langle p_y,v\rangle)$$
$$+ (3\dot{\phi}\ddot{\phi} + \phi\dddot{\phi})/\alpha(y)\,\langle p_y,u\rangle\langle p_y,v\rangle\langle p_y,w\rangle,$$

where K_y is the *angular metric* of $a = \langle\cdot,\cdot\rangle$ given by

$$K_y(u,v) = \langle u,v\rangle - \langle y,u\rangle\langle y,v\rangle/\alpha^2(y),$$

or, in coordinates, $K_{ij} = F \cdot F_{y^iy^j} = g_{ij} - g_{ip}y^p g_{iq}y^q/F^2$. The (α,β)-norms form a rich class of computable Minkowski norms and play an important role in Finsler geometry, see [BCS, Mat, SS, YZ]. Consider some particular cases of (α,β)-norm.

(i) For $\phi(s) = 1 + s$, $|s| < b < 1$ with $b := \alpha(\beta)$, we get Minkowski norm $F = \alpha + \beta$ (i.e., Euclidean norm shifted by a small vector), introduced by a physicist G. Randers[2] to study the unified field theory. We have $\rho = 1 + s$, $\rho_0 = 1$ and $\rho_1 = 1$.

(ii) The (α,β)-norms $F = \alpha^{l+1}/\beta^l$ ($l > 0$), i.e., $\phi(s) = 1/s^l$ ($s > 0$), are called *generalized Kropina metrics*, and have applications in general dynamical systems. The *Kropina metric*, i.e., $l = 1$, first introduced by L. Berwald in con-

---

[2] Randers G. On an asymmetrical metric in the four-space of general relativity, Phys. Rev. 59 (1941), 195–199.

nection with a Finsler plane with rectilinear extremal, and investigated by
V.K. Kropina.[3] We have $\rho = 2/s^2$, $\rho_0 = 3/s^4$ and $\rho_1 = -4/s^3$.

(iii) The (α, β)-norm $F = \frac{\alpha^2}{\alpha - \beta}$, i.e., $\phi(s) = \frac{1}{1-s}$ with $|s| < 1$, (called *slope-metric*)
was introduced by M. Matsumoto to study the time it takes to negotiate any
given path on a hillside. We have $\rho = \frac{1-2s}{(1-s)^3}$, $\rho_0 = \frac{3}{(1-s)^4}$ and $\rho_1 = \frac{1-4s}{(1-s)^4}$.
Kropina and slope norms are not Minkowski norms: their F's are not defined
on the whole $\mathbb{R}^n \setminus 0$ (have singularities) and their g_y's are not positive definite.

(iv) A Finsler metric is a *polynomial* (α, β)-*norm* if $\phi(s) = \sum_{i=0}^k C_i s^i$, $C_0 = 1$, $C_k \neq 0$. The *quadratic metric* $F = (\alpha + \beta)^2/\alpha$, i.e., $\phi(s) = (1 + s)^2$ with $|s| < 1$,
appears in many geometrical problems. We have $\rho = (1 - s)(1 + s)^3$, $\rho_0 = 6(1 + s)^2$ and $\rho_1 = 2(1 - 2s)(1 + s)^2$.

(v) Define by $\phi(s) = e^{s/k}$, $|s| < b < |k|$, the *exponential metric* $F = \alpha\, e^{\beta/(k\alpha)}$. Con-
dition (1.21) reads as a quadratic inequality $s^2 + ks - (b^2 + k^2) < 0$. Taking $s = b$
in (1.21) yields $k(s - k) < 0$ when $|s| < |k|$. Thus, (1.21) is satisfied for arbitrary
numbers s and b such that $|s| \leq b < |k|$. We have $\rho = e^{2s/k}(k - s)/k > 0$, $\rho_0 = 2\, e^{2s/k}/k^2$ and $\rho_1 = e^{2s/k}(k - 2s)/k^2$.

1.3 Basic of the Extrinsic Geometry of Foliations

Extrinsic geometry of a foliated Riemannian manifold deals with the properties,
which can be expressed in terms of the second fundamental form of the leaves and its
invariants (e.g., the principal curvatures). The mixed sectional curvature, the partial
Ricci and mixed scalar curvatures play an important role in the extrinsic geometry
of foliations. Several authors investigated, whether on a given Riemannian manifold
there exists a totally geodesic foliation (i.e., of the simplest extrinsic geometry).
One of the principal problems of extrinsic geometry of foliations reads as follows,
see [CW, Gl, LaW, R1, RW2]: *Given a foliated manifold* $(M, \mathscr{F})$ *and a geometric
property* (P) *of a submanifold, find a Riemannian metric* g *on* M, *for instance, in a
given conformal class, such that* $\mathscr{F}$ *enjoys* (P) *with respect to g.* Such problems of
the existence and classification of metrics with a given geometry on foliations were
studied already in the 1970s when they provided a topological tautness condition for
a foliation [Su2], equivalent to the existence of a Riemannian metric making all the
leaves minimal. The extrinsic geometry of foliations concerns the following topics:

- Integral formulas, which provide geometrical obstructions for existence of foliations or compact leaves of them, and under some conditions yield splitting
  results.
- Variational formulae for geometric quantities by change of metric, which are
  used in minimization of actions on foliations and exploring extrinsic geometric
  flows.
- Prescribing geometric quantities of foliations by $\mathscr{F}$-conformal change of metric.

---

[3] V.K. Kropina, On projective two-dimensionnal Finsler spaces with special metric. Trudy Sem.
Vector Tenzor Anal., 11 (1961), 277–292.

1.3.1 Fundamental Tensors

Let $\text{Sym}^2(M)$ be the space of all symmetric $(0,2)$-tensors tangent to M. A *pseudo-Riemannian metric* of index q is an element $g \in \text{Sym}^2(M)$ (of the space of symmetric $(0,2)$-tensor fields) such that each g_x $(x \in M)$ is a non-degenerate bilinear form of index q on the tangent space T_xM. When $q = 0$, g is a Riemannian metric, and a *Lorentz metric* when $q = 1$, see [BF, Ch2].

Let M^{n+p} be a connected smooth manifold with a pseudo-Riemannian metric g of index q with the Levi-Civita connection ∇ and complementary orthogonal non-degenerate distributions $\widetilde{\mathscr{D}}$ and $\mathscr{D}$ (subbundles of the tangent bundle TM) of ranks $\dim_{\mathbb{R}} \widetilde{\mathscr{D}}_x = n$ and $\dim_{\mathbb{R}} \mathscr{D}_x = p$ for every $x \in M$. A distribution $\widetilde{\mathscr{D}}$ is *non-degenerate*, if g_x is a non-degenerate bilinear form on $\widetilde{\mathscr{D}}_x \subset T_xM$ for every $x \in M$; in this case, $\mathscr{D}$ is also non-degenerate. A pair $(\widetilde{\mathscr{D}}, \mathscr{D})$ of complementary orthogonal distributions is called an *almost product structure* on (M, g).

Let $^\top$ and $^\perp$ be g-orthogonal projections onto $\widetilde{\mathscr{D}}$ and $\mathscr{D}$, respectively; for any $X \in \mathfrak{X}_M$ we write $X = X^\top + X^\perp$. Thus, $g = g^\top + g^\perp$, where

$$g^\perp(X,Y) = g(X^\perp, Y^\perp), \quad g^\top(X,Y) = g(X^\top, Y^\top), \qquad X,Y \in \mathfrak{X}_M.$$

Let $\mathfrak{X}_M$ (resp., $\mathfrak{X}_{\widetilde{\mathscr{D}}}$) be the module over $C^\infty(M)$ of vector fields on M (resp. evaluated in $\widetilde{\mathscr{D}}$). The *fundamental tensors* $\mathfrak{P}$ and $\mathscr{O}$ were introduced in [Gr1] by

$$\mathfrak{P}_Y X = (\nabla_{Y^\perp}(X^\perp))^\top + (\nabla_{Y^\perp}(X^\top))^\perp, \quad \mathscr{O}_Y X = (\nabla_{Y^\top}(X^\top))^\perp + (\nabla_{Y^\top}(X^\perp))^\top,$$
$$(1.25)$$

where $X,Y \in \Gamma(TM)$. In the book, we define several tensors for one of the distributions (say, $\mathscr{D}$; similar tensors for $\widetilde{\mathscr{D}}$ will be denoted using $^\top$ or $^\sim$ notation). The *co-nullity (splitting) operator* $C : \mathscr{D} \times \widetilde{\mathscr{D}} \to \widetilde{\mathscr{D}}$ of a distribution $\widetilde{\mathscr{D}}$ is defined by

$$C_\xi(X) = -(\nabla_X \xi)^\top, \quad X \in \Gamma(\widetilde{\mathscr{D}}), \ \xi \in \Gamma(\mathscr{D}). \tag{1.26}$$

Hence, $C = 0$ if and only if $\widetilde{\mathscr{D}}$ is integrable with totally geodesic integral manifolds.

Remark 1.33. One may show that fundamental tensors can be expressed through co-nullity tensors, e.g.,

$$\langle \mathscr{O}_Y X, \xi \rangle = -\langle \nabla_Y \xi, X \rangle = \langle C_\xi Y, X \rangle, \quad X,Y \in \Gamma(\widetilde{\mathscr{D}}), \ \xi \in \Gamma(\mathscr{D}).$$

The *2nd fundamental form* $h : \widetilde{\mathscr{D}} \times \widetilde{\mathscr{D}} \to \mathscr{D}$ and the *integrability tensor* $T : \widetilde{\mathscr{D}} \times \widetilde{\mathscr{D}} \to \mathscr{D}$ of $\widetilde{\mathscr{D}}$ are defined by

$$2h(X,Y) = (\nabla_X Y + \nabla_Y X)^\perp,$$
$$2T(X,Y) = (\nabla_X Y - \nabla_Y X)^\perp = [X,Y]^\perp. \tag{1.27}$$

The (self-adjoint) *shape operator* A_ξ and the skew-symmetric operator $T_\xi^\sharp$, where $\xi \in \mathscr{D}$, on $\widetilde{\mathscr{D}}$ are dual to the tensors h and T, respectively:

$$g(A_\xi X, Y) = \langle h(X,Y), \xi \rangle, \quad \langle T_\xi^\sharp X, Y \rangle = \langle T(X,Y), \xi \rangle. \tag{1.28}$$

The distribution $\widetilde{\mathscr{D}}$ is *totally geodesic* if and only if $h = 0$ (this is the case when *any geodesic of M that is tangent to $\widetilde{\mathscr{D}}$ at one point is tangent to $\widetilde{\mathscr{D}}$ at all its points*), and $\widetilde{\mathscr{D}}$ is tangent to a foliation if and only if $T = 0$. The case $h = T = 0$ corresponds to a *totally geodesic foliation*. The *mean curvature vector* of $\widetilde{\mathscr{D}}$ is $H = \text{trace}_g h$.

Given vector field $\xi \in \Gamma(\mathscr{D})$, $C_\xi = A_\xi + T_\xi^\sharp$ is a (1,1)-tensor in $\widetilde{\mathscr{D}}$. If $\widetilde{\mathscr{D}}$ is tangent to a foliation $\mathscr{F}$, then $C_\xi = A_\xi$. Observe that

$$\langle (C_\xi - C_\xi^*)X, Y \rangle = \langle [X,Y], \xi \rangle, \quad X,Y \in \widetilde{\mathscr{D}}, \ \xi \in \mathscr{D}, \tag{1.29}$$

$$\langle (C_\xi + C_\xi^*)X, Y \rangle = 2\langle h_{X,Y}, \xi \rangle, \quad X,Y \in \widetilde{\mathscr{D}}, \ \xi \in \mathscr{D},$$

where $*$ is the conjugation of a (1,1)-tensor. We have the following identities on $\widetilde{\mathscr{D}}$:

$$A_\xi = (C_\xi + C_\xi^*)/2, \qquad T_\xi^\sharp = (C_\xi - C_\xi^*)/2. \tag{1.30}$$

A pseudo-Riemannian manifold may admit many kinds of geometrically interesting foliations, e.g. $\mathscr{F}$ (with $T\mathscr{F} = \widetilde{\mathscr{D}}$) is *totally umbilical, harmonic*, or *totally geodesic*, if $h = (1/n) Hg^\top$, $H = 0$, or $h = 0$, respectively (see survey in [R1]). Totally geodesic foliations appear naturally in the study of manifolds with degenerate differential forms and curvature-like tensors. Parallel circles or winding lines on a flat torus and a Hopf field of great circles on a sphere S^3 are simple examples of geodesic foliations. Codimension-one totally geodesic foliations of closed non-negatively curved space forms are completely understood: they are given by parallel hyperplanes in the case of a flat torus and they don't exist for spheres. If the codimension is bigger than one, examples of geometrically distinct totally geodesic foliations on space forms are abundant. However, a compact space form $M(c)$ with $c < 0$ does not admit a totally geodesic foliation, see [Ze].

A foliation $\mathscr{F}$ (with $T\mathscr{F} = \widetilde{\mathscr{D}}$) is *conformal, transversely harmonic*, or *Riemannian*, if, respectively, $\tilde{h} = (1/p)\tilde{H}g^\perp$, $H^\perp = 0$, or $\tilde{h} = 0$. In other words, a foliation is Riemannian [Mol] if it respects a bundle-like metric: that is, locally the distance between leaves is constant. Using the natural representation of $O(p) \times O(n)$ on the tangent bundle TM, they described thirty-six different classes of Riemannian almost-product manifolds [Na]; some of them being foliations, e.g. harmonic, totally umbilical and totally geodesic. Following this line of research, several geometers completed the geometric interpretation, and gave examples for each class, e.g. using hypersurfaces and products of manifolds [Miq, S1]. There exist foliations with harmonic and nowhere totally geodesic leaves on Lie groups with left-invariant (bundle-like) metrics [TY].

For a simply connected manifold M, the *splitting* (i.e., the local metric decomposition) is equivalent to a global product structure. On the other hand, the geodesic foliation of a Klein bottle with flat metric splits only locally. If a complete and simply connected Riemannian manifold M has two complementary orthogonal totally geodesic foliations, then de Rham's Decomposition Theorem (see e.g. [Pe]) asserts that M is isometric to the product of the two leaves through a point $m \in M$.

The following convention is adopted for the range of indices:

$$a,b,c \ldots \in \{1 \ldots n\}, \quad i,j,k \ldots \in \{1 \ldots p\}.$$

It will be convenient to use orthonormal frames with certain nice properties.

Remark 1.34. One may show that a local adapted orthonormal frame $\{E_a, \mathscr{E}_i\}$, where $\{E_a\} \subset \widetilde{\mathscr{D}}$, and $\varepsilon_i = g(\mathscr{E}_i, \mathscr{E}_i) \in \{-1, 1\}$, $\varepsilon_a = g(E_a, E_a) \in \{-1, 1\}$, always exists on M. Of course, take a local frame (E_a) of the first distribution and complete it with vector fields (W_i) to get a local frame of TM. Then, apply the procedure called *Gram-Schmidt orthonormalization* to the frame $(V_1, \ldots, V_n, W_1, \ldots, W_p)$.

We will use the following convention for various tensors: $T_i^\sharp = T_{\mathscr{E}_i}^\sharp$, $A_i = A_{\mathscr{E}_i}$, etc. The *partial divergence* $\nabla_{\widetilde{\mathscr{D}}}^*$ of a $(1, j+1)$-tensor S on M is the $(1, j)$-tensor

$$(\nabla_{\widetilde{\mathscr{D}}}^* S)(Y_1, \ldots, Y_j) = -\sum_{i \leq n} \varepsilon_i (\nabla_i S)(\mathscr{E}_i, Y_1, \ldots, Y_j), \tag{1.31}$$

see (1.11a). For a $(1, 2)$-tensor S, define a $(0, 2)$-tensor $\operatorname{div}^\perp S$ (similarly, $\operatorname{div}^\top S$) by

$$(\operatorname{div}^\perp S)(X, Y) = \sum_i \varepsilon_i \langle (\nabla_i S)(X, Y), \mathscr{E}_i \rangle, \quad X, Y \in \mathfrak{X}_M,$$

see (1.11b). For a $\mathscr{D}$-valued $(0, 2)$-tensor S, using $\langle S, H \rangle (X, Y) := \langle S(X, Y), H \rangle$ we get

$$\operatorname{div}^\perp S = \operatorname{div} S + \langle S, H \rangle. \tag{1.32}$$

For example, $\operatorname{div}^\perp h := \sum_i \varepsilon_i \langle \nabla_i h, \mathscr{E}_i \rangle$ is a symmetric bilinear form on $\widetilde{\mathscr{D}}$. The $\mathscr{D}$-*divergence* of $X \in \mathfrak{X}_M$ is defined by

$$\operatorname{div}^\perp X = \sum_i \varepsilon_i \langle \nabla_i X, \mathscr{E}_i \rangle.$$

Thus, $\operatorname{div} X = \operatorname{div}^\perp X + \operatorname{div}^\top X$. For $X \in \mathfrak{X}_{\mathscr{D}}$ we get $\operatorname{div}^\perp X = \operatorname{div} X + \langle X, H \rangle$.

The $\widetilde{\mathscr{D}}$-Laplacian of $f \in C^2(M)$ is $\Delta^\top f = \operatorname{div}^\top (\nabla^\top f)$, where $\nabla^\top f := (\nabla f)^\top$ is the projection of the gradient of f onto $\widetilde{\mathscr{D}}$. Similarly, $\Delta^\perp f = \operatorname{div}^\perp (\nabla^\perp f)$ is the $\mathscr{D}$-Laplacian of f.

Applying Green's Theorem to the identity

$$\operatorname{div} X = \operatorname{div}^\perp X - \langle X, H \rangle,$$

for $X \in \Gamma(\mathscr{D})$ with compact support on (M, g), we obtain

$$\int_M (\operatorname{div}^\perp X) \, d\operatorname{vol} = \int_M \langle H, X \rangle \, d\operatorname{vol}. \tag{1.33}$$

So, $\int_M (\operatorname{div}^\perp X) d\operatorname{vol} = 0$ for compactly supported $X \in \Gamma(\mathscr{D})$ if and only if $H = 0$.

In analogy with the fact that on a closed connected Riemannian manifold, every harmonic function is constant, one may show that *if* $\Delta^\top f = H(f)$ *then* $\nabla^\perp f = 0$.

1.3.2 The Mixed Curvature

There are three kinds of sectional curvature for a pseudo-Riemannian manifold (M, g) endowed with a smooth distribution $\widetilde{\mathscr{D}}$: *tangential, transversal* and *mixed* (denoted by K_{mix}). The *mixed plane* is spanned by two vectors such that the first (second) vector is tangent (orthogonal) to $\widetilde{\mathscr{D}}$. *Mixed curvatures* stand for the sectional curvatures of mixed planes. The mixed curvature of a foliated manifold is encoded in the Riccati equation (also called the Raychaudhuri equation in relativity) and controls the deviation of leaves along the leaf geodesics, see e.g. [R1]. In the language of mechanics it measures the relative acceleration of two particles moving forward on neighboring geodesics. In relativity, it measures the tidal force experienced by a rigid body moving along a geodesic. For constant K_{mix} the solutions of equations mentioned above (and the relative behavior of geodesics on nearby leaves) are well-known. Riemannian foliations (and submersions) with totally geodesic leaves (fibers) have $K_{\mathrm{mix}} \geq 0$. One of the simplest curvature invariants of $(M, g, \widetilde{\mathscr{D}})$ is the *mixed scalar curvature*, i.e., an averaged sectional curvature of planes that non-trivially intersect $\widetilde{\mathscr{D}}$ and its orthogonal complement denoted by $\mathscr{D}$:

$$S_{\mathrm{mix}} = \sum_{a,i} \varepsilon_a \varepsilon_i \langle R(E_a, \mathscr{E}_i) E_a, \mathscr{E}_i \rangle, \qquad (1.34)$$

see [R1, W4]. To compute S_{mix} on (M, g) we only need to fix one of the distributions $(\widetilde{\mathscr{D}}, \mathscr{D})$, then the second distribution is defined as its g-orthogonal complement.

If either $\mathscr{D}$ or $\widetilde{\mathscr{D}}$ is spanned by a vector N such that $g(N, N) = \varepsilon_N \in \{-1, 1\}$, then S_{mix} is simply the Ricci curvature $\varepsilon_N \mathrm{Ric}_{N,N}$. For a surface (M^2, g), foliated by curves, we get $S_{\mathrm{mix}} = K$—the Gaussian curvature.

Definition 1.35 (See [R6]). The symmetric $(0,2)$-tensor

$$\mathrm{Ric}^\perp_{X,Y} := \sum_a \varepsilon_a \langle R(X^\perp, E_a) Y^\perp, E_a \rangle \qquad (1.35)$$

is called the *partial Ricci curvature* of $\widetilde{\mathscr{D}}$. Also $\mathrm{Ric}^\top_{X,Y} := \sum_i \varepsilon_i \langle R(X^\top, \mathscr{E}_i) Y^\top, \mathscr{E}_i \rangle$. The dual to (1.35) symmetric $(1,1)$-tensor is called the *partial Ricci tensor*,

$$\mathrm{Ric}^\perp(X) = \sum_a \varepsilon_a (R(E_a, X^\perp) E_a)^\perp.$$

For a foliation spanned by a unit vector N, we have $\mathrm{Ric}^\perp_{X,Y} = \varepsilon_N R(N, X, N, Y)$ for all $X, Y \in TM$, and $\mathrm{Ric}^\perp = R_N := \varepsilon_N R(N, \cdot)N$ is called the *Jacobi operator* for N. We consider the understanding (and prescribing, see Sections 1.4 and 5.3.8) of the partial Ricci curvature as a fundamental problem in extrinsic geometry of foliations.

One may show that $S_{\mathrm{mix}} = \mathrm{trace}_g\, \mathrm{Ric}^\perp$, and that definition (1.35) does not depend on the choice of $\{E_i\}$; thus, the partial Ricci curvature, $\mathrm{Ric}^\perp_{X,X}$, for a unit vector $X \in \mathscr{D}$ is the "mean value" of sectional curvatures over all mixed planes containing X.

Define the tensor $R^\perp_{\xi_1,\xi_2}(X) = \left(R(\xi_1,X^\perp)\xi_2\right)^\perp$ and the self-adjoint operator $R^\perp_\xi := R^\perp_{\xi,\xi}$, where $\xi_1,\xi_2 \in \widetilde{\mathscr{D}}$. Then we have $(\mathrm{Ric}^\perp)^\sharp(X) = \sum_{i=1}^p R^\perp_{\mathscr{E}_i}(X)$.

Define the self-adjoint $(1,1)$-tensors (*Casorati type operators*) $\mathscr{A} = \sum_i \varepsilon_i A_i^2$ and $\mathscr{T} = \sum_i \varepsilon_i (T_i^\sharp)^2$. The $\mathscr{D}$-*deformation* $\mathrm{Def}_{\mathscr{D}} Z$ of a vector field Z (e.g., $Z = H$) is the symmetric part of ∇Z restricted to $\mathscr{D}$,

$$\mathrm{Def}_{\mathscr{D}} Z(X,Y) = (1/2)\left(\langle \nabla_X Z, Y\rangle + \langle \nabla_Y Z, X\rangle\right), \quad X,Y \in \mathscr{D}; \tag{1.36}$$

it measures the degree to which the flow of a vector field Z distorts the metric.

Curvatures of the distributions are derived from the curvature of M, using fundamental tensors. The *Ricci equation* for distributions reads as, see [Gr1],

$$\langle R(X,\xi)Y,\eta\rangle = \langle(\nabla_\xi \mathscr{O})_X Y + \mathscr{O}(\mathscr{O}_X \xi, Y), \eta\rangle + \langle(\nabla_X \mathfrak{P})_\xi \eta + \mathfrak{P}(\mathfrak{P}_\xi X, \eta), Y\rangle. \tag{1.37}$$

From the *Codazzi equation* for a pair of complementary distributions, see [Gr1],

$$\langle R(X,Y)Z,\xi\rangle = \langle(\nabla_Y \mathscr{O})_X Z - (\nabla_X \mathscr{O})_Y Z + \mathfrak{P}(\mathscr{O}_X Y, Z) - \mathfrak{P}(\mathscr{O}_Y X, Z), \xi\rangle \tag{1.38}$$

we obtain the following identity with co-nullity operator:

$$(\nabla_X C)_\xi(Y) - (\nabla_Y C)_\xi(X) = -(R(X,Y)\xi)^\top + (\nabla_{[X,Y]^\perp}\xi)^\top. \tag{1.39}$$

Lemma 1.36. *For any* $X,Y \in \mathscr{D}$, $\xi_i \in \widetilde{\mathscr{D}}$, *we have*

$$R(\xi_1,X,\xi_2,Y) = \langle((\nabla_{\xi_1} C)_{\xi_2} - C_{\xi_2} C_{\xi_1})(X), Y\rangle + \langle((\nabla_X A)_Y - A_Y A_X)(\xi_1), \xi_2\rangle, \tag{1.40}$$

$$\mathrm{Ric}^\perp(X,Y) = \mathrm{div}_{\mathscr{F}}\, \tilde{h}(X,Y) - \langle(\mathscr{A}+\mathscr{T})(X), Y\rangle + \mathrm{Def}_{\mathscr{D}} H(X,Y) - \langle A_X, A_X\rangle. \tag{1.41}$$

Proof. By Remark 1.33, (1.40) follows from (1.37). Note that

$$\sum_i \langle(\nabla_X A)_Y(\mathscr{E}_i), \mathscr{E}_i\rangle = \sum_i \nabla_X \langle A_Y(\mathscr{E}_i), \mathscr{E}_i\rangle = \nabla_X \langle \sum_i h(\mathscr{E}_i, \mathscr{E}_i), Y\rangle = \langle \nabla_X H, Y\rangle.$$

Tracing (1.40) on $\widetilde{\mathscr{D}}$ yields

$$\mathrm{Ric}^\perp(X,Y) = -\langle \nabla^*_{\mathscr{F}} C(X), Y\rangle - \langle \sum_i C_i^2(X), Y\rangle + \langle \nabla_X H, Y\rangle - \mathrm{trace}(A_Y A_X).$$

The symmetric part of this is (1.41), and the antisymmetric part is

$$\langle \mathrm{div}_{\widetilde{\mathscr{D}}}\, T^\sharp(X), Y\rangle - \sum_i \langle(A_i T_i^\sharp + T_i^\sharp A_i)(X), Y\rangle = d^\top H(X,Y),$$

where $d^\top H(X,Y) = \frac{1}{2}\left[\langle \nabla_X H, Y\rangle - \langle \nabla_Y H, X\rangle\right]$ is the antisymmetric part of $\nabla^\perp H$. $\square$

One may show that for a *totally geodesic* foliation $\mathscr{F}$, we have

$$R^{\perp}_{\xi_1,\xi_2} = (\nabla_{\xi_1} C)_{\xi_2} - C_{\xi_2} C_{\xi_1}, \quad \mathrm{Ric}^{\perp} = \mathrm{div}_{\mathscr{F}} \tilde{h} - (\widetilde{\mathscr{A}} + \widetilde{\mathscr{T}})^{\flat}, \tag{1.42}$$

and that the symmetric and antisymmetric parts of $(1.42)_1$ with $\xi_1 = \xi_2 = \xi$ are

$$(R^{\perp}_{\xi})^{\sharp} = \nabla_{\xi} A_{\xi} - A^2_{\xi} - (T^{\sharp}_{\xi})^2, \quad \nabla_{\xi} T^{\sharp}_{\xi} = A_{\xi} T^{\sharp}_{\xi} + T^{\sharp}_{\xi} A_{\xi}. \tag{1.43}$$

Of course, the first equation in (1.42) follows from (1.36) with $A_X = A_Y = 0$, and the second equation in (1.42) follows from (1.41) with $H = A_X = 0$.

In contrast to scalar curvature, S_{mix} is related to the extrinsic geometry of foliations. The next formula [W4], being the trace of (1.41), has many applications:

$$S_{\mathrm{mix}} = \mathrm{div}(H + \tilde{H}) - \langle h, h \rangle - \langle \tilde{h}, \tilde{h} \rangle + \langle H, H \rangle + \langle \tilde{H}, \tilde{H} \rangle + \langle T, T \rangle + \langle \tilde{T}, \tilde{T} \rangle, \tag{1.44}$$

where $\langle h, h \rangle = \sum_{i,j} \varepsilon_i \varepsilon_j \langle h(\mathscr{E}_i, \mathscr{E}_j), h(\mathscr{E}_i, \mathscr{E}_j) \rangle$ and $\langle T, T \rangle = \sum_{i,j} \varepsilon_i \varepsilon_j \langle T(\mathscr{E}_i, \mathscr{E}_j), T(\mathscr{E}_i, \mathscr{E}_j) \rangle$,
which shows how S_{mix} is built of invariants of the distributions, see also [ARW, R1, RW2]. Applying to (1.44) the Divergence Theorem for a closed manifold, yields the integral formula, which provides decomposition criteria for foliations with an integrable normal distribution under constraints on the sign of S_{mix}. For example:

- If $\mathscr{F}$ is a compact harmonic foliation of a Riemannian manifold (M, g) with an integrable normal distribution $\mathscr{D}$ and $S_{\mathrm{mix}} \geq 0$, then M splits along the foliations.
- If $\mathscr{F}$ and $\mathscr{F}^{\perp}$ are complementary orthogonal totally umbilical foliations of a closed oriented (M, g) with $S_{\mathrm{mix}} \leq 0$, then M splits along the foliations.

Definition 1.37. The difference, called the *extrinsic curvature* of $\widetilde{\mathscr{D}}$,

$$R_{\mathrm{ex}}(X, Y, Z, W) = \langle h(X^{\top}, Z^{\top}), h(Y^{\top}, W^{\top}) \rangle - \langle h(X^{\top}, W^{\top}), h(Z^{\top}, Y^{\top}) \rangle$$

is useful in the study of extrinsic geometry of foliations. The traces (along $\widetilde{\mathscr{D}}$)

$$\mathrm{Ric}_{\mathrm{ex}} = \sum_a \varepsilon_a R_{\mathrm{ex}}(\cdot, E_a, \cdot, E_a), \quad S_{\mathrm{ex}} = \sum_b \varepsilon_b \mathrm{Ric}_{\mathrm{ex}}(E_b, E_b)$$

are called the *extrinsic Ricci curvature* and *extrinsic scalar curvature* of $\widetilde{\mathscr{D}}$.

From Definition 1.37 it follows that the extrinsic scalar curvatures of $\widetilde{\mathscr{D}}$ and $\mathscr{D}$ obey

$$S_{\mathrm{ex}} = \langle H, H \rangle - \langle h, h \rangle, \quad \widetilde{S}_{\mathrm{ex}} = \langle \tilde{H}, \tilde{H} \rangle - \langle \tilde{h}, \tilde{h} \rangle, \tag{1.45}$$

respectively. Thus, (1.44) is equivalent to the formula

$$S_{\mathrm{mix}} = S_{\mathrm{ex}} + \widetilde{S}_{\mathrm{ex}} + \langle T, T \rangle + \langle \tilde{T}, \tilde{T} \rangle + \mathrm{div}(H + \tilde{H}). \tag{1.46}$$

Example 1.38. Assume that $n = 1$. Then $\widetilde{\mathscr{D}}$ is spanned by a nonsingular vector field N. An example of such foliations is provided by a circle action $S^1 \times M \to M$ without fixed points. A flow of a unit vector field N is *geodesic* if $h = 0$ (the orbits are geodesics), and is *Riemannian* if $\tilde{h} = 0$ (the orbits are locally equidistant) A nonsingular Killing vector clearly defines a Riemannian flow; moreover, a Killing vector of unit length generates a geodesic Riemannian flow.

Put $g^\top = \varepsilon_N N^\flat \otimes N^\flat$, where $\varepsilon_N = \langle N, N \rangle \in \{-1, 1\}$. Then $S_{\mathrm{mix}} = \varepsilon_N \mathrm{Ric}_{N,N}$, where $\mathrm{Ric}_{N,N} = \sum_i \varepsilon_i \langle R_N(\mathscr{E}_i), \mathscr{E}_i \rangle$, and the partial Ricci curvature has a simple form:

$$\mathrm{Ric}^\top = \varepsilon_N \mathrm{Ric}_{N,N} \, g^\top, \quad \mathrm{Ric}^\perp = \varepsilon_N (R_N)^\flat.$$

Here $R_N = R(N, \cdot)N$ is the Jacobi operator. For a foliation by N-curves we have $T = 0$, while $\langle \tilde{T}, \tilde{T} \rangle = \varepsilon_N \langle \tilde{T}_N^\sharp, \tilde{T}_N^\sharp \rangle$. We have $\tilde{h} = \tilde{h}_{sc} N$, where $\tilde{h}_{sc} = \varepsilon_N \langle \tilde{h}, N \rangle$ is the scalar second fundamental form of $\widetilde{\mathscr{D}}$, and $\tilde{H} = \varepsilon_N \tilde{\tau}_1 N$. Let $\tilde{A}_N$ be the shape operator associated to $\tilde{h}_{sc}$ and $\tilde{\tau}_i = \mathrm{trace}\, \tilde{A}_N^i$ $(i \geq 0)$. We have $S_{\mathrm{ex}} = 0$, $\tilde{S}_{\mathrm{ex}} = \tilde{\tau}_1^2 - \tilde{\tau}_2$ and

$$\mathrm{div}\, N = \sum_i \varepsilon_i \langle \nabla_{\mathscr{E}_i} N, \mathscr{E}_i \rangle = -\langle N, \sum_i \varepsilon_i \nabla_{\mathscr{E}_i} \mathscr{E}_i \rangle = -\langle N, \tilde{H} \rangle = -\tilde{\tau}_1,$$
$$\mathrm{div}(\tilde{\tau}_1 N) = N(\tilde{\tau}_1) + \tilde{\tau}_1 \mathrm{div}\, N = N(\tilde{\tau}_1) - \tilde{\tau}_1^2.$$

The curvature vector of the flow lines is $H = \varepsilon_N \nabla_N N$. It follows that $h = H g^\top$ and

$$\langle h, h \rangle = \langle H, H \rangle, \quad \langle \tilde{H}, \tilde{H} \rangle = \tilde{\tau}_1^2, \quad \langle \tilde{h}, \tilde{h} \rangle = \tilde{\tau}_2, \tag{1.47}$$

where $\tilde{\tau}_i$ is the trace of the ith power of $\tilde{A}_N$. By (1.32), we obtain

$$\mathrm{div}\, \tilde{h} = \nabla_N \tilde{h}_{sc} - \tilde{\tau}_1 \tilde{h}_{sc}, \quad \mathrm{div}\, h = (\mathrm{div}\, H) g^\top. \tag{1.48}$$

The identities of Lemma 1.36 are simplified, see [RW2, RZ2], to

$$\varepsilon_N \big(R_N + (\tilde{A}_N)^2 + (\tilde{T}_N^\sharp)^2 \big)^\flat = N(\tilde{h}_{sc}) - H^\flat \otimes H^\flat + \mathrm{Def}_{\mathscr{D}} H,$$
$$\varepsilon_N \mathrm{Ric}_{N,N} = \mathrm{div}\, H + \varepsilon_N (N(\tilde{\tau}_1) - \tilde{\tau}_2) + \langle \tilde{T}, \tilde{T} \rangle. \tag{1.49}$$

Observe that $(1.49)_2$ is simply the trace of $(1.49)_1$.

Definition 1.39. The *mixed scalar curvature* of the curvature tensor $\overline{R}$ of a connection $\overline{\nabla}$ on an almost product pseudo-Riemannian manifold (M, g) is the function

$$\overline{S}_{\mathrm{mix}} = \frac{1}{2} \sum_{a,i} \varepsilon_a \varepsilon_i \big(\langle \overline{R}(E_a, \mathscr{E}_i) E_a, \mathscr{E}_i \rangle + \langle \overline{R}(\mathscr{E}_i, E_a) \mathscr{E}_i, E_a \rangle \big), \tag{1.50}$$

compare with (1.34). The definition (1.50) does not depend on the order of distributions and on the choice of a local frame.

Let $M = B \times F$ be the product of pseudo-Riemannian manifolds (B, g_B) and (F, g_F) with the canonical projections $\pi: M \to B$ and $\sigma: M \to F$. Given positive differentiable functions $u, v: M \to \mathbb{R}$ $(i = 1, 2)$, the metric of a *doubly twisted product* $B \times_{(u,v)} F$ is defined by $g = (u^2 g_B) \oplus (v^2 g_F)$, i.e.,

$$g(X, Y) = u^2 g_B(\pi_* X, \pi_* Y) + v^2 g_F(\sigma_* X, \sigma_* Y), \quad X, Y \in TM.$$

The leaves $B \times \{y\}$ (tangent to $\widetilde{\mathscr{D}}$) and the fibers $\{x\} \times F$ (tangent to $\mathscr{D}$) are totally umbilical in (M,g), and this property characterizes doubly twisted products, [PR]. If $v = 1$ then we have a *twisted product*, Figure 1.7, and a *warped product* if also u depends on F only, see also [Ch3]. Some examples are provided by rotational symmetric metrics, which appear on rotation surfaces in space forms.

Lemma 1.40. *Let $M = B \times_\varphi F$ be a twisted product with $\varphi : B \to \mathbb{R}_+$. Then $\mathrm{Ric}^\perp = -(\Delta_{\mathscr{F}} \varphi/\varphi) g^\perp$, and the following conditions are equivalent:*

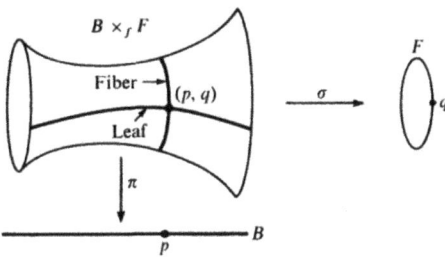

Fig. 1.7 Twisted product

> (a) $\mathrm{Hess}_\varphi^{\mathscr{F}} = -\lambda \varphi \, \mathrm{id}_{\mathscr{F}}$ *for some function* $\lambda : F \to \mathbb{R}$,
> (b) K_{mix} *depends on a point on M only.*

Proof. For a twisted product we have $h = 0$, $T = 0$, $\tilde{h} = -\nabla^{\mathscr{F}} (\log \varphi) g^\perp$ and $\tilde{H} = -n \nabla^{\mathscr{F}} (\log \varphi)$. The sectional curvature is $K(X, \xi) = -\frac{1}{\varphi} \xi(\xi(\varphi))$ for all local unit vector fields $\xi \in \mathfrak{X}_{\widetilde{\mathscr{D}}}$ and $X \in \mathscr{D}$, see [PR]. Note that $\langle \mathrm{Hess}_\varphi^{\mathscr{F}} (\xi), \xi \rangle = \xi(\xi(\varphi))$. Using (1.35), we then find the partial Ricci tensor

$$\mathrm{Ric}^\perp = -(\Delta_{\mathscr{F}} \varphi/\varphi) g^\perp.$$

From the above the last claim follows. □

Definition 1.41. A *doubly twisted product* of metric-affine spaces $(B^n, g_B, \mathfrak{T}_B)$ and $(F^p, g_F, \mathfrak{T}_F)$ with positive warping functions $u, v \in C^\infty(B \times F)$ is a manifold $M = B \times F$ with metric $g = v^2 g^\top + u^2 g^\perp$ and the contorsion tensor $\mathfrak{T} = u^2 \mathfrak{T}^\top + v^2 \mathfrak{T}^\perp$, where

$$g^\top(X,Y) = v^2 g_B(X^\top, Y^\top), \quad g^\perp(X,Y) = u^2 g_F(X^\perp, Y^\perp),$$
$$\mathfrak{T}_X^\top Y = u^2 (\mathfrak{T}_B)_{X^\top} Y^\top, \quad \mathfrak{T}_X^\perp Y = v^2 (\mathfrak{T}_F)_{X^\perp} Y^\perp.$$

For $v = 1$ we have the metric-affine *twisted product*, and a metric-affine *warped product* when also $u \in C^\infty(B)$.

Observe that $\mathfrak{T}^* = \mathfrak{T}^{*\top} + \mathfrak{T}^{*\perp}$, where $(\mathfrak{T}^{*\top})_X Y = (\mathfrak{T}_B^*)_{X^\top} Y^\top$ and $(\mathfrak{T}^{*\perp})_X Y = (\mathfrak{T}_F^*)_{X^\perp} Y^\perp$. Let $\mathscr{D}$ be tangent to the *fibers* $\{x\} \times F$ and $\widetilde{\mathscr{D}}$ tangent to the *leaves* $B \times \{y\}$. The second fundamental forms and mean curvature vectors of $B \times_{(v,u)} F$ are

$$\tilde{h} = -\nabla^{\top}(\log u)g^{\perp}, \quad h = -\nabla^{\perp}(\log v)g^{\top}, \quad \tilde{H} = -p\nabla^{\top}(\log u), \quad H = -n\nabla^{\perp}(\log v).$$

Thus, the leaves and the fibers are totally umbilical with respect to $\overline{\nabla}$ and ∇.

Example 1.42. One may show that if the factors of the doubly twisted product of metric-affine manifolds $(B^n, g_B, \mathfrak{T}_B)$ and $(F^p, g_F, \mathfrak{T}_F)$ are Riemann–Cartan manifolds then $(M, g, \overline{\nabla} = \nabla + \mathfrak{T})$ is also a Riemann–Cartan space. Since

$$\text{div}\,\tilde{H} = -p(\Delta^{\top}u)/u - (p^2 - p)\langle\nabla^{\top}u, \nabla^{\top}u\rangle/u^2,$$
$$\langle\tilde{H},\tilde{H}\rangle - \langle\tilde{h},\tilde{h}\rangle = (p^2 - p)\langle\nabla^{\top}u, \nabla^{\top}u\rangle/u^2,$$
$$\text{div}\,H = -n(\Delta^{\perp}v)/v - (n^2 - n)\langle\nabla^{\perp}v, \nabla^{\perp}v\rangle/v^2,$$
$$\langle H, H\rangle - \langle h, h\rangle = (n^2 - n)\langle\nabla^{\perp}v, \nabla^{\perp}v\rangle/v^2,$$

the formula (1.44) reduces to

$$S_{\text{mix}} = -p(\Delta^{\top}u)/u - n(\Delta^{\perp}v)/v.$$

From the above, (1.50) and (2.55) we get the linear elliptic PDE along a leaf,

$$-\Delta^{\top}u - (\beta + \overline{S}_{\text{mix}}/p)u + u^2\langle\text{trace}\,\mathfrak{T}^{\top}, \nabla u\rangle = 0, \tag{1.51}$$

where $\beta = (n/p)(v^{-1}\Delta^{\perp}v - v\langle\text{trace}\,\mathfrak{T}^{\perp}, \nabla v\rangle$ and u is unknown function. Let B be a closed manifold with $g_B > 0$ and trace $\mathfrak{T}_B = 0$. Hence, trace $\mathfrak{T}^{\top} = 0$, and (1.51) becomes the eigenvalue problem for the Schrödinger operator $\mathcal{H}$ with $\beta = (n/p)v^{-1}\Delta^{\perp}v$. Thus, $B \times_{(v,e_0)} F$ has leafwise constant $\overline{S}_{\text{mix}} = p\lambda_0$, where λ_0 is the least eigenvalue of $\mathcal{H}$ and $e_0 > 0$ is its simple eigenvector. For $\mathfrak{T}_B = 0 = \mathfrak{T}_F$ we get a Riemannian doubly twisted product of leafwise constant S_{mix}.

1.4 The Partial Ricci Curvature

We consider the understanding of the partial Ricci curvature as a fundamental problem in extrinsic geometry of foliations. In Section 1.4.1, we give examples of structures with $\widetilde{\mathscr{D}}$-conformal tensor $\text{Ric}^{\perp}$, which provide fixed points of the flow (5.52b) with certain values of Φ on structures with positive partial Ricci curvature. Section 1.4.2 is devoted to the following question: *Given a partial Ricci candidate T on $(M, \mathscr{F})$, is there a metric g satisfying $\text{Ric}^{\perp}(g) = T$?* Sections 1.4.3 and 1.4.4 deal with the weighted mixed curvature.

1.4.1 Structures with Constant Partial Ricci Curvature

Almost Contact Structure Contact manifolds come with a natural foliation given by the flowlines of the Reeb field. They also admit an *associated metric* with well known properties, [Bl]. An *almost contact manifold* (M, ϕ, ξ, η) is a $(2n+1)$-dimensional manifold M, which carries a $(1,1)$-tensor field ϕ, a vector field ξ, called characteristic or *Reeb vector field*, and a 1-form η satisfying

$$\phi^2 = -\mathrm{id}_{TM} + \eta \otimes \xi, \quad \eta(\xi) = 1.$$

One may show that $\phi\xi = 0$, $\eta \circ \phi = 0$ and ϕ has rank $2n$, see [Bl, Theorem 4.1]. We get an *almost contact metric structure*, if there is metric g (in general, non-unique) such that

$$g(\phi X, \phi Y) = g(X, Y) - \eta(X)\eta(Y).$$

Setting $Y = \xi$ we have $\eta(X) = g(X, \xi)$. A manifold M^{2n+1} with a 1-form η such that $d\eta(\xi, X) = 0$ for $X \in TM$ and $\eta(\xi) = 1$, is called a *contact manifold*, and ξ is called the *characteristic vector field*. A Riemannian metric g on a contact manifold (M^{2n+1}, η) is *associated* if there exists a $(1,1)$-tensor ϕ such that for all $X, Y \in TM$

$$\eta(X) = g(\xi, X), \quad d\eta(X, Y) = g(X, \phi(Y)), \quad \phi^2 = -I + \eta \otimes \xi. \tag{1.52}$$

The above (ϕ, ξ, η, g) is called a *contact metric structure* on M. An almost contact metric manifold (M, ϕ, ξ, η, g) is *normal* if

$$N_\phi + 2d\eta \otimes \xi = 0,$$

where

$$N_\phi(X, Y) := \phi^2[X, Y] + [\phi X, \phi Y] - \phi[\phi X, Y] - \phi[X, \phi Y]$$

is the *Nijenhuis torsion* of ϕ. A *Sasakian structure* is a normal contact metric structure, whose fundamental 2-form Φ defined by $\Phi(X, Y) = g(X, \phi Y)$ is $d\eta$.

Remark 1.43. One may show that on a contact metric manifold the integral curves of ξ are geodesics, i.e., $\nabla_\xi \xi = 0$. Of course, for a contact metric structure we have (e.g., [Bl, Theorem 4.5])

$$0 = (\mathscr{L}_\xi \eta)(X) = \xi\langle X, \xi\rangle - \langle \nabla_\xi X - \nabla_X \xi, \xi\rangle = \langle X, \nabla_\xi \xi\rangle$$

for all X; hence, the integral curves of ξ are geodesics.

Contact structures on M^{2n+1} are defined by $\omega = 0$ with $\omega \wedge (d\omega)^n \neq 0$ for a 1-form ω. A contact metric structure, for which ξ is Killing is called *K-contact*, see [Bl]. A manifold with a *geodesic Riemannian ξ-flow* (i.e., the integral curves of ξ are geodesics forming a Riemannian foliation) is called *Sasakian* if the curvature of sections containing ξ equals one, in other words, its curvature satisfies

$$R(X, \xi)Y = \langle \xi, Y\rangle X - \langle X, Y\rangle \xi.$$

A Sasakian metric can be defined using the construction of the Riemannian cone. Given a Riemannian manifold (M, g), its Riemannian cone is a product of M with a half-line $\mathbb{R}_{>0}$, equipped with the cone metric $t^2 g + dt^2$. A contact Riemannian manifold is Sasakian, if its Riemannian cone with the cone metric is a Kähler manifold.

For all contact (metric) manifolds we consider in this work, $\mathscr{D}$ is spanned by ξ and $\widetilde{\mathscr{D}}$ denotes the orthogonal complement of $\mathscr{D}$. If $n = 1$, $\widetilde{\mathscr{D}} = \ker \omega$ for a one-form ω satisfying $\omega \wedge d\omega > $ (meaning non-zero and positive oriented); distributions $\widetilde{\mathscr{D}} = \ker \omega$ with $\omega \wedge d\omega \geq 0$ are called *confoliations*, see [ET].

Remark 1.44. One can make from a Riemannian contact manifold (M, η, g) a pseudo-Riemannian contact manifold by setting $\bar{g} = g - 2\eta \otimes \eta$ as the new metric [BP]. Then $-\bar{g}(X, \xi) = \eta(X)$ for all $X \in TM$ and the remaining equations of (1.52) hold for $\bar{g}$ without changes. This transformation does not invalidate our main results.

Almost Contact 3-Structure In higher dimensions, an (almost) contact metric structure can be replaced by an (almost) contact 3-structure, introduced by Y. Kuo and C. Udrişte, as a set of three (almost) contact structures, $(M, \phi_i, \xi_i, \eta^i)$,

$$\phi_i^2 = -\operatorname{id}_{TM} + \eta^i \otimes \xi_i, \quad \eta^i(\xi_j) = \delta_j^i$$

with the same compatible metric g: $g(\phi_i X, \phi_i Y) = g(X, Y) - \eta^i(X)\eta^i(Y)$, obeying

$$\phi_k = \phi_i \circ \phi_j - \eta^i \otimes \xi_j = -\phi_j \circ \phi_i + \eta^j \otimes \xi_i$$

for any cyclic permutation (i, j, k) of $(1, 2, 3)$, see [Bl]. The dimension of M with an almost contact 3-structure is $4n + 3$. The quaternionic-like relations force restrictive geometric conditions. In particular, if each of (ϕ_i, ξ_i, η^i) is Sasakian, i.e., $g(X, \phi_i Y) = d\eta^i(X, Y)$, then we get a *3-Sasakian structure*; its characteristic distribution defines a totally geodesic Riemannian foliation. Let a distribution $\widetilde{\mathscr{D}}$ be spanned by 3 characteristic vector fields (ξ_i), and $\widetilde{\mathscr{D}}$ its orthogonal complement. For a 3-Sasakian structure we have $[\xi_i, \xi_j] = c\,\xi_k$ for some $c \in \mathbb{R}$ and any cyclic permutation (i, j, k) of $(1, 2, 3)$; thus, $\widetilde{\mathscr{D}}$ is integrable; moreover, it defines a totally geodesic Riemannian foliation with the property $T_i^\sharp = \phi_i$ along $\mathscr{D}$. Hence $\operatorname{Ric}^\perp = 3\,\operatorname{id}^\perp$ and the metric g is a fixed point of the flow (5.52b) with $\Phi = 3$.

An f-Structure An *f-structure* (due to Yano, [Yan]) on a manifold M is a non null $(1, 1)$-tensor field f on M of constant rank such that $f^3 + f = 0$. A manifold M equipped with an f-structure is called an *f-manifold*, and it is known that TM splits into two complementary subbundles $\widetilde{\mathscr{D}} = f(TM)$ and $\mathscr{D} = \ker f$, and that the restriction of f to $\widetilde{\mathscr{D}}$ determines a complex structure on it and the rank of f is even. This structure generalizes the almost complex and the almost contact structures. An interesting case of f-structure occurs when $\mathscr{D}$ is parallelizable for which there exist global vector fields ξ_i, $i \in \{1, \ldots, p\}$, with their dual 1-forms η^i, satisfying

$$f^2 = -\operatorname{id}_{TM} + \sum_i \eta^i \otimes \xi_i, \quad \eta^i(\xi_j) = \delta_j^i,$$

e.g., [FIP]. A manifold M endowed with such structure is denoted by (M, f, ξ_i, η^i); the vector fields ξ_i, $i = 1, \ldots, p$, are called characteristic vector fields. An f-structure, on a manifold M, is called *normal* if the following tensor field vanishes:

$$N_f + 2\sum_i d\eta^i \otimes \xi_i = 0.$$

A pseudo-Riemannian metric g on (M, f, ξ_i, η^i) is *compatible*, if

$$g(f(X), f(Y)) = g(X, Y) - \sum_i \varepsilon_i \eta^i(X) \eta^i(Y), \qquad (1.53)$$

where $\varepsilon_i = g(\xi_i, \xi_i) = \pm 1$. From (1.53) we get

$$g(X, \xi_i) = \varepsilon_i \eta^i(X), \qquad g(X, f(Y)) = -g(f(X), Y), \quad X, Y \in TM.$$

Define the 2-form Φ putting $\Phi(X, Y) = g(X, f(Y))$ for any $X, Y \in TM$. If $d\eta^i = \Phi$ for any i, then (M, f, ξ_i, η^i) with a compatible metric g is called an *almost S-manifold*. A normal almost S-manifold is called *S-manifold*; in this case, the sectional curvature is $K(X, \xi_i) = g(X, X)$. Hence, $\mathrm{Ric}^\perp = p\,\mathrm{id}^\perp$ and the metric g is a fixed point of the flow (5.52b) with $\Phi = p$.

A Para-f-Structure A smooth manifold M^{2n+p} is called an *almost para-f-manifold* [BM] if it admits a $(1,1)$-tensor field f of rank $2n$, satisfying $f^3 - f = 0$. The kernel distribution $\mathscr{D} = \ker f$ has dimension p and ± 1 eigen-distributions of f, denoted by $\mathscr{D}^+$ and $\mathscr{D}^-$, respectively, have the same dimension equal to n. Set $\widetilde{\mathscr{D}} := f(TM) = \mathscr{D}^+ \oplus \mathscr{D}^-$. This structure generalizes the almost product and the almost paracontact structures in a similar way to f-structures of corank greater than one generalize almost contact structures. Using known classification for metric almost product manifolds, they obtained thirty six classes of metric para-f-manifolds with parallelizable kernel, see [Tar]. If there exist global (structure) vector fields ξ_i $(1 \le i \le p)$, and 1-forms η^i, satisfying the compatibility conditions

$$f^2 = \mathrm{id}_{TM} - \sum_i \eta^i \otimes \xi_i, \quad f(\xi_i) = 0, \quad \eta^i \circ f = 0, \quad \eta^i(\xi_j) = \delta_j^i,$$

then M is an *almost para-f-manifold with complemented frames*. If an almost para-f-manifold admits a (*compatible*) pseudo-Riemannian metric g such that

$$g(fX, fY) = -g(X, Y) + \sum_i \eta^i(X) \eta^i(Y),$$

then it is called a *metric almost para-f-manifold*. Such metric g is always of signature $(n + p, n)$, an example is, see [FPM],

$$g(X, Y) = (1/2) \left(\bar{g}(X, Y) - \bar{g}(fX, fY) + \sum_i \eta^i(X) \eta^i(Y) \right),$$

where $\bar{g}(X, Y) = G(f^2 X, f^2 Y) + \sum_i \eta^i(X) \eta^i(Y)$ and G is any given metric on M. An almost para-f-manifold is *normal* if the following tensor field vanishes:

$$N_f - 2 \sum_i d\eta^i \otimes \xi_i = 0.$$

On a metric almost para-f-manifold, we define a 2-form by $F(X, Y) = g(X, f(Y))$. If $F = d\eta^i$ for any i, then M is called a *para-f-manifold*. A *para-S-manifold* is a normal para-f-manifold. In this case,

$$R(X, Y)\xi_i = \sum_j \left[\eta^j(X) f^2(Y) - \eta^j(Y) f^2(X) \right]$$

for all i, see [FPM, OD]. For $X = \xi_i$, $Y \in \widetilde{\mathscr{D}}$ we get $R(\xi_i, Y)\xi_i = Y$. Hence, $\text{Ric}^{\perp} = p\,\text{id}^{\perp}$ and the metric g is a fixed point of the flow (5.52b) with $\Phi = p$.

1.4.2 Prescribing the Partial Ricci Curvature

Different aspects of the problem of finding a Riemannian metric g, whose Ricci tensor is a given second-order symmetric tensor T, were considered by many authors, e.g., [Bes, PT2, PP, Pu]. We solve a similar problem for the partial Ricci curvature.

In Section 1.4.2 we use the following notations. $\mathscr{D}_1$ and $\mathscr{D}_2$ with $\dim \mathscr{D}_i = p_i \geq 2$ $(i = 1,2)$, $p_1 + p_2 = n$ are smooth complementary orthogonal distributions on a Riemannian manifold (M^n, g) with the Riemannian curvature tensor R, and an adapted local orthonormal frame $e_1, \ldots, e_n$, i.e., $e_i \in \mathscr{D}_1$ for $i \leq p_1$ and $e_\alpha \in \mathscr{D}_2$ for $\alpha > p_1$. The *partial Ricci curvature related to* $\mathscr{D}_1$ (and similarly for $\mathscr{D}_2$) is defined here as the symmetric bilinear form

$$\text{Ric}_1(X,Y) = \sum_{\alpha > p_1} \langle R(e_\alpha, X)Y, e_\alpha \rangle$$

on TM, compare with (1.35). Therefore, $S_{\text{mix}} = \text{trace}_g \text{Ric}_{i \mid \mathscr{D}_i}$ for $i = 1, 2$.

We find necessary and sufficient conditions on T for the existence of metrics $\tilde{g} = (1/\phi^2)g$ (conformal to the metric g), which solve the equations

$$\widetilde{\text{Ric}}_{i \mid \mathscr{D}_i} = T_{\mid \mathscr{D}_i} \quad (i = 1, 2). \tag{1.54}$$

The compatibility condition for (1.54) is $\text{trace}_g T_{\mid \mathscr{D}_1} = \text{trace}_g T_{\mid \mathscr{D}_2}$ with the traces equal to S_{mix}. We start with the solution to (1.54) at a point $x \in M$. Let M be a neighborhood of the origin O in $\mathbb{R}^n = \mathbb{R}^{p_1} \times \mathbb{R}^{p_2}$, and $\mathscr{D}_i$ $(i = 1, 2)$ be tangent to the i-th factor.

Example 1.45. Let a diagonal n-by-n matrix T obey $K = \sum_{i \leq p_1} T_{ii} = \sum_{\alpha > p_1} T_{\alpha\alpha}$. Then the metric $g = \sum_{a=1}^{n} \left(1 - \sum_{b \leq n} c_{bb} x_b^2\right) dx^a \otimes dx^a$, where

$$c_{ii} = T_{ii}/p_2 \ (1 \leq i \leq p_1), \quad c_{\alpha\alpha} = (T_{\alpha\alpha} - K/p_2)/p_1 \ (p_1 < \alpha \leq n),$$

defined in a neighborhood of O in $\mathbb{R}^n = \mathbb{R}^{p_1} \times \mathbb{R}^{p_2}$, satisfies (1.54) at the origin. Of course, since $g_{i\alpha} \equiv 0$, the partial Ricci curvature in local coordinates is

$$\begin{aligned}
\text{Ric}_1(g)_{ij} &= -\tfrac{1}{2}\sum_{\alpha\beta} g^{\alpha\beta}(g_{ij,\alpha\beta} + g_{\alpha\beta,ij}) + Q_1, \\
\text{Ric}_2(g)_{\alpha\beta} &= -\tfrac{1}{2}\sum_{ij} g^{ij}(g_{\alpha\beta,ij} + g_{ij,\alpha\beta}) + Q_2,
\end{aligned} \tag{1.55}$$

where Q_i is a function of g and its derivatives, and is homogeneous of degree 2 in the first derivatives of g. Substituting in (1.55)

$$g = \sum_{a=1}^{n} \left(1 - \sum_{b \leq n} c_{bb} x_b^2\right) dx^a \otimes dx^a,$$

we get at O the linear system of rank $< n$ (of n equations and n variables),

$$p_2 c_{ii} + \sum_\alpha c_{\alpha\alpha} = T_{ii} \ (1 \le i \le p_1), \quad p_1 c_{\alpha\alpha} + \sum_i c_{ii} = T_{\alpha\alpha} \ (p_1 < \alpha \le n).$$

We present the formulae with the partial Ricci and the mixed scalar curvatures for conformally related metrics on $(M, \mathscr{D}_1, \mathscr{D}_2)$. The conformal change of a metric g keeps the orthogonality of $\mathscr{D}_1$ and $\mathscr{D}_2$. Given a smooth function ϕ on M, we call

$$\Delta^{(1)}\phi = \sum_{i \le p_1} h_\phi(e_i, e_i), \quad \Delta^{(2)}\phi = \sum_{\alpha > p_1} h_\phi(e_\alpha, e_\alpha)$$

the $\mathscr{D}_1$- and $\mathscr{D}_2$-*Laplacian* of ϕ, respectively, where h_ϕ is the hessian of ϕ. Certainly, $\Delta^{(1)} + \Delta^{(2)} = \Delta$. Let $\mathbb{R}^n = \mathbb{R}^{p_1} \times \mathbb{R}^{p_2}$ $(p_1, p_2 > 0)$ be a decomposition of Euclidean space. The partial Laplacians are then $\Delta^{(1)} = \sum_{i \le p_1} \partial^2/\partial x_i^2$, and $\Delta^{(2)} = \sum_{\alpha > p_1} \partial^2/\partial x_\alpha^2$.

Proposition 1.46. *The partial Ricci curvatures and the mixed scalar curvature transform under conformal change of a metric* $\tilde{g} = (1/\phi^2)g$ *by the formulae*

$$\widetilde{\mathrm{Ric}}_1 = \mathrm{Ric}_1 + \left(p_2 \phi\, h_\phi + (\phi \Delta^{(2)}\phi - p_2 \|\nabla\phi\|^2)g\right)/\phi^2, \tag{1.56a}$$

$$\widetilde{\mathrm{S}}_{\mathrm{mix}} = \phi^2 \mathrm{S}_{\mathrm{mix}} + \phi\,(p_1 \Delta^{(2)}\phi + p_2 \Delta^{(1)}\phi) - p_1 p_2 \|\nabla\phi\|^2. \tag{1.56b}$$

Proof. Define a metric $\tilde{g}$ on M by $\tilde{g} = e^\psi g$, where $\psi = -2\log\phi$. The connections ∇ and $\tilde{\nabla}$ of g and $\tilde{g}$ are related by

$$\tilde{\nabla}_X Y = \nabla_X Y + \frac{1}{2}(X(\psi)Y + Y(\psi)X - \langle X, Y\rangle\nabla\psi),$$

see [GKM]. For the curvature tensor $\tilde{R}$ of $\tilde{g}$ we obtain

$$\tilde{R}(X,Y)Y = R(X,Y)Y + \frac{1}{2}(\langle h_\psi(X), Y\rangle Y - \langle h_\psi(Y), Y\rangle X - \|Y\|^2 h_\psi(X)) + \frac{1}{4}((Y(\psi)^2$$
$$- \|Y\|^2\|\nabla\psi\|^2)X + (X(\psi)Y(\psi) - Y(\psi)\|\nabla\psi\|^2)Y + X(\psi)\|Y\|^2\nabla\psi), \tag{1.57}$$

where $X \in \mathscr{D}_2, Y \in \mathscr{D}_1$, thus $\langle X, Y\rangle = 0$. Note that if Z is a g-unit vector then $\tilde{Z} = Ze^{-\psi/2}$ is a $\tilde{g}$-unit vector. Let $\nabla^{(2)}$ be the $\mathscr{D}_2$-gradient of a function. From this and (1.57), we get the relation between partial Ricci curvatures in both metrics

$$\widetilde{\mathrm{Ric}}_1 = \mathrm{Ric}_1 - \frac{1}{2}((\Delta^{(2)}\psi)g + p_2 h_\psi) + \frac{1}{4}((\|\nabla^{(2)}\psi\|^2 - p_2\|\nabla\psi\|^2)g + p_2\|\nabla\psi\|^2). \tag{1.58}$$

To shorten the formulae, we turn back to $e^\psi = 1/\phi^2$. Substituting $\nabla\psi = -\frac{2}{\phi}\nabla\phi$,

$$h_\psi(e_A, e_B) = \frac{2}{\phi^2}\|\nabla_{e_A}\phi\|^2 - \frac{2}{\phi}h_\phi(e_A, e_B), \quad \Delta^{(2)}\psi = \frac{2}{\phi^2}\|\nabla^{(2)}\phi\|^2 - \frac{2}{\phi}\Delta^{(2)}\phi,$$

in (1.58), we obtain (1.56a). By $\widetilde{\mathrm{S}}_{\mathrm{mix}} = \phi^2 \sum_{i \le p_1} \widetilde{\mathrm{Ric}}_1(e_i, e_i)$, (1.56b) is the result of the trace operation applied to (1.56a). The formula for Ric_2 is dual to Ric_1. $\qquad\square$

Let $\bar{g} = u^{4/(n-2)}g$ be a conformal metric, where $u > 0$ is a function on M. Then the mixed scalar curvature $\hat{S}_{\mathrm{mix}}$ of $\bar{g}$ satisfies the PDE, see [R6, Corollary 1],

$$-\frac{n}{n-2}\Delta u + S_{\mathrm{mix}}\, u = \hat{S}_{\mathrm{mix}}\, u^{\frac{n+2}{n-2}}, \tag{1.59}$$

when $p_1 = p_2$, similar to the classical PDE for the scalar curvature, see [Bes].

We will consider (1.54) for a domain V of a locally conformally flat space. Let $(x_1,\ldots,x_n)$ be coordinates on V with the metric $g_{ij} = \delta_{ij}/F^2$, where $F > 0$ is a differentiable function on M. We will fix on V canonical foliations

$$\mathscr{F}_1 = \{x_\alpha = c_\alpha,\ \alpha > p_1\}, \quad \mathscr{F}_2 = \{x_i = c_i,\ i \le p_1\}, \quad c_i, c_\alpha \in \mathbb{R},$$

consisting of coordinate submanifolds. Let $\mathscr{D}_2 = T\mathscr{F}_1$ and $\mathscr{D}_1 = T\mathscr{F}_2$ be their tangent distributions.

Theorem 1.47. *Let $f_1, f_2 \in C^1(V)$ and let T be a symmetric $(0,2)$-tensor such that*

$$T_{ij} = f_1\delta_{ij}, \quad T_{\alpha\beta} = f_2\delta_{\alpha\beta}, \quad T_{i\alpha} = 0 \quad (i,j \le p_1,\ \alpha,\beta > p_1).$$

Then, there is a metric $\tilde{g} = (1/\phi^2)\bar{g}$ solving (1.54) if and only if

$$\phi F = \sum\nolimits_{i \le p_1}(a_1 x_i^2 + b_i x_i) + \sum\nolimits_{\alpha > p_1}(a_2 x_\alpha^2 + b_\alpha x_\alpha) + c,$$
$$f_1 = -p_2(\phi F)^{-2}[\lambda - 2(a_2 - a_1)\mu], \quad f_2 = -p_1(\phi F)^{-2}[\lambda - 2(a_2 - a_1)\mu].$$

Here $a_1, a_2, b_k, c \in \mathbb{R}$, and

$$\lambda = \left(\sum\nolimits_k b_k^2\right) - 2(a_1 + a_2)c, \quad \mu = \sum\nolimits_{i \le p_1}(a_1 x_i^2 + b_i x_i) + \sum\nolimits_{\alpha > p_1}(a_2 x_\alpha^2 + b_\alpha x_\alpha).$$

Proof. The compatibility condition is $p_1 f_1 = p_2 f_2$. Observe that $\tilde{g} = \bar{g}/\phi^2 = g/(\phi F)^2 = g/\varphi^2$, where g is a Euclidean metric, and $\varphi = \phi F$. In view of $\mathrm{Ric}_1 = \mathrm{Ric}_2 = 0$ for g, we have

$$\widetilde{\mathrm{Ric}}_1 = \left[p_2\varphi h_\varphi + \left(\varphi\Delta^{(2)}\varphi - p_2\|\nabla\varphi\|^2\right)g\right]/\varphi^2,$$

see (1.56a), and similarly for $\widetilde{\mathrm{Ric}}_2$. Since $T_{|\mathscr{D}_k} = \widetilde{\mathrm{Ric}}_{k|\mathscr{D}_k}$ $(k = 1,2)$, we obtain

$$\varphi^2 f_1 g = p_2\varphi h_\varphi + \left(\varphi\Delta^{(2)}\varphi - p_2\|\nabla\varphi\|^2\right)g \quad \text{on } \mathscr{D}_1,$$

and similarly for f_2. Thus, the problem is reduced to studying the system of PDE's

$$p_2\varphi_{,x_i x_i} = \varphi f_1 - \Delta^{(2)}\varphi + p_2\|\nabla\varphi\|^2/\varphi, \quad i \le p_1$$
$$p_1\varphi_{,x_\alpha x_\alpha} = \varphi f_2 - \Delta^{(1)}\varphi + p_1\|\nabla\varphi\|^2/\varphi, \quad \alpha > p_1, \tag{1.60}$$
$$\varphi_{,x_k x_m} = 0, \quad 1 \le k \ne m \le n.$$

From $(1.60)_3$ we conclude that $\varphi = \sum_{k=1}^n \phi_k(x_k)$. From $(1.60)_{1,2}$ we find $\phi_i''(x_i) = 2a_1 \in \mathbb{R}$ for all $i \le p_1$ and $\phi_\alpha''(x_\alpha) = 2a_2 \in \mathbb{R}$ for all $\alpha > p_1$. Therefore, $\varphi =$

$\sum_{i \le p_1}(a_1 x_i^2 + b_i x_i) + \sum_{\alpha > p_1}(a_2 x_\alpha^2 + b_\alpha x_\alpha) + c$, where $b_k, c \in \mathbb{R}$. We also have $\Delta^{(1)} \varphi = 2a_1 p_1$ and $\Delta^{(2)} \varphi = 2a_2 p_2$. Hence $(1.60)_{1,2}$ are reduced to

$$2p_2(a_1 + a_2) = \varphi f_1 + p_2 \|\nabla \varphi\|^2/\varphi, \quad 2p_1(a_1 + a_2) = \varphi f_2 + p_1 \|\nabla \varphi\|^2/\varphi.$$

Comparing them, we see that the equality $p_1 f_1 = p_2 f_2 (= \phi^2 \widetilde{S}_{\mathrm{mix}})$ is necessary for the solution existence. In view of $\|\nabla \varphi\|^2 - 2(a_1 + a_2)\varphi = \lambda - 2(a_2 - a_1)\mu$, we get f_1 and f_2, as required. If $a_1 a_2 \le 0$ and $a_1^2 + a_2^2 + \sum_k b_k^2 > 0$ then the set $\{\phi = 0\}$ of singularities of $\tilde{g}$ is non-empty and can be explicitly described. If $a_1 a_2 > 0$ then the inequality $\phi > 0$ means that the discriminant of a quadratic equation is negative. $\square$

Example 1.48.

(i) One may show that if $a_i = a$, then $\widetilde{S}_{\mathrm{mix}} = -p_1 p_2 \lambda$ defined in Theorem 1.47 for $M = \mathbb{R}^{p_1} \times \mathbb{R}^{p_2}$, and the singularity set of $\tilde{g}$ can be described in terms of λ. Of course, if $\lambda < 0$ then a non-complete metric $\tilde{g}$ is defined on $\mathbb{R}^n$, and if $\lambda \ge 0$ then, excluding the homothety, the singularity set of $\tilde{g}$ consists of a point if $\lambda = 0$; a hyperplane if $\lambda > 0$ and $a = 0$; and an $(n-1)$-sphere if $\lambda > 0$ and $a \ne 0$.

(ii) Let M be the hyperbolic space $(\mathbb{H}^n, \bar{g})$, represented by the half space model $\mathbb{R}^n_+$, $x_n > 0$, and $\bar{g}_{ij} = \delta_{ij}/x_n^2$. For any pair of integers $p_1, p_2 > 0$, $p_1 + p_2 = n$, denote by $\mathcal{F}$ a foliation by p_2-planes $\{x\} \times \mathbb{R}^{p_2}$, where $x = (x_1, \ldots, x_{p_1}, 0, \ldots, 0)$. Let $\mathcal{D}_2$ be the distribution tangent to $\mathcal{F}$ and $\mathcal{D}_1$ its orthogonal complement. Using $F = x_n$, one may show that $\phi x_n = \sum_{i \le p_1}(a_1 x_i^2 + b_i x_i) + \sum_{\alpha > p_1}(a_2 x_\alpha^2 + b_\alpha x_\alpha) + c$. The singularity sets for different values of parameters are described in [R6, Example 2].

Remark 1.49. One may extend Theorem 1.47 for tensors T such that, see [R6],

(i) $T_{|\mathcal{D}_1} = \sum_{i \le p_1} f_i(x_k) dx_i^2$ and $T_{|\mathcal{D}_2} = \sum_{\alpha > p_1} f_\alpha(x_k) dx_\alpha^2$, with a fixed index $k \le p_1$. Then there is a metric $\tilde{g} = \bar{g}/\phi^2$ solving the problem (1.54) if and only if there is a differentiable function $U(x_k)$ on V such that $\phi F = e^U$ and

$$f_k = p_2 U'', \quad f_i = -p_2 U'^2 \ (i \ne k), \quad f_\alpha = U'' - (p_1 - 1)U'^2 \ (\alpha > p_1).$$

(ii) $T_{|\mathcal{D}_1} = \sum_{i \le p_1} f_i(x_k, x_\delta) dx_i^2$ and $T_{|\mathcal{D}_2} = \sum_{\alpha > p_1} f_\alpha(x_k, x_\delta) dx_\alpha^2$ with fixed indices $k \le p_1$ and $\delta > p_1$. Then there is a metric $\tilde{g} = \bar{g}/\phi^2$ solving the problem (1.54) if and only if there are differentiable functions $v(x_k), w(x_\delta)$ such that $\phi F = v + w$, where

$$f_k = [(p_2 v'' + w'')(v + w) - p_2(v'^2 + w'^2)]/(v + w)^2,$$
$$f_\delta = [(v'' + p_1 w'')(v + w) - p_1(v'^2 + w'^2)]/(v + w)^2,$$
$$f_\alpha = f_\delta - p_1 w''/(v + w) \ (\forall \alpha \ne \delta), \quad f_i = f_k - p_2 v''/(v + w) \ (\forall i \ne k).$$

As a consequence of above results for $u = (\phi F)^{1/(p-1)}$ (when $p_1 = p_2 = p$), we find C^∞ solutions to the PDE's of the type (1.59). For certain functions $\check{K}$ there exist conformally flat metrics $\tilde{g}$, whose mixed scalar curvature is $\check{K}$.

Corollary 1.50. *Assume that $p_1 = p_2$ and $\widetilde{K}$ is defined by any of conditions (i)–(iii):*

(i) $-p_1 p_2 [\lambda + 2(a_2 - a_1)\mu]$, *where* $a_1, a_2, b_k, c \in \mathbb{R}$, *and* λ, μ *as in Theorem 1.47.*
(ii) $p_2 e^{2U} [U'' - (p_1 - 1)U'^2]$, *where* $U(x_k)$ *is differentiable, for some* $k \le p_1$.
(iii) $(v + w)(p_2 v'' + p_1 w'') - p_1 p_2 (v'^2 + w'^2)$, *where* $v(x_k)$, $w(x_\delta)$ *are differentiable functions, for some* $k \le p_1, \delta > p_1$.

Then (1.59) has a solution, globally defined on $\mathbb{R}^n$, *given by*

(i) $u = \left[\sum_{i \le p_1} (a_1 x_i^2 + b_i x_i) + \sum_{\alpha > p_1} (a_2 x_\alpha^2 + b_\alpha x_\alpha) + c \right]^{2/(n-2)}$.
(ii) $u = e^{2U/(n-2)}$.   *(iii)* $u = (v + w)^{2/(n-2)}$.

Proof. For all cases (i)–(iii) we define $u = \varphi^{2/(n-2)}$.

(i) By Theorem 1.47, the mixed scalar curvature for the metric $\tilde{g}$ is $\widetilde{S}_{\mathrm{mix}} = -p_1 p_2 [\lambda + 2(a_2 - a_1)\mu]$. Substituting in (1.59) with $K = 0$, we get (1.59).
(ii) It follows that $\widetilde{S}_{\mathrm{mix}} = p_2 e^{2U} (U'' - (p_1 - 1)U'^2)$ for the metric of case (ii) in Remark 1.49. Similarly to (i), we obtain (1.59).
(iii) It follows that $\widetilde{S}_{\mathrm{mix}} = (v + w)(p_2 v'' + p_1 w'') - p_1 p_2 (v'^2 + w'^2)$ for the metric of case (iii) in Remark 1.49. Similarly to (i), we obtain (1.59). □

1.4.3 The Weighted Mixed Curvature

The dimension of a compact totally geodesic foliation with $K_{\mathrm{mix}} > 0$ can be easily estimated from above by half of the dimension of the manifold itself: the idea (by T. Frankel [F]) is that two compact totally geodesic submanifolds (for instance, large spheres in a round sphere) in a Riemannian space of positive curvature necessarily intersect each other if the sum of their dimensions is not less than the dimension of the whole space. On the other hand, if the radius of the circle of S^2 is "small," then there exists a large circle of S^2 that is "far" from it. J. Morvan [Mrv] started from this elementary fact and gave an upper bound for the distance between two submanifolds of a Riemannian space with positive sectional curvature. The generalization of the results for the weighted curvature as the version for foliations is given below.

The "weighted" curvature is defined for a Riemannian manifold endowed with a vector field X, e.g., when X is a gradient of a density function. The *weighted scalar curvature* appeared in Perelman's functionals for the Ricci flow. The *weighted Ricci curvature* was first studied by Lichnerowicz, and later by Bakry–Emery and many others. The study of *weighted Ricci tensor*,

$$\mathrm{Ric}_X^d = \mathrm{Ric} + \frac{1}{2}\mathscr{L}_X g - \frac{1}{d} X^\flat \otimes X^\flat \tag{1.61}$$

was motivated by the *curvature-dimension* inequality **CD**(c,d): $\mathrm{Ric}_X^d \ge c$, for a brief overview see [Ca]. Here d is an upper bound of the "generalized dimension" of the weighted manifold, c is a lower bound of the Ricci tensor, and $\mathscr{L}$ is the Lie deriva-tive. The property (1.61) arises for the Ric of a warped product of M of dimension $d > 0$ with a manifold B, the warping function $\phi = -(1/d)\log f$ and $X = \nabla f$.

Let (M^{n+p}, g) be equipped with a vector field X and complementary orthogonal distributions $\widetilde{\mathscr{D}}$ and $\mathscr{D}$ of ranks $\dim\widetilde{\mathscr{D}} = p > 0$ and $\dim\mathscr{D} = n > 0$. We consider three kinds of *weighted mixed curvature* (sectional, qth Ricci and scalar), introduce the notions of "curvature dimension" of a distribution and the "mixed-curvature-dimension" inequality, and obtain natural generalizations of several results known for the case of $X = 0$. We define several functions on $(M, g, \widetilde{\mathscr{D}}, \mathscr{D}, X)$, which "in-terpolate" between the weighed sectional and Ricci curvatures; such functions on (M, g) for $X = 0$ were studied by many geometers, see surveys in [R1, R2]. Let W^q be a subspace of $\widetilde{\mathscr{D}}_m$ spanned by $q \le p$ orthonormal vectors $\{x_1, \ldots, x_q\}$ at a point $m \in M$, and $y \in \mathscr{D}_m$ a unit vector. Set

$$\widetilde{\mathrm{Ric}}_q(y, W) := \sum_{i=1}^{q} K(y, x_i).$$

Riemannian manifolds of $\widetilde{\mathrm{Ric}}_q > 0$ form a lager class than ones of $K_{\mathrm{mix}} > 0$.

Definition 1.51. We define the *weighted mixed qth Ricci curvature* of $\{y, W^q\}$ by

$$\widetilde{\mathscr{Ric}}_q^{\mathfrak{d}}(y, W) := \widetilde{\mathrm{Ric}}_q(y, W) + \frac{q}{2}(\mathscr{L}_{X/p}g)(y, y) + \frac{qp}{\mathfrak{d}}\langle X/p, y\rangle^2, \tag{1.62}$$

where $W^q \subset \widetilde{\mathscr{D}}_m$, $y \in \mathscr{D}_m$, and $\mathfrak{d} \in \mathbb{R}$ is called the *curvature dimension of $\widetilde{\mathscr{D}}$*. By the *mixed-curvature-dimension* inequality $\mathbf{CD}^\top(c, \mathfrak{d}, q)$ for $\widetilde{\mathscr{D}}$ we mean the inequality

$$\widetilde{\mathscr{Ric}}_q^{\mathfrak{d}} \ge c. \tag{1.63}$$

$\mathscr{Ric}_q^d(x, W)$ for $W^q \subset \mathscr{D}_m$, $x \in \widetilde{\mathscr{D}}_m$, and $\mathbf{CD}^\perp(c, d, q)$ for $\mathscr{D}$ are defined similarly.

Example 1.52. Let $M^{k+3} = S^3 \times \hat{M}^k$ $(k > 0)$ be the product of a unit 3-sphere and a Riemannian manifold. Suppose that Y is a lift of any unit vector field on S^3 and the Killing field X is a lift of Hopf field on S^3 (corresponding to standard complex structure on $\mathbb{R}^4 = \mathbb{C}^2$). Set $\mathscr{D} = \mathrm{span}(Y)$. Then $\widetilde{\mathscr{Ric}}_{k+1}^{\mathfrak{d}}(Y, \cdot) \ge 1$ for $\mathfrak{d} > 0$.

The *weighted mixed sectional curvature* is the weighted sectional curvature of the planes that non-trivially intersect each of the distributions,

$$\widetilde{\mathscr{K}}^{\mathfrak{d}}(y, x) := K(y, x) + \left(\frac{1}{2}(\mathscr{L}_{X/p}g)(y, y) + \frac{p}{\mathfrak{d}}\langle X/p, y\rangle^2\right)\|x\|^2, \tag{1.64}$$

see (1.62) for $q = 1$ and $W = \{x\}$. Similarly, we define $\mathscr{K}^d(x, y)$. The x and y in (1.64) are placed in asymmetric way; generally, we have $\mathscr{K}^n(x, y) \ne \widetilde{\mathscr{K}}^p(y, x)$. Observe that $\widetilde{\mathscr{Ric}}_q^{\mathfrak{d}}(y, W) = \sum_{i=1}^{q} \widetilde{\mathscr{K}}^{\mathfrak{d}}(y, x_i)$. The weighted sectional curvature ap-pears in the formula for the second variation of energy of a path.

Lemma 1.53. *Let $\mathscr{F}$ be a p-dimensional Riemannian foliation of (M,g,X), and $\bar{\gamma}: [a,b] \times (-\varepsilon,\varepsilon) \to M$ a variation of the geodesic $\gamma(t) = \bar{\gamma}(t,0)$ and the variation field on γ, $x(t) = \partial_s \bar{\gamma}|_{s=0}$ belongs to $T\mathscr{F}$. Then the index form on a geodesic γ is*

$$\mathscr{I}(x,x) = \int_a^b \left(\|\dot{x} - \langle \dot{\gamma}, X \rangle x\|^2 - \widetilde{\mathscr{K}^p}(\dot{\gamma}, x)\|x\|^2 \right) dt + \langle \dot{\gamma}, X \rangle \|x\|^2 \big|_a^b. \qquad (1.65)$$

Proof. A geodesic started orthogonally to a leaf of a Riemannian foliation remains to be normal to leaves. Thus, the proof is as the proof of [Wy, Proposition 5.1]. □

Remark 1.54. Let V_1, V_2 be subspaces in $\mathbb{R}^l$, $\dim V_1 = \dim V_2$. Then there exist orthonormal bases $\{a_i\} \subset V_1$, $\{b_i\} \subset V_2$ (which correspond to extremal values of angle between given subspaces) with the property $a_i \perp b_j$ $(i \neq j)$.

Let $\operatorname{diam}\mathscr{F}$ be the *maximal distance between the leaves of a foliation $\mathscr{F}$*.

Theorem 1.55. *Let (M^{n+p}, g, X) be endowed with a p-dimensional Riemannian foliation $\mathscr{F}$ $(T\mathscr{F} = \mathscr{D})$ with compact leaves of the second fundamental form h. If $CD^\top(c,\mathfrak{d},q)$ holds for some $\mathfrak{d} \geq p$, $c > 0$ and $1 \leq q \leq p$, then*

$$(\operatorname{diam}\mathscr{F})^2 \leq \frac{2q\|X^\perp\|}{c + q\|X^\perp\|^2} + \begin{cases} \dfrac{2q}{c}\|h\| + \dfrac{\pi^2}{4} & \text{if } p \leq n-1, \\[2mm] \dfrac{2q}{c}\|h\| + (q-p+n-1)\dfrac{\pi^2}{4c} & \text{if } n-1 < p < n+q-1, \\[2mm] \dfrac{2q}{c}\|h\| & \text{if } p \geq n+q-1. \end{cases}$$

Proof. Consider two leaves L_1, L_2 with distance $l = \operatorname{dist}(L_1, L_2)$, which is reached at points $m_1 \in L_1$ and $m_2 \in L_2$. The shortest geodesic $\gamma(t)$ $(0 \leq t \leq 1)$ with length l between m_1, m_2 is orthogonal to L_1 and L_2. Since $\mathscr{F}$ is a Riemannian foliation, γ intersect the leaves orthogonally for all $t \in (0,1)$.

Assume the second case: $n-1 < p < n-1+q$ (the other two cases are similar). Then the parallel displacement of $T_{m_1}L_1$ along γ will intersect $T_{m_2}L_2$ by q'−dimensional subspace V_2, where $p - n + 1 \leq q' < q$. The inverse image of V_2 in $T_{m_1}L_1$ we denote by V_1. For small l, let $T_{m_1}L_1 = V_1 \oplus V_1'$ be the orthogonal decomposition where the parallel image of V_1' is uniquely projected onto $T_{m_2}L_2$ (denote its orthogonal projection in $T_{m_2}L_2$ by V_2'). Let vectors $e_1, \ldots, e_{q'}$ form an orthonormal basis of V_1 and continue them to parallel vector fields $\bar{e}_1, \ldots, \bar{e}_{q'}$ along γ. Note that $\bar{e}_1(m_2), \ldots, \bar{e}_{q'}(m_2)$ belong to V_2. Let vectors $a_1, \ldots, a_s$ (where $s = q - q' = \dim V_1'$) form an orthonormal basis of V_1' and vectors $b_1, \ldots, b_s$ form an orthonormal basis of V_2', and continue them to parallel vector fields $\bar{a}_1, \ldots, \bar{a}_s$ and $\bar{b}_1, \ldots, \bar{b}_s$ along γ. Consider the field of parallel planes $\sigma_i(t)$ along γ, spanned by vectors $\bar{a}_i(t), \bar{b}_i(t)$. Assume, that $\{a_i\}$, $\{b_i\}$ correspond to extremal angles between V_1' and parallel image of V_2', see Remark 1.54. Then $\sigma_i(t) \perp \sigma_j(t)$ for $i \neq j$. We take the unit vector $\tilde{b}_i(t) \in \sigma_i(t)$ such that $\langle \bar{a}_i, \tilde{b}_i(t) \rangle = 0$. One may choose b_i and $\tilde{b}_i(t)$ with the properties $\langle \bar{a}_i, \bar{b}_i \rangle \geq 0$ and $\langle \bar{b}_i, \tilde{b}_i(t) \rangle \geq 0$. Let us introduce the unit vector fields $x_i(t) = (\cos \theta_i t)\bar{a}_i + (\sin \theta_i t)\tilde{b}_i(t)$ along γ, where $\theta_i = \arccos(\bar{a}_i, \bar{b}_i) \in [0, \frac{\pi}{2}]$. Note that $\langle x_i(t), x_j(t) \rangle = 0$ $(i \neq j)$, and $\langle \dot{x}_i(t), x_i(t) \rangle = 0$. We have $q' + s = q$. Using the 2nd variation of $\mathscr{E}$ on γ, (1.65), along $x_i(t)$ and $\bar{e}_j$, we obtain

$$\mathscr{E}_{x_i}''(0) = \langle h(b_i,b_i),\dot{\gamma}(1)/l\rangle - \langle h(a_i,a_i),\dot{\gamma}(0)/l\rangle + \theta_i^2$$

$$- l^2 \int_0^1 \left(\widetilde{\mathscr{K}^p}(\dot{\gamma},x_i(t)) + \langle \dot{\gamma}/l,X\rangle^2 \right) dt + 2\langle \dot{\gamma}/l,X\rangle\,|_0^1 \geq 0,$$

$$\mathscr{E}_{\bar{e}_j}''(0) = \langle h(\bar{e}_j,\bar{e}_j),\dot{\gamma}(1)/l\rangle - \langle h(e_j,e_j),\dot{\gamma}(0)/l\rangle$$

$$- l^2 \int_0^1 \left(\widetilde{\mathscr{K}^p}(\dot{\gamma},\bar{e}_j) + \langle \dot{\gamma}/l,X\rangle^2 \right) dt + 2\langle \dot{\gamma}/l,X\rangle\,|_0^1 \geq 0. \tag{1.66}$$

Since $s = q - q' \leq q - p + n - 1$, $\sum_i \theta_i^2 \leq (\pi^2/4)\,s$, we have

$$\sum_{i=1}^{q'} \| \langle h(b_i,b_i),\dot{\gamma}(1)/l\rangle - \langle h(a_i,a_i),\dot{\gamma}(0)/l\rangle \| \leq 2q'\|h\|,$$

$$\sum_{j=1}^{s} \| \langle h(\bar{e}_j,\bar{e}_j),\dot{\gamma}(1)/l\rangle - \langle h(e_j,e_j),\dot{\gamma}(0)/l\rangle \| \leq 2s\|h\|.$$

By (1.63) and condition $\eth \geq p$, we get $\widetilde{\mathscr{Ric}}_q^p(\dot{\gamma},W) \geq c$ for W spanned by $x_i(t)$ and $\bar{e}_j$; hence,

$$\sum_{i=1}^{q'} \widetilde{\mathscr{K}^p}(\dot{\gamma},x_i(t)) + \sum_{j=1}^{q-q'} \widetilde{\mathscr{K}^p}(\dot{\gamma},\bar{e}_j) \geq c.$$

Then by (1.66), $l^2(c + q\|X^\perp\|^2) \leq 2q\|h\| + (q - p + n - 1)\pi^2/4 + 2q\|X^\perp\|$. $\qquad\square$

Corollary 1.56. *Let (M^{n+p},g) be endowed with a compact totally geodesic foliation $\mathscr{F}^p$ and a vector field X tangent to the leaves. Suppose that inequality $\widetilde{\mathscr{Ric}}_q^0 > 0$ is satisfied for some $\eth \geq p$ and $1 \leq q \leq p$. Then $p < n + q - 1$.*

1.4.4 Toponogov Conjecture

Denote by $\rho(n) - 1$ the *maximal number of point-wise linearly independent vector fields on a sphere S^{n-1}*, such vector fields are built using orthogonal multiplications on $\mathbb{R}^n$. The topological invariant $\rho(n)$ [A] (*Adams number*) is $\rho((\text{odd})\,2^{4b+c}) = 8b + 2^c$ for any $b \geq 0$, $0 \leq c \leq 3$, see Table 1.1, and $\rho(n) \leq 2\log_2 n + 2 \leq n$.

Table 1.1 The number of vector fields on the $(n-1)$-sphere

$n-1$	1	3	5	7	9	11	13	15	17	19	21	23	25	27	29
$\rho(n)-1$	1	3	1	7	1	3	1	8	1	3	1	7	1	3	1

There are no much theorems in differential geometry, which involve $\rho(n)$.

D. Ferus [Fe] found a topological obstruction for existence of totally geodesic foliations. For a p-dimensional totally geodesic foliation of a closed Riemannian manifold (M^{n+p},g), whose sectional curvature, K_{mix}, has the same positive value for all mixed planes, he proved that

$$p \leq \rho(n) - 1. \tag{1.67}$$

Idea of Proof (of Ferus's Theorem) As a rule, a linear operator defined on an even-dimensional vector space has no eigenvectors. The situation changes when we consider a p-parameter family of linear operators. Let $\{e_i\}$ be an orthonormal basis of $T_m\mathscr{F}$. Then the following p continuous vector fields:

$$w_i(y) := C(e_i, y) - \langle C(e_i, y), y \rangle y, \ 1 \le i \le p,$$

are tangent to the unit sphere $S^{n-1} \subset T_m^\perp \mathscr{F}$. If $p \ge \rho(n)$ then these vector fields are linearly dependent at some point $y \in S^{n-1}$ with weights λ_i, i.e., $\sum_i \lambda_i w_i(y) = 0$. Then the co-nullity operator has real eigenvector: $C(x, y) = \lambda y$, where $x = \sum_i \lambda_i e_i$ and $\lambda = \langle C(x, y), y \rangle$. The proof of the theorem is based on the above argument on eigenvectors and analysis of the matrix Riccati equation

$$\dot{C}_{\dot\gamma} + (C_{\dot\gamma})^2 + R_{\dot\gamma} = 0$$

on a leaf geodesic γ ($\dot\gamma = x$), where $C_x = C(x, \cdot)$ for short. If $R_{\dot\gamma} = k\,\mathrm{id}^\perp$, then the eigenvectors of the solution $C_{\dot\gamma}$ do not depend on t; every eigenfunction $\lambda(t)$ satisfies the scalar Riccati equation $\lambda'(t) + \lambda^2(t) + k = 0$, and hence, for $k > 0$, a solution cannot be extended to the whole real line $t \in \mathbb{R}$.   □

Since $\rho(n) - 1 = 0$ for an odd n, Ferus's theorem prohibits the existence of a geodesic foliation of an even-dimensional manifold with positive constant mixed curvatures. In the case of $p = 1$ and an even n, the manifold is foliated by complete geodesics. Hopf's fiber bundle $\pi : S^3 \to S^2$, the sphere S^3 being equipped with the standard metric, gives a simple example of such a foliation for odd $n + p = 3$.

Among Toponogov's many contributions to global Riemannian geometry (see webpage http://math.haifa.ac.il/ROVENSKI/toponogov_e.html) there is the following, see survey in [R1, p. 30]:

Conjecture 1.57. Ferus's theorem can be generalized by replacing the hypothesis "all mixed curvatures are equal to a positive constant" with the weaker one: "all mixed curvatures are positive".

The conjecture is still open for closed foliated manifolds. Using variation of geodesics for a locally given foliation, there were proven (i)–(ii) in [R1]:

(i) The exactness of estimate (1.67) and necessity of more conditions, because of local counterexample: let $n = 2m + 1$, $R_{2m+1} = [1, \ldots, 1, 4]$ a diagonal matrix, and

$$\underset{m\ blocks}{C(t)} = \begin{pmatrix} C_2 & \cdots & 0 & 0 \\ \vdots & \ddots & \vdots & \vdots \\ 0 & \cdots & C_2 & 0 \\ 0 & \cdots & 0 & C_3 \end{pmatrix}, C_2 = \begin{pmatrix} 0 & 1 \\ -1 & 0 \end{pmatrix}, C_3 = \begin{pmatrix} \sin 2t & \cos 2t & -\sin t \\ \cos 2t & -\sin 2t & -\cos t \\ 4\sin t & 4\cos t & 0 \end{pmatrix}.$$

The above (co-nullity operator) $C(t)$ solves the matrix Riccati equation $\dot{C} + C^2 + R_{2m+1} = 0$; hence, corresponds to a geodesic foliation of an open domain on CP^{m+1}, while $p = 1 = \rho(n)$ — (1.67) is not valid.

(ii) Conjecture 1.57 for the special case, when M is a ruled submanifold of a sphere. Namely, *let M^{n+p} be a ruled submanifold with complete rulings $\{L\}$ in a sphere $S^N(1)$ satisfying condition*

$$(A_\xi X)^2 + (A_\xi Y)^2 \le 1 \quad X, Y \in TL, \ X \perp Y, \ |X| = |Y| = 1.$$

If $p \ge \max\{\rho(t) : t \le n\}$ then M^{n+p} is locally the product $L^p \times \tilde{M}^n$. In particular, if such $M^{n+p} \subset S^N(1)$ satisfies $K_{\mathrm{mix}} > 0$ then (1.67) is valid.

Approach using the flows of metrics to Conjecture 1.57 is given in Section 5.3. We introduce the *weighted Jacobi operator* $\mathscr{R}_x$ $(x \in \widetilde{\mathscr{D}})$ by

$$\mathscr{R}_x := R_x + \big((\mathscr{L}_{X/n} g)(x,x)/2 + \langle X/n, x\rangle^2 \big)\,\mathrm{id}^\perp, \tag{1.68}$$

and similarly define $\widetilde{\mathscr{R}}_y$ $(y \in \mathscr{D})$. Set $\mathscr{R}ic(x,x) = \mathrm{trace}_g\,\mathscr{R}_x$ and $\widetilde{\mathscr{R}ic}(y,y) = \mathrm{trace}_g\,\widetilde{\mathscr{R}}_y$. Thus, $\mathscr{R}ic(x,x) = \mathscr{R}ic_n^p(x, \mathscr{D}_m)$, see (1.62).

In this section we study the following Toponogov type conjecture, see [R11].

Conjecture 1.58. *The inequality $p < \rho(n)$ is valid for a closed manifold (M^{n+p}, g) endowed with a totally geodesic foliation $\mathscr{F}^p$ and a vector field X such that $\mathscr{R}_x > \|X^\top/n\|^2\,\mathrm{id}^\perp$ for all unit vectors $x \in T\mathscr{F}$.*

Since $\mathscr{R}_x > 0$ $(x \ne 0)$ yields the inequality

$$R_x > -(1/2)\,(\mathscr{L}_{X/n} g)(x,x)\,\mathrm{id}^\perp,$$

then R_x (and, hence, K_{mix}) in the above conjecture can be negative somewhere.

Much is known about foliations with $K_{\mathrm{mix}} = \mathrm{const}$. Some examples of such foliations are provided by (1) *k-nullity* foliations on manifolds with degenerate curvature tensor (the certain metrics are called partially hyperbolic, parabolic or elliptic); (2) *relative nullity* foliations of curvature-invariant submanifolds M (e.g., of space forms). Submanifolds with positive relative nullity index, $\mu(m) = \dim\ker h(m)$ for all $m \in M$ (introduced in [CK]) have a ruled developable structure. The weighted modification of the co-nullity operator of a totally geodesic foliated manifold (M, g) equipped with a vector field X, is defined to be

$$\mathscr{B}_x = C_x - \langle X/n, x\rangle\,\mathrm{id}^\perp.$$

Then the following "weighted" Riccati equation is valid along leaf geodesics:

$$\dot{\mathscr{B}}_{\dot\gamma} + (\mathscr{B}_{\dot\gamma})^2 + 2\,\langle X/n, \dot\gamma\rangle\,\mathscr{B}_{\dot\gamma} + \mathscr{R}_{\dot\gamma} = 0. \tag{1.69}$$

Next theorem and corollary with constant $\mathscr{R}_x > 0$ generalize Ferus's theorem.

Theorem 1.59. *Let (M^{n+p}, g, X) be endowed with a totally geodesic foliation $\mathscr{F}^p$. Suppose that there exist $k = \mathrm{const} > 0$ and a point $m \in M$ such that $\mathscr{R}_{\dot\gamma} = k\,\mathrm{id}^\perp$ is valid along any leaf geodesic $\gamma : [0, \pi/\sqrt{k}] \to M$ with $\gamma(0) = m$, and*

$$\langle X/n, \dot{\gamma} \rangle^2 \leq k. \tag{1.70}$$

Then $p < \rho(n)$.

Proof. Assume the contrary, $p \geq \rho(n)$. Then there exist unit vectors $x \in T_m \mathcal{F}$ and $y \in T_m^{\perp} \mathcal{F}$ and $\lambda \leq 0$ such that $\mathcal{B}_x y = \lambda y$ for a geodesic $\gamma(t)$ with initial velocity $\dot{\gamma}(0) = x$ (see Idea of proof of Ferus's theorem). Let $\bar{y}(t)$ ($\bar{y}(0) = y$) be a parallel vector field along γ. The eigenvectors of the solution $\mathcal{B}_{\dot{\gamma}}$ of (1.69) with $\mathcal{R}_{\dot{\gamma}} = k \, \mathrm{id}^{\perp}$ do not depend on t. Then $\mathcal{B}_{\dot{\gamma}} \bar{y}(t) = \lambda(t) \bar{y}(t)$ for certain eigenfunction $\lambda(t)$, which satisfies the scalar Riccati equation $\dot{\lambda} + \lambda^2 + 2\lambda \langle X/n, \dot{\gamma} \rangle + k = 0$. By (1.70), solution $\lambda(t)$ of above ODE cannot be extended to $[0, \pi/\sqrt{k}]$, a contradiction. $\square$

The *relative nullity space* of the second fundamental form h of a submanifold $M \subset \bar{M}$ at $m \in M$ is $\ker h(m) = \{x \in T_m M : h(x, y) = 0 \text{ for all } y \in T_m M\}$. A submanifold $M \subset \bar{M}$ is *curvature-invariant* if the curvature tensor of $\bar{M}$ satisfies $(\bar{R}(x, y)z)^{\perp} = 0$, $(x, y, z \in TM)$. Such submanifolds with positive index of relative nullity $\mu(M) = \min_{m \in M} \mu(m)$ have a ruled developable structure. The *extrinsic qth Ricci curvature* is

$$\mathrm{Ric}_h^q(x_0, x_1, \ldots, x_q) = \sum_{i=1}^{q} \left(\langle h(x_0, x_0), h(x_i, x_i) \rangle - \langle h(x_0, x_i), h(x_0, x_i) \rangle \right).$$

Corollary 1.60. *Let M^n be a complete curvature-invariant submanifold in $(\bar{M}^{n+p}, \bar{g})$ endowed with a vector field X. Suppose that there exists real $k > 0$ such that for any unit vector $Z \in \ker h$, the weighted Jacobi operator of $\bar{M}$ obeys (1.70) and $\bar{\mathcal{R}}_Z = k \, \mathrm{id}^{\perp}$. Then M is totally geodesic if any of conditions a), b) is satisfied:*

(a) $\mu(M) > \nu(n) := \max\{t : t < \rho(n-t)\}$, (b) $\mathrm{Ric}_h^q \leq 0$ and $2p < n - \nu(n) - q + \delta_{1q}$.

Conjecture 1.57 has been studied in [R1] for a foliation given near a complete leaf: the necessity of additional assumptions in local case was shown. The proof was based on estimates of the length of associated Jacobi field of "extremal" geodesics, and it examines conditions when co-nullity operator has no real eigenvectors, providing $p < \rho(n)$. Here we extend the above method to study Conjecture 1.58.

A smooth $(1,1)$-tensor field $Y(t) : T_{\gamma(t)}^{\perp} \mathcal{F} \to T_{\gamma(t)}^{\perp} \mathcal{F}$ on a leaf geodesic γ is called a *Jacobi tensor* if it satisfies the equation $\ddot{Y} + R_{\dot{\gamma}} Y = 0$, and $\ker Y(t) \cap \ker \dot{Y}(t) = \{0\}$ for all t; hence, the action of Y on linearly independent parallel sections of $T_{\gamma}^{\perp} \mathcal{F}$ gives rise to linearly independent Jacobi vector fields. We have $C_{\dot{\gamma}} = \dot{Y} Y^{-1}$.

Lemma 1.61 (See Lemma 4.7 in [R1]). *Let a solution $y(t) \subset \mathbb{R}^n$ of the Jacobi ODE*

$$\ddot{y} + R(t)y = 0 \quad (0 \leq t \leq \pi/\sqrt{k}), \tag{1.71}$$

be written in the form $y(t) = \bar{y}(t) + u(t)$, where $\bar{y}(t) = y(0)\cos(\sqrt{k}t) + \frac{y'(0)}{\sqrt{k}}\sin(\sqrt{k}t)$ and the norm $\|R(t) - k\,\mathrm{id}\| \leq \varepsilon_1 < k/2$. Then

$$\|u(t)\| \leq \frac{\varepsilon_1}{k - (1 - \cos(\sqrt{k}t))\varepsilon_1} \int_0^t \sqrt{k}\,|\bar{y}(s)|\sin(\sqrt{k}(t-s))\,ds.$$

The *turbulence of a leaf* L of a totally geodesic foliation is the rotational component of the co-nullity operator, see [R1],

$$a(L) = \sup\{\langle C_x(y), z\rangle : x \in TL, \ y, z \in TL^{\perp}, \ y \perp z, \ \|x\| = \|y\| = \|z\| = 1\}.$$

If $a(L) = 0$ for all leaves then $T^{\perp}\mathscr{F}$ is tangent to a totally umbilical foliation.

The following theorem (with $\mathscr{R}_x > 0$) and corollary (with $\mathscr{R}_x \geq 0$) generalize [R1, Theorems 4.10 and 4.16] when $X = 0$. Theorem 1.62 is not true without condition (1.72b), but the coefficient 0.3 is obtained by the method for proving.

Theorem 1.62. *Let $\mathscr{F}^p$ be a totally geodesic foliation of (M^{n+p}, g, X), and there is $m \in M$ such that on any leaf geodesic $\gamma : [0, \frac{\pi}{\sqrt{k}}] \to L$ ($\gamma(0) = m$) we have (1.70) and*

$$0 < k_1 \operatorname{id}^{\perp} \leq \mathscr{R}_{\dot\gamma} \leq k_2 \operatorname{id}^{\perp}, \tag{1.72a}$$

$$(k_2 - k_1 + 2\varepsilon)\max\{a(L)^2, k\} \leq 0.3\, k(k_2 + \varepsilon), \tag{1.72b}$$

where $k = \frac{1}{2}(k_1 + k_2)$ and $\varepsilon := \|\langle \nabla_{\dot\gamma}(X/n), \dot\gamma\rangle + \langle X/n, \dot\gamma\rangle^2\| < k_1$. Then $p < \rho(n)$.

Sketch of Proof It is sufficient to show that linear operators $\mathscr{B}_x : T_m^{\perp}\mathscr{F} \to T_m^{\perp}\mathscr{F}$, ($x \neq 0$), have no real eigenvalues. Suppose the opposite, i.e., there exist unit vectors $x_0 \in T_m\mathscr{F}$, $y_0 \in T_m^{\perp}\mathscr{F}$ and $\lambda \leq 0$ with the property $\mathscr{B}_{x_0}(y_0) = \lambda y_0$. Let $\gamma(t) : [0, \pi/\sqrt{k}] \to M$, $\dot\gamma(0) = x_0$ be a leaf geodesic, and $y(t) : \gamma \to T_{\dot\gamma}^{\perp}\mathscr{F}$ a Jacobi vector field on γ through the vector y_0. Hence (1.71) holds with

$$\|R(t) - k \operatorname{id}\| \leq \frac{k_2 - k_1}{2} + \varepsilon,$$

see (1.68), where $\dot y = \nabla_{\dot\gamma} y$ and $\ddot y = \nabla_{\dot\gamma}\nabla_{\dot\gamma} y$. The Jacobi vector field $y(t)$ has the form $y(t) = (\cos(\sqrt{k}t) + (\lambda/\sqrt{k})\sin(\sqrt{k}t))y_0 + u(t)$, where $u(0) = u'(0) = 0$. (For $k_2 = k_1$, we have $u(t) = 0$, hence $y(t)$ vanishes at $t_0 = \operatorname{arcctg}(-\lambda/\sqrt{k})/\sqrt{k}$, see the proof of Theorem 1.59). Using Lemma 1.61, we show for (1.72a) and $k_1 - \varepsilon \geq 0.582(k_2 + \varepsilon)$, that $|y(t)|$—the length of the Jacobi vector field $y(t)$—has a local minimum at t_m in the interval $(0, \pi/\sqrt{k})$, Figure 1.8.

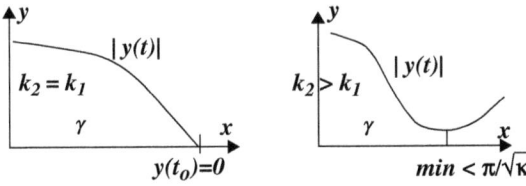

Fig. 1.8 The $\mathscr{F}$-parallel Jacobi field along the "extremal" geodesic

Our second observation is that the function $V(t)$, the area of a parallelogram, whose sides are the vectors $y(t)$ and $y'(t)$, varies "slowly" along a geodesic γ. (This function is constant when $k_2 = k_1$.) By Lemma 1.61 and (1.72b), these will yield a contradiction, because $V(t)$ cannot increase from zero value $V(0)$ to a "large" value $V(t_m)$ on a given interval with length $t_m < \pi/\sqrt{k}$. □

Notice that $\varepsilon = 0$ (see Theorem 1.62) is provided by $X^\top = 0$. If $k_1 \geq 0$ in Theorem 1.62 and $p \geq \rho(n)$ then $k_1 = k_2 = 0$, and we get the following splitting result.

Corollary 1.63. *Let $\mathscr{F}$ be a p-dimensional totally geodesic foliation of (M^{n+p}, g, X) with compact leaves. Suppose that (1.70) and the following inequalities hold:*

$$0 \leq k_1 \, \mathrm{id}^\perp \leq \mathscr{R}_x \leq k_2 \, \mathrm{id}^\perp \quad (x \in T\mathscr{F}, \|x\| = 1),$$
$$(k_2 - k_1 + 2\varepsilon) \cdot \max\{a(L)^2, k\} \leq 0.3 \, (k_2 + \varepsilon) \, k, \tag{1.73}$$

for any leaf L, $k = \frac{1}{2}(k_1 + k_2)$ and $\varepsilon := \left\| \langle \nabla_x(X/n), x \rangle + \langle X/n, x \rangle^2 \right\| < k_1$. If $p \geq \rho(n)$ then M splits along $\mathscr{F}$.

Theorem 1.62 and Corollary 1.63 are false without conditions (1.72b) and (1.73)$_2$, but their coefficient 0.3 is obtained by the method for proving.

1.5 Appendix

Here, just for the readers' convenience, we provide a concise review on differential geometry of manifolds. Few lemmas, propositions, theorems are extracted from the text. Sentences looking like statements are distinguished by italic fonts only. Minimum of proves are given. Only a few examples can be found. Readers not familiar enough with manifolds are referred to any of hundreds of books on the topic: one can choose his/her favorite book arbitrarily between "α" [AG] and "ω" [War].

1.5.1 Tensors and Differential Forms

Let M be a smooth manifold of dimension n and x—a point of M. Vectors tangent to M at x are defined as linear operators $v : C^\infty(M) \to \mathbb{R}$ which satisfy the following *Leibniz condition*:

$$v(f_1 \cdot f_2)(x) = f_1(x) \cdot v(f_2) + v(f_1) \cdot f_2(x). \tag{1.74}$$

Certainly, any linear combination of operators satisfying (1.74) satisfies (1.74) too, so the space T_xM of all the vectors tangent to M at x carries the natural structure of a vector space over $\mathbb{R}$; T_xM is the tangent space of M at x.

If $\phi = (\phi_1, \ldots \phi_n)$ is a chart defined in a neighborhood of x, then the formula $f \to \frac{\partial}{\partial x_i}(f \circ \phi^{-1})(\phi(x))$ defines a vector $(\partial/\partial\phi(i))(x)$ of T_xM. It is easy to see that the vectors $\partial/\partial\phi_i(x)$, $i = 1, \ldots n$, are linearly independent and span T_xM. Therefore, $\dim T_xM = n$. In particular, the space tangent to an open subset U of $\mathbb{R}^n$ at any its point x is canonically isomorphic to $\mathbb{R}^n$: the canonical isomorphism $\iota(a)(f) = (t \to f(x + ta))'(0)$ is determined for all elements a of $\mathbb{R}^n$ and smooth functions f on U.

If a map $F : M \to N$ is smooth and $x \in M$, then the formula

$$dF(x)(v)(f) = v(f \circ F),$$

where $f \in C^\infty(N)$ and $v \in T_xM$, defines a linear map $dF(x) : T_xM \to T_{F(x)}N$. The map $dF(x)$ (denoted often by F_{*x}) is called the *differential* of F at x. In particular, the differential $dc(t)$ of a smooth curve $c(t)$ on a manifold M is determined by $dc(t)(d/dt)(f) = (f \circ c)'(t)$ for $f \in C^\infty(M)$. The vector $\dot{c}(t) = dc(t)(d/dt)$ is said to be *tangent to the curve* c. From the definition, it follows directly that the differential of a composition of smooth maps coincides with the composition of their differentials: $d(F \circ G)(x) = dF(G(x)) \circ dG(x)$ for all x in the domain of G. Also, $d\,\mathrm{id}_M(x) = \mathrm{id}_{T_xM}$.

A smooth map $F : M \to N$ is called an *immersion* (resp., a *submersion*), whenever its differential F_{*x} is injective (resp., surjective) for all $x \in M$. If F is an immersion, then $m = \dim M \le \dim N = n$ and for any $x \in M$ there exist charts ϕ on M and ψ on N such that $x \in D_\phi$, $f(x) \in D_\psi$ and $(\psi \circ F \circ \phi^{-1})(u_1, \ldots, u_m) = (u_1, \ldots, u_n, 0, \ldots, 0)$ for all $(u_1, \ldots, u_m)$ in the domain of $\psi \circ F \circ \phi^{-1}$. Similarly, for a submersion $m \ge n$ holds and there exist such charts that satisfy $(\psi \circ F \circ \phi^{-1})(u_1, \ldots, u_m, \ldots, u_n) = (u_1, \ldots, u_m)$. Roughly speaking, up to transformation by charts, immersions look like canonical embeddings $\mathbb{R}^m \times \{0\} \subset \mathbb{R}^m \times \mathbb{R}^{n-m} = \mathbb{R}^n$, while submersions look like canonical projections $\mathbb{R}^m \to \mathbb{R}^n$.

The differential of a diffeomorphism is an isomorphism of tangent spaces. Conversely, *if the differential $dF(x)$ of a smooth map $F : M \to N$ is an isomorphism, then there exist open neighborhoods U of x and V of $F(x)$ such that F maps U onto V diffeomorphically.* Consequently, if $dF(x)$ maps all tangent spaces T_xM isomorphically onto $T_{F(x)}N$, then F is a *local diffeomorphism*, that is, M admits an open covering $\mathscr{U}$ such that F maps each of $U \in \mathscr{U}$ diffeomorphically onto and open subset $F(U)$ of N.

If $F : M \to N$ is an immersion, then $F(M)$ is an immersed *submanifold* of M. A subset $N \subset M$ is a *submanifold* when N carries a differential structure for which $\mathrm{id}_N : N \to M$ is an immersion. In other words, $N \subset M$ is a submanifold, whenever for any $x \in N$ there exists a chart $\phi = (\phi_1, \ldots, \phi_m)$ $(m = \dim M)$ on M such that $x \in D_\phi$ and the connected components of $N \cap D_\phi$ containing x are given by equations $\phi_{n+1} = \ldots = \phi_m = 0$. Then, $\dim N = n$ and the number $m - n$ is called the *codimension* of N. In general, the manifold topology of N is stronger than that induced from M. If these two topologies coincide, then N is *embedded* in M. *Any compact submanifold is embedded.* If $F : M \to N$ is a submersion, then the fibres $F^{-1}(x)$, $x \in M$ are submanifolds of M. More generally, if $R \subset M \times M$ is an equivalence relation (that is, R is reflexive, symmetric and transitive) such that R is a submanifold of $M \times M$ and the natural projections $\pi_i : R \to M$ $(i = 1, 2)$ are submersions, then *the quotient space M/R (equipped with the quotient topology) has a unique structure of a smooth manifold for which both projections π_1 and π_2 become submersions.*

Let TM be the disjoint union of all the tangent spaces $T_x M$, $x \in M$. One can define the canonical projection $\pi : TM \to M$ which maps all the elements of $T_x M$ onto x. For any chart $\phi = (\phi_1, \ldots, \phi_n)$, $n = \dim M$, on M one can define a map $\tilde{\phi} = (\tilde{\phi}_1, \ldots, \tilde{\phi}_{2n}) : \pi^{-1}(D_\phi) \to \mathbb{R}^n$ by $\tilde{\phi}_j = \phi_j \circ \pi$ and $\tilde{\phi}_{n+j} = d\phi_j$ for $j = 1 \ldots, n$. The map $\tilde{\phi}$ is a bijection between $\pi^1(D_\phi)$ and $G_\phi \times \mathbb{R}^n$. If π and ψ are two charts on M, then the composition $\tilde{\psi} \circ \tilde{\phi}^{-1}$ occurs to be a diffeomorphism between open subsets of $\mathbb{R}^n$. Therefore, TM carries the unique topology for which the domains of maps $\tilde{\phi}$ are open and all these maps become homeomorphisms. The atlas $\tilde{\mathscr{A}} = \{\tilde{\phi} : \phi \in \mathscr{A}\}$, $\mathscr{A}$ being an atlas determining the differential structure of M, provides on TM the structure of a smooth manifold. This manifold is called the *tangent bundle* of M.

Vector fields on a manifold M can be defined as *sections* of the tangent bundle TM, that is such smooth maps $X : M \to TM$ that $\pi \circ X = \mathrm{id}_M$. Therefore, for any $x \in M$, $X(x) \in T_x M$ is a linear transformation of $C^\infty(M)$ satisfying (1.74). Given such an X and $f \in C^\infty(M)$ one can define a function $X(f)$ by $X(f)(x) := X(x)(f)$. Smoothness of X and the construction of the differential structure of TM implies that $X(f) \in C^\infty(M)$. Therefore, vector fields can be defined alternatively as $\mathbb{R}$-linear transformations X of $C^\infty(M)$ satisfying condition (1.74). The set $\mathfrak{X}_M$ of all such transformations carries the algebraic structure of a modulus over the ring $C^\infty(M)$:

$$(X + Y)(f) = X(f) + Y(f), \quad (hX)(f) = h \cdot X(f)$$

for $X, Y \in \mathfrak{X}_M$ and $h, f \in C^\infty(M)$. The *Lie bracket* $[\cdot, \cdot] : \mathfrak{X}_M \times \mathfrak{X}_M \to \mathfrak{X}_M$, defined by

$$[X, Y](f) = X(Y(f)) - Y(X(f)), \quad X, Y \in \mathfrak{X}_M, \ f \in C^\infty(M),$$

is bilinear over $\mathbb{R}$ and $[X, Y] = -[Y, X]$. Moreover, $[X, f \cdot Y] = f \cdot [X, Y] + X(f) \cdot Y$ for $f \in C^\infty(M)$ and $[\cdot, \cdot]$ satisfies the *Jacobi identity*:

$$[X, [Y, Z]] + [Y, [Z, X]] + [Z, [X, Y]] = 0.$$

For any chart ϕ one has $[\partial/\partial\phi_i, \partial/\partial\phi_j] = 0$ for all i and j. The space $\mathfrak{X}_M$ equipped with the Lie bracket becomes an algebra over $\mathbb{R}$.

A curve $c : (a, b) \to M$ is called an *integral curve* of a vector field $X \in \mathfrak{X}_M$, whenever $\dot{c}(t) = X(c(t))$ for all $t \in (a, b)$. The theory of ordinary differential equations implies that for any $x \in M$ there exists an integral curve $c : I \to M$ of X such that $0 \in I$ and $c(0) = x$. Such curve is unique (up to domain), therefore for any $x \in M$ there exists a unique maximal (with respect to the domain) integral curve $c_x : I_x \to M$ of X. Each such c_x is called a *trajectory* of X. The set $W_X = \{(x, t) : t \in I_x\}$ is open in $M \times \mathbb{R}$ and the mapping $\phi : W_X \to M$ given by $\phi(x, t) = c_x(t)$ is smooth. If $\phi_t(x) = \phi(x, t)$, then ϕ_t occurs to be a diffeomorphism between open subsets of M, $\phi_0 = \mathrm{id}_M$ and $\phi_{t+s} = \phi_t \circ \phi_s$ wherever defined. The mapping ϕ (or, the family ϕ_t) is called the *flow* of X.

In general, $W_X \neq M \times \mathbb{R}$ meaning that some trajectories of X "die" in finite time. A vector field X is said to be *complete*, whenever $W_X = M \times \mathbb{R}$. In this case, all the trajectories of X are defined on the whole real line and all the members ϕ_t of the flow of X map diffeomorphically M onto itself. As in the case of real valued functions, the support of a vector field X is defined as the closure of the set $\{x \in M : X(x) \neq 0\}$. *Any vector field with compact support occurs to be complete.* Consequently, *all the vector fields on a compact manifold M are complete.*

The dual space $T_x^*M = \text{Lin}(T_xM, \mathbb{R})$ is called the *cotangent space* of M at $x \in M$, while its elements—the *covectors*. Similarly to TM, the disjoint union T^*M of all the cotangent spaces carries the structure of a smooth manifold called the *cotangent bundle* of M. Its sections ω are called 1-*forms* and can be considered as $C^\infty(M)$-linear transformations of $\mathfrak{X}_M$ into $C^\infty(M)$:

$$\omega(X)(x) = \omega(x)(X(x)), \quad X \in \mathfrak{X}_M, \quad x \in M.$$

If $F : M \to N$ is smooth and $x \in M$, then the differential $F_{*x} : T_xM \to T_{F(x)}N$ induces the dual linear transformation $F_x^* : T_{F(x)}^*N \to T_x^*M$:

$$F_x^*(\omega)(v) := \omega(F_{*x}(v)), \quad v \in T_xM, \quad \omega \in T_{F(x)}^*N.$$

The map T_x^*F is called the *co-differential* of X (at x). If $f \in C^\infty(M)$, then df is a 1-form. Similarly to $\mathfrak{X}_M$, the set of all 1-forms carries the structure of a $C^\infty(M)$-module. Any 1-form ω can be expressed in the form $\omega = \sum_i h_i \cdot df_i$ for some smooth functions f_i and h_i. Any 1-form ω supported in the domain D_ϕ of a chart $\phi = (\phi_1, \ldots, \phi_n)$ can be expressed uniquely as $\omega = \sum_{i=1}^n \omega(\partial/\partial\phi_i) \cdot d\phi_i$.

For any point x of a smooth manifold M and any $r, s \geq 0$ one can define the space $T_x^{r,s}M = \mathbb{R} \otimes T_xM \otimes \ldots \otimes T_xM \otimes T_x^*M \otimes \ldots \otimes T_x^*M$ of *tensors* of type (r,s), where the number of T_xM-factors equals r and that of T_x^*M-factors equals s. In particular, $(0,0)$-tensors are real numbers, $(1,0)$-tensors are tangent vectors and $(0,1)$-tensors are covectors. Differential structures on TM and on T^*M induce in a natural way the structure of a smooth manifold on $T^{r,s}M$, the disjoint union of all the spaces $T_x^{r,s}M$ $(x \in M)$; $T^{r,s}M$ is the *tensor bundle* of type (r,s) of M. The sections of $T^{r,s}M$, that is smooth maps $S : M \to T^{r,s}M$ which satisfy $S(x) \in T_x^{r,s}M$ for all $x \in M$, are called *tensor fields* of type (r,s). In particular, $(0,0)$-tensor fields are just smooth functions, $(1,0)$-tensor fields are vector fields are $(0,1)$-tensor fields are 1-forms. As before, the space of all the tensors of a given type has the algebraic structure of a $C^\infty(M)$-module. We denote by $\otimes$ the product of tensors and use the symmetrization operator to define the symmetric product of tensors:

$$B \odot C = \text{Sym}(B \otimes C) = \frac{1}{2}(B \otimes C + C \otimes B).$$

The canonical isomorphism ι between $T_xM \otimes T_x^*M$ and $\text{Lin}(T_xM, T_xM)$ given by

$$\iota(v \otimes \omega)(w) = \omega(w) \cdot v, \quad v, w \in T_xM, \quad \omega \in T_x^*M,$$

allows us to consider $(1,1)$-tensors as endomorphisms of tangent spaces and $(1,1)$-tensor fields as endomorphism of the module of vector fields. Moreover, $(0,s)$- and $(1,s)$-tensors can be considered as s-linear maps on T_xM with values in $\mathbb{R}$ and T_xM, respectively. Similarly, $(0,s)$- and $(1,s)$-tensor fields can be identified with s-linear (over the ring $C^\infty(M)$) maps on $\mathfrak{X}_M$ valued in $C^\infty(M)$ and $\mathfrak{X}_M$, respectively.

Another canonical isomorphisms, that is between $T_x M \otimes T_x^* M$ and $T_x^* M \otimes T_x M$, allows to define the tensor multiplication $\otimes : T_x^{r,s} M \times T_x^{r',s'} M \to T_x^{r+r',s+s'} M$:

$$(v_1 \otimes \ldots \otimes v_r \otimes \omega_1 \otimes \ldots \otimes \omega_s) \otimes (v_1 \otimes \ldots \otimes v_{r'} \otimes \omega_1 \otimes \ldots \otimes \omega_{s'})$$
$$= (v_1 \otimes \ldots \otimes v_{r+r'} \otimes \omega_1 \otimes \ldots \otimes \omega_{s+s'}),$$

and the similar multiplication of tensor fields. Up to canonical isomorphisms, this multiplications are associative in the *tensor algebra* $T_x^{**} M = \bigoplus_{r,s \geq 0} T_x^{r,s} M$ of M at x and in the analogous $C^\infty(M)$-algebra of all the tensor fields on M.

Now, let X be a vector field on M and (ϕ_t) its flow. The differentials ϕ_{t*} and co-differentials ϕ_t^* together define linear transformations $\phi_t^\sharp$ of spaces of tensors (and, tensor fields) of given types (r,s): $\phi_t^\sharp = \phi_{t*}$ when $r = 1, s = 0$, and $\phi_t^\sharp = \phi_t^*$ when $r = 0, s = 1$ and $\phi_t^\sharp(S_1 \otimes S_2) = \phi_t^\sharp(S_1) \otimes \phi_t^\sharp(S_1)$ for arbitrary tensors (tensor fields) S_1 and S_2 of arbitrary types. The *Lie derivative* $\mathscr{L}_X$ is a transformation of spaces of tensor fields (of certain types) given by

$$\mathscr{L}_X S(x) = \lim_{t \to 0} \frac{1}{t} \left(S(x) - \phi_t^\sharp(S(\phi_{-t}(x))) \right) \tag{1.75}$$

for any tensor field S of any type (r,s). In particular, $\mathscr{L}_X f = X(f)$ when $r = s = 0$ and $\mathscr{L}_X Y = [X,Y]$ when $r = 1$, $s = 0$,

$$(\mathscr{L}_X S)(Y_1, \ldots, Y_s) = X(S(Y_1, \ldots, Y_s)) - \sum_{i=1}^s S(Y_1, \ldots, [X,Y_i], \ldots, Y_s), \quad Y_i \in \mathfrak{X}_M,$$

when $r = 0$ and

$$(\mathscr{L}_X S)(Y_1, \ldots, Y_s) = [X, S(Y_1, \ldots, Y_s)] - \sum_{i=1}^s S(Y_1, \ldots, [X,Y_i], \ldots, Y_s), \quad Y_i \in \mathfrak{X}_M,$$

when $r = 1$. For other types of tensor fields, the formula for $\mathscr{L}_X S$ can be obtained from the above particular cases and the *Leibniz condition*

$$\mathscr{L}_X(S_1 \otimes S_2) = \mathscr{L}_X S_1 \otimes S_2 + S_1 \otimes \mathscr{L}_X S_2,$$

which follows directly from (1.75) and the properties of $\phi_t^\sharp$ mentioned before. Note that $\mathscr{L}_X S = 0$ if and only if S is invariant under the flow of X that is when $\phi_t^\sharp S = S$ for any t. In particular, $[X,Y] = 0$ if and only if the flows (ϕ_t) and (ψ_s) of X and Y commute that is when $\phi_t \circ \psi_s = \psi_s \circ \phi_t$ for any t and s. Finally, we have the identity $\mathscr{L}_{[X,Y]} = [\mathscr{L}_X, \mathscr{L}_Y]$, where the bracket $[\cdot, \cdot]$ on the RHS denotes the commutator of linear operators: $[A,B] = A \circ B - B \circ A$.

A differential k-*form* on a manifold M is a skew-symmetric tensor field ω of type $(0,k)$. Skew symmetry of ω means that $\omega(X_{\sigma_1}, \ldots, X_{\sigma_k}) = \operatorname{sgn} \sigma \cdot \omega(X_1, \ldots, X_k)$ for arbitrary vector fields $X_1, \ldots X_k \in \mathfrak{X}_M$ and any permutation $\sigma = (\sigma_1, \ldots, \sigma_k)$ of the set $\{1, \ldots, k\}$. As before, the space $\Lambda^k(M)$ of all k-forms carries the algebraic structure of a module over the ring $C^\infty(M)$ and that of the real vector space. Certainly,

$\Lambda^k(M) = 0$ when $k > n = \dim M$. Given one-forms $\omega_1, \ldots, \omega_k$, the exterior (wedge) product $\omega_1 \wedge \ldots \wedge \omega_k$ is defined as the skew-symmetrization of the tensor product:

$$\omega_1 \wedge \ldots \wedge \omega_k = \frac{1}{k!} \sum_\sigma \operatorname{sgn} \sigma \cdot \omega_{\sigma_1} \otimes \ldots \otimes \omega_{\sigma_k},$$

where σ ranges over the set of all permutations of k elements, becomes a k-form. If a k-form ω is supported in the domain D_ϕ of a chart $\phi = (\phi_i, \ldots, \phi_n)$, then

$$\omega = \sum_{1 \le i_1 < \ldots < i_k \le n} \omega_{i_1, \ldots, i_k} \cdot d\phi_{i_1} \wedge \ldots \wedge d\phi_{i_k} \tag{1.76}$$

for some smooth functions $\omega_{i_1, \ldots, i_k}$ on D_ϕ. The formula

$$(\omega_1 \wedge \ldots \wedge \omega_k) \wedge (\omega_{k+1} \wedge \ldots \wedge \omega_{k+l}) = \omega_1 \wedge \ldots \wedge \omega_{k+l}$$

for arbitrary 1-forms ω_i determines the exterior (wedge) product $\wedge$ of forms of arbitrary degrees: $\wedge : \Lambda^k(M) \times \Lambda^l(M) \to \Lambda^{k+l}(M)$. If ω and η are differential forms of degree, respectively, k and l, then

$$\eta \wedge \omega = (-1)^{kl} \omega \wedge \eta.$$

The direct sum $\Lambda^*(M) = \bigoplus_{k=0}^n \Lambda^k(M)$ equipped with $\wedge$ becomes an algebra, called *exterior algebra* of M, over the ring of smooth functions on M.

For any $\omega \in \Lambda^k(M)$ the formula

$$d\omega(X_1, \ldots, X_{k+1}) = \sum_{i=1}^{k+1} (-1)^{i+1} X_i \omega(X_1, \ldots, \hat{X}_i \ldots, X_{k+1})$$
$$+ \sum_{i<j} (-1)^{i+j} \omega([X_i, X_j], X_1, \ldots, \hat{X}_i, \ldots, \hat{X}_j, \ldots, X_{k+1}),$$

where $X_1, \ldots, X_{k-1} \in \mathfrak{X}_M$ and $\hat{\ }$ denotes the operator of omission, defines a $(k+1)$-form $d\omega$ called the *exterior differential* of ω. The differential operator $d : \Lambda^*(M) \to \Lambda^*(M)$ is $\mathbb{R}$-linear, $d^2 = d \circ d = 0$ and

$$d(\omega \wedge \eta) = d\omega \wedge \eta + (-1)^k \omega \wedge d\eta$$

when ω is a k-form. In particular,

$$d\omega = \sum_{1 \le i_1 < \ldots < i_k \le n} d\omega_{i_1, \ldots, i_k} \wedge d\phi_{i_1} \wedge \ldots \wedge d\phi_{i_k}$$

when ω is given by (1.76). A form ω satisfying $d\omega = 0$ is said to be *closed*. Forms of the view $d\eta$ are called *exact*. The identity $d^2 = 0$ implies the following: *any exact form is closed*. The converse is not true: the quotients $H^k(M, \mathbb{R}) = Z^k/B^k$, Z^k and B^k being, respectively, the vector spaces of closed and exact k-forms, are called *de Rham cohomology* groups of M and describe, to some extent, the topology of M. The numbers $b_k = \dim H^k(M, \mathbb{R})$ are called *Betti numbers* of a manifold M and the sum

$$\chi(M) = \sum_k (-1)^k b_k$$

is its *Euler number*, or, *Euler characteristic*. If M is closed and orientable, then *a nowhere-vanishing vector field on M exists if and only if $\chi(M) = 0$.*

Throughout the book, we use also the operator of *contraction* ι_X: if ω is a k-form and X is a vector field, then $\iota_X \omega$ is a $(k-1)$-form given by

$$\iota_X \omega(Y_1, \ldots, Y_{k-1}) = \omega(X, Y_1, \ldots, Y_{k-1}), \quad Y_i \in \mathfrak{X}_M.$$

Certainly $\iota_X^2 = 0$. The Lie derivative $\mathscr{L}_X \omega$ of a k-form ω is a k-form as well. The operators d, ι_X and $\mathscr{L}_X$ are related by

$$\mathscr{L}_X = d \circ \iota_X + \iota_X \circ d \quad \text{and} \quad \iota_{X,Y} = [\mathscr{L}_X, \iota_Y].$$

A nowhere-vanishing differential form Ω of degree $n = \dim M$ on M is called the *volume form*. A form like that exists if and only if M is orientable. Given a volume form Ω one can define the corresponding *divergence* operator div. Namely, at any point $x \in M$, the form $\mathscr{L}_X \Omega(x)$ is a multiple of $\Omega(x)$ (with a real coefficient depending smoothly on x and the vector field X) and we may put

$$\mathscr{L}_X \Omega = \operatorname{div} X \cdot \Omega. \tag{1.77}$$

From what we said before it follows that X *is divergence free* $(\operatorname{div} X = 0)$ *if and only if the flow* (ϕ_t) *of X preserves volume*, that is, when $\phi_t^* \Omega = \Omega$ for any t.

1.5.2 Frobenius Theorem

Arbitrary, integrable or not, distributions (subbundles of the tangent bundle) on a manifold appear in various situations, such as fields of tangent planes of foliations or kernels of tensors, on contact and f- manifolds and in sub-Riemannian geometry, see [BF]. Some problems about foliations naturally extend to distributions.

If $\mathscr{F}$ is a C^r-foliation of M and $r \geq 1$, then the set $E = T\mathscr{F}$ of all the vectors $v \in TM$ tangent to the leaves of $\mathscr{F}$ becomes a C^{r-1}-subbundle of TM. This subbundle (distribution) is *involutive* in the following sense: If $\mathfrak{X}_E$ denotes the family of all the E-valued vector fields on M, then $X, Y \in \mathfrak{X}_E$ provides $[X, Y] \in \mathfrak{X}_E$, where $[\cdot, \cdot]$ is the Lie bracket on M. The classical Frobenius Theorem says the following:

Theorem 1.64. *For any involutive C^r-subbundle E of a manifold M there exists a C^r-foliation $\mathscr{F}$, for which $E = T\mathscr{F}$.*

Proof. Here, we sketch the proof. The details can be found in several books, for instance in [Tam]. We perform induction with respect to p, the dimension of fibres of E. For $p = 1$, the theorem is obvious (compare Example 1.67).

Assume that $p > 1$, that our statement is true for bundles of fibre dimension $p-1$ and take an involutive subbundle $E \subset TM$ of fibre dimension p. Then, E is generated locally by linearly independent vector fields $X_1, \ldots, X_p$ such that

$$[X_i, X_j] = \sum_{k=1}^{p} a_{ij}^k X_k, \quad i, j = 1, \ldots, p,$$

for some smooth functions a_{ij}^k defined on an open subset U of M. Diminishing U if necessary, one can find a chart $(x_1, \ldots, x_n)$ on U for which $X_1 = \partial/\partial x_1$. Modifying $X_2, \ldots, X_p$ if necessary, one can assume that $X_k = \sum_{i=2}^{n} a_k^j \partial/\partial x_j$ for another smooth functions a_k^j on U. The fields $X_2, \ldots, X_p$ are linearly independent and generate an involutive plane field $\tilde{E}$. By inductive assumption, $\tilde{E}$ defines locally (say, on U) a foliation $\tilde{\mathscr{F}}$. There exists a chart $(y_1, \ldots, y_n)$ such that the leaves of $\tilde{\mathscr{F}}$ are given by $y_1 = c_1, y_{p+1} = c_{p+1}, \ldots, y_n = c_n$ for some constants $c_1, c_{p+1}, \ldots, c_n$. Define functions z_i, $i = 1 \ldots, n$, by $z_1 = x_1, z_i = y_i$ for $i > 1$. The system $(z_1, \ldots, z_n)$ satisfies the condition $\det\left[\frac{\partial z_i}{\partial x_j}\right]_{i,j \leq n} \neq 0$, thus, it defines locally (say, again on U) a chart. In this chart, E occurs to be generated by $\partial/\partial z_1, \ldots, \partial/\partial z_p$, or, equations $z_{p+1} = c_{p+1}, \ldots, z_n = c_n$, where c_j, $p < j \leq n$, are constants, define submanifolds, the leaves of a foliation $\mathscr{F}_U$ on U for which $E|_U = T\mathscr{F}_U$. Gluing together leaves (plaques) of such $\mathscr{F}_U$ over U belonging to an open cover of M, we get a global foliation $\mathscr{F}$ of M for which $E = T\mathscr{F}$. $\square$

Remark 1.65. Theorem 1.64 still holds if one removes the constant-rank assumption, but the involutivity assumption does not suffice. A sufficient condition, in the case when E is spanned by a family $\{Y_1, \ldots, Y_p\}$ of vector fields, is the finite-type condition $[Y_i, Y_j] = \sum_k a_{ij}^k Y_k$, where the coefficients a_{ij}^k are locally bounded. The above condition is not a property of the distribution, but of the spanning family. Indeed, let $x = (x_1, x_2) \in \mathbb{R}^2$ and $E = \text{span}(Y_1, Y_2)$, where $Y_1 = e^{-1/\|x\|^2} \partial_1$ and $Y_2 = \|x\|^2 \partial_2$. We have

$$[Y_1, Y_2] = 2x_2/\|x\|^2 Y_1 + 2x_1/\|x\|^2 e^{-1/\|x\|^2} Y_2$$

with unbounded (in any neighborhood of the origin) function $x \mapsto 2x_2/\|x\|^2$.

Similarly, let $D(M) = \bigoplus_{k=0}^{n} D^k(M)$ be the graded algebra of differential forms, $d : D(M) \to D(M)$ the exterior differentiation and $\mathscr{J} \subset D(M)$ an ideal such that $d\mathscr{J} \subset \mathscr{J}$. If for all $x \in M$, $\ker \mathscr{J}(x) = \{v \in T_x M : \omega(v) = 0 \text{ for all } \omega \in \mathscr{J}^1\}$ has the same dimension, say p, then the subbundle $\ker \mathscr{J}$ of TM is involutive and determines a foliation $\mathscr{F}$ of codimension $q = n - p$.

If there is a form $\omega \in D^k(M)$ such that $T\mathscr{F} = \ker \omega$ then, by the Frobenius involutivity condition, there is a form $\eta \in D^1(M)$ such that

$$d\omega = \omega \wedge \eta. \tag{1.78}$$

Such η is not unique: two forms satisfying (1.78) differ by a scalar multiple of ω.

Remark 1.66. A known result of multilinear algebra tells us that if v is a nonzero element of a vector space V and $\zeta \in \wedge^2 V$ satisfies $v \wedge \zeta = 0$, then $\zeta = v \wedge w$ for some $w \in V$. Consequently, the condition $\omega \wedge d\omega = 0$ implies locally $d\omega = \omega \wedge \eta$ for some 1-form η. By Frobenius Theorem, the equation $\omega = 0$, where ω is a nowhere-vanishing 1-form satisfying $\omega \wedge d\omega = 0$, defines a foliation of codimension one. This happens, for instance, when ω is closed ($d\omega = 0$). For such 1-form ω, condition $\omega \wedge d\omega = 0$ is equivalent to (1.78).

In codimension $q > 1$, $\mathcal{J}^1$ is generated locally (and globally when the bundle $TM/T\mathcal{F}$ is trivial) by q linearly independent 1-forms $\omega_1, \ldots, \omega_q$. Involutivity of the corresponding distribution, intersection of the kernels of all ω_i's, can be expressed by the following system of conditions for some 1-forms η_{ij} $(i, j = 1, \ldots, q)$:

$$d\omega_i = \sum_j \eta_{ij} \wedge \omega_j, \quad i = 1, \ldots, q.$$

Example 1.67. Any one-dimensional subbundle of the tangent bundle is involutive, and determines a foliation. In particular, any nowhere vanishing vector field X on M determines a foliation of M with leaves being orbits of its flow. A simple example of a noninvolutive subbundle is provided by the kernel of $\omega = dz + x \cdot dy$ in $\mathbb{R}^3$; this form ω defines the standard *contact structure* on $\mathbb{R}^3$, see Figure 1.9.

Note that foliations exist on several manifolds. For instance, in [Th2], the following important result has been established.

Theorem 1.68. *Every* $(n-1)$*-dimensional plane field on a closed manifold M of dimension n is homotopic to a smooth, involutive one. Consequently, any closed manifold of Euler characteristic zero carries smooth, codimension-one foliations.*

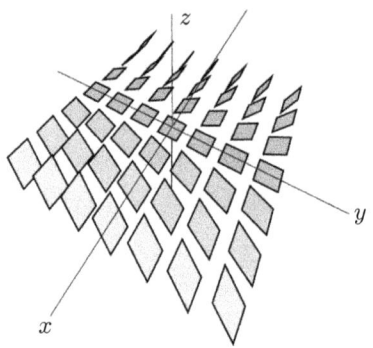

Fig. 1.9 A contact structure $\ker(dz + xdy)$ in $\mathbb{R}^3$

The article [Th2] contains a version of the above theorem for compact manifolds with boundary, and [Th3] is devoted to the existence of foliations of codimension greater than one. Finally, let us recall that a vector bundle E with fibers of dimension k is *orientable*, whenever its exterior power $\wedge^k E$ is trivial, that is, has a global section. Therefore, a manifold M is orientable in the sense of Section 1.2 if and only if its tangent bundle TM is orientable in the above sense. A foliation $\mathcal{F}$ is said to be *orientable* or *transversely orientable*, whenever the bundle $T\mathcal{F}$ or, respectively, the quotient $TM/T\mathcal{F}$ is orientable. If M is orientable, then any foliation of M is orientable if and only if it is transversely otientable. If codim $\mathcal{F} = 1$, then $\mathcal{F}$ is transversely orientable if and only if it is given by equation $\omega = 0$, where ω is a globally defined and nowhere vanishing 1-form on M.

1.5.3 The Elementary Symmetric Functions

The *elementary symmetric functions* σ_j of an $n \times n$-matrix C (or a linear transformation of $\mathbb{R}^n$) are given by

$$\sum_{j=0}^{n} \sigma_j(C)t^j = \det(\mathrm{id}_n + tC),$$

and are called *generalized mean curvatures* in the case of shape operator. Thus, $\sigma_0 = 1$, $\sigma_1 = \mathrm{trace}\, C$, ..., $\sigma_m = \det C$. The *power sums* of the principal curvatures $k_1, \ldots, k_n$ of a codimension-one foliation $\mathscr{F}$ (the eigenvalues of the shape operator) are given by $\tau_j = k_1^j + \ldots + k_n^j$ ($j \geq 0$). Denote by $\tau_j(C) = \mathrm{trace}\, C^j$ for $j \in \mathbb{N}$ the *power sums* symmetric functions of C, and set $\vec{\tau} = (\tau_1, \ldots, \tau_n)$. By the Cayley–Hamilton theorem, one may express C^b, using the lower powers C^j ($j < b$). By the Newton formulas,

$$\begin{aligned}
\tau_j - \tau_{j-1}\sigma_1 + \ldots + (-1)^{j-1}\tau_1\sigma_{j-1} + (-1)^j j\sigma_j &= 0 \quad (1 \leq j \leq n), \\
\tau_j - \tau_{j-1}\sigma_1 + \ldots + (-1)^n \tau_{j-n}\sigma_n &= 0 \quad (j > n),
\end{aligned} \tag{1.79}$$

the functions τ_j are polynomials of elementary symmetric functions $\sigma_1, \ldots, \sigma_n$. Many authors investigated higher-order mean curvatures of hypersurfaces using the Newton transformations of the shape operator. Newton transformations of the shape operator have been applied successfully to foliations, see [AKN, AW1, R5, RW3].

Definition 1.69. The *Newton transformations* $T_r(C)$ of the $n \times n$-matrix C are defined inductively or explicitly by

$$T_0(C) = \mathrm{id}, \quad T_r(C) = \sigma_r \, \mathrm{id} - C \, T_{r-1}(C) \quad (0 < r \leq n), \tag{1.80a}$$

$$T_r(C) = \sum_{j=0}^{r}(-1)^j \sigma_{r-j} C^j = \sigma_r \, \mathrm{id} - \sigma_{r-1} C + \ldots + (-1)^r C^r. \tag{1.80b}$$

For example, $T_1(C) = \sigma_1 \, \mathrm{id} - C$. Notice that C and $T_r(C)$ commute. By the Cayley–Hamilton Theorem, $T_n(C) = 0$. Let $C(t)$ be a smooth family of n-by-n matrices with the symmetric functions $\tau_j(t) = \mathrm{trace}\, C^j(t)$. Using $\dot{C}^k = C\dot{C}^{k-1} + \dot{C}C^{k-1}$ ($k > 0$), by induction we get $\dot{C}^k = \sum_{i=1}^{k} C^{i-1}\dot{C}C^{k-i}$. By $\mathrm{trace}(AB) = \mathrm{trace}(BA)$, and that the trace commutes with derivative, we get

$$\dot{\tau}_k = \mathrm{trace}(\dot{C}^k) = k\,\mathrm{trace}(C^{k-1}\dot{C}). \tag{1.81}$$

Remark 1.70. From the following equalities: $\dot{\sigma}_r(C) = \sigma_{1,r-1}(\dot{C}, C)$, see [RW2, Lemma 1.1] and $\sigma_{r,1}(B, C) = \mathrm{trace}(T_{r-1}(B)C)$, see [RW2, Lemma 1.2]. one may prove the identity for elementary symmetric functions of $C(s)$ (see [RW2, p. 7]):

$$\dot{\sigma}_r = \sum_{j=0}^{r-1}(-1)^r \sigma_{r-j-1}\,\mathrm{trace}(C^j\dot{C}) = \mathrm{trace}(T_{r-1}(C)\dot{C}), \quad r = 1, \ldots, n. \tag{1.82}$$

Lemma 1.71. *We have*

$$\text{trace}\, T_r(C) = (n-r)\,\sigma_r, \quad \text{trace}(C\,T_r(C)) = (r+1)\,\sigma_{r+1},$$

$$\text{trace}(C^2\,T_r(C)) = \sigma_1\,\sigma_{r+1} - (r+2)\,\sigma_{r+2}, \quad \text{trace}(T_{r-1}(C)\,\dot{C}) = \dot{\sigma}_r.$$

Proof. The first three algebraic properties follow directly from the Newton formulae (1.79). The last identity, follows from the formula (1.82). □

Next, we study the invariants $\sigma_\lambda(C_1, \ldots, C_m)$ of a set of matrices that generalize the elementary symmetric functions of a symmetric matrix C. Given $n \times n$-matrices $C_1, \ldots C_m$ and a unit matrix I_n one can consider the determinant $\det(I_n + t_1 C_1 + \ldots + t_m C_m)$ and express it as a polynomial of real variables $t = (t_1, \ldots, t_m)$. Given $\lambda = (\lambda_1, \ldots \lambda_m)$, a sequence of nonnegative integers with $|\lambda| := \lambda_1 + \ldots + \lambda_m \leq n$, we shall denote by $\sigma_\lambda(C_1, \ldots, C_m)$, see [RW5], its coefficient at $t^\lambda = t_1^{\lambda_1} \cdots t_m^{\lambda_m}$:

$$\det(I_n + t_1 C_1 + \ldots + t_m C_m) = \sum_{|\lambda| \leq n} \sigma_\lambda(C_1, \ldots C_m) t^\lambda. \tag{1.83}$$

Certainly, $\sigma_i(C)$ is the i-th elementary symmetric function of a single matrix C. We collect properties of these invariants, used in the sequel, in the following.

Lemma 1.72. *For any $\lambda = (\lambda_1, \ldots \lambda_m)$ and any $n \times n$ matrices C_i, C and B one has*
(I) $\sigma_\lambda(0, C_2, \ldots, C_m) = 0$ *if $\lambda_1 > 0$ and* $\sigma_{0, \hat{\lambda}}(C_1, \ldots, C_m) = \sigma_{\hat{\lambda}}(C_2, \ldots, C_m)$,
   where $\hat{\lambda} = (\lambda_2, \ldots, \lambda_m)$,
(II) $\sigma_\lambda(C_{s(1)}, \ldots C_{s(m)}) = \sigma_{\lambda \circ s}(C_1, \ldots C_m)$, *where $s \in S_m$ and $\lambda \circ s = (\lambda_{s(1)}, \ldots \lambda_{s(m)})$,*
(III) $\sigma_\lambda(I_n, C_2, \ldots C_m) = \binom{n - |\hat{\lambda}|}{\lambda_1} \sigma_{\hat{\lambda}}(C_2, \ldots C_m)$,
(IV) $\sigma_{\lambda_1, \lambda_2, \hat{\lambda}}(C, C, C_3, \ldots C_m) = \binom{\lambda_1 + \lambda_2}{\lambda_1} \sigma_{\lambda_1 + \lambda_2, \hat{\lambda}}(C, C_3, \ldots C_m)$,
(V) $\sigma_{1, \hat{\lambda}}(C + B, C_2, \ldots, C_m) = \sigma_{1, \hat{\lambda}}(C, C_2, \ldots C_m) + \sigma_{1, \hat{\lambda}}(B, C_2, \ldots C_m)$ *and*
   $\sigma_\lambda(aC_1, C_2, \ldots, C_m) = a^{\lambda_1} \sigma_\lambda(C_1, C_2, \ldots C_m)$ *if $a \in \mathbb{R} \setminus 0$.*

Proof. All these properties are easy to prove, see [RW5]. Here, we include a short proof of one of them just for the convenience of the reader. Equality (IV) arises directly from comparison of the coefficients on both sides of

$$\sum_\lambda \sigma_\lambda(C, C, C_3, \ldots C_n) t^\lambda = \det(I_n + (t_1 + t_2)C + t_3 C_3 + \ldots + t_m C_m)$$

$$= \sum_\mu \sigma_\mu(C, C_3, \ldots C_m)(t_1 + t_2)^{\mu_1} \cdot \hat{t}^{\hat{\mu}}$$

$$= \sum_\mu \sum_{k=0}^{\mu_1} \binom{\mu_1}{k} \sigma_\mu(C, C_3, \ldots C_m) t_1^k \cdot t_2^{\mu_1 - k} \cdot \hat{t}^{\hat{\mu}}. \quad \square$$

By the First Fundamental Theorem of Matrix Invariants [Gu], all σ_λ can be expressed in terms of the traces of the matrices involved and their products. The following remark provides a tool for writing down the explicit formulae for all σ_λ's in terms of traces (of given matrices and their products).

Remark 1.73. For any $n \times n$-matrices C_i, $\lambda = (\lambda_1, \ldots, \lambda_m)$ and $k \geq 0$, we have

$$\sigma_{(\lambda,k)}(C_1, \ldots, C_m, C_{m+1}) = \sigma_\lambda(C_1, \ldots, C_m)\,\sigma_k(C_{m+1})$$
$$- \sum_\mu \sigma_\mu(C_1, \ldots, C_m, C_{m+1}, C_1 C_{m+1}, \ldots, C_m C_{m+1}),$$

where the sum ranges over all the sequences $\mu = (\lambda_1 - j_1, \ldots \lambda_m - j_m, k - j_1 - \ldots - j_m, j_1, \ldots, j_m)$ with $j_1 \leq \lambda_1$, $\ldots$, $j_m \leq \lambda_m$, $0 < j_1 + \ldots + j_m \leq k$. Thus, $|\mu| = |\lambda| - |j|$. For example, $\sigma_{k,1}(B,C) = \sigma_k(B)\,\mathrm{trace}(C) - \sigma_{k-1,1}(B, BC)$.

These invariants can be used in calculation of the determinant of a matrix $B(t)$ expressed as a power series $B(t) = \sum_{i=0}^\infty t^i B_i$. The coefficient at t^j in the series of $\det(B(t))$ depends only on the part $\sum_{i \leq j} t^i B_i$ of $B(t)$. Set $\|\lambda\| = \lambda_1 + 2\lambda_2 + \ldots + k\lambda_k$.

Lemma 1.74. *If $B(t)$ is $n \times n$ matrix given by $B(t) = \sum_{i=0}^\infty t^i B_i$, then*

$$\det(B(t)) = 1 + \sum_{k=1}^\infty \left(\sum_{\lambda, \|\lambda\| = k} \sigma_\lambda(B_1, \ldots B_k) \right) t^k. \tag{1.84}$$

Since $\det : \mathcal{M}(n) \to \mathbb{R}$, $\mathcal{M}(n) \approx \mathbb{R}^{n^2}$ being the space of all $n \times n$-matrices, is a polynomial function, the series in (1.84) is convergent for all $t \in (-r_0, r_0)$, where $r_0 = 1/\limsup_{k \to \infty} \|B_k\|^{1/k}$ is the radius of convergence of the series $B(t)$.

Chapter 2
Integral Formulas

Integral formulas provide a powerful tool for obtaining global results in analysis and geometry. The integral formulas for foliations (and non-integrable distributions) of Riemannian, Finsler and metric-affine manifolds, to which this chapter is devoted, are useful for solving many problems (see, e.g. [AW1, R1, RW3, To, W4] and surveys in [ARW, RW2]): (i) the existence and characterization of foliations, whose leaves have a given geometric property, such as being totally geodesic, totally umbilical or minimal; (ii) prescribing the higher mean curvatures of the leaves of a foliation; (iii) minimizing functionals like volume and energy defined for tensor fields on a foliated manifold. We consider also distributions and tensors defined outside of a "singularity set" under additional assumption of convergence of certain integrals. Integral formulas for foliations are obtained by two different methods:

- applying the Divergence Theorem to appropriate vector fields;
- calculating the determinant of the Jacobi tensor for a family of diffeomorphisms expressed as a power series of the parameter and then integrated.

2.1 Codimension One Foliations of Riemannian Manifolds

The Gauss-Bonnet integral formula provides an obstruction in the problem "prescribing the curvature of a manifold". G. Reeb, in a short paper [Re1], proved that the integral of the mean curvature $H = \sigma_1$ of the leaves of a codimension-one foliation $\mathscr{F}$ of a closed Riemannian manifold (M, g) is equal to zero,

$$\int_M H \, d\,\mathrm{vol} = 0; \qquad (2.1)$$

thus, either $H \equiv 0$ or $H(x)H(x') < 0$ for some points $x \neq x'$ on M. The proof of (2.1) is based on the Divergence Theorem and the identity $\mathrm{div}\, N = -H$ with N a unit

© Springer Nature Switzerland AG 2021
V. Rovenski, P. Walczak, *Extrinsic Geometry of Foliations*, Progress
in Mathematics 339, https://doi.org/10.1007/978-3-030-70067-6_2

normal to the leaves. The second formula in the series of total σ_k's (the elementary symmetric functions of principal curvatures k_i of the leaves, see Section 1.5.3) says that the total σ_2 is a half of the *total Ricci curvature* in the N-direction, [N]:

$$\int_M (2\sigma_2 - \mathrm{Ric}_{N,N}) \, d\,\mathrm{vol} = 0, \tag{2.2}$$

which is a consequence of Green's theorem applied to $\nabla_N N + \sigma_1 \cdot N$.

Remark 2.1. One may prove (2.2) applying the following equality:

$$\mathrm{div}(\sigma_1 \cdot N + \nabla_N N) = \mathrm{Ric}_{N,N} - 2\sigma_2. \tag{2.3}$$

Next, using $\langle \nabla_N N, N \rangle = 0$, we obtain

$$\begin{aligned}
\mathrm{div}(\sigma_1 \cdot N + \nabla_N N) &= \mathrm{div}_{M'}(\sigma_1 \cdot N + \nabla_N N) + \langle \nabla_N(\sigma_1 \cdot N + \nabla_N N), N \rangle \\
&= \mathrm{div}_{M'}(\sigma_1 \cdot N + \nabla_N N) + N(\sigma_1) - \|\nabla_N N\|^2.
\end{aligned}$$

Thus, using (2.3), we get

$$\mathrm{div}_{M'}(\sigma_1 \cdot N + \nabla_N N) = \mathrm{Ric}_{N,N} - 2\sigma_2 - N(\sigma_1) + \|\nabla_N N\|^2.$$

From this and the Divergence Theorem for (M', g'), we obtain the counterpart of (2.2) for a closed leaf (M', g'):

$$\int_{M'} (2\sigma_2 - \mathrm{Ric}_{N,N} - \sigma_1^2 - N(\sigma_1) - \|\nabla_N N\|^2) \, d\,\mathrm{vol}_{g'} = 0. \tag{2.4}$$

For $n = 1$, formula (2.2) reduces to the integral of Gaussian curvature, $\int_M K \, dV_M = 0$.

Moreover, (2.2) admits a leaf-wise counterpart for a closed leaf M' with induced metric g', which has been used to estimate the energy of a vector field and to show that codimension-one foliations with negative Ricci curvature are far from being totally umbilical, [BW, LaW].

It was shown in [BLR] (generalizing result in [Asi]) that total σ_k's of a codimension-one foliation $\mathscr{F}$ on a compact space form $M^{n+1}(c)$ do not depend on $\mathscr{F}$: they depend on n, k, c and volume of M only,

$$\int_M \sigma_k \, d\,\mathrm{vol} = \begin{cases} c^{k/2} \binom{n/2}{k/2} \mathrm{Vol}(M, g), & n \text{ and } k \text{ even}, \\ 0, & \text{either } n \text{ or } k \text{ odd.} \end{cases} \tag{2.5}$$

In what follows we present two ways to derive the series of integral formulas for a codimension-one foliation of a closed Riemannian manifold, generalizing (2.5).

Example 2.2. Here is an amazing consequence of (2.5) for any sufficiently smooth codimension-one foliation of the round unit sphere S^3 (communicated to the authors by D. Asimov). By S. Novikov's theorem, any such foliation contains a leaf diffeomorphic to a torus. So, by Gauss-Bonnet theorem, there is a point with zero

Gaussian leaf curvature $K_{\mathscr{F}}$. By (2.5) for $n = j = 2$ and $c = 1$, the average of σ_2 is 1. Because $K_{\mathscr{F}} = 1 + \sigma_2$, there is a point, where $K_{\mathscr{F}} > 1 + 1 = 2$. Hence, the set of values of the function $K_{\mathscr{F}} : S^3 \to \mathbb{R}$ contains the interval $[0, 2 + \varepsilon]$ for some $\varepsilon > 0$.

Another version of the above example is as follows. Let a complete Riemannian manifold (M, g) with $\mathrm{Ric} \geq 2c > 0$ be endowed with a codimension-one foliation. Then the image of the function $\sigma_2 : M \to \mathbb{R}$ contains interval $[0, c + \varepsilon]$ for some $\varepsilon > 0$. Indeed, by Myers Theorem, e.g., [Bes, Theorem 6.51], M is compact, and by Reeb formula (2.1), $\sigma_1(x) = 0$ at some $x \in M$. Then $\sigma_2(x) = -\sum_i k_i^2 \leq 0$. By (2.2), $\int_M \sigma_2 \, d\mathrm{vol}_g \geq c \, \mathrm{Vol}(M, g)$. Hence, there exists $y \in M$ such that $\sigma_2(y) > c$ (otherwise, $\sigma_2 \leq c$ on M, then $\sigma_2 \equiv c > 0$ on M—a contradiction to $\sigma_2(x) \leq 0$).

Our problem concerns "prescribing the higher mean curvatures σ_i of a codimension-one foliated manifold". In this section, we prove a series of integral formulas for such symmetric functions σ_i. Our hypothesis is that *these integral formulas are the main obstructions for recovering metrics by higher mean curvatures of codimension-one foliations.*

2.1.1 Using a Family of Diffeomorphisms

Integral formulae for codimension-one foliations on a Riemannian manifold of finite volume (which are especially nice for a locally symmetric space, i.e., $\nabla R = 0$) can be obtained by calculating the determinant of the Jacobi tensor for a family of diffeomorphisms, see [BLR, RW5]. The following theorem generalizes (2.5); namely, the restriction that the ambient space has a constant sectional curvature is dropped.

Theorem 2.3. *Let $\mathscr{F}$ be a codimension-one foliation (with the unit normal N) on a complete locally symmetric space (M^{n+1}, g) of finite volume. Assume that $\mathscr{F}$ has bounded extrinsic geometry, i.e.,*

$$\sup_M \|A_N\| < \infty, \tag{2.6}$$

and set $B_{2k} = \frac{(-1)^k}{(2k)!} R_N^k$, $B_{2k+1} = \frac{(-1)^k}{(2k+1)!} R_N^k A_N$. Then for any $m > 0$

$$\int_M \sum_{\|\lambda\|=m} \sigma_\lambda \left(B_1, \ldots, B_m \right) d\mathrm{vol} = 0. \tag{2.7}$$

Proof. We call $\exp_x : T_x M \to M$, as usual, the *exponential map* at $x \in M$. Let $\gamma_x : t \to \exp_x(tN)$ be a unique geodesic in M with $\gamma_x(0) = x$ and $\gamma_x'(0) = N$. Choose a positively oriented orthonormal frame $(e^1, \ldots e^n)$ of $T_x \mathscr{F}$ and extend it by parallel translation to the frame $(E_x^1, \ldots, E_x^n)$ of vector fields along γ_x. Denote also by E_x^{n+1} the parallel field along γ_x with $E_x^{n+1}(0) = N_x$. For any $i \leq n$, denote by $Y_x^i(t)$ the Jacobi field along γ_x satisfying $Y_x^i(0) = e^i$ and $(Y_x^i)'(0) = A_x(e^i)$, where A_x is the Weingarten operator (of a leaf at x) relative to N_x. Denote by R_N the curvature operator $X \mapsto R(X, N)N$ in $T\mathscr{F}$ and by $R_x(t)$ the matrix with entries $\langle R(E_x^i(t), E_x^{n+1}(t)) E_x^{n+1}(t), E_x^j(t) \rangle$ (*Jacobi operator*). Denote also by $Y_x(t)$ the

$n \times n$ matrix consisting of the scalar products $\langle Y_x^i(t), E_x^j(t) \rangle$ (*Jacobi tensor*). Then $Y_x(0) = \mathrm{id}_n$ and $Y_x'(0) = A_x$. Consider the maps $\{\phi_t : M \to M, \ t \in (-\varepsilon, \varepsilon)\}$ defined by $\phi_t(x) = \exp_x(tN)$. One may show (see, for instance, [Kl, Lemma 3.1.17]) that

$$|d\phi_t(x)| = \det Y_x(t).$$

Note that $\|R_N\| \leq \|R\|$ and $\|R\| = \mathrm{const}$, because $\nabla R = 0$. In view of (2.6), there exists $\varepsilon > 0$ such that for all $x \in M$ there are no focal points of a leaf L_x along $\gamma_x(t)$, $t \in [0, \varepsilon)$. Hence, $\{\phi_t : M \to M\}_{t \in (-\varepsilon, \varepsilon)}$ are diffeomorphisms. Let $|d\phi_t(x)|$ be the Jacobian of ϕ_t at a point x of M. For short, write $R_N := R_x(0)$. The Jacobi equation $Y_x'' = -R_x(t)Y_x$ and assumption that (M, g) is locally symmetric ($\nabla R = 0$) imply that

$$Y_x^{(2m)}(0) = (-R_N)^m, \quad Y_x^{(2m+1)}(0) = (-R_N)^m A_x, \quad m = 0, 1, 2, \ldots$$

Hence, our Jacobi tensor $Y_x = \sum_{m=0}^{\infty} Y_x^{(m)}(0) \frac{t^m}{m!}$ has the form

$$Y_x(t) = \mathrm{id}_n + tA_x - \frac{t^2}{2!}R_N - \frac{t^3}{3!}R_N A_x + \frac{t^4}{4!}R_N^2 + \ldots \tag{2.8}$$

Certainly, the radius of convergence of the series in (2.8) is uniformly bounded from below on M (by $1/\|R\| > 0$). Therefore—by Lebesque Dominated Convergence Theorem—its integration together with Exchange Variable Theorem yield the equality for arbitrary $t > 0$ small enough

$$\mathrm{Vol}(M, g) = \int_M \det\left(\mathrm{id}_n + tA_x - \frac{t^2}{2!}R_N - \frac{t^3}{3!}R_N A_x + \frac{t^4}{4!}R_N^2 + \ldots\right) d\,\mathrm{vol}. \tag{2.9}$$

All the integral coefficients at t^k, $k > 0$, in (2.9) vanish, therefore, (2.9) together with Lemma 1.74 imply (2.7). $\qquad \square$

Example 2.4. For initial values $m = 1, 2, 3, 4$, the equations (2.7) give (2.1), (2.2) and

$$\int_M \left(\sigma_3(A) - \frac{1}{2}\mathrm{Ric}_{N,N}\,\sigma_1(A) + \frac{1}{3}\sigma_1(R_N A)\right) d\,\mathrm{vol} = 0,$$

$$\int_M \left(\sigma_4(A) + \frac{1}{4}\sigma_2(R_N) - \frac{1}{6}\sigma_{1,1}(A, R_N A) + \frac{1}{24}\sigma_1(R_N^2) - \frac{1}{2}\sigma_{2,1}(A, R_N)\right) d\,\mathrm{vol} = 0.$$

For $K_{\mathrm{mix}} = \mathrm{const}$, (2.7) reduces to (2.5). Thus, there are no codimension-one foliations of $K_{\mathrm{mix}} = \mathrm{const} \neq 0$ on a compact even-dimensional manifold.

Remark 2.5. Similar to (2.7) formulae can be derived on arbitrary Riemannian manifolds. They are more complicated since they contain terms, which depend on covariant derivatives of R_N. More precisely, (2.7) contains just terms of the form $R_N^{(k)}$ with $k \leq m - 2$, where $R_N^{(1)} = \nabla_N R_N$, $R_N^{(2)} = \nabla_N \nabla_N R_N$, and so on.

Corollary 2.6. *Let $\mathscr{F}$ be a codimension-one foliation of a complete Riemannian manifold (M, g) of finite volume with $K_{\mathrm{mix}} = c = \mathrm{const}$ and $\sup_M \|A_\xi\| < \infty$. Then (2.5) are valid.*

Proof. Let (M^{n+1}, g) be of constant mixed sectional curvature, i.e., $R_N = c\,\mathrm{id}_n$.

(a) For compact M with zero mixed sectional curvature, $R_N = 0$, and we obtain the Jacobi tensor Y_x of a simple form, linear in t: $Y_x(t) = \mathrm{id}_n + tA_x$. Then (2.9) reduces to $\mathrm{Vol}(M, g) = \int_M \det(\mathrm{id}_n + tA_x)\,d\,\mathrm{vol}$. This yields the case $c = 0$ of (2.5), i.e.,

$$\int_M \sigma_k(A_x)\,d\,\mathrm{vol} = 0, \quad k > 0.$$

(b) Assume now that the mixed curvature of M is positive. Then the matrix R_N is positive definite and, since it is also symmetric, one can consider its square root $\sqrt{R_N}$ and write the Taylor series (2.8) for Y_x in the form

$$Y_x(t) = \cos(t\sqrt{R_N})\big(\mathrm{id}_n + \tan(t\sqrt{R_N})(\sqrt{R_N})^{-1}A_x\big).$$

Here and below, the trigonometric functions of matrices are defined by the same Taylor series that hold for the trigonometric functions of real numbers. Exchange Variable Theorem for integration implies that the equality

$$\mathrm{Vol}(M, g) = \int_M \det\big(\cos(t\sqrt{R_N})\big) \det\big(\mathrm{id}_n + \tan(t\sqrt{R_N})(\sqrt{R_N})^{-1}A_x\big)\,d\,\mathrm{vol} \tag{2.10}$$

is valid for arbitrary t small enough. If (M, g) has positive constant mixed curvature $c = 1$, then $R_N = \mathrm{id}_n$ and (2.10) reduces to

$$\mathrm{Vol}(M, g) = \cos^n(t) \int_M \det\big(\mathrm{id}_n + \tan(t)A_x\big)\,d\,\mathrm{vol}.$$

One can use the substitution $\tan(t) \to \tilde{t}$ and the identity $\cos^2 t = (1 + \tilde{t}^2)^{-1}$ for further derivations.

(c) The case of negative constant mixed curvature of M is similar to the case (b), i.e., $R_N = -\mathrm{id}_n$. According to the above (a)–(c), we have (2.5). $\qquad\square$

2.1.2 Applications

2.1.2.1 Totally Geodesic and Umbilical Foliations

From (2.7) one may get obstructions for existence of codimension-one totally geodesic and umbilical foliations. If (M, g) is locally symmetric and the leaves of $\mathscr{F}$ are totally geodesic, then $A_x \equiv 0$, $Y_x^{(2m+1)}(0) = 0$ and $Y_x^{(2m)}(0) = (-R_N)^m$. By (2.8), we get the Jacobi tensor $Y_x(t) = \mathrm{id}_n - \frac{t^2}{2!}R_N + \frac{t^4}{4!}R_N^2 + \dots$, and (2.7) reads

$$\int_M \sum_{\|\lambda\|=m} \sigma_\lambda \left(-\frac{1}{2!}R_N, \frac{1}{4!}R_N^2, \ \ldots \ \frac{(-1)^m}{(2m)!}R_N^m \right) d\,\text{vol} = 0. \qquad (2.11)$$

The initial members of (2.11) are

$$\int_M \text{Ric}_{N,N}\, d\,\text{vol} = 0, \qquad \int_M \left(\sigma_2(R_N) + \frac{1}{6}\sigma_1(R_N^2) \right) d\,\text{vol} = 0, \qquad (2.12)$$

and so on. Equalities (2.12) imply directly the following statement.

Corollary 2.7. *The curvature tensor R_N in the direction N normal to a codimension-one totally geodesic foliation $\mathscr{F}$ on a locally symmetric Riemannian manifold vanishes identically provided that either Ricci curvature is everywhere non-negative (or, non-positive), or $\sigma_2(R_N)$ is non-negative.*

Our integral formulae (2.12) for σ_i with $i > 2$ provide conditions for the mean curvature $H = \sigma_1$ of $\mathscr{F}$. For a totally umbilical foliation with $A = (H/n)\,\text{id}_n$ on a locally symmetric manifold, the matrices in (2.7) reduce to $B_{2k} = \frac{(-1)^k}{(2k)!}R_N^k$ and $B_{2k+1} = \frac{(-1)^k}{(2k+1)!}(H/n)R_N^k$, and such conditions can be derived from Example 2.4 with the use of Remark 1.72 and read simply as

$$\int_M \left(\frac{(n-1)(n-2)}{n^2}H^2 - \text{Ric}_{N,N} \right) d\,\text{vol} = 0, \qquad (2.13a)$$

$$\int_M H \left(\frac{n(n-1)(n-2)}{(3n-2)n^2}H^2 - \text{Ric}_{N,N} \right) d\,\text{vol} = 0, \qquad (2.13b)$$

$$\int_M \left(\binom{n}{4}\frac{H^4}{n^4} - \frac{(n-1)(3n-4)}{12n^2}\text{Ric}_{N,N}H^2 + \frac{1}{4}\sigma_2(R_N) + \frac{1}{24}\sigma_1(R_N^2) \right) d\,\text{vol} = 0, \quad (2.13c)$$

etc. These integrals for n even, e.g., (2.13a,c), contain polynomials in H^2. In case of positive coefficients these polynomials are positive for all values of H and one may get obstructions for existence of such totally umbilical foliations.

2.1.2.2 Foliations with Conformally Defined Metric

Given $x \in M^{n+1}$, let $k_i(x)$ be the principal curvatures at x of the leaf L_x of $\mathscr{F}$ (the eigenvalues of A_x). The non-negative function $\phi(x) = \sum_{i<j}(k_i(x) - k_j(x))^2$ on M can be considered as a measure of non-umbilicity of the leaves of $\mathscr{F}$. From $\phi(x) = n\,\text{trace}(A_x^2) - \text{trace}^2(A_x)$ and $2\sigma_2 = \text{trace}^2(A_x) - \text{trace}(A_x^2)$, one may express $\phi(x)$ using elementary symmetric polynomials of the eigenvalues $k_i(x)$ as

$$\phi(x) = (n-1)\sigma_1^2 - 2n\,\sigma_2.$$

Assume that $\mathscr{F}$ has no umbilical points and define a new Riemannian metric g^c on M by $g^c = \mu^2 g$, where $\mu^2 = \phi$. If $\tilde{g}$ is another metric on M, conformally equivalent to g, and $\tilde{g}^c$ is obtained from $\tilde{g}$ by the above procedure, then $\tilde{g}^c = g^c$. The metric

g^c is called the *conformally defined metric* on $(M,\mathcal{F},g)$, [W5]. Its volume form is $\text{vol}^c = \mu^{n+1} \text{vol}$, where vol is the Riemannian volume form on (M,g). If X is a g-unit vector, then $X^c = X/\mu$ is a g^c-unit vector. It follows that

$$\text{Ric}^c_{N^c,N^c} = \frac{1}{\mu^2}\text{Ric}_{N,N} - \frac{\Delta\mu}{\mu^3} - \frac{n-1}{\mu^3}h_\mu(N,N) - \frac{n-2}{\mu^4}||\nabla\mu||^2 + \frac{2(n-1)}{\mu^4}\langle\nabla\mu,N\rangle^2.$$

Here, h_μ is the Hessian of μ. The operators A and A^c (at x) are related by

$$A^c = \frac{1}{\mu}\left(A - \frac{1}{\mu}\langle\nabla\mu,N\rangle\,\text{id}\right).$$

A similar relation is valid between the mean curvature functions

$$\sigma_1^c = \frac{1}{\mu}\left(\sigma_1 - \frac{1}{n\mu}\langle\nabla\mu,N\rangle\right),$$

$$\sigma_2^c = \frac{1}{\mu^2}\sigma_2 - \frac{1}{\mu^3}(n-1)\sigma_1\langle\nabla\mu,N\rangle + \frac{1}{2\mu^4}n(n-1)\langle\nabla\mu,N\rangle^2,$$

and so on. We say that $\mathcal{F}$ is *conformally harmonic* if $\sigma_1^c = 0$. By the above, this is equivalent to the condition $\mu\sigma_1 = \langle\nabla\mu,N\rangle$.

Using the integral formula (2.1) applied to σ_1^c, one may show that if $\mathcal{F}$ is a codimension one umbilics free foliation of a closed manifold (M,g), then either $\mathcal{F}$ is conformally harmonic or there exist points x and y in M such that $\sigma_1(x) > \langle\nabla\log\mu,N\rangle(x)$ and $\sigma_1(y) < \langle\nabla\log\mu,N\rangle(y)$.

Furthermore, we get more integral formulae for a compact manifold with a codimension-one foliation without umbilical points. These involve also derivatives of the principal curvatures k_i's (that is $\nabla\mu$). For example,

$$\int_M \frac{1}{\mu}\left(\sigma_1 - \frac{1}{n\mu}\langle\nabla\mu,N\rangle\right)d\,\text{vol} = 0,$$

$$\int_M \Big(2\mu^{n-1}\sigma_2 - \frac{2(n-1)}{n}\mu^{n-2}\sigma_1\langle\nabla\mu,N\rangle - \mu^{n-1}\text{Ric}_{N,N}$$
$$+(n-1)(n-2)\mu^{n-3}\langle\nabla\mu,N\rangle^2 + (n-1)\mu^{n-2}h_\mu(N,N)\Big)d\,\text{vol} = 0.$$

However, constructing umbilics free foliations on some specific Riemannian manifolds can be not so obvious. For instance, the problem of existence of such foliations on the standard sphere S^{2n+1} seems to be open and interesting. For $n = 1$, it reduces to the problem of existence of umbilics free Reeb components in the standard $\mathbb{R}^3$.

2.1.3 Using the Divergence Theorem

Here, we apply the Divergence Theorem to a suitable vector field to represent a series of integral formulae for a codimension-one foliation $\mathcal{F}$ with unit normal N on a closed Riemannian manifold (M,g). We omit proofs in this section, (see [RW2] and recent article [R16]) because more general results with proofs are given in Section 2.3.2.

For short, set $Z = \nabla_N N$, $\sigma_k = \sigma_k(A_N)$, $\tau_k = \tau_k(A_N)$, $A = A_N$, $\mathscr{A} = \mathscr{A}_N$, $\sigma_1 = \langle H, N \rangle$, etc. Given functions f_j on $\mathbb{R}^n$ $(0 \le j < n)$, define a $(1,1)$-tensor field

$$\mathscr{A} := \sum_{j=0}^{n-1} f_j(\vec{\tau}(A)) A^j, \quad \text{where} \quad \vec{\tau}(A) = \tau_1(A), \dots, \tau_n(A). \tag{2.14}$$

The degree of $\mathscr{A}$ is $\deg \mathscr{A} = \max\{j : f_j \not\equiv 0\}$. The choice of the RHS for $\mathscr{A}$ in (2.14) is natural for the following reasons:

- the powers A^j are the only $(1,1)$-tensors, obtained algebraically from A, while $\tau_1, \dots, \tau_n$, or, equivalently, $\sigma_1, \dots, \sigma_n$, generate all scalar invariants of A.
- the Newton transformation $T_r(A)$ $(r < n)$ of (1.80b) depends on all A^j $(j \le r)$.

Define a linear operator $R(X) : \widetilde{\mathscr{D}} \to \widetilde{\mathscr{D}}$ (for $X \in \Gamma(TM)$) by

$$R(X) : Z \to (R(Z,X)N)^\top, \quad Z \in \Gamma(\widetilde{\mathscr{D}}),$$

and denote $R_N = R(N)$, i.e., $R_N : Z \to (R(Z,N)N)^\top$.

For clarity and simplicity, we restrict ourselves to the case $\mathscr{A} = T_r(A)$, see [LuW].

Proposition 2.8. *We have*

$$\operatorname{div}_{\mathscr{F}}(T_r(A)Z) = \underline{\langle -\nabla_{\mathscr{F}}^* T_r(A), Z \rangle} + \operatorname{trace}(T_r(A)R_N) + \langle T_r(A)Z, Z \rangle - N(\sigma_{r+1})$$
$$- (r+2)\sigma_{r+2} + \sigma_1 \sigma_{r+1},$$

where the underlined term is given by

$$\langle \nabla_{\mathscr{F}}^* T_r(A), Z \rangle = -\sum_{j \le r} \operatorname{trace}\left(T_{r-j}(A)R((-A)^{j-1}Z)\right).$$

Theorem 2.9. *For any compact leaf L we have*

$$\int_L \left(\sum_{1 \le j \le r} \operatorname{trace}\left(T_{r-j}(A)R((-A)^{j-1}Z)\right) - N(\sigma_{r+1}) + \sigma_1 \sigma_{r+1} - (r+2)\sigma_{r+2}\right.$$
$$\left. + \operatorname{trace}(T_r(A)R_N) + \langle T_r(A)Z, Z \rangle\right) d\operatorname{vol}_L = 0.$$

Example 2.10. For $r = 0$, by Proposition 2.8,

$$\operatorname{div}_{\mathscr{F}} Z = \tau_2 - N(\tau_1) + \operatorname{Ric}_{N,N} + \langle Z, Z \rangle \ge 0,$$

see also Remark 2.1. Thus, if $\operatorname{Ric}_{N,N} \ge 0$ then $A = 0$, $\operatorname{Ric}_{N,N} = 0$ and $Z = 0$ on any compact leaf with the property $N(\tau_1) \le 0$.

Note that $\operatorname{div}(\mathscr{A}Z) = -\nabla_{\mathscr{F}}^*(\mathscr{A}Z) - \langle \mathscr{A}Z, Z \rangle$. By the above, for the Newton transformations of A we have

$$\operatorname{div}\left(T_r(A)Z + \sigma_{r+1}N\right) = -\langle \nabla_{\mathscr{F}}^* T_r(A), Z \rangle - (r+2)\sigma_{r+2} + \operatorname{trace}(T_r(A)R_N).$$

Theorem 2.11. *For a closed Riemannian manifold (M, g) we have*

$$\int_M \left(\underline{\langle -\nabla^*_{\mathscr{F}} T_r(A), Z \rangle} - (r+2)\,\sigma_{r+2} + \mathrm{trace}(T_r(A) R_N) \right) d\,\mathrm{vol} = 0, \quad (2.15)$$

where the underlined terms are given in Proposition 2.8.

For $r = 0$, (2.15) reads as (2.2), see also [AW1], and for $r = 1$, (2.15) reduces to

$$\int_M \left(3\,\sigma_3 - \sigma_1 \mathrm{Ric}_{N,N} + \sigma_1(AR_N) - \mathrm{Ric}_{N,Z} \right) d\,\mathrm{vol} = 0.$$

One may show, see [AW1], that for integrable $\mathscr{D}$, (2.15) coincides with

$$\int_M \left((r+2)\sigma_{r+2} - \sigma_1(T_r(A)R_N) - \sum_{j=1}^r \mathrm{trace}(R((-A)^{j-1}Z)T_{r-j}(A)) \right) d\,\mathrm{vol} = 0.$$

The *total higher-order mean curvatures* and *total power sums* are integrals

$$I_{\sigma,j} = \int_M \sigma_j \, d\,\mathrm{vol}, \quad I_{\tau,j} = \int_M \tau_j \, d\,\mathrm{vol}, \quad j \in \mathbb{N}.$$

Using (2.15) and the identity trace $T_r(A) = (n-r)\,\sigma_r$ (Lemma 1.71), we get $I_{\sigma,r+2} = c\,\frac{n-r}{r+2} I_{\sigma,r}$ on $M^{n+1}(c)$, where $I_{\sigma,0} = \mathrm{Vol}(M, g)$ and $I_{\sigma,1} = 0$. Then (2.5) follows by induction, and similarly, $I_{\tau,0} = n\,\mathrm{Vol}(M, g)$, $I_{\tau,2} = -c\,I_{\tau,0}$, etc.

Let $\mathscr{F}$ be a harmonic foliation (i.e., the mean curvature $H = 0$), and let N define a geodesic foliation of a closed manifold (M, g). One may show (see also Corollary 2.49 and two lines after it) that if $R_N = c\,\widehat{\mathrm{id}}$ (i.e., the mixed sectional curvature is constant), and n is even, then for any $s > 0$,

$$I_{\tau,2s+1} = 0, \quad I_{\tau,2s} = (-c)^s\,n\,\mathrm{Vol}(M, g). \quad (2.16)$$

When (M, g) is an Einstein manifold with $\dim M > 2$,

$$\mathrm{Ric}_{X,Y} = nc\,\langle X, Y \rangle \quad \text{for all } X, Y \text{ and some } c \in \mathbb{R}$$

and $\mathscr{F}$ is umbilical (i.e., $A = \lambda\,\mathrm{id} = (\tau_1/n)\,\mathrm{id}$), we get formulae similar to (2.5).

Corollary 2.12. *Let $\mathscr{F}$ $(\dim \mathscr{F} = n > 1)$, be a totally umbilical foliation of a closed Einstein manifold (M^{n+1}, g) (with $\mathrm{Ric} = nc \cdot g$ for some $c \in \mathbb{R}$), then we have*

$$I_{\sigma,k} = \begin{cases} c^{\frac{k}{2}} \binom{n/2}{k/2} \mathrm{Vol}(M, g), & n \text{ and } k \text{ even} \\ 0, & \text{either } n \text{ or } k \text{ odd.} \end{cases}$$

Proof. Under our assumptions, $\mathrm{Ric}_{N,Z} = 0$ and $\mathrm{Ric}_{N,N} = nc$. In this case, $T_r(A) = \frac{n-r}{n}\,\sigma_r\,\mathrm{id}$. Hence, $\langle \mathrm{div}_{\mathscr{F}} T_r(A), Z \rangle = 0$ and $\mathrm{trace}(T_r(A) R_N) = c\,\frac{n-r}{r}\,\sigma_r$, see Proposition 2.8 and (2.15). Then (2.15) reads:

$$\int_M \left((r+2)\sigma_{r+2} - nc\,\frac{n-r}{n}\,\sigma_r \right) d\,\mathrm{vol} = 0 \quad \Rightarrow \quad I_{\sigma,r+2} = c\,\frac{n-r}{r+2} I_{\sigma,r},$$

From the above, the claim follows by induction.   □

The study of hypersurfaces with constant higher-order mean curvatures has been of increasing interest in recent years, see [AW1, BKO]. Now, we apply integral formulae to provide some results for foliations whose leaves have constant σ_2.

Proposition 2.13. *Let (M,g) be a closed Einstein manifold with non-negative sectional curvature and $\mathscr{F}$ a foliation of M whose leaves have constant σ_2. Then σ_2 must be constant all over M.*

Proof. By Proposition 2.15 below, either σ_2 is constant on M (so the assertion of our proposition is satisfied) or there exists a closed leaf L of $\mathscr{F}$ with the property

$$\sigma_{2|L} = \alpha, \tag{2.17}$$

where $\alpha = \max_M \sigma_2$. On the other hand, M has non-negative sectional curvature, thus $\mathrm{Ric}_{N,N} \geq 0$, and (2.2) implies that $\int_M \sigma_2 \, d\,\mathrm{vol} \geq 0$. If σ_2 is not constant on M then $\alpha > 0$ and $\sigma_2 > 0$ on L. Then, $\sigma_1^2 \geq 2\sigma_2 > 0$ and consequently $\sigma_1 \neq 0$ on L. Without loss of generality, we may assume that $\sigma_1 > 0$ on L. As the eigenvalues of $T_1(A)$ are of the form $\sigma_1 - k_i$, k_i being the principal curvatures of the leaves, and

$$\sigma_1^2 = \sum_{i=1}^n k_i^2 + 2\sigma_2 > k_i^2,$$

we infer that $T_1(A)$ is positive definite on L. This and the assumption of sectional curvature give $\mathrm{trace}(T_1(A)R_N) \geq 0$ and $\langle T_1(A)Z,Z \rangle \geq 0$. From (2.17) we conclude that the derivative $N(\sigma_2)$ vanishes on L. On the other hand, we also have $\mathrm{Ric}_{N,Z} = 0$, so that from Theorem 2.47 with $r = 1$ we get

$$0 = \int_L \left(\mathrm{trace}(A^2 T_1(A)) + \mathrm{trace}(R_N T_1(A)) + \langle T_1(A)Z,Z \rangle \right) d\,\mathrm{vol}_L > 0.$$

Thus, we arrived at a contradiction, which shows σ_2 is constant on M.   □

Similarly, we get the following.

Proposition 2.14. *Let (M,g) be an Einstein (not necessarily compact) manifold with non-negative sectional curvature. A leaf of a foliation of M, whose leaves have the same constant $\sigma_2 > 0$, cannot be compact.*

Proof. Assume that a foliation with the above-mentioned properties has a closed leaf L. As before, we obtain that $T_1(A)$ is positive definite on M. As the leaves have the same constant σ_2, then $N(\sigma_2) = 0$. Moreover, $\mathrm{Ric}_{N,Z} = 0$, thus the Divergence Theorem and the Proposition 2.8 applied to L yield a contradiction.   □

Proposition 2.15 (See [BKO]). *Let $\mathscr{F}$ be a codimension-one C^3-foliation of a connected Riemannian manifold M and let $f : M \to \mathbb{R}$ be a continuous function which is constant along the leaves of $\mathscr{F}$. If f is not constant on M, then the set $G = \{x \in M : f(x) = \max_M f(x)\}$ contains at least one compact leaf.*

Proof. It is clear that G is a closed, thus compact, subset of M. Since f is constant along the leaves of $\mathscr{F}$ then G is a union of leaves, i.e., G is saturated. Since G is compact and saturated, G contains a minimal set $\mathscr{M}$. Since $\mathscr{M}$ is closed and M is compact then $\mathscr{M}$ is compact. If $\mathscr{M}$ has nonempty interior, then $\mathscr{M}$ must coincide with M (see [CN, page 53]). It can't happen, otherwise, f would be constant on M, which contradicts our hypothesis. It follows that $\mathscr{M}$ is either a compact leaf or a union of exceptional leaves (see [CN, Theorems 4 and 7]). If the last possibility occurs, then, according to [HH, Theorem 4.1.1], there is an open saturated set U containing $\mathscr{M}$ such that for any leaf L of $\mathscr{F}$ in U, the closure $\overline{L}$ contains $\mathscr{M}$. It follows that U is contained in the interior of G. Thus, either G contains a compact leaf or we may find an open saturated set $\mathscr{V}$ containing all minimal sets in G, with $\mathscr{V}$ contained in the interior of G. Since G is saturated compact and different from M, then $G \setminus \mathscr{V} \neq \emptyset$ is compact and saturated. Hence, it still contains a minimal set. This contradiction shows that G must contain a compact leaf. □

2.2 Foliations and Singularities

In the foliation theory, singular foliations have been defined usually as families $\mathscr{F}$ of maximal integral submanifolds (leaves) of generalized distributions $\mathscr{D}$ (e.g. [St, Su]) on manifolds M defined as functions which assign to each $x \in M$ a linear subspace $\mathscr{D}_x$ of the tangent space T_xM in such a way that for any $x \in M$ and $v \in \mathscr{D}_x$ there exists a smooth vector field V defined in a neighborhood U of x and such that $V(x) = v$ and $V(y) \in \mathscr{D}_y$ for all y of U. *A priori*, the dimension $p(x)$ of $\mathscr{D}_x$ depends on $x \in M$. Such foliations play a particular role in the theory of Riemannian foliations [Mol] and analysis of nonlinear systems in mechanics [Is]. In terms of parameterizations, such singular distributions can be represented as the images of TM under smooth endomorphisms $P : TM \to TM$, see [RP], and a formula similar to (1.44) is valid. Here, for our purposes, we use a different, very simple notion of singular foliations.

Definition 2.16. A foliation $\mathscr{F}$ of the complement $M \setminus \Sigma$ of a union Σ of finitely many closed codimension ≥ 2 submanifolds of a manifold M is called a *singular foliation* of M; Σ itself is called the *singular locus* of $\mathscr{F}$.

The singular case is important because many manifolds admit no smooth (e.g. codimension-one) distributions, while they admit such distributions or foliations outside some set. An example of singular foliations is given by "open book decompositions" of manifolds.

Example 2.17.

(i) Certainly, the restriction $\mathscr{F}_\Sigma$ of a regular foliation $\mathscr{F}$ of M to $M \setminus \Sigma$ is a singular foliation with the singular locus Σ; in this case, one can consider Σ as the set of "removable singularities".

(ii) The tori $T_\theta = \{(w,z) \in \mathbb{C}^2 : |w| = \cos\theta, |z| = \sin\theta\}$, $\theta \in (0, \pi/2)$, form a singular foliation of the unit sphere S^3 with singular locus Σ consisting of two great circles, geodesic lines on S^3.

(iii) In general, given Σ as in Definition 2.16 on a Riemannian manifold (M,g), the hypersurfaces Σ_r (with $r > 0$ and small enough) consisting of all the points in distance r from Σ form a singular foliation of a tubular neighborhood $\mathscr{N}$ of Σ.

2.2.1 Adapted Singular Foliations

Let $\Sigma \subset M^{n+1}$ be a compact embedded oriented submanifold of dimension q, and d_M the distance function on M. Following [Gr2], denote by $T(\Sigma,t)$ and $\Sigma(t)$ the *tube of radius t* around Σ and its boundary set, i.e.,

$$T(\Sigma,t) = \{x \in M : d_M(x,\Sigma) \le t\}, \quad \Sigma(t) = \{x \in M : d_M(x,\Sigma) = t\}.$$

The hypersurfaces $\Sigma(t)$ $(t < r)$, form a Riemannian foliation on $T(\Sigma,r)$ with a singular leaf Σ. Denote by $T\Sigma^\perp$ the normal bundle of Σ in M. The *exponential map of a normal bundle* $T\Sigma^\perp$ is defined by

$$\exp_\Sigma^\perp(x,\xi) = \exp_x(\xi), \quad \xi \in T_x\Sigma^\perp.$$

For small $t > 0$ the map $\exp_\Sigma^\perp : \{(x,\xi) \in T\Sigma^\perp : |\xi| \le t\} \to T(\Sigma,t)$ is a diffeomorphism, and the sets $\Sigma(t)$ are compact embedded oriented hypersurfaces. Let γ be a unit speed geodesic with $\gamma(0) = x \in \Sigma$ and $\gamma'(0) = \xi \in T\Sigma^\perp$. Hence $\gamma(t) = \exp_x(t\xi)$. The second fundamental tensor $S_\xi(t)$ of the tubular hypersurface $\Sigma(t)$ in the direction $\xi_t := \gamma'(t)$ satisfies the matrix Riccati equation $S'_\xi(t) = S_\xi^2(t) + R(t)$, where $R(t) = R(\cdot,\gamma'(t))\gamma'(t)$ is the restriction to $\Sigma(t)$ of the curvature operator. Set

$$V_\Sigma^M(t) = \text{the } (n+1)\text{-dimensional volume of } T(\Sigma,t),$$
$$W_\Sigma^M(t) = \text{the } n\text{-dimensional volume of } \Sigma(t).$$

Let O_Σ be the largest neighborhood of the zero section of $T\Sigma^\perp$ for which $\exp^\perp : O_\Sigma \to \exp^\perp(O_\Sigma)$ is a diffeomorphism. Denote by ω the (globally defined) volume form of M, and by $\omega^\perp$ the volume form of a normal bundle $T\Sigma^\perp$. So $\exp^{\perp*}(\omega)$ is a multiple of $\omega^\perp$. The function $\mathrm{chvol} : \exp^\perp(O_\Sigma) \to \mathbb{R}$ is the everywhere non-negative function given by

$$\exp^{\perp*}(\omega) = (\mathrm{chvol} \circ \exp^\perp) \cdot \omega^\perp.$$

The *infinitesimal change of volume function of* Σ *in the direction* ξ is the function defined for $(x,t\xi) \in O(\Sigma)$ by $\theta_\xi(t) = (\mathrm{chvol} \circ \exp^\perp)(x,t\xi)$, [Gr2]. This function measures the extent to which $\exp_\Sigma^\perp$ distorts volumes in the direction ξ. It satisfies the fundamental ODE that expresses the infinitesimal change of volume in terms of the shape operator. Namely, along γ for $t > 0$ we have

$$-\frac{d}{dt}\theta_\xi(t)/\theta_\xi(t) = \frac{n-q}{t} + \operatorname{trace} S_\xi(t).$$

For Euclidean space $M = \mathbb{R}^{n+1}$ we have just $\theta_\xi(t) = \det(\mathrm{id}_n - tS_\xi)$ as long as the RHS of this formula is nonnegative. One may show that

$$\frac{d}{dt} V_\Sigma^M(t) = W_\Sigma^M(t) = t^{n-q} \int_\Sigma \int_{S^{n-q}(1)} \theta_\xi(t) \, d\xi \, d\,\mathrm{vol},$$

and hence

$$V_\Sigma^M(t) = \alpha t^{n-q+1} + o(t^{n-q+1}),$$

that is, the Taylor expansion for $V_\Sigma^M(t)$ starts with a t^{n-q+1}-term.

For some submanifolds Σ (totally geodesic, etc.) in symmetric spaces M of rank 1 the function $V_\Sigma^M(t)$ has been explicitly calculated or at least estimated, see [Gr2].

Suppose that a unit vector field N is not defined on a union of compact embedded connected submanifolds $\Sigma_i \subset M^{n+1}$, $i = 1,\ldots,s$ of codim $\Sigma_i \geq 2$.

Definition 2.18. A vector field N (e.g. a unit normal to the leaves of a foliation $\mathscr{F}$) is called *adapted* to a submanifold Σ_i if there exists $\varepsilon_i > 0$ such that either N or $-N$ restricted on $T(\Sigma_i, \varepsilon_i)$ is the outward normal $\xi_i(t)$ to the tube hypersurfaces $\Sigma_i(t)$, $0 < t < \varepsilon_i$. Let v_{Σ_i} be the sign of the scalar product $\langle N, \xi_i \rangle$.

Examples of foliations with adapted normal field N are singular Riemannian foliations (the standard definition of a singular Riemannian foliation is that in [Mol, Chapter 6]), e.g., tube foliations of compact rank one symmetric spaces. For more general notion of singularity, see Section 2.2.3. We shall show that the formula (2.9) can be extended (with some additional terms) to foliations adapted to their singularities Σ_i. Note that our maps ϕ_t are defined on $M \setminus \bigcup_i \Sigma_i$ only.

Lemma 2.19. Let N be a unit vector field on a compact (M^{n+1}, g) and $\phi_t(x) = \exp_x(tN)$ the flow from (5.22) defined outside the singular locus. Let N be adapted to the components Σ_i (being compact connected embedded submanifolds of M) of its singular locus. Then for sufficiently small $t \geq 0$, i.e., $t < \min_i \varepsilon_i$, we have

$$\int_{M \setminus \bigcup_i \Sigma_i} |d\phi_t(x)| \, d\,\mathrm{vol} = \mathrm{Vol}(M, g) - \sum_i v_{\Sigma_i} V_{\Sigma_i}^M(t).$$

Proof. Denote by Σ_+ the union all singular components Σ_i with $v_{\Sigma_i} = 1$ and by Σ_- the union of the remaining ones. The map ϕ_t is a diffeomorphism from $M \setminus \{T(\Sigma_-, t) \bigcup \Sigma_+\}$ to $M \setminus \{T(\Sigma_+, t) \bigcup \Sigma_-\}$ for all $t < \min_i \varepsilon_i$. Hence

$$\int_{M \setminus \{T(\Sigma_-, t) \bigcup \Sigma_+\}} |d\phi_t(x)| \, d\,\mathrm{vol} = \mathrm{Vol}(M, g) - V_{\Sigma_+}^M(t).$$

This yields the required formula. $\qquad\square$

If the dimension of singular components Σ_i is small, then the term $\sum_i v_{\Sigma_i} V_{\Sigma_i}^M(t)$ does not contain t^m-terms with $m \leq n - \max_i \dim \Sigma_i$. Hence, we obtain

Theorem 2.20. Let $(M^{n+1}, \mathscr{F}, g)$ be a compact foliated locally symmetric space, and N be a unit vector field on M orthogonal to $\mathscr{F}$ and adapted to the components

Σ_i *(being compact connected embedded submanifolds of M) of its singular locus. Then (2.7) is valid for any* $0 < m \le n - \max_i \dim \Sigma_i$.

Corollary 2.21. *Let* $(M^{n+1}, \mathscr{F}, g)$ *be a singular Riemannian foliation of a compact manifold, and the singular components* Σ_i *are compact connected embedded submanifolds of M. Then (2.7) is valid for any* $0 < m \le n - \max_i \dim \Sigma_i$. *In particular, if* $\max_i \{\dim \Sigma_i\} \le n - 2$ *then*

$$\int_{M \setminus \bigcup_i \Sigma_i} \sigma_1 \, d\mathrm{vol} = 0, \qquad \int_{M \setminus \bigcup_i \Sigma_i} \left(2\sigma_2 - \mathrm{Ric}_{N,N} \right) d\mathrm{vol} = 0.$$

There exist foliations of compact rank one symmetric spaces, for which the singular locus consists of arbitrarily many components. One can start with an arbitrary nonsingular foliation of S^3 and perform the turbulization (see [CC, W1]) along disjoint loops Γ_i, $i \le m$; replacing all Reeb components of obtained foliation by families of "parallel" tori we get a foliation with the singular locus $\Gamma = \bigcup_{i=1}^m \Gamma_i$. Also, ([CC, Chapter 11]) for all nontrivial nonsplit links Γ in S^3 there exist codimension-one foliations of S^3 such that all their Reeb components are built around the components Γ_i of Γ; working as before we get a foliation with the singular locus Γ.

2.2.2 Improper Integrals

Given a submanifold Σ of an n-dimensional oriented manifold M and an n-form Ω on $M \setminus \Sigma$, one can define the improper integral of Ω by

$$\int_M \Omega \, d\mathrm{vol} = \lim_{U \to \Sigma} \int_{M \setminus U} \Omega \, d\mathrm{vol}, \tag{2.18}$$

whenever the limit in (2.18) exists; here, U ranges over the family of all open neighborhoods of Σ. One may show (see, Theorem 4.4 in,[1] for instance) that if M is compact and codim $\Sigma \ge 2$, then compactly supported forms on $M \setminus \Sigma$ are dense in the Sobolev space $H^1(M \setminus \Sigma)$ of differential forms on $M \setminus \Sigma$ which are L²-integrable with respect to their C¹-norms. Therefore, by Stokes' Theorem,

If ω *is an* $(n-1)$-*form of the space* $H^1(M \setminus \Sigma)$, *then* $\int_M d\omega = 0$. *Also, if X is a vector field on* $M \setminus \Sigma$, *M is closed oriented and Riemannian manifold and the norms* $\|X\|$ *and* $\|\nabla X\|$, ∇ *being—as usual—the Levi-Civita connection on M, are* L²-*integrable with respect to the Riemannian volume, then*

$$\int_M (\mathrm{div}\, X) \, d\mathrm{vol} = 0. \tag{2.19}$$

If M is equipped with a Riemannian metric g, then one can restrict attention to tubular neighborhoods $\mathscr{N}_r$ of radius r and put

---

[1] J. Brüning and M. Lesch, Hilbert complexes, J. Func. Anal. 108 (1992), 88–132.

$$\int_M \Omega\,\mathrm{d}\mathrm{vol} = \lim_{r\to 0}\int_{M\setminus\mathcal{N}_r}\Omega\,\mathrm{d}\mathrm{vol} \tag{2.20}$$

whenever the limit in (2.20) exists. Certainly, existence of the limit in (2.18) implies that in (2.20), however one cannot expect the converse in general.

Since here we work with Riemannian manifolds all the time, we shall use definition (2.20) (rather than (2.18)) everywhere throughout the book.

Let f be a nonnegative real function defined on $M\setminus\Sigma$, M being a closed oriented Riemannian manifold and Σ—a closed submanifold of M of codimension k. Given a sufficiently small $r > 0$, denote by $\mathcal{N}_r$ the tube of radius r around Σ and by Σ_r its boundary, the tubular surface at distance r. Also, let X be a vector field defined on $M\setminus\Sigma$. The following lemmas generalize those of [LuW, W6].

Lemma 2.22. *If $p > 1$, $(k-1)(p-1) \geq 1$ and $\liminf\limits_{r\to 0^+}\int_{\Sigma_r} f\,\mathrm{d}\sigma > 0$, then*

$$\int_M f^p\,\mathrm{d}\mathrm{vol} = \infty.$$

Proof. Since M is compact, its geometry is bounded and there exist positive constants r_0, c and ε such that $\mathrm{Vol}(\Sigma_r, g') \leq c\cdot r^{k-1}$ and $\int_{\Sigma_r} f\,\mathrm{d}\sigma \geq \varepsilon$ for all $r \leq r_0$. Hölder's inequality implies that

$$\int_{\Sigma_r} f\,\mathrm{d}\sigma \leq \Big(\int_{\Sigma_r} f^p\,\mathrm{d}\sigma\Big)^{1/p}\mathrm{Vol}(\Sigma_r, g')^{1/q},$$

where $\frac{1}{p} + \frac{1}{q} = 1$. Thus

$$\int_{\Sigma_r} f^p\,\mathrm{d}\sigma \geq \frac{\varepsilon^p}{c^{p/q}}\cdot r^{p(1-k)/q}$$

for all $r \leq r_0$. Finally, if $r \leq r_0$, then

$$\int_M f^p\,\mathrm{d}\mathrm{vol} \geq \int_{\mathcal{N}_r} f^p\,\mathrm{d}\mathrm{vol} = \int_0^r\Big(\int_{\Sigma_t} f^p\,\mathrm{d}\sigma\Big)\mathrm{d}t \geq \frac{\varepsilon^p}{c^{p/q}}\cdot\int_0^r t^{(1-k)p/q}\,\mathrm{d}t = \infty$$

when $(1-k)p/q \leq -1$, equivalently, when $(k-1)(p-1) \geq 1$. $\square$

Lemma 2.23. *If $(k-1)(p-1) \geq 1$ and $\int_M \|X\|^p\,\mathrm{d}\mathrm{vol} < \infty$, then equality (2.19) is valid.*

Proof. Let v_r be a suitably oriented unit vector field orthogonal to the tubular surface Σ_r. By Stokes' theorem and Lemma 2.22 applied to $f = \|X\|$, we obtain

$$\Big|\int_{M\setminus\mathcal{N}_r}(\mathrm{div}X)\mathrm{d}\mathrm{vol}\Big| = \Big|\int_{\Sigma_r}\langle X, v_r\rangle\,\mathrm{d}\sigma\Big| \leq \int_{\Sigma_r}\|X\|\,\mathrm{d}\sigma \to 0$$

as $r \to 0$. $\square$

We give first applications of the above lemma. As mentioned in Section 2.1, the mean curvature $H = \sigma_1$ of the leaves of a transversely oriented codimension-one foliation $\mathscr{F}$ of a Riemannian manifold M coincides (up to its sign) with the divergence of the positively oriented unit normal N. Since $\|N\| = 1$, it is integrable over M when M is compact. Therefore, Lemma 2.23 implies directly the following.

Theorem 2.24. *If $\mathscr{F}$ is a singular codimension-one foliation of a closed manifold M and the singular locus Σ of $\mathscr{F}$ is of codimension ≥ 2, then (2.1) is valid.*

If M is closed, then (2.3) implies (2.2) directly; and again, since

$$\|\sigma_1 \cdot N + \nabla_N N\|^2 = \sigma_1^2 + \kappa^2,$$

$\kappa = \|\nabla_N N\|$ being the curvature of the curves orthogonal to the leaves of $\mathscr{F}$, our Lemma 2.23 yields the following.

Theorem 2.25. *If $\mathscr{F}$ is a singular codimension-one foliation of a closed manifold M and the singular locus Σ of $\mathscr{F}$ is of codimension ≥ 2 and*

$$\int_M (\sigma_1^2 + \kappa^2)\, d\,\mathrm{vol} < \infty,$$

then (2.2) is valid.

2.2.3 Civilized Foliations

In [BLR], the authors distinguished a class of singular foliations called *civilized* (*civilisé* in French). Roughly speaking, $\mathscr{F}$ is *civilized* when, for r small enough, the tangent spaces to the leaves are close to the spaces tangent to the tubular hypersurfaces Σ_r consisting of points in distance r from Σ. In [BLR], this property has been expressed for $\mathscr{F}$ with Σ of codimension 2 in terms of a one-form ω defining $\mathscr{F}$: one should be able to extend such ω locally to a neighborhood of any point p of Σ and the germ of ω restricted to a normal geodesic disc at any $p \in \Sigma$ should be reversible. Here, we shall define them in terms of the unit normal N, in the way convenient for our purposes. So, let $\mathscr{F}$ be a singular codimension-one foliation of M, Σ—its singular locus, and $\mathscr{N}$—a tubular neighborhood of Σ, the union of Σ and surfaces Σ_r for $0 < r < r_0$. Let N and v be unit normals on $\mathscr{N} \setminus \Sigma$ for $\mathscr{F}$ and for the foliation of $\mathscr{N} \setminus \Sigma$ by Σ_r's.

Definition 2.26. The foliation $\mathscr{F}$ will be called C^k-*civilized* ($k \geq 2$) when there exists a smooth vector field Z on M and $r_1 \in (0, r_0)$ such that on Σ_r for any $r \in (0, r_1)$,

$$N = v + r^k \cdot Z. \tag{2.21}$$

Example 2.27. Foliations of Example 2.17 (ii) and (iii) are C^k-civilized for any k, while that of (i) is not. Singular Riemannian foliations [Mol] satisfy (2.21) with $Z = 0$. Moreover, foliations satisfying (2.21) with $Z = 0$ were considered in Section 2.2.1, where they are called adapted to Σ.

Equality (2.21) implies directly the following one:

$$\nabla N = \nabla v + d(r^k) \otimes Z + r^k \cdot \nabla Z, \qquad (2.22)$$

where ∇ is the Levi-Civita connection on M. Consequently, the norms $\|(N-v)(p)\|$ and $\|\nabla(N-v)(p)\|$ converge to 0 uniformly as $p \to \Sigma$.

Tracing both sides of (2.22) and taking into account the definition

$$\sigma_1 = \operatorname{trace} A = -\operatorname{trace} \nabla N = -\operatorname{div} N$$

of the mean curvature of $\mathscr{F}$ and the similar equality $h_r = -\operatorname{trace} \nabla v = -\operatorname{div} v$ for the mean curvature h_r of Σ_r, we obtain the formula

$$\sigma_1 = h_r - Z(r^k) - r^k \operatorname{div} Z,$$

which implies the following estimate:

$$|\langle \sigma_1 \cdot N, v \rangle| \le (|h_r| + kr^{k-1}\|Z\| + r^k|\operatorname{div} Z|) \cdot (1 + r^k\|Z\|) \qquad (2.23)$$

on Σ_r with r small enough. Moreover, (2.21) implies that

$$\nabla_N N = \nabla_v v + r^k \nabla_Z v + v(r^k)Z + r^k \nabla_v Z + r^k Z(r^k)Z + r^{2k}\nabla_Z Z. \qquad (2.24)$$

Since $v = \partial/\partial r$, $\|v\| = 1$ and $\nabla_v v = 0$, (2.24) provides us with another estimate:

$$|\langle \nabla_N N, v \rangle| \le kr^{k-1}\|Z\| + kr^{2k-1}\|Z\|^2 + r^k\|\nabla Z\| + r^{2k}\|Z\| \cdot \|\nabla Z\|. \qquad (2.25)$$

Since locally, in a neighborhood of Σ, extrinsic geometry of Σ_r's is quasi-isometric to that of cylinders of revolution in the Euclidean space, the mean curvature h_r of Σ_r satisfies, for r small enough, the inequality

$$|h_r - f_r/r| \le C, \qquad (2.26)$$

where C is a constant depending on the extrinsic geometry of Σ, and f_r a function on Σ_r that is uniformly close to 1 when r is close to 0. Set $m = \operatorname{codim} \Sigma$ and denote by ω_{m-1} the volume of the unit sphere of dimension $m - 1$. Put also

$$C_Z = \max_M \|Z\|, \quad C_{\nabla Z} = \max_M \|\nabla Z\|.$$

With this notations, estimates (2.23), (2.25) and (2.26) imply the existence of constants $\varepsilon_r > 0$ such that $\varepsilon_r \to 0$ when $r \to 0$ and

$$\left| \int_{\Sigma_r} \langle H + \tilde{H}, v \rangle d\sigma_r \right| \le \omega_{m-1}(r^{m-1} + \varepsilon_r) \cdot \operatorname{Vol}(\Sigma, g')\big(1/r + C$$
$$+ 2kr^{k-1}C_Z + kr^{2k-1}C_Z^2 + (1+n)r^k C_{\nabla Z} + r^{2k}C_Z C_{\nabla Z}\big).$$

Passing to the limit as $r \to 0$ we end up with following result.

Theorem 2.28. *If $\mathscr{F}$ is a C^k- civilized ($k \geq 2$) foliation of $M \setminus \Sigma$, then*

$$\int_M (\mathrm{Ric}_{N,N} - 2\sigma_2)\, d\, \mathrm{vol} = \begin{cases} 2\pi \cdot \mathrm{Vol}(\Sigma, g') & \text{when } \mathrm{codim}\, \Sigma = 2 \\ 0 & \text{when } \mathrm{codim}\, \Sigma > 2. \end{cases}$$

For manifolds M of constant sectional curvature c we obtain the following.

Corollary 2.29 (Théorème 3.8 from [BLR]). *If $\mathscr{F}$ is a civilized foliation of $M \setminus \Sigma$, codim $\Sigma = 2$ and the sectional curvature of M is constant equal to c, then*

$$\int_M \sigma_2\, d\, \mathrm{vol} = \frac{c(n-1)}{2} \mathrm{Vol}(M, g) - \pi \cdot \mathrm{Vol}(\Sigma, g').$$

Comparing Theorems 2.25 and 2.28 one gets also

Corollary 2.30. *If $\mathscr{F}$ is a civilized foliation of $M \setminus \Sigma$, $\dim M \geq 3$ and codim $\Sigma = 2$, then*

$$\int_M (\sigma_1^2 + \kappa^2)\, d\, \mathrm{vol} = \infty. \tag{2.27}$$

Example 2.31. Let us find the mean curvature of tori T_θ from Example 2.17(ii) and show that the foliation of S^3 by these tori satisfies (2.27). Tori T_θ coincide with the products of two orthogonal circles of radii $\cos\theta$ and $\sin\theta$, therefore, of curvatures $1/\cos\theta$ and $1/\sin\theta$. Projecting the corresponding curvature vectors in the direction orthogonal to S^3 we conclude that the mean curvature of T_θ equals $\pm 2\cos 2\theta / \sin 2\theta$ (with the sign depending on the choice of a transverse orientation). Its volume amounts to $4\pi^2 \sin\theta \cos\theta$, the product of lengths of the two circles mentioned above. Integrating the product of these quantities over the interval $(0, \pi/2)$ and using the equality $\kappa = 0$ (that is the fact that all the curves orthogonal to our foliation are geodesic lines on S^3), we arrive at the expected equality (infinity of the integral).

Remark 2.32.

(i) In [BLR], it has been conjectured (Conjecture 3.10) that results similar to those for σ_2 can be obtained for other mean curvatures σ_k with $k > 2$. Integral formulae of Section 2.3 (in particular, of Section 2.1) seem to provide tools suitable for obtaining such results. Also [RW5, Lemma 4] could be useful for such purposes. Note that [RW5, Theorem 2 and Corollary 4] provide a partial solution of that conjecture for locally symmetric spaces.

(ii) Several integral formulae which appear in this book in the context of regular foliations of closed manifolds can be easily generalized to singular foliations, civilized or not, under suitable integrability conditions which allow to use our technical lemmas of Section 2.2.2.

(iii) A *singular distribution* $\mathscr{D}$ on a manifold M assigns to each point $x \in M$ a linear subspace $\mathscr{D}_x$ of the tangent space $T_x M$ in such a way that, for any $v \in \mathscr{D}_x$, there exists a smooth vector field V defined in a neighborhood U of x and such that $V(x) = v$ and $V(y) \in \mathscr{D}_y$ for all y of U. A priori, the dimension $\dim \mathscr{D}_x$

depends on $x \in M$. If $\dim \mathscr{D}_x = \text{const}$, then we obtain a regular distribution. *Singular foliations* are defined as families of maximal integral submanifolds (leaves) of integrable singular distributions (certainly, regular foliations correspond to integrable regular distributions). The Bochner technique works for tensors lying in the kernel of Hodge Laplacian $\Delta_H = d\delta + \delta d$ on a closed manifold, e.g., [Jo, Pe]: using maximum principles, they prove that such tensors are parallel. In [RP, RPS, RPS2], a manifold endowed with a singular or regular distribution, determined as the image of the tangent bundle under a smooth endomorphism (e.g., the orthogonal projection) is studied and the Bochner's technique is generalized to the case of a distribution.

2.3 Foliations of Arbitrary Codimension

In further generalizations of integral formulas (discussed in Section 2.1) for foliations of arbitrary codimension [AW1, AW2, BLR, BN, LuW, R5, W4], the integrand depends on second fundamental forms, integrability tensors and mixed curvature.

Applying the Stokes' theorem to (1.44), yields a general integral formula for complementary orthogonal distributions $(\mathscr{D}, \widetilde{\mathscr{D}})$ (see notations in Section 1.3.1) on a closed Riemannian manifold (M, g), obtained in [W4],

$$\int_M \left(S_{\text{mix}} + \|h\|^2 + \|\tilde{h}\|^2 - \|H\|^2 - \|\tilde{H}\|^2 - \|T\|^2 - \|\tilde{T}\|^2 \right) d \operatorname{vol} = 0. \qquad (2.28)$$

One may show, using calculations of Example 1.38, that for a codimension-one integrable distribution with a unit normal N, (2.28) reduces to (2.2).

In this section, we study two methods to derive a series of integral formulae for a $(n + p)$-dimensional Riemannian manifold (M, g) endowed with a transversely oriented p-dimensional foliation $\mathscr{F}$ with the tangent distribution $\widetilde{\mathscr{D}}$ and the n-dimensional normal distribution $\mathscr{D}$. Our integral formulae involve the shape (or conullity) operator, and certain components of the curvature tensor and their products.

Integral formulas of Section 2.3 can be extended
– for foliations defined outside of a "singularity set" Σ.
– for holomorphic foliations of complex Riemannian manifolds, e.g., [Sv].

2.3.1 Using a Family of Diffeomorphisms

In what follows (M, g) denotes a compact (or of finite volume) Riemannian manifold with the Levi-Civita connection ∇, and the curvature tensor R; $\mathscr{F}$ a transversely oriented foliation of M, $T\mathscr{F}$ $(\mathscr{D})$ its tangent (normal) distribution,

$$R_\xi^{\text{mix}}(y) = (R(y, \xi)\,\xi)^\perp \quad (\xi \in T\mathscr{F},\ y \perp T\mathscr{F}),$$

the mixed curvature operator, K_{mix} the mixed sectional curvature, $C_\xi y = -(\nabla_y \xi)^\perp$ the co-nullity operator (for unit vectors ξ), $\sigma_i(C_\xi)$ the i-th mean curvature, see Section 1.5.3. For the case of codimension-one foliations, ξ is a positive oriented, unit normal vector field.

The results in Section 2.1 suggest the existence of similar to (2.1) and (2.2) formulae for σ_i's with $i > 2$. Here, we provide such formulae: they generalize (2.5) (the constancy of curvature condition is rejected) and relate total σ_i's with integrals involving some algebraic invariants of the co-nullity operator, the mixed curvature operator and their products. We write several such formulae, on locally symmetric spaces as well as on arbitrary Riemannian manifolds, where they involve also covariant derivatives of the mixed curvature operator.

2.3.1.1 Algebraic Preliminaries

Let $T : (\mathbb{R}^p)^h \times \mathbb{R}^n \to \mathbb{R}^n$ be a multilinear map. For each $\xi \in \mathbb{R}^p$ define a linear operator $T_\xi := T(\xi, \dots, \xi; \cdot) : \mathbb{R}^n \to \mathbb{R}^n$. In coordinate form $\xi = \sum_i \xi_i e_i$, we have $T_\xi = \sum T_{i_1, \dots i_p} \xi_{i_1} \cdot \dots \cdot \xi_{i_p}$, where $T_{i_1, \dots i_p} = T(e_{i_1}, \dots e_{i_p}; \cdot)$. Let $d\omega_{p-1}$ be the Lebesgue measure on the unit sphere S^{p-1} in $\mathbb{R}^p$.

Lemma 2.33. *If h and m are odd, then $\int_{\|\xi\|=1} \sigma_m(T_\xi) \, d\omega_{p-1} = 0$.*

Proof. By Remark 1.72 (V), $\sigma_m(T_{-\xi}) = \sigma_m((-1)^h T_\xi) = (-1)^{hm} \sigma_m(T_\xi)$. □

Set $I_\lambda := \int_{\|\xi\|=1} \xi^\lambda \, d\omega_{p-1}$, where $\xi^\lambda = \prod_{i \le p} \xi_i^{\lambda_i}$. One may show that

$$I_\lambda = \frac{2}{\Gamma(p/2 + (1/2)\sum_i \lambda_i)} \prod_{i \le p} \frac{1 + (-1)^{\lambda_i}}{2} \Gamma\left(\frac{1 + \lambda_i}{2}\right),$$

see [PBM], where Γ is the Gamma function. For example,

$$I_{0,\dots 0} = \frac{2\pi^{p/2}}{\Gamma(p/2)} = \mathrm{vol}(S_1^{p-1}), \quad I_{2\lambda_1,0,\dots 0} = 2\pi^{\frac{p-1}{2}} \frac{\Gamma(1/2 + \lambda_1)}{\Gamma(p/2 + \lambda_1)}.$$

Lemma 2.34. *If $h = 1$, then $\int_{\|\xi\|=1} \sigma_{2k-1}(\sum_{i \le p} \xi_i T_i) \, d\omega_{p-1} = 0$ and*

$$\int_{\|\xi\|=1} \sigma_{2k}\left(\sum_{i \le p} \xi_i T_i\right) d\omega_{p-1} = \sum_{|\lambda|=k} I_{2\lambda} \, \sigma_{2\lambda}(T_1, \dots T_p). \tag{2.29}$$

Proof. Using (1.83), one obtains for $T_\xi = \sum_{i \le p} \xi_i T_i$

$$\sum_j \sigma_j(T_\xi) t^j = \det\left(\mathrm{id} + \sum_{i \le p}(t \, \xi_i) T_i\right) = \sum_j t^j \sum_{|\lambda|=j} \sigma_\lambda(T_1, \dots T_p) \xi^\lambda.$$

Hence,

$$\sigma_j(T_\xi) = \sum_{|\lambda|=j} \sigma_\lambda(T_1, \dots T_p) \xi^\lambda. \tag{2.30}$$

From this and definition of $I_{2\lambda}$, (2.29) follows. □

Example 2.35. For $k = 1, 2, \ldots, (2.29)$ read as

$$\int_{\|\xi\|=1} \sigma_2(T_\xi) \, d\omega_{p-1} = I_{2,0,\ldots 0} \sum_{i \leq p} \sigma_2(T_i),$$

$$\int_{\|\xi\|=1} \sigma_4(T_\xi) \, d\omega_{p-1} = I_{4,0,\ldots 0} \sum_{i \leq p} \sigma_4(T_i) + I_{2,2,0,\ldots 0} \sum_{i,j \leq p} \sigma_{2,2}(T_i, T_j),$$

and so on, where $I_{2,0\ldots 0} = \frac{\pi^{p/2}}{\Gamma(p/2+1)}$, $I_{4,0\ldots 0} = \frac{3\pi^{p/2}}{2\Gamma(p/2+2)}$, $I_{2,2,0\ldots 0} = \frac{\pi^{p/2}}{2\Gamma(p/2+2)}$.

If $T_\xi = \sum_{i,j} \xi_i \xi_j T_{ij}$, then, similarly to (2.30), one has

$$\sigma_k(T_\xi) = \sum_{|\lambda|=k} \sigma_{(\lambda_{11},\ldots \lambda_{ij},\ldots \lambda_{pp})}(T_{11}, \ldots, T_{ij}, \ldots, T_{pp}) \prod_{i,j \leq p} (\xi_i \xi_j)^{\lambda_{ij}}, \quad (2.31)$$

and so on. For $k = 1, 2$, (2.31) reads as $\sigma_1(T_\xi) = \sum_{i,j} \sigma_1(T_{ij}) \xi_i \xi_j$ and

$$\sigma_2(T_\xi) = \sum_{i,j} \sigma_2(T_{ij}) \xi_i \xi_j + \sum_{i,j,s,l} \sigma_{1,1}(T_{ij}, T_{sl}) \xi_i \xi_j \xi_s \xi_l.$$

The integrals of $\sigma_k(T_\xi)$ are similar to (2.29), e.g.,

$$\int_{\|\xi\|=1} \sigma_1(T_\xi) \, d\omega_{p-1} = \frac{\pi^{p/2}}{\Gamma(p/2+1)} \sum_{i \leq p} \sigma_1(T_{ii}).$$

2.3.1.2 The Integral Formulas

Let $\mathscr{F} = \{L\}$ be a transversely oriented foliation of M. We call $\overline{\mathscr{F}} = \{TL\}$ the *lift foliation* of $T\mathscr{F}$. The metric $\bar{g}$ in $T\mathscr{F}$ is the restriction of the Sasaki metric of TM (see Section 1.2.1). We call $S\mathscr{F} = \{(x, \xi) \in T\mathscr{F} : |\xi| = 1\}$ the *tangent sphere bundle of radius 1* of $\mathscr{F}$, a hypersurface in $T\mathscr{F}$ consisting of all pairs (x, ξ), where $x \in M$ and $\xi \in TL$ is a vector of length 1. We call $\widetilde{\mathscr{F}} = \{SL\}$ the *lift foliation of* $S\mathscr{F}$. The vertical vector field $\bar{\xi}$ is outward normal to $S\mathscr{F}$ in $T\mathscr{F}$ at each point $(x, \xi) \in S\mathscr{F}$. The *tangential lift* of y is a vector field $\tilde{y}^t$ tangent to $S\mathscr{F}$ and defined by $\tilde{y}^t = \bar{y}_V - \bar{g}(\bar{y}_V, \bar{\xi}) \bar{\xi}$. Using Sasaki metric $\bar{g}$, we define metric $\tilde{g}$ on $S\mathscr{F} \subset T\mathscr{F}$,

$$\tilde{g}(\tilde{z}_H, \tilde{y}_H) = \bar{g}(\bar{z}_H, \bar{y}_H), \quad \tilde{g}(\tilde{z}_H, \tilde{y}^t) = 0, \quad \tilde{g}(\tilde{z}^t, \tilde{y}^t) = \bar{g}(\bar{z}_V, \bar{y}_V) - \bar{g}(\bar{z}_V, \bar{\xi}) \bar{g}(\bar{y}_V, \bar{\xi}).$$

The natural projection $\tilde{\pi} : S\mathscr{F} \to M$ is a Riemannian submersion with totally geodesic fibers S_1^{p-1}, the spheres of a radius 1. Hence, the volume $\text{Vol}(S\mathscr{F})$ is the product of $\text{Vol}(S_1^{p-1})$ and $\text{Vol}(M, g)$, see [Ber, Note 7.1.1.1].

Lemma 2.36. *Let $\mathscr{F} = \{L\}$ be a p-dimensional totally geodesic foliation of M. Then $\widetilde{\mathscr{F}} = \{SL\}$, the $(2p-1)$-dimensional lift foliation of $T_1\mathscr{F}$, is totally geodesic.*

Proof. This follows from the fact that TL is totally geodesic in TM if and only if L is a totally geodesic submanifold in M. □

The tangent bundle to $S\mathscr{F}$ is orthogonally decomposed into the sum of *horizontal* and *vertical* subbundles: $T(S\mathscr{F}) = \widetilde{V} \oplus \widetilde{H}$. In view of isomorphism $\tilde{\pi}_* : \widetilde{H} \to TM$, the almost product structure $TM = T\mathscr{F} \oplus T\mathscr{F}^\perp$ induces the orthogonal decomposition of $\widetilde{H}$ into the sum of subbundles: $\widetilde{H} = \widetilde{H}^\top \oplus \widetilde{H}^\perp$.

Definition 2.37. The *characteristic vector field* $\mathscr{N} \subset \widetilde{H}^\top$ (velocity of the geodesic flow on $S\mathscr{F}$) at a point $\tilde{x} = (x, \xi)$ is defined by $\tilde{\pi}_*(\mathscr{N}) = \xi$. The *characteristic Weingarten* and *Jacobi operators* on $T_1\mathscr{F}$ are defined by

$$\tilde{A}_\mathscr{N} : \tilde{y} \to \widetilde{\nabla}_{\tilde{y}}\mathscr{N} - \langle \widetilde{\nabla}_{\tilde{y}}\mathscr{N}, \mathscr{N} \rangle \mathscr{N}, \quad \tilde{R}_\mathscr{N} : \tilde{y} \to \tilde{R}(\tilde{y}, \mathscr{N})\mathscr{N}.$$

For totally geodesic $\mathscr{F}$, the restriction of $\mathscr{N}$ on any leaf SL generates the geodesic flow on this leaf, and the distribution $\widetilde{H}^\perp$ is $\widetilde{\nabla}$-parallel along $\mathscr{N}$.

Theorem 2.38. *Let $\mathscr{F}^p$ be a totally geodesic foliation of a complete locally symmetric manifold (M^{p+n}, g) of finite volume. Set*

$$B_{\xi,2k} = \frac{1}{(2k)!}(-R_\xi^{\mathrm{mix}})^k, \quad B_{\xi,2k+1} = \frac{1}{(2k+1)!}(-R_\xi^{\mathrm{mix}})^k C_\xi,$$

and assume that the normal distribution $T^\perp\mathscr{F}$ has bounded extrinsic geometry, i.e.,

$$\sup{}_{\|\xi\|=1} \|C_\xi\| < \infty.$$

Then for any $m > 0$ one has

$$\int_M \left(\int_{\|\xi\|=1} \left(\sum{}_{\|\lambda\|=m} \sigma_\lambda(B_{\xi,1}, \ldots, B_{\xi,m})\right) d\omega\right) d\mathrm{vol}_g = 0. \tag{2.32}$$

Proof. Let $\tilde{y}$ be a vector field on $T_1\mathscr{F}$ and $y = \pi_*(\tilde{y})$ the projection on M. Then

$$\tilde{A}_\mathscr{N}\tilde{y} = \widetilde{\nabla}_{\tilde{y}}\mathscr{N} \subset \widetilde{V} \oplus \widetilde{H}^\top \qquad (\tilde{y} \subset \widetilde{V} \oplus \widetilde{H}^\top),$$
$$\tilde{A}_\mathscr{N}\tilde{y} = \widetilde{\nabla}_{\tilde{y}}\mathscr{N} = \widetilde{(\nabla_y \xi)^\perp} \subset \widetilde{H}^\perp \quad (\tilde{y} \subset \widetilde{H}^\perp),$$
$$\tilde{R}_\mathscr{N}\tilde{y} \subset \widetilde{V} \oplus \widetilde{H}^\top \qquad (\tilde{y} \subset \widetilde{V} \oplus \widetilde{H}^\top),$$
$$\tilde{R}_\mathscr{N}\tilde{y} = \widetilde{R_\xi^{\mathrm{mix}}y} \subset \widetilde{H}^\perp \quad (\tilde{y} \subset \widetilde{H}^\perp, \ \xi = \tilde{\pi}_*(\mathscr{N})).$$

By the above, the Weingarten and Jacobi operators, $\tilde{A}_\mathscr{N}$, $\tilde{R}_\mathscr{N}$, and hence the *Jacobi operator* $\tilde{Y}_\mathscr{N}$ have 2×2 block form:

$$\tilde{A}_\mathscr{N} = \begin{pmatrix} \tilde{A}_\mathscr{N}^L & * \\ 0 & C_\xi \end{pmatrix}, \quad \tilde{R}_\mathscr{N} = \begin{pmatrix} \tilde{R}_\mathscr{N}^L & 0 \\ 0 & R_\xi^{\mathrm{mix}} \end{pmatrix}, \quad \tilde{Y}_\mathscr{N}(t) = \begin{pmatrix} Y_\mathscr{N}^L(t) & * \\ 0 & Y_\xi^\perp(t) \end{pmatrix}$$

where $\tilde{A}_\mathscr{N}^L$ and $\tilde{R}_\mathscr{N}^L$ are the restrictions of $\tilde{A}_\mathscr{N}$ and $\tilde{R}_\mathscr{N}$ on SL. By Schur identity, we have $\det \tilde{Y}_\mathscr{N}(t) = \det Y_\mathscr{N}^L(t) \det Y_\xi^\perp(t)$. Recall that $\mathscr{N}$ generates the geodesic flow on each SL and $\det Y_\mathscr{N}^L(t)$ presents the volume element in L. Hence, by Liouville's Theorem, $\det Y_\mathscr{N}^L(t) = 1$ for all t. Finally, we have $\det \tilde{Y}_\mathscr{N}(t) = \det Y_\xi^\perp(t)$ for all t.

Note that $\|R_\xi^{\mathrm{mix}}\| \le \|R\|$ ($|\xi| = 1$) and $\|R\| = \mathrm{const}$, because $\nabla R = 0$. Given $\xi \in T_x L$, choose a positive oriented orthonormal frame $(e_1, \ldots, e_n)$ of $T_x L^\perp$ and extend it by parallel translation to the frame $(E_1, \ldots, E_n)$ of vector fields along the geodesic $\gamma : t \to \exp_x(t\,\xi)$. Hence, for the Jacobi tensor $Y_\xi^\perp(t)$ we get

$$Y_\xi^\perp(t) = \mathrm{id}_n + t\,C_\xi - \frac{t^2}{2!} R_\xi^{\mathrm{mix}} - \frac{t^3}{3!} R_\xi^{\mathrm{mix}} C_\xi + \frac{t^4}{4!} (R_\xi^{\mathrm{mix}})^2 + \ldots$$

The diffeomorphisms $\phi_t(x) = \exp_x(t\,\tilde{N}(x))$ of $S\mathscr{F}$ are defined for $t \in \mathbb{R}$ (since leaves of $\mathscr{F}$ are complete). The Jacobian of ϕ_t satisfies $|d\phi_t(x)| = \det \tilde{Y}_{\tilde{N}}(t)$ [RW5]. The rest of proof is similar to codimension-one case (compare Section 2.1.1). $\qquad\square$

Remark 2.39. The expressions for $B_{\xi,i}$ ($i > 2$) in Theorem 2.38 are more complex without assumption that (M, g) is locally symmetric, since they also contain terms with covariant derivatives of the mixed curvature operator. For example,

$$B_{\xi,3} = -\frac{1}{6}(R_\xi^{\mathrm{mix}} C_\xi + \nabla_\xi R_\xi^{\mathrm{mix}}), \quad B_{\xi,4} = \frac{1}{4!}((R_\xi^{\mathrm{mix}})^2 - \nabla_\xi^2 R_\xi^{\mathrm{mix}} - 2(\nabla_\xi R_\xi^{\mathrm{mix}}) C_\xi).$$

Note that $\|\lambda\| = \mathrm{odd}$ implies $\int_{\|\xi\|=1} \sigma_\lambda\,(B_{\xi,1}(x), \ldots, B_{\xi,m}(x))\,d\omega_{p-1} = 0$.

For few initial values of m, $m = 2, 4$, (2.32) read as follows:

$$\int_M \Big(\int_{\|\xi\|=1} \big(\sigma_2(C_\xi) - \frac{1}{2}\sigma_1(R_\xi^{\mathrm{mix}}) \big)\,d\omega \Big)\,d\mathrm{vol}_g = 0,$$

$$\int_M \Big(\int_{\|\xi\|=1} \big(\sigma_4(C_\xi) + \frac{1}{4}\sigma_2(R_\xi^{\mathrm{mix}}) + \frac{1}{24}\sigma_1((R_\xi^{\mathrm{mix}})^2)$$

$$-\frac{1}{6}\sigma_{1,1}(C_\xi, R_\xi^{\mathrm{mix}} C_\xi) \big)\,d\omega \Big)\,d\mathrm{vol}_g = 0. \qquad (2.33)$$

One may reduce (2.33) to integrals over M, e.g., $(2.33)_1$ reduces to (2.28).

Corollary 2.40. *Let $\mathscr{F}^p$ be a totally geodesic foliation of a complete Riemannian manifold M^{p+n} of finite volume with $K_{\mathrm{mix}} = c \ge 0$ and $\sup_{\|\xi\|=1} \|C_\xi\| < \infty$. Then*

$$\int_{\xi \in S^\perp} \sigma_k(C_\xi)\,d\omega = \begin{cases} \dfrac{2\pi^{p/2}}{\Gamma(p/2)} \dbinom{n/2}{k/2} c^{k/2}\,\mathrm{Vol}(M, g), & n \text{ and } k \text{ even}, \\[2mm] 0, & \text{either } n \text{ or } k \text{ odd.} \end{cases} \qquad (2.34)$$

Proof. If $K_{\mathrm{mix}} = c = \mathrm{const}$, then $\nabla_\xi R_\xi^{\mathrm{mix}} = 0$ ($\xi \in T\mathscr{F}$). By the proof of Theorem 2.38 we have

$$\sum_{m \le k} z^m \int_{\xi \in S^\perp} \sigma_m(C_\xi)\,d\omega = \mathrm{vol}(S\mathscr{F})\,(1 \pm |c|\,z^2)^{k/2},$$

hence, (2.34). For k odd we obtain $c = 0$ (the case $c \ne 0$ leads to a contradiction). $\qquad\square$

For a codimension-one distribution $\widetilde{\mathscr{D}}$ (i.e., $p = 1$), the projection $\pi : S^\perp \to M$ is a double covering. One can show that if $\mathscr{D}$ is transversally orientable, then (2.34) reduces to (2.5) with doubled RHS.

Definition 2.41. Given orthonormal frames $\{e_i\}_{i \leq p}$ of $\widetilde{\mathscr{D}}_x$ and $\{e_\alpha\}_{\alpha \leq n}$ of $\mathscr{D}_x = (T_x\mathscr{F})^\perp$ at $x \in M$, set $C_{\alpha,\beta}^i = \langle C_{e_i}e_\alpha, e_\beta\rangle$. The *extrinsic 2k-curvature* of $\mathscr{D}_x$ is given by

$$\gamma_{2k}(\mathscr{D}_x) = \frac{1}{(2k)!\,2^k}\sum_{\alpha,\beta}\varepsilon_{\alpha,\beta}\Big[\sum_{i_1=1}^p (C_{\alpha_1,\beta_1}^{i_1}C_{\alpha_2,\beta_2}^{i_1} - C_{\alpha_1,\beta_2}^{i_1}C_{\alpha_2,\beta_1}^{i_1}) \times$$
$$\ldots \times \sum_{i_k=1}^p (C_{\alpha_{2k-1},\beta_{2k-1}}^{i_k}C_{\alpha_{2k},\beta_{2k}}^{i_k} - C_{\alpha_{2k-1},\beta_{2k}}^{i_k}C_{\alpha_{2k},\beta_{2k-1}}^{i_k})\Big],$$

where $\varepsilon_{\alpha,\beta} = \varepsilon_{\alpha_1,\ldots,\alpha_{2k};\beta_1,\ldots,\beta_{2k}}$ looks like a product, we have to say 1 (or, -1) if $\alpha = (\alpha_1,\ldots,\alpha_{2k})$ are $2k$ distinct integers and $\beta = (\beta_1,\ldots,\beta_{2k})$ are even (or, odd) permutations of α; otherwise it is 0. The sum is taken over all α_i, β_i between 1 and n.

For a totally geodesic foliation $\mathscr{F}$ ($T\mathscr{F} = \widetilde{\mathscr{D}}$) on a closed space form $M^{n+p}(c)$ with $c \geq 0$, the total $\gamma_{2k}(\mathscr{D}_x)$ depends on k, p, n, c and $\mathrm{Vol}(M,g)$ only [BN]:

$$\gamma_{2k}(\mathscr{D}) = \begin{cases} \dfrac{\binom{p+2k-1}{2k}}{\binom{(p+2k-1)/2}{k}}\binom{n/2}{k}c^k\,\mathrm{vol}(M,g) & \text{if } n \text{ even, } p \text{ odd;} \\[3mm] \dfrac{2^{2k}(k!)^2}{(2k)!}\binom{p/2+k-1}{k}\binom{n/2}{k}c^k\,\mathrm{vol}(M,g) & \text{if } n, p \text{ are even;} \\[3mm] 0 & \text{if } n \text{ odd.} \end{cases} \tag{2.35}$$

Remark 2.42.

(a) For $p = 1$ the equality $\gamma_{2k}(\mathscr{D}_x) = \sigma_{2k}(\mathscr{D}_x)$ is valid, moreover, (2.34) and (2.35) coincide when $p = 1$. For $p > 1$, we have $\gamma_{2k}(\mathscr{D}_x) \neq \sigma_{2k}(\mathscr{D}_x)$.

(b) There are no totally geodesic foliations $\mathscr{F}^p$ ($p > 0$) with $K_{\mathrm{mix}} = \mathrm{const} \neq 0$ on a closed manifold (M^{p+n}, g) for n odd. Of course, let $\widetilde{\mathscr{D}}$ be tangent to a totally geodesic foliation of a closed M and the mixed curvature be constant $c \in \mathbb{R}$. Notice that $\sigma_{2s+1}(\mathscr{D}) = 0$. By (2.44) and the equality $\mathrm{trace}\,T_r(C_N) = (n-r)\sigma_r(N)$, we get

$$\sigma_{r+2}(\mathscr{D}) = \frac{c(n-r)}{r+2}\sigma_r(\mathscr{D}), \qquad \sigma_0(\mathscr{D}) = \int_{S^\perp} d\omega^\perp = \frac{2\pi^{p/2}}{\Gamma(p/2)}\,\mathrm{Vol}(M).$$

By induction, $\sigma_{2k}(\mathscr{D}) \neq 0$ when $c \neq 0$, that contradicts to (2.34) for n odd.

2.3.2 Using the Divergence Theorem

Here, we extend results of Section 2.1.3 for foliations of any codimension. Assume, for simplicity, that $\widetilde{\mathscr{D}}$ is integrable and tangent to a foliation $\mathscr{F}$. Let $S^\perp = \{\xi \in \mathscr{D} : \langle \xi, \xi\rangle = 1\}$ be the unit sphere bundle with the Sasaki metric and the volume form $d\omega$. The natural projection $\pi : S^\perp \to M$ is a Riemannian submersion with totally geodesic fibers $\{S_x^\perp\}_{x \in M}$ (unit spheres). Then the volume form is decomposed as $d\,\mathrm{vol}_{S^\perp} = d\,\mathrm{vol}_{S_x^\perp} \cdot d\,\mathrm{vol}_M$, see [Ber, Note 7.1.1.1]. Hence, $d\,\mathrm{vol}(S^\perp, d\omega)$ is the product of $d\,\mathrm{vol}(S_1^{n-1})$ and $d\,\mathrm{vol}_g$, and the differentiating along M commutes with the integration along the fibers $S_x^\perp$.

Given functions f_j on $\mathbb{R}^p$ $(0 \le j < p)$, define a (1,1)-tensor field $\mathscr{A}$ by

$$\mathscr{A} = \int_{\xi \in S_x^\perp} \mathscr{A}_\xi \, d\omega, \quad \text{where } \mathscr{A}_\xi := \sum_{j=0}^{p-1} f_j(\vec{\tau}(\xi)) A_\xi^j \quad \text{and } x \in M,$$

compare with (2.14), where A_ξ is the shape operator for the normal ξ. Evidently, $\langle A_\xi(X), Y \rangle = \langle h_{X,Y}, \xi \rangle$ for $X, Y \in \mathscr{D}$. Define a linear operator $R(X,Y) : \widetilde{\mathscr{D}} \to \widetilde{\mathscr{D}}$ by

$$R(X,Y) : Z \to (R(Z,X)Y)^\top, \quad Z \in \Gamma(\widetilde{\mathscr{D}}), \ X, Y \in \Gamma(TM),$$

and denote $R_\xi = R(\xi, \xi)$, i.e., $R_\xi : Z \to (R(Z,\xi)\xi)^\top$ for $Z \in \widetilde{\mathscr{D}}$.

2.3.2.1 The Leaf-Wise Divergence

Let $\{e_i\}$, $(1 \le i \le p)$ be a local orthonormal frame of $T\mathscr{F}$.

Lemma 2.43. *The $\mathscr{F}$-divergence of $(A_\xi)^k$ for $k > 0$ is given by the formula*

$$-(\nabla_\mathscr{F}^* A_\xi^k) = \sum_{1 \le j \le k} \left((k-j+1)^{-1} A_\xi^{j-1}(\nabla \tau_{k-j+1}(A_\xi))^\top\right.$$
$$\left. - \sum_i A_\xi^{j-1}(R(\xi, A_\xi^{k-j} e_i)e_i)^\top\right). \tag{2.36}$$

Proof. We will calculate at a point $x \in M$. One may assume $(\nabla_{e_i} \xi)^\perp = 0$ at x. Decomposing $(A_\xi)^k = A_\xi(A_\xi)^{k-1}$ for $k > 1$, we get at a point x, see (1.31),

$$-(\nabla_\mathscr{F}^* A_\xi^k) = -A_\xi(\nabla_\mathscr{F}^* A_\xi^{k-1}) + \sum_i (\nabla_{e_i}^\mathscr{F} A_\xi) A_\xi^{k-1} e_i. \tag{2.37}$$

Since (2.37) is tensorial, it is valid for any point of M. Using (1.29), (1.39) and symmetries of the curvature tensor, we compute for $X \in T\mathscr{F}$,

$$\sum_i \langle (\nabla_{e_i}^\mathscr{F} A_\xi)(A_\xi^{k-1} e_i), X \rangle = \sum_i \langle A_\xi^{k-1} e_i, (\nabla_{e_i}^\mathscr{F} A_\xi) X \rangle$$
$$= \sum_i \langle A_\xi^{k-1} e_i, (\nabla_X^\mathscr{F} A)_\xi \, e_i - R(e_i, X)\xi + \nabla_{[e_i,X]^\perp}\xi \rangle$$
$$= \text{trace}(A_\xi^{k-1}(\nabla_X^\mathscr{F} A)_\xi) - \sum_i \langle R(\xi, A_\xi^{k-1} e_i)e_i, X \rangle.$$

Here, we used $[e_i, X]^\perp = 0$. For $X \in T\mathscr{F}$, (2.37) gives us

$$-\langle (\nabla_\mathscr{F}^* A_\xi^k), X \rangle = -\langle A_\xi(\nabla_\mathscr{F}^* A_\xi^{k-1}), X \rangle + \text{trace}(A_\xi^{k-1}(\nabla_X^\mathscr{F} A_\xi))$$
$$- \sum_i \langle R(\xi, A_\xi^{k-1} e_i)e_i, X \rangle.$$

The above and the identity $k \cdot \text{trace}(A_\xi^{k-1}(\nabla_X^\mathscr{F} A_\xi)) = X(\tau_k)(A_\xi)$ for $k > 0$ yield

$$-(\nabla_\mathscr{F}^* A_\xi^k) = -A_\xi(\nabla_\mathscr{F}^* A_\xi^{k-1}) + \frac{1}{k}(\nabla \tau_k(A_\xi))^\top - \sum_i (R(\xi, A_\xi^{k-1} e_i)e_i)^\top, \tag{2.38}$$

see (1.81). By induction, (2.36) follows from (2.38).   □

Example 2.44. By Lemma 2.43 with $f_j = (-1)^j \sigma_{r-j}(A_\xi)$ and $j \leq r$, we have for the Newton transformations of A_ξ (see Section 2.1.3 for codimension-one case):

$$\nabla_{\mathscr{F}}^* T_r(A_\xi) = -\sum_{1 \leq j \leq r} \left(\sum_i (-A_\xi)^{j-1} (R(\xi, T_{r-j}(A_\xi)e_i)e_i) \right)^\top. \qquad (2.39)$$

Of course, by inductive definition (1.80a), we obtain

$$-(\nabla_{\mathscr{F}}^* T_r)(A_\xi) = (\nabla \sigma_r(A_\xi))^\top + A_\xi(\nabla_{\mathscr{F}}^* T_{r-1}(A_\xi)) - \sum_i (\nabla_{e_i}^{\mathscr{F}} A)_\xi T_{r-1}(A_\xi)e_i.$$

Then, using the identity $X(\sigma_r)(A_\xi) = \text{trace}(T_{r-1}(A_\xi)(\nabla_X^{\mathscr{F}} A)_\xi)$ for any $X \in T\mathscr{F}$, see Lemma 1.71 or (1.82), and Codazzi equation (1.39), we get

$$-\nabla_{\mathscr{F}}^* T_r(A_\xi) = A_\xi(\nabla_{\mathscr{F}}^* T_{r-1}(A_\xi)) + \sum_i (R(\xi, T_{r-1}(A_\xi)e_i)e_i)^\top.$$

By induction, one can may show that $\nabla_{\mathscr{F}}^*(T_r(A_\xi))$ for $r > 0$ is given by (2.39).

2.3.2.2 The Integral Formulas

For a vector field ξ in $S^\perp$, put $Z_\xi = (\nabla_\xi \xi)^\top$ for short. Recall that $f_k = f_k(\vec{\tau}(\xi))$.

Lemma 2.45. *Let $\{e_i, e_\alpha\}$ be a local orthonormal frame of TM adapted for $(\widetilde{\mathscr{D}}, \mathscr{D})$ such that $\nabla_X^{\mathscr{F}} e_i(x) = 0$ and $(\nabla_X e_\alpha(x))^\perp = 0$ for any vector $X \in T_x M$. Then for any unit vector $\xi = \sum_{\alpha \leq n} y_\alpha e_\alpha \in S_x^\perp$ with $y_\alpha \in \mathbb{R}$ we have at the point $x \in M$*

$$\langle \nabla_{e_i}^{\mathscr{F}} \nabla_\xi \xi, e_j \rangle = (A_\xi^2)_{ij} + \langle R(e_i, \xi)\xi, e_j \rangle - (\nabla_\xi^{\mathscr{F}} A_\xi)_{ij} + \sum_{\alpha \leq n} \langle \nabla_\xi e_\alpha, e_i \rangle \langle \nabla_{e_\alpha} \xi, e_j \rangle.$$

Proof. First, observe that

$$-\langle Z_\xi, \nabla_{e_i} e_j \rangle = \langle \nabla_{e_i} Z_\xi, e_j \rangle + \langle \nabla_{e_i} \xi, \nabla_\xi e_j \rangle + \langle \xi, \nabla_{e_i} \nabla_\xi e_j \rangle. \qquad (2.40)$$

We have

$$(\nabla_\xi^{\mathscr{F}} A_\xi)_{ij} = \nabla_\xi \langle \xi, \nabla_{e_i} e_j \rangle = \langle Z_\xi, \nabla_{e_i} e_j \rangle + \langle \xi, \nabla_\xi \nabla_{e_i} e_j \rangle.$$

Therefore, we obtain at x

$$(A_\xi^2)_{ij} + \langle R(e_i, \xi)\xi, e_j \rangle - (\nabla_\xi^{\mathscr{F}} A_\xi)_{ij} = (A_\xi^2)_{ij} - \langle R(e_i, \xi)e_j, \xi \rangle + \xi \langle \nabla_{e_i} \xi, e_j \rangle$$
$$= (A_\xi^2)_{ij} - \langle Z_\xi, \nabla_{e_i} e_j \rangle - \langle \nabla_{e_i} \nabla_\xi e_j, \xi \rangle + \langle \nabla_{[e_i, \xi]} e_j, \xi \rangle. \qquad (2.41)$$

Using (2.40), conditions at x,

$$\nabla_{e_i} \xi = \sum_{\alpha \leq n} \langle \nabla_{e_i} \xi, e_\alpha \rangle e_\alpha, \quad \nabla_\xi e_i = \sum_{\alpha \leq n} \langle \nabla_\xi e_i, e_\alpha \rangle e_\alpha,$$

and $(A_\xi^2)_{ij} = -\sum_{\alpha \le n} \langle \nabla_{e_i}\xi, e_\alpha \rangle \langle \nabla_{e_\alpha}e_j, \xi \rangle$, we simplify the last line in (2.41) as $\langle \nabla_{e_i}Z_\xi, e_j \rangle - \sum_{\alpha \le n} \langle \nabla_\xi e_\alpha, e_i \rangle \langle \nabla_{e_\alpha}\xi, e_j \rangle$. From the above, the claim follows. $\qquad\square$

Applying Lemma 2.45, and $f_j = (-1)^j \sigma_{r-j}(A_\xi)$ $(j \le r)$ and (2.39), we obtain

Proposition 2.46. *For any $x \in M$,*

$$\mathrm{div}_{\mathscr{F}}\Big(\int_{S_x^\perp} T_r(A_\xi)Z_\xi \, d\omega\Big) = \int_{S_x^\perp} \big(\langle -\nabla_{\mathscr{F}}^* T_r(A_\xi), Z_\xi \rangle - \xi(\sigma_{r+1})(A_\xi) - (r+2)\,\sigma_{r+2}(A_\xi)$$
$$+\sigma_1(A_\xi)\,\sigma_{r+1}(A_\xi) + \mathrm{trace}(T_r(A_\xi)R_\xi) + \sum_{\alpha \le n}\langle T_r(A_\xi)((\nabla_{e_\alpha}\xi)^\top), \nabla_\xi e_\alpha \rangle\big)\,d\omega,$$

where $\langle \nabla_{\mathscr{F}}^* T_r(A_\xi), Z_\xi \rangle = -\sum_{1 \le j \le r} \mathrm{trace}\,\big(T_{r-j}(A_\xi)R((-A_\xi)^{j-1}Z_\xi, \xi)\big)$.

Proof. This is technical, see [R5], and we omit details. $\qquad\square$

Applying the Divergence Theorem to M or to any compact leaf, we get

Theorem 2.47. *On any closed Riemannian manifold (M, g) we have*

$$\int_{S^\perp} \big(\langle -\nabla_{\mathscr{F}}^* T_r(A_\xi), Z_\xi \rangle - (r+2)\,\sigma_{r+2}(A_\xi) - \langle T_r(A_\xi)Z_\xi, \tilde{H} \rangle$$
$$+ \mathrm{trace}(T_r(A_\xi)R_\xi) + \sum_{\alpha \le n}\langle T_r(A_\xi)((\nabla_{e_\alpha}\xi)^\top), \nabla_\xi e_\alpha \rangle\big)\,d\omega = 0. \quad (2.42)$$

For any compact leaf L of $\mathscr{F}$ we have

$$\int_{S_L^\perp} \big(\sum_{1 \le j \le r} \mathrm{trace}\,\big(T_{r-j}(A_\xi)R((-A)_\xi^{j-1}Z_\xi, \xi)\big) - \xi(\sigma_{r+1})(A_\xi) - (r+2)\sigma_{r+2}(A_\xi)$$
$$+\sigma_1(A_\xi)\sigma_{r+1}(A_\xi) + \mathrm{trace}(T_r(A_\xi)R_\xi) + \langle T_r(A_\xi)Z_\xi, Z_\xi \rangle\big)\,d\omega_L^\perp = 0.$$

Remark 2.48. The integrals over $S_x^\perp$ when $n > 1$ can be found explicitly, see Sect. 2.3.1.1.

We shall find integral formulae on manifolds with a pair of complementary distributions $\mathscr{D}$ and $\widetilde{\mathscr{D}}$. The idea is to compute the divergence of a vector field $X = \int_{S_x^\perp} A_\xi(\nabla_\xi \xi)\,d\omega$ where $x \in M$. For a codimension-one distribution (see Section 2.1.3), this is simply $X = A_\xi(\nabla_\xi \xi)$.

We look at first members of series of (2.42). For $r = 0$, we get

$$\int_{S^\perp} \big(-2\sigma_2(A_\xi) + \mathrm{trace}\,R_\xi - \langle Z_\xi, \tilde{H} \rangle + \sum_{\alpha \le n}\langle (\nabla_{e_\alpha}\xi)^\top, \nabla_\xi e_\alpha \rangle\big)\,d\omega = 0.$$

One can reduce this formula to (2.28).
For $r = 2$, (2.42) gives us

$$\int_{S^\perp} \big(-4\sigma_4(A_\xi) - \langle T_2(A_\xi)Z_\xi, \tilde{H} \rangle + \mathrm{trace}(T_2(A_\xi)R_\xi) + \mathrm{trace}\,R(A_\xi Z_\xi, \xi)$$
$$-T_1(A_\xi)R(Z_\xi, \xi) + \sum_{\alpha \le n}\langle T_2(A_\xi)((\nabla_{e_\alpha}\xi)^\top), \nabla_\xi e_\alpha \rangle\big)\,d\omega = 0. \quad (2.43)$$

If $\mathscr{D}$ is tangent to a totally geodesic foliation, then (2.43) shortens to the formula

$$\int_{S^\perp} \big(4\,\sigma_4(A_\xi) - \mathrm{trace}(T_2(A_\xi)R_\xi)\big)\,d\omega = 0.$$

2.3.2.3 Totally Geodesic and Umbilical Foliations

A totally geodesic distribution $\mathscr{D}$ is characterized by the property

$$(\nabla_\xi \xi)^\top = 0, \quad \xi \in \mathscr{D}.$$

Corollary 2.49. *If $\mathscr{D}$ is tangent to a totally geodesic foliation of a closed M, then*

$$\int_{S^\perp} \big((r+2)\,\sigma_{r+2}(A_\xi) - \mathrm{trace}(T_r(A_\xi)R_\xi)\big)\,d\omega = 0. \tag{2.44}$$

The *total k-th mean curvature σ_k* and the total τ_k related to $\widetilde{\mathscr{D}}$ are defined by

$$\sigma_k(\widetilde{\mathscr{D}}) = \int_{S^\perp} \sigma_k(A_\xi)\,d\omega, \quad \tau_k(\widetilde{\mathscr{D}}) = \int_{S^\perp} \tau_k(A_\xi)\,d\omega.$$

Note that $\sigma_{2s+1}(\widetilde{\mathscr{D}}) = \tau_{2s+1}(\widetilde{\mathscr{D}}) = 0$. For manifolds of constant curvature our formulae reduce (with a simple choice of f_j) to (2.34).

Remark 2.50. One may show that (2.44) itself yields (2.34). Of course, let $\mathscr{D}$ be tangent to a totally geodesic foliation with constant mixed curvature $c \ge 0$. By $\mathrm{trace}\, T_r(A_\xi) = (p-r)\,\sigma_r(A_\xi)$ and (2.44), we get $\sigma_{r+2}(\widetilde{\mathscr{D}}) = \frac{c(p-r)}{r+2}\,\sigma_r(\widetilde{\mathscr{D}})$ and $\sigma_0(\widetilde{\mathscr{D}}) = \frac{2\pi^{n/2}}{\Gamma(n/2)}\mathrm{Vol}(M,g)$. Then (by induction), (2.34) follows from the above.

Note that for $r > 0$, (2.44) is different from the result of [RW3, Theorem 3.2] in general, but gives the same (2.34), when the mixed curvature is constant $c \ge 0$.

The total *extrinsic mean curvatures $\gamma_r(\widetilde{\mathscr{D}})$* (of Definition 2.41) satisfy, see [AW2],

$$\gamma_{r+2}(\widetilde{\mathscr{D}}) = \frac{c(p-r)(n+r)}{(r+2)(r+1)}\,\gamma_r(\widetilde{\mathscr{D}}), \quad \gamma_0(\widetilde{\mathscr{D}}) = \sigma_0(\widetilde{\mathscr{D}}).$$

Thus, $\sigma_{2s}(\widetilde{\mathscr{D}}) = \prod_{i=1}^s \frac{n+2i-2}{2i-1} \cdot \gamma_{2s}(\widetilde{\mathscr{D}})$. Indeed, $\sigma_r(\widetilde{\mathscr{D}}) = \gamma_r(\widetilde{\mathscr{D}})$ for $n = 1$. Similarly, we have $\tau_0(\widetilde{\mathscr{D}}) = p\frac{2\pi^{n/2}}{\Gamma(n/2)}\mathrm{Vol}(M,g)$, $\tau_2(\widetilde{\mathscr{D}}) = -c\,\tau_0(\widetilde{\mathscr{D}})$, etc.

Using Remark 2.48, by induction we obtain

Corollary 2.51. *Let $\mathscr{F}$ be a harmonic foliation, $\mathscr{D}$ determines a totally geodesic foliation of a closed M, and the mixed sectional curvature $K(X,\xi) = c = \mathrm{const}$ for $X \in \mathscr{D}$ and $\xi \in T\mathscr{F}$. If p even and $n > 1$, then for any $s > 0$ we have*

$$\tau_{2s}(\widetilde{\mathscr{D}}) = \frac{2\pi^{n/2}}{\Gamma(n/2)}\,(-c)^s\,p\,\mathrm{Vol}(M,g).$$

For $n = 1$, since the projection $\pi : S^\perp \to M$ is a double covering, the above formula reduces to $\tau_{2s}(\widetilde{\mathscr{D}}) = (-c)^s\,p\,\mathrm{Vol}(M,g)$, see (2.16).

Remark 2.52. Let $\pi : P \rightarrow M$ be the principal bundle of orthonormal frames (oriented orthonormal frames, respectively) of $\widetilde{\mathscr{D}}$ with the structure group G, which is either the full orthogonal group or the special orthogonal group. Each element $(x, e) = (e_1, \ldots, e_q) \in P_x$, $x \in M$, induces the system of endomorphisms $A(x, e) = (A_1(x, e), \ldots, A_q(x, e))$ of $\mathscr{D}_x$, where $A_\alpha(x, e)$ is the shape operator corresponding to (x, e), i.e., $A_\alpha(x, e)(X) = -(\nabla_X e_\alpha)^\top$ for $X \in \mathscr{D}_x$. Equality (1.82) is the starting point to define *generalized Newton transformation*, see [AKN]. Namely, $T_{(0,\ldots,0)} = 1$ and

$$T_\lambda = \sigma_\lambda \, \mathrm{id} - \sum_\alpha A_\alpha T_{\alpha_\flat(\lambda)} = \sigma_\lambda \, \mathrm{id} - \sum_\alpha T_{\alpha_\flat(\lambda)} A_\alpha, \quad \text{if } |\lambda| \geq 1,$$

where $\alpha_\flat(i_1, \ldots, i_q) = (i_1, \ldots, i_{\alpha-1}, i_\alpha - 1, i_{\alpha+1}, \ldots, i_q)$. These transformations T_λ are called the *generalized Newton transformations*.

Let $T_\lambda(x, e)$ be the generalized Newton transformation of the operator $A(x, e)$. Averaging over a fiber we obtain a set of globally defined functions $\widehat{\sigma}_\lambda$

$$\widehat{\sigma}_\lambda(x) = \int_{P_x} \sigma_\lambda(x, e) \, \mathrm{d}e = \int_G \sigma_\lambda(x, e_0 a) \, \mathrm{d}a,$$

called *extrinsic curvatures* of a distribution $\mathscr{D}$. As in the codimension-one case, one may express total extrinsic curvatures in terms of generalized Newton transformations, see Section 1.5.3, and the second fundamental forms of $\mathscr{D}$ and $\widetilde{\mathscr{D}}$. In the case of integrable totally geodesic orthogonal distribution $\widetilde{\mathscr{D}}$, we have

$$|\lambda| \sigma_\lambda^M = \sum_{\alpha, \beta} \int_P \mathrm{trace}(R(\alpha, \beta) T_{\beta_\flat, \alpha_\flat(u)});$$

if in addition, sectional curvature c of M is constant, they reduce to the following recurrence formula, which implies (2.35):

$$|\lambda| \sigma_\lambda^M = c \sum_\alpha \int_P \mathrm{trace}(T_{\alpha_\flat^2(\lambda)}) = c \sum_\alpha \int_P (n - |\lambda| + 2) \sigma_{\alpha_\flat^2(\lambda)}$$
$$= c(n - |\lambda| + 2) \sum_\alpha \sigma_{\alpha_\flat^2(\lambda)}^M.$$

The initial conditions are $\sigma_{(0,\ldots,0)}^M = \mathrm{Vol}(P)$ and $\sigma_{\alpha^\sharp(0,\ldots,0)}^M = 0$, where

$$\alpha^\sharp(i_1, \ldots, i_q) = (i_1, \ldots, i_{\alpha-1}, i_\alpha + 1, i_{\alpha+1}, \ldots, i_q).$$

In this case, the total extrinsic curvatures do not depend on the geometry of $\mathscr{D}$.

Certainly, there are several other integral formulas and their consequences than we have indicated here, and there exist more applications of Theorem 2.47 (and its consequences). The reader is welcomed to find his own.

2.3.3 Splitting of Weighted Generalized Products

Here, we use the notion of weighted curvature (see Sections 1.4.3 and 1.4.4) to prove splitting theorems for the case of Riemannian almost product manifolds of nonnegative or nonpositive weighted mixed scalar curvature.

We say that $(M, g, \mathscr{D})$ *splits* if both distributions $\mathscr{D}$ and $\widetilde{\mathscr{D}}$ are integrable and M is locally isometric to a product with foliations tangent to $\mathscr{D}$ and $\widetilde{\mathscr{D}}$. Recall, e.g., [BF, Section 4.4], that if a simply connected manifold splits, then it is the product.

Definition 2.53. The *weighted mixed scalar curvature* of a Riemannian almost product manifold $(M, g, \widetilde{\mathscr{D}}, \mathscr{D})$ endowed with a vector field X is defined by

$$\mathscr{S}^{d,\eth} = S_{\text{mix}} + \frac{1}{2}\operatorname{div} X + \frac{1}{2d}\|X^\top\|^2 + \frac{1}{2\eth}\|X^\perp\|^2, \tag{2.45}$$

where $d, \eth \in \mathbb{R}$ are curvature dimensions of $\mathscr{D}$ and $\widetilde{\mathscr{D}}$, respectively.

Based on (1.34) and (1.62), we find that $\mathscr{S}^{d,\eth} = \frac{1}{2}\operatorname{trace}_g\left(\mathscr{R}ic_n^d + \widetilde{\mathscr{R}ic}_p^\eth\right)$.

Modifying Stokes' theorem on a complete open Riemannian manifold (M, g) gives the following, see [CSC, Proposition 1].

Lemma 2.54. *Let (M, g) be a complete open Riemannian manifold endowed with a vector field X such that $\operatorname{div} X \geq 0$. If the norm $\|X\|_g \in L^1(M, g)$, then $\operatorname{div} X \equiv 0$.*

Proof. Let ω be a $(n-1)$-form on M ($\dim M = n$) given by $\omega = \iota_X d\operatorname{vol}_g$, i.e., the contraction of the volume form $d\operatorname{vol}_g$ in the direction of X. If $\{e_1, \dots, e_n\}$ is an orthonormal frame on an open set $U \subset M$, with coframe $\omega_1, \dots, \omega_n$, then

$$\iota_X d\operatorname{vol}_g = \sum\nolimits_{i=1}^n (-1)^{i-1}\langle X, e_i\rangle \omega_1 \wedge \dots \wedge \hat{\omega}_i \wedge \dots \wedge \omega_n.$$

Since the $(n-1)$-forms $\omega_1 \wedge \dots \wedge \hat{\omega}_i \wedge \dots \wedge \omega_n$ are orthonormal in $\Omega^{n-1}(M)$, we get

$$\|\omega\|_g^2 = \sum\nolimits_{i=1}^n \langle X, e_i\rangle^2 = \|X\|_g^2.$$

Thus, $\|\omega\|_g \in L^1(M)$ and $d\omega = d(\iota_X d\operatorname{vol}_g) = (\operatorname{div} X)d\operatorname{vol}_g$. There exist domains B_i on M such that $M = \bigcup_{i\geq 1} B_i$, $B_i \subset B_{i+1}$ and $\lim_{i\to\infty}\int_{B_i} d\omega = 0$, see [Yau]. Then $\int_{B_i}(\operatorname{div} X)d\operatorname{vol}_g = \int_{B_i} d\omega \to 0$. Since $\operatorname{div} X \geq 0$ on M, then $\operatorname{div} X = 0$ on M. $\qquad \square$

In the next two theorems we consider harmonic distributions with $\mathscr{S}^{N,\mathscr{N}} \geq 0$.

Theorem 2.55. *Let (M, g) be a complete open (or closed) Riemannian manifold endowed with complementary orthogonal integrable distributions $(\widetilde{\mathscr{D}}, \mathscr{D})$ and a vector field $X \in \mathfrak{X}^\top$ satisfying conditions $\langle X, \tilde{H}\rangle = 0$ and $\|X_{|M'}\|_{g'} \in L^1(M', g')$ for all leaves (M', g') of $\widetilde{\mathscr{D}}$. Suppose that $\widetilde{\mathscr{D}}$ is harmonic and $\mathscr{S}^{N,\mathscr{N}} \geq 0$ for some negative N and $\mathscr{N} \neq 0$. Then $X = 0$ and M splits.*

Proof. By conditions, (2.45) and (1.44), we have

$$\operatorname{div}^{\top}\left(\tilde{H}+\frac{1}{2}X\right) = \mathscr{S}^{N,\mathscr{N}} + \|\tilde{h}\|^2 + \|h\|^2 - \frac{1}{2N}\|X\|^2.$$

By Lemma 2.54 for each leaf, $\operatorname{div}^{\top}(\tilde{H}+\frac{1}{2}X) = 0$ when $\mathscr{S}^{N,\mathscr{N}} \geq 0$ and $N < 0$; thus, $h = 0 = \tilde{h}$ and $X = 0$. By de Rham decomposition theorem, (M,g) splits. $\square$

Corollary 2.56. *Let (M,g) be endowed with two complementary orthogonal integrable distributions $(\widetilde{\mathscr{D}},\mathscr{D})$ and a vector field $X \in \mathfrak{X}^{\top}$. If $\mathscr{D}$ is harmonic, then $\widetilde{\mathscr{D}}$ has no compact harmonic leaves M' with $\mathscr{S}^{N,\mathscr{N}}|_{M'} > 0$ for some $N < 0$ and $\mathscr{N} \neq 0$.*

Theorem 2.57. *Let (M,g) be a complete open (or closed) Riemannian manifold endowed with complementary orthogonal harmonic foliations and a vector field X such that $\|X\|_g \in L^1(M,g)$. If $\mathscr{S}^{N,\mathscr{N}} \geq 0$ for some negative N and $\mathscr{N}$, then $X = 0$ and M splits.*

Proof. Under conditions, we obtain

$$\frac{1}{2}\operatorname{div}X = \mathscr{S}^{N,\mathscr{N}} + \|\tilde{h}\|^2 + \|h\|^2 - \frac{1}{2N}\|X^{\top}\|^2 - \frac{1}{2\mathscr{N}}\|X^{\perp}\|^2.$$

By Lemma 2.54, we get $\operatorname{div}X = 0$ when $\mathscr{S}^{N,\mathscr{N}} \geq 0$ and $N,\mathscr{N} < 0$. Thus, $h = 0 = \tilde{h}$ and $X = 0$. By de Rham decomposition theorem, (M,g) splits. $\square$

If $\mathscr{D}$ is totally umbilical, then $\|\tilde{h}\|^2 - \|\tilde{H}\|^2 = -\frac{n-1}{n}\|\tilde{H}\|^2$, and similarly, for $\widetilde{\mathscr{D}}$. In the next theorem we consider totally umbilical distributions with $\mathscr{S}^{N,\mathscr{N}} \leq 0$.

Theorem 2.58. *Let (M,g) be a closed (or a complete open) Riemannian manifold endowed with complementary orthogonal totally umbilical distributions $\widetilde{\mathscr{D}}$ and $\mathscr{D}$ and a vector field X obeying $\|\xi_{|M}\|_g \in L^1(M,g)$, where $\xi = \tilde{H} + H + \frac{1}{2}X$. Suppose that $\mathscr{S}^{N,\mathscr{N}} \leq 0$ for some positive N and $\mathscr{N}$. Then $X = 0$ and M splits.*

Proof. Let $\dim M = n + p$ and $\dim \mathscr{D} = n$. Under conditions, we get

$$\operatorname{div}\xi = \mathscr{S}^{N,\mathscr{N}} - \|T\|^2 - \|\tilde{T}\|^2 - \frac{n-1}{n}\|\tilde{H}\|^2 - \frac{p-1}{p}\|H\|^2 - \frac{1}{2N}\|X^{\top}\|^2 - \frac{1}{2\mathscr{N}}\|X^{\perp}\|^2.$$

From this and Lemma 2.54 and since $\mathscr{S}^{N,\mathscr{N}} \leq 0$ for $N,\mathscr{N} > 0$, we get $\operatorname{div}\xi = 0$. Thus $T,\tilde{T},H,\tilde{H}$ and X vanish. By de Rham theorem, (M,g) splits. $\square$

2.3.4 Multi-Product Structures

A Riemannian manifold (M,g) endowed with $k > 2$ pairwise orthogonal complementary n_i-dimensional distributions $\mathscr{D}_i$ $(1 \leq i \leq k)$ with $\sum n_i = \dim M$ (called here a *Riemannian almost multi-product manifold*) appears in such topics as multiply warped products, the webs composed of several foliations, and Dupin hypersurfaces of real space-forms (i.e., the number k of distinct principal curvatures is constant). We study the mixed scalar curvature of such structure and prove integral formulae and splitting theorems with this kind of curvature, which generalize (1.44) for $k = 2$.

There always exists on M a local adapted orthonormal frame $\{E_1, \ldots, E_n\}$, where

$$\{E_1, \ldots, E_{n_1}\} \subset \mathscr{D}_1, \quad \{E_{n_1+1}, \ldots, E_{n_2}\} \subset \mathscr{D}_2, \quad \ldots \quad \{E_{n_{k-1}+1}, \ldots, E_{n_k}\} \subset \mathscr{D}_k.$$

Definition 2.59 (See [R14]). The function on (M, g) with $k > 2$ orthogonal complementary distributions,

$$S_{\mathscr{D}_1, \ldots, \mathscr{D}_k} = \sum_{i<j} S(\mathscr{D}_i, \mathscr{D}_j)$$

will be called the *mixed scalar curvature* of $(M, g; \mathscr{D}_1, \ldots, \mathscr{D}_k)$, where

$$S(\mathscr{D}_i, \mathscr{D}_j) = \sum_{n_{i-1} < a \leq n_i, \, n_{j-1} < b \leq n_j} \langle R(E_a, E_b) E_a, E_b \rangle, \quad i \neq j.$$

Let h_i, H_i, T_j be related to the distribution $\mathscr{D}_i$ as usual, and $P_i : TM \to \mathscr{D}_i$ be ortho-projectors. Let also h_{ij}, H_{ij}, T_{ij} be the second fundamental forms, their mean curvature vector fields and integrability tensors related to the distributions $\mathscr{D}_i \oplus \mathscr{D}_j$. Note that $H_i = \sum_{j \neq i} P_j H_i$, etc. By definition (2.59), we get the decomposition formula

$$2S_{\mathscr{D}_1, \ldots, \mathscr{D}_k} = \sum_i S_{\mathscr{D}_i, \mathscr{D}_i^\perp}. \tag{2.46}$$

Note that for the scalar curvature $S : M \to \mathbb{R}$ of (M, g) we have

$$S = 2S_{\mathscr{D}_1, \ldots, \mathscr{D}_k} + \sum_i S(\mathscr{D}_i),$$

where $S(\mathscr{D}_i)$ are scalar curvatures of suitable distributions (functions on M).

Let $S(r, k)$ be the set of all r-combinations (i.e., subsets of r distinct elements) of $\{1, \ldots, k\}$. For example, $S(k-1, k)$ contains k elements. For any $q \in S(r, k)$ we may assume $q_1 < \ldots < q_r$. Set $h_q = h_{q_1, \ldots, q_r}$ and $T_q = T_{q_1, \ldots, q_r}$ for $\mathscr{D}_{q_1} \oplus \ldots \oplus \mathscr{D}_{q_r}$.

Theorem 2.60 ([R14]). *For $(M, g; \mathscr{D}_1, \ldots, \mathscr{D}_k)$ we have*

$$\operatorname{div} \sum_i (H_i + H_i^\perp) = 2S_{\mathscr{D}_1, \ldots, \mathscr{D}_k}$$
$$+ \sum_i (\|h_i\|^2 - \|H_i\|^2 - \|T_i\|^2 + \|h_i^\perp\|^2 - \|H_i^\perp\|^2 - \|T_i^\perp\|^2), \tag{2.47}$$

and for a closed manifold M we have the following integral formula:

$$\int_M \left(2S_{\mathscr{D}_1, \ldots, \mathscr{D}_k} + \sum_{r \in \{1, k-1\}} \sum_{q \in S(r,k)} (\|h_q\|^2 - \|H_q\|^2 - \|T_q\|^2)\right) d\operatorname{vol}_g = 0. \tag{2.48}$$

Proof. For $k = 2$ we have (1.44). To illustrate the proof for $k > 2$, first consider the case of $k = 3$. Using (1.44) for the distributions $\mathscr{D}_1$ and $\mathscr{D}_1^\perp = \mathscr{D}_2 \oplus \mathscr{D}_3$, we get

$$\operatorname{div}(H_1 + H_1^\perp) = 2S_{\mathscr{D}_1, \mathscr{D}_1^\perp}$$
$$+ \|h_1\|^2 - \|H_1\|^2 - \|T_1\|^2 + \|h_1^\perp\|^2 - \|H_1^\perp\|^2 - \|T_1^\perp\|^2, \tag{2.49}$$

and similarly for $(\mathscr{D}_2, \mathscr{D}_2^{\perp})$ and $(\mathscr{D}_3, \mathscr{D}_3^{\perp})$. Summing 3 copies of (2.49), we obtain (2.47) for $k = 3$. Next, for any $k > 2$, we apply (1.44) for pairs of distributions $(\mathscr{D}_i, \mathscr{D}_i^{\perp})$ and get k equalities. After summation and using (2.46), we get (2.47). Using Stokes' theorem for (2.47) on a closed manifold M yields (2.48). $\qquad\square$

Example 2.61. From (2.47) for $k = 3$ with $h_3 = T_3 = 0$ it follows that if $k = 3$ and $\mathscr{D}_3$ is tangent to a totally geodesic foliation, then for a closed Riemannian almost multi-product manifold we have

$$\int_M \left(S_{\mathscr{D}_1, \mathscr{D}_2, \mathscr{D}_3} + \|h_1\|^2 + \|h_2\|^2 - \|H_1\|^2 - \|H_2\|^2 - \|T_1\|^2 - \|T_2\|^2 \right) d\operatorname{vol} = 0.$$

Example 2.62. Let (M^n, g) admits n pairwise orthogonal codimension-one foliations $\mathscr{F}_i$, and let N_i be unit vector fields orthogonal to $\mathscr{F}_i$. Writing down (2.2) for each N_i, summing for $i = 1, \ldots, n$, and using the equality $S = \sum_i \operatorname{Ric}_{N_i, N_i}$, yields the integral formula with the scalar curvature S of (M, g),

$$\int_M \left(2 \sum_i \sigma_2(\mathscr{F}_i) - S \right) d\operatorname{vol}_g = 0. \tag{2.50}$$

Two immediately consequences of (2.50): 1) if $S < 0$ then each foliation $\mathscr{F}_i$ cannot be totally umbilical; 2) if $S > 0$ then each foliation $\mathscr{F}_i$ cannot be harmonic.

The next result allows us to reorganize some terms in the formula (2.47).

Theorem 2.63 ([R14]). *For $(M, g; \mathscr{D}_1, \ldots, \mathscr{D}_k)$ and any $r \in \{2, \ldots, k-1\}$, we have*

$$\operatorname{div} X = \sum_{q \in S(r,k)} \left(\|H_q\|^2 + \langle \textstyle\sum_{i=1}^r H_{qi} - r H_q, \sum_{j \notin q} H_j \rangle \right), \tag{2.51}$$

where $X = \sum_{q \in S(r,k)} H_q - C_{k-2}^{r-1} \sum_i H_i$. On a closed manifold M we have the following integral formula:

$$\int_M \sum_{q \in S(r,k)} \left(\|H_q\|^2 + \langle \textstyle\sum_{i=1}^r H_{qi} - r H_q, \sum_{j \notin q} H_j \rangle \right) d\operatorname{vol}_g = 0. \tag{2.52}$$

Proof. Using equality $H_{1\ldots r} = P_{r+1\ldots k}(H_1 + \ldots + H_r)$ for $q = \{1, \ldots, r\}$, we find

$$\operatorname{div} H_{1\ldots r} = \operatorname{div}_{r+1\ldots k} H_{1\ldots r} - \|H_{1\ldots r}\|^2$$
$$= \operatorname{div}_{r+1\ldots k}(H_1 + \ldots + H_r) + \langle H_1 + \ldots + H_r, H_{r+1\ldots k} \rangle - \|H_{1\ldots r}\|^2,$$

and similarly for all C_k^r cases of $q \in S(r,k)$. Summing the above, we use equalities

$$\mathrm{div}_{r+1\ldots k} H_1 = \sum_{j>r} \mathrm{div}_j H_1, \quad \mathrm{div}_{2\ldots k} H_1 = \mathrm{div}\, H_1 + \|H_1\|^2, \quad \text{etc.}$$

Using induction, we write (2.51) for $k-1$ distributions $\mathcal{D}_1, \ldots, \mathcal{D}_{k-2}$ and $\mathcal{D}_{k-1,k}$, and write similar formulas for other choices of a pair of distributions. Summing these C_k^2 equations, we get equation of the form (2.51) with k, comparing coefficients yields (2.51). Using Stokes' theorem for (2.51) on a closed manifold M yields (2.52). $\qquad\square$

Remark 2.64. One may show that (2.51) for $k = 3$ reads as

$$\mathrm{div}\left(\sum_{i<j} H_{ij} - \sum_i H_i\right) = \sum_i \|H_i\|^2 + \sum_{i<j} \left(2\langle H_i, H_j\rangle - \|H_{ij}\|^2\right). \qquad (2.53)$$

If $H_3 = 0$, then (2.53) reads as $\langle H_1, H_2\rangle = 0$. Applying (2.53) to (2.49) we obtain the formula, proved in an article[2] by long calculations,

$$\mathrm{div}\left(\sum_i H_i\right) = \mathrm{S}_{\mathcal{D}_1,\mathcal{D}_2,\mathcal{D}_3} - \sum_i \|H_i\|^2 - \sum_{i<j} \langle H_i, H_j\rangle$$
$$+ \frac{1}{2} \sum_i \left(\|h_i\|^2 - \|T_i\|^2\right) + \frac{1}{2} \sum_{i<j} \left(\|h_{ij}\|^2 - \|T_{ij}\|^2\right).$$

This yields the following integral formula on a closed (M, g):

$$\int_M \Big[\mathrm{S}_{\mathcal{D}_1,\mathcal{D}_2,\mathcal{D}_3} - \sum_i \|H_i\|^2 - \sum_{i<j} \langle H_i, H_j\rangle$$
$$+ \frac{1}{2} \sum_i \left(\|h_i\|^2 - \|T_i\|^2\right) + \frac{1}{2} \sum_{i<j} \left(\|h_{ij}\|^2 - \|T_{ij}\|^2\right)\Big]\, \mathrm{d\,vol} = 0.$$

Just (2.47) for $k = 3$, see Remark 2.64, one can find many geometrical applications, because an almost 3-product structure appears in different cases:

(1) almost para-f-manifolds, see Section 1.4.1.
(2) multiply warped product manifolds, e.g., [Ch3, Section 3.6].
(3) on orientable 3-manifolds, since they admit 3 linearly independent vector fields.
(4) the theory of webs composed of three generic foliations, e.g., [AG].
(5) hypersurfaces in space forms with 3 distinct principal curvatures, see [CR, R14].

Next, we use the above results to prove some splitting and non-existence of immersions results for Riemannian almost multi-product manifolds. We say that $(M, g; \mathcal{D}_1, \ldots, \mathcal{D}_k)$ *splits* if all distributions $\mathcal{D}_i$ are integrable and M is locally the direct product $M_1 \times \ldots \times M_k$ with foliations tangent to $\mathcal{D}_i$. Recall that if a simply connected manifold splits then it is the direct product. In the following definition we apply the submanifolds theory to Riemannian almost multi-product manifolds.

---

[2] M. Banaszczyk, R. Majchrzak, An integral formula for a Riemannian manifold with three orthogonal distributions. Acta Sci. Math., 54 (1990), 201–207.

Definition 2.65. A pair $(\mathscr{D}_i, \mathscr{D}_j)$ with $i \neq j$ of distributions on $(M, g; \mathscr{D}_1, \ldots, \mathscr{D}_k)$ with $k > 2$ is called

(a) *mixed totally geodesic*, if $h_{ij}(X, Y) = 0$ for all $X \in \mathscr{D}_i$ and $Y \in \mathscr{D}_j$.
(b) *mixed integrable*, if $T_{ij}(X, Y) = 0$ for all $X \in \mathscr{D}_i$ and $Y \in \mathscr{D}_j$.

By mathematical induction we have the following. If each pair $(\mathscr{D}_i, \mathscr{D}_j)$ on $(M, g; \mathscr{D}_1, \ldots, \mathscr{D}_k)$ with $k > 2$ is

(a) mixed totally geodesic, then $h_q(X, Y) = 0$; (b) mixed integrable, then $T_q(X, Y) = 0$

for all $q \in S(r, k)$, $2 < r < k$ and $X \in \mathscr{D}_{q_1}, Y \in \mathscr{D}_{q_2}$.

Theorem 2.66. *Let a Riemannian almost multi-product manifold* $(M, g; \mathscr{D}_1, \ldots, \mathscr{D}_k)$ *with* $k > 2$ *have integrable harmonic distributions* $\mathscr{D}_1, \ldots, \mathscr{D}_k$. *If* $S_{\mathscr{D}_1, \ldots, \mathscr{D}_k} \geq 0$ *and each pair* $(\mathscr{D}_i, \mathscr{D}_j)$ *is mixed integrable, then M splits.*

Proof. From $H_{1\ldots r} = P_{r+1\ldots k}(H_1 + \ldots + H_r)$ it follows that $H_i = 0$ for all $i \in \{1, \ldots, k\}$, then $H_q = 0$ for all $q \in S(r, k)$ and $2 \leq r < k$. Similarly, if $T_{ij} = 0$ for all $i \in \{1, \ldots, k\}$, then $T_q = 0$ for all $q \in S(r, k)$ and $2 \leq r < k$. By conditions and (2.47),

$$2\,S_{\mathscr{D}_1, \ldots, \mathscr{D}_k} + \sum\nolimits_{r \in \{1, k-1\}} \sum\nolimits_{q \in S(r,k)} (\|h_q\|^2 = 0.$$

Thus, $h_q = 0$ for all $q \in S(r, k)$ with $r \in \{1, k-1\}$, in particular, $h_i = 0$ $(1 \leq i \leq k)$. By well-known de Rham decomposition theorem, (M, g) splits. □

Theorem 2.67. *Let a complete open Riemannian almost multi-product manifold* $(M, g; \mathscr{D}_1, \ldots, \mathscr{D}_k)$ *with* $k > 2$ *have totally umbilical distributions such that each pair* $(\mathscr{D}_i, \mathscr{D}_j)$ *is mixed totally geodesic and* $\langle H_i, H_j \rangle = 0$ *for* $i \neq j$. *If* $S_{\mathscr{D}_1, \ldots, \mathscr{D}_k} \leq 0$ *and* $\|H_i\| \in L^1(M, g)$ *for* $1 \leq i \leq k$, *then M splits.*

Proof. By assumptions, from (2.47) we get

$$\operatorname{div} \xi = 2\,S_{\mathscr{D}_1, \ldots, \mathscr{D}_k} + \sum\nolimits_{r \in \{1, k-1\}} \sum\nolimits_{q \in S(r,k)} (\|h_q\|^2 - \|H_q\|^2 - \|T_q\|^2), \qquad (2.54)$$

where $\xi = \sum_{r \in \{1, k-1\}} \sum_{q \in S(r,k)} H_q$. By conditions, and since $\|H_q\| \leq \sum_{i=1}^k \|H_{q_i}\|$, we get $\|H_q\| \in L^1(M, g)$. Since $\|\xi\| \leq \sum_{r \in \{1, k-1\}} \sum_{q \in S(r,k)} \|H_q\|$, then also $\|\xi\| \in L^1(M, g)$. By conditions, for any $q = (q_1, \ldots, q_r) \in S(r, k)$ with $r \geq 1$ we have

$$\|h_q\|^2 - \|H_q\|^2 = -\sum\nolimits_{i=1}^r \frac{n_{q_i} - 1}{n_{q_i}} \|P_q^\perp H_{q_i}\|^2 \leq 0,$$

where $P_q^\perp$ is the orthoprojector on the distribution $\bigoplus_{j \neq q} \mathscr{D}_j$. Hence, from the inequality $S_{\mathscr{D}_1, \ldots, \mathscr{D}_k} \leq 0$ and (2.54) we get $\operatorname{div} \xi \leq 0$. By conditions and Lemma 2.54, we get $\operatorname{div} \xi = 0$. By (2.54), $S_{\mathscr{D}_1, \ldots, \mathscr{D}_k} = 0$ and T_i and h_i vanish. By de Rham decomposition theorem, (M, g) splits. □

Example 2.68. Totally umbilical integrable distributions appear on multiply twisted products. Let $(F_1, g_1)\ldots,(F_k, g_k)$ be Riemannian manifolds. A *multiply twisted product* $F_1 \times_{u_2} F_2 \times \ldots \times_{u_k} F_k$ is the product $M = F_1 \times \ldots \times F_k$ with the metric $g = g_{F_1} \oplus u_2^2 g_{F_2} \oplus \ldots \oplus u_k^2 g_{F_k}$, where $u_i : F_1 \times F_i \to (0, \infty)$ for $i = 2, \ldots, k$ are smooth functions. Twisted products ($k = 1$) and *multiply warped product*, i.e., $u_i : F_1 \to (0, \infty)$, are special cases of multiply twisted products. Let $\mathscr{D}_i$ be the distribution on M of the vectors tangent to (the horizontal lifts of) F_i, e.g., [Ch3]. The leaves (i.e., tangent to $\mathscr{D}_i$, $i \geq 2$) are totally umbilical submanifolds, with $H_i = -n_i P_1 \nabla(\log u_i)$, and the fibers (i.e., tangent to $\mathscr{D}_1$) are totally geodesic submanifolds. Since

$$\operatorname{div} H_i = -n_i (\Delta_1 u_i)/u_i - (n_i^2 - n_i)\|P_1 \nabla u_i\|^2/u_i^2,$$

where Δ_1 is the Laplacian on $C^2(F_1)$, and we have

$$S_{\mathscr{D}_1,\ldots,\mathscr{D}_k} = \sum_{i \geq 2} n_i (\Delta_1 u_i)/u_i.$$

Corollary 2.69. *Let a multiply twisted product (M, g) be complete open and the equalities $\langle H_i, H_j \rangle = 0$ hold for $i \neq j$. If $S_{\mathscr{D}_1,\ldots,\mathscr{D}_k} \leq 0$ and $\|H_i\| \in L^1(M, g)$ for $2 \leq i \leq k$, then (M, g) is the direct product.*

2.4 Foliations of Metric-Affine Manifolds

Here, we apply the approach of Section 2.3.2 to vector fields $Z_k = (A_H)^k \tilde{H} + (\tilde{A}_{\tilde{H}})^k H$ for $k = 0, 1$. The formulas, under some conditions yield splitting of ambient metric-affine manifolds (including Riemannian manifolds, submersions and twisted products). We also work with a closed manifold equipped with vector fields and distributions, defined on the complement to the "singularity set" Σ, as in Definition 2.16.

Define the "mean curvature type" vector fields related to contorsion tensor $\mathfrak{T}$ by

$$H_{\mathfrak{T}} := \sum_a \varepsilon_a \mathfrak{T}_a E_a, \quad \tilde{H}_{\mathfrak{T}} := \sum_i \varepsilon_i \mathfrak{T}_i \mathscr{E}_i,$$

where, $\{E_a, \mathscr{E}_i\}$ is a local adapted orthonormal frame, see Remark 1.34. For projections of these vectors we will use notations $H_{\mathfrak{T}}^\perp = (H_{\mathfrak{T}})^\perp$, $\tilde{H}_{\mathfrak{T}}^\top = (\tilde{H}_{\mathfrak{T}})^\top$, etc.

2.4.1 Integral Formulas with the Mixed Scalar Curvature

Here, we calculate the divergence of $Z_0 = \tilde{H} + H$, see (1.44), for metric-affine manifolds and generalize integral formula (2.28) obtained for Riemannian manifolds. Using (1.17) and (2.28), we represent $\bar{S}_{\mathrm{mix}}$ of $\bar{\nabla} = \nabla + \mathfrak{T}$ in the form

$$\overline{S}_{mix} = \frac{1}{2}\sum_{a,i}\varepsilon_a\varepsilon_i\big(\langle(\nabla_i\mathfrak{T})_a E_a, \mathscr{E}_i\rangle - \langle(\nabla_a\mathfrak{T})_i E_a, \mathscr{E}_i\rangle + \langle(\nabla_a\mathfrak{T})_i\mathscr{E}_i, E_a\rangle$$
$$- \langle(\nabla_i\mathfrak{T})_a\mathscr{E}_i, E_a\rangle + \langle[\mathfrak{T}_i, \mathfrak{T}_a]E_a, \mathscr{E}_i\rangle + \langle[\mathfrak{T}_a, \mathfrak{T}_i]\mathscr{E}_i, E_a\rangle\big) + S_{mix}. \quad (2.55)$$

Denote by $\langle B, C\rangle_{|V}$ the inner product of tensors restricted on the subbundle

$$V = (\mathscr{D} \times \widetilde{\mathscr{D}}) \oplus (\widetilde{\mathscr{D}} \times \mathscr{D}) \subset TM \times TM.$$

Lemma 2.70. *We have*

$$\mathrm{div}\,\big(H^{\perp}_{\mathfrak{T}-\mathfrak{T}^*} + \tilde{H}^{\top}_{\mathfrak{T}-\mathfrak{T}^*}\big) = 2\,(\overline{S}_{mix} - S_{mix}) - 2\bar{Q}, \quad (2.56)$$

where

$$2\bar{Q} = \langle H_{\mathfrak{T}}, \tilde{H}_{\mathfrak{T}^*}\rangle + \langle\tilde{H}_{\mathfrak{T}}, H_{\mathfrak{T}^*}\rangle + \langle H_{\mathfrak{T}-\mathfrak{T}^*} + \tilde{H}_{\mathfrak{T}^*-\mathfrak{T}}, H - \tilde{H}\rangle$$
$$+ \langle\mathfrak{T} - \mathfrak{T}^* + \widehat{\mathfrak{T}} - \widehat{\mathfrak{T}^*}, \tilde{A} - \tilde{T}^{\sharp} + A - T^{\sharp}\rangle - \langle\widehat{\mathfrak{T}^*}, \widehat{\mathfrak{T}}\rangle_{|V}.$$

Proof. Using (2.55) we get $\overline{S}_{mix} - S_{mix} = \frac{1}{2}(q_1 + q_2)$, where

$$q_1 = \sum_{a,i}\varepsilon_a\varepsilon_i\,[\,\langle(\nabla_i\mathfrak{T})_a E_a, \mathscr{E}_i\rangle - \langle(\nabla_a\mathfrak{T})_i E_a, \mathscr{E}_i\rangle + \langle[\mathfrak{T}_i, \mathfrak{T}_a]E_a, \mathscr{E}_i\rangle\,],$$
$$q_2 = \sum_{a,i}\varepsilon_a\varepsilon_i\,[\,\langle(\nabla_a\mathfrak{T})_i\mathscr{E}_i, E_a\rangle - \langle(\nabla_i\mathfrak{T})_a\mathscr{E}_i, E_a\rangle + \langle[\mathfrak{T}_a, \mathfrak{T}_i]\mathscr{E}_i, E_a\rangle\,].$$

Assume $(\nabla_i E_a)^{\top} = 0$ and $(\nabla_a\mathscr{E}_i)^{\perp} = 0$ at a point $x \in M$. Thus,

$$\nabla_i E_a = -\sum_j\varepsilon_j\langle(\tilde{A}_a + \tilde{T}_a^{\sharp})\mathscr{E}_i, \mathscr{E}_j\rangle\mathscr{E}_j, \quad \nabla_a\mathscr{E}_i = -\sum_b\varepsilon_b\langle(A_i + T_i^{\sharp})E_a, E_b\rangle E_b.$$

We calculate three terms of q_1 at $x \in M$:

$$\sum_{a,i}\varepsilon_a\varepsilon_i\langle(\nabla_i\mathfrak{T})_a E_a, \mathscr{E}_i\rangle = \sum_{a,i}\varepsilon_a\varepsilon_i\langle\mathfrak{T}_i E_a + \mathfrak{T}_a\mathscr{E}_i, (\tilde{A}_a - \tilde{T}_a^{\sharp})\mathscr{E}_i\rangle + \mathrm{div}^{\perp} H_{\mathfrak{T}},$$
$$\sum_{a,i}\varepsilon_a\varepsilon_i\langle(\nabla_a\mathfrak{T})_i E_a, \mathscr{E}_i\rangle = \sum_{a,i}\varepsilon_a\varepsilon_i\langle\mathfrak{T}_a^*\mathscr{E}_i + \mathfrak{T}_i^* E_a, (A_i - T_i^{\sharp})E_a\rangle + \mathrm{div}^{\top} \tilde{H}_{\mathfrak{T}^*},$$
$$\sum_{a,i}\varepsilon_a\varepsilon_i\langle[\mathfrak{T}_i, \mathfrak{T}_a]E_a, \mathscr{E}_i\rangle = \langle H_{\mathfrak{T}}, \tilde{H}_{\mathfrak{T}^*}\rangle - \sum_{a,i}\varepsilon_a\varepsilon_i\langle\mathfrak{T}_i E_a, \mathfrak{T}_a^*\mathscr{E}_i\rangle,$$

and find

$$q_1 = \mathrm{div}^{\perp} H_{\mathfrak{T}} - \mathrm{div}^{\top} \tilde{H}_{\mathfrak{T}^*} + \sum_{a,i}\varepsilon_a\varepsilon_i\,[\,\langle\mathfrak{T}_i E_a + \mathfrak{T}_a\mathscr{E}_i, (\tilde{A}_a - \tilde{T}_a^{\sharp})\mathscr{E}_i\rangle$$
$$- \langle\mathfrak{T}_a^*\mathscr{E}_i + \mathfrak{T}_i^* E_a, (A_i - T_i^{\sharp})E_a\rangle - \langle\mathfrak{T}_i E_a, \mathfrak{T}_a^*\mathscr{E}_i\rangle\,] + \langle H_{\mathfrak{T}}, \tilde{H}_{\mathfrak{T}^*}\rangle. \quad (2.57)$$

Replacing the roles of $\widetilde{\mathscr{D}}$ and $\mathscr{D}$ in (2.57), one can find q_2. From the above, using

$$\mathrm{div}^{\perp} H_{\mathfrak{T}} = \mathrm{div}\,H^{\perp}_{\mathfrak{T}} + \langle H_{\mathfrak{T}}, H - \tilde{H}\rangle, \quad \mathrm{div}^{\top} \tilde{H}_{\mathfrak{T}^*} = \mathrm{div}\,\tilde{H}^{\top}_{\mathfrak{T}^*} - \langle\tilde{H}_{\mathfrak{T}^*}, H - \tilde{H}\rangle,$$
$$\mathrm{div}^{\perp} H_{\mathfrak{T}^*} = \mathrm{div}\,H^{\perp}_{\mathfrak{T}^*} + \langle H_{\mathfrak{T}^*}, H - \tilde{H}\rangle, \quad \mathrm{div}^{\top} \tilde{H}_{\mathfrak{T}} = \mathrm{div}\,\tilde{H}^{\top}_{\mathfrak{T}} - \langle\tilde{H}_{\mathfrak{T}}, H - \tilde{H}\rangle,$$

we obtain

$$\text{div}\left(H^{\perp}_{\mathfrak{T}-\mathfrak{T}^*}+\tilde{H}^{\top}_{\mathfrak{T}-\mathfrak{T}^*}\right)=2\left(\overline{S}_{\text{mix}}-S_{\text{mix}}\right)$$
$$-\langle H_{\mathfrak{T}},\tilde{H}_{\mathfrak{T}^*}\rangle-\langle \tilde{H}_{\mathfrak{T}},H_{\mathfrak{T}^*}\rangle-\langle H_{\mathfrak{T}-\mathfrak{T}^*}+\tilde{H}_{\mathfrak{T}^*-\mathfrak{T}},H-\tilde{H}\rangle$$
$$-\sum_{a,i}\varepsilon_a\varepsilon_i\left[\langle(\mathfrak{T}_i-\mathfrak{T}_i^*)E_a+(\mathfrak{T}_a-\mathfrak{T}_a^*)\mathscr{E}_i,(\tilde{A}_a-\tilde{T}_a^{\sharp})\mathscr{E}_i+(A_i-T_i^{\sharp})E_a\rangle\right.$$
$$\left.-\langle \mathfrak{T}_a\mathscr{E}_i,\mathfrak{T}_i^* E_a\rangle-\langle\mathfrak{T}_a^*\mathscr{E}_i,\mathfrak{T}_i E_a\rangle\right],$$

which is equivalent to (2.56).   □

Notice that for statistical manifolds Lemma 2.70 yields

$$\overline{S}_{\text{mix}}-S_{\text{mix}}-\langle H_{\mathfrak{T}},\tilde{H}_{\mathfrak{T}}\rangle+(1/2)\langle\mathfrak{T},\mathfrak{T}\rangle_{|V}=0.$$

Theorem 2.71. *Let* $(M,g,\overline{\nabla})$ *be a closed metric-affine space and* $\widetilde{\mathscr{D}}$ *a distribution defined on the complement to the "singularity set"* Σ. *If* $\|\xi\|_g\in L^2(M,g)$, *where* $\xi=\tilde{H}+H+\frac{1}{2}H^{\perp}_{\mathfrak{T}-\mathfrak{T}^*}+\frac{1}{2}\tilde{H}^{\top}_{\mathfrak{T}-\mathfrak{T}^*}$, *then the following integral formula is valid:*

$$\int_M\left\{\overline{S}_{\text{mix}}-\langle T,T\rangle-\langle\tilde{T},\tilde{T}\rangle+\langle h,h\rangle+\langle\tilde{h},\tilde{h}\rangle-\langle H,H\rangle-\langle\tilde{H},\tilde{H}\rangle-\bar{Q}\right\}d\,\text{vol}=0,$$

where $\bar{Q}$ *is defined in Lemma 2.70.*

Proof. By (1.44) and Lemma 2.70, we have on $M\setminus\Sigma$:

$$\text{div}\,\xi=\overline{S}_{\text{mix}}-\langle T,T\rangle-\langle\tilde{T},\tilde{T}\rangle+\langle h,h\rangle+\langle\tilde{h},\tilde{h}\rangle-\langle H,H\rangle-\langle\tilde{H},\tilde{H}\rangle-\bar{Q}.\quad(2.58)$$

Thus, the claim follows from (2.58) and Lemma 2.23.   □

Corollary 2.72. *Let* $(M,g,\overline{\nabla})$ *be a closed statistical manifold and* $\widetilde{\mathscr{D}}$ *a distribution on the complement to a "singularity set"* Σ. *If* $\|\tilde{H}+H\|_g\in L^2(M,g)$, *then*

$$\int_M\left\{\overline{S}_{\text{mix}}-\langle T,T\rangle-\langle\tilde{T},\tilde{T}\rangle+\langle h,h\rangle+\langle\tilde{h},\tilde{h}\rangle-\langle H,H\rangle-\langle\tilde{H},\tilde{H}\rangle\right.$$
$$\left.-\langle H_{\mathfrak{T}},\tilde{H}_{\mathfrak{T}}\rangle+(1/2)\langle\mathfrak{T},\mathfrak{T}\rangle_{|V}\right\}d\,\text{vol}=0.$$

Definition 2.73. We say that (M',g') is a *leaf of a distribution* $\widetilde{\mathscr{D}}$ on (M,g) if M' is a submanifold of M with induced metric g' and $T_xM'=\widetilde{\mathscr{D}}_x$ for any $x\in M'$.

The following condition for $\mathfrak{T}^*=\mathfrak{T}$ reduces to $H=0$, i.e., $\widetilde{\mathscr{D}}$ is harmonic:

$$2H=H^{\perp}_{\mathfrak{T}^*-\mathfrak{T}}.\quad(2.59)$$

Theorem 2.74. *Suppose that a distribution* $\widetilde{\mathscr{D}}$ *on a metric-affine space* $(M,g,\overline{\nabla})$ *has a compact leaf* (M',g') *with the condition* (2.59) *on a neighborhood of* M'. *Then the following integral formula along the leaf is valid:*

$$\int_{M'}\left\{\overline{S}_{\text{mix}}-\langle\tilde{T},\tilde{T}\rangle+\langle\tilde{h},\tilde{h}\rangle+\langle h,h\rangle-\bar{Q}'\right\}d\,\text{vol}_{g'}=0,\quad(2.60)$$

where

$$2\bar{Q}' = \langle H_{\mathfrak{T}}, \tilde{H}_{\mathfrak{T}^*}\rangle + \langle \tilde{H}_{\mathfrak{T}}, H_{\mathfrak{T}^*}\rangle - \langle H_{\mathfrak{T}-\mathfrak{T}^*}, \tilde{H}\rangle - \langle \tilde{H}_{\mathfrak{T}-\mathfrak{T}^*}, H\rangle$$
$$+ \langle \mathfrak{T} - \mathfrak{T}^* + \widehat{\mathfrak{T}} - \widehat{\mathfrak{T}^*}, \tilde{A} - \tilde{T}^{\sharp} + A\rangle - \langle \mathfrak{T}^*, \widehat{\mathfrak{T}}\rangle_{|V}.$$

Proof. Using $T = 0$ along M', (1.44), Lemma 2.70 and equalities

$$\mathrm{div}^{\perp}\tilde{H} = -\langle \tilde{H}, \tilde{H}\rangle, \quad \mathrm{div}^{\top}H = -\langle H, H\rangle,$$
$$\mathrm{div}^{\perp}\tilde{H}^{\top}_{\mathfrak{T}-\mathfrak{T}^*} = -\langle \tilde{H}_{\mathfrak{T}-\mathfrak{T}^*}, \tilde{H}\rangle, \quad \mathrm{div}^{\top}H^{\perp}_{\mathfrak{T}-\mathfrak{T}^*} = -\langle H_{\mathfrak{T}-\mathfrak{T}^*}, H\rangle,$$

we obtain

$$\mathrm{div}^{\top}\big(\tilde{H} + \tfrac{1}{2}\tilde{H}^{\top}_{\mathfrak{T}-\mathfrak{T}^*}\big) + \mathrm{div}^{\perp}\big(H + \tfrac{1}{2}H^{\perp}_{\mathfrak{T}-\mathfrak{T}^*}\big)$$
$$= \overline{S}_{\mathrm{mix}} + \langle \tilde{h}, \tilde{h}\rangle + \langle h, h\rangle - \langle \tilde{T}, \tilde{T}\rangle - \bar{Q}'. \tag{2.61}$$

By conditions (2.59), the $\mathrm{div}^{\perp}$-term in (2.61) vanishes along M'. Thus, (2.60) follows from the Divergence theorem for $\xi = \tilde{H} + \tfrac{1}{2}\tilde{H}^{\top}_{\mathfrak{T}-\mathfrak{T}^*}$ on M'. $\qquad\square$

Corollary 2.75. *Let a distribution $\widetilde{\mathscr{D}}$ on a statistical manifold $(M, g, \overline{\nabla})$ admit a compact leaf (M', g') with the condition $H = 0$ on a neighborhood of M'. Then*

$$\int_{M'}\big\{\overline{S}_{\mathrm{mix}} - \langle \tilde{T}, \tilde{T}\rangle + \langle \tilde{h}, \tilde{h}\rangle + \langle h, h\rangle - \langle H_{\mathfrak{T}}, \tilde{H}_{\mathfrak{T}}\rangle - \langle \mathfrak{T}, \mathfrak{T}\rangle_{|V}\big\}\mathrm{d}\,\mathrm{vol}_{g'} = 0.$$

Remark 2.76. For $\mathscr{D}$ spanned by a unit vector field N, from Theorems 2.71 and 2.74 there follow integral formulas, which for $\mathfrak{T} = 0$ abbreviate to (2.2) and (2.4).

2.4.2 Integral Formula with the Ricci Curvature

The divergence of the vector field $Z_1 = A_H\tilde{H} + \tilde{A}_{\tilde{H}}H$ on a Riemannian manifold was calculated in [LuW]:

$$\mathrm{div}(A_H\tilde{H} + \tilde{A}_{\tilde{H}}H) = \mathrm{Ric}_{H,\tilde{H}} + Q_1, \tag{2.62}$$

where

$$Q_1 = \langle H, \nabla_{\tilde{H}}H\rangle + \langle \tilde{H}, \nabla_H\tilde{H}\rangle + \langle \mathrm{trace}^{\perp}_g(\nabla_\cdot T)(\cdot, \tilde{H}), H\rangle + \langle \mathrm{trace}^{\top}_g(\nabla_\cdot\tilde{T})(\cdot, H), \tilde{H}\rangle$$
$$+ \langle A_H, \nabla.\tilde{H}\rangle + \langle \tilde{A}_{\tilde{H}}, \nabla.H\rangle - \langle A_{\tilde{H}}H, H\rangle - \langle \tilde{A}_H\tilde{H}, \tilde{H}\rangle + 2\sum_a \varepsilon_a\big(\langle A_{(\nabla_a H)^{\perp}}\tilde{H}, E_a\rangle$$
$$+ \langle \nabla_{T(\tilde{H}, E_a)}E_a, H\rangle\big) + 2\sum_i \varepsilon_i\big(\langle \tilde{A}_{(\nabla_i\tilde{H})^{\top}}H, \mathscr{E}_i\rangle + \langle \nabla_{\tilde{T}(H, \mathscr{E}_i)}\mathscr{E}_i, \tilde{H}\rangle\big), \tag{2.63}$$

and if the distributions are totally umbilical, integrable and have constant mean curvature, then $Q_1 = -(\tfrac{1}{n} + \tfrac{1}{p})\|H\|^2 \cdot \|\tilde{H}\|^2$. On a closed manifold (M, g) we have

$$\int_M (\mathrm{Ric}_{H,\tilde{H}} + Q_1)\,d\,\mathrm{vol} = 0. \tag{2.64}$$

Here, we extend (2.62) and the above integral formula for metric-affine manifolds.

Lemma 2.77. *For the metric-affine case we have*

$$\mathrm{div}(\mathfrak{T}_{\tilde{H}}H + \mathfrak{T}_H\tilde{H}) = -(\overline{\mathrm{Ric}}_{H,\tilde{H}} - \mathrm{Ric}_{H,\tilde{H}}) + \bar{Q}_1, \tag{2.65}$$

where $\overline{\mathrm{Ric}}_{H,\tilde{H}} = \mathrm{Sym}\big(\sum_a \varepsilon_a \langle \bar{R}_{\tilde{H},E_a}H, E_a\rangle + \sum_i \varepsilon_i \langle \bar{R}_{H,\mathscr{E}_i}\tilde{H}, \mathscr{E}_i\rangle\big)$ *and*

$$
\begin{aligned}
\bar{Q}_1 = \mathrm{Sym}\big(&\langle \bar{\nabla}_{\tilde{H}}H, H_{\mathfrak{T}^*}\rangle + \langle \bar{\nabla}_H\tilde{H}, \tilde{H}_{\mathfrak{T}^*}\rangle - \langle \mathfrak{T}_{\tilde{H}}H, \tilde{H}\rangle - \langle \mathfrak{T}_H\tilde{H}, H\rangle \\
&- \sum_a \varepsilon_a \langle \nabla_{\tilde{H}}(\mathfrak{T}_a H) - \mathfrak{T}_{(h+T)(\tilde{H},E_a)+\nabla_a\tilde{H}}H + \mathfrak{T}_{\tilde{H}}(\nabla_a H), E_a\rangle \\
&- \sum_i \varepsilon_i \langle \nabla_H(\mathfrak{T}_i\tilde{H}) - \mathfrak{T}_{(\tilde{h}+\tilde{T})(H,\mathscr{E}_i)+\nabla_i H}\tilde{H} + \mathfrak{T}_H(\nabla_i\tilde{H}), \mathscr{E}_i\rangle\big). \tag{2.66}
\end{aligned}
$$

Proof. Using (1.17), we calculate

$$
\begin{aligned}
\sum_a \varepsilon_a &\langle \bar{R}_{\tilde{H}\,E_a}H - R(\tilde{H}, E_a)H, E_a\rangle \\
&= \mathrm{Sym}\big(\mathrm{div}^\top(\mathfrak{T}_{\tilde{H}}H) + \langle \nabla_{\tilde{H}}H, H_{\mathfrak{T}^*}\rangle + \langle \mathfrak{T}_{\tilde{H}}H, H_{\mathfrak{T}^*}\rangle \\
&\quad - \sum_a \varepsilon_a [\langle \nabla_{\tilde{H}}(\mathfrak{T}_a H), E_a\rangle - \langle \mathfrak{T}_{(h+T)(\tilde{H},E_a)}H, E_a\rangle \\
&\quad - \langle \mathfrak{T}_{\nabla_a\tilde{H}}H, E_a\rangle + \langle \mathfrak{T}_{\tilde{H}}(\nabla_a H), E_a\rangle + \langle \mathfrak{T}_{\tilde{H}}(\mathfrak{T}_a H), E_a\rangle]\big). \tag{2.67}
\end{aligned}
$$

Summing (2.67) with a similar formula for $\sum_i \varepsilon_i \langle \bar{R}_{\mathscr{E}_i,H}\tilde{H} - R(H,\mathscr{E}_i)\tilde{H}, \mathscr{E}_i\rangle$ and using

$$\mathrm{div}^\perp(\mathfrak{T}_H\tilde{H}) = \mathrm{div}(\mathfrak{T}_H\tilde{H}) - \langle H, \mathfrak{T}_H\tilde{H}\rangle, \quad \mathrm{div}^\top(\mathfrak{T}_{\tilde{H}}H) = \mathrm{div}(\mathfrak{T}_{\tilde{H}}H) - \langle \tilde{H}, \mathfrak{T}_{\tilde{H}}H\rangle,$$

yield (2.65) and (2.66), where Sym means the symmetrization of a tensor. $\qquad\square$

Theorem 2.78 (See [R9]). *Let* $(M, g, \bar{\nabla})$ *be a closed metric-affine space and* $\mathscr{D}$ *a distribution defined on the complement to the "singularity set"* Σ. *If* $\|\xi\|_g \in$ $\mathrm{L}^2(M, g)$, *where* $\xi = A_H\tilde{H} + \tilde{A}_{\tilde{H}}H + \mathfrak{T}_H\tilde{H} + \mathfrak{T}_{\tilde{H}}H$, *then*

$$\int_M \big\{\overline{\mathrm{Ric}}_{H,\tilde{H}} + Q_1 + \bar{Q}_1\big\}d\,\mathrm{vol} = 0. \tag{2.68}$$

Proof. From (2.62) and (2.65) we obtain

$$\mathrm{div}(A_H\tilde{H} + \tilde{A}_{\tilde{H}}H + \mathfrak{T}_{\tilde{H}}H + \mathfrak{T}_H\tilde{H}) = -\overline{\mathrm{Ric}}_{H,\tilde{H}} + Q_1 + \bar{Q}_1.$$

Applying Lemma 2.23 to (2.63), (2.65) and (2.66), we obtain (2.68). $\qquad\square$

For $\mathfrak{T} = 0$, we have $\bar{Q}_1 = 0$, and (2.68) reduces to (2.64). One may get a number of formulas from (2.68). Recall that $\mathscr{D}$ has constant mean curvature whenever its mean curvature vector H obeys $\nabla^\perp H = 0$, where $\nabla^\perp$ is the connection in $\mathscr{D}$ induced by the Levi-Civita connection on M.

In conditions of Theorem 2.78, let distributions $\mathscr{D}$ and $\widetilde{\mathscr{D}}$ be totally umbilical, integrable and have constant mean curvature. Then (2.68) reads as (see [R9])

$$
\int_M \Big\{ \overline{\mathrm{Ric}}_{H,\tilde{H}} - (\frac{1}{n} + \frac{1}{p})\, \|H\|^2 \cdot \|\tilde{H}\|^2 + \mathrm{Sym}\big(\langle \overline{\nabla}_H \tilde{H}, \tilde{H}_{\mathfrak{T}^*} \rangle + \langle \overline{\nabla}_{\tilde{H}} H, H_{\mathfrak{T}^*} \rangle
$$
$$
- \langle \tilde{H}, \mathfrak{T}_{\tilde{H}-H} H \rangle - \langle H, \mathfrak{T}_{H-\tilde{H}} \tilde{H} \rangle - \sum_a \varepsilon_a \langle \overline{\nabla}_{\tilde{H}} (\mathfrak{T}_a H) - \mathfrak{T}_{\nabla_a \tilde{H}} H + \mathfrak{T}_{\tilde{H}}(\nabla_a H),\ E_a \rangle
$$
$$
- \sum_i \varepsilon_i \langle \overline{\nabla}_H (\mathfrak{T}_i \tilde{H}) - \mathfrak{T}_{\nabla_i H} \tilde{H} + \mathfrak{T}_H (\nabla_i \tilde{H}),\ \mathscr{E}_i \rangle \big) \Big\}\, d\,\mathrm{vol} = 0.
$$

2.4.3 Splitting Results

Here, we use integral formulas of Section 2.4.1 (either on the whole metric-affine space $(M, g, \overline{\nabla}; \mathscr{D}, \widetilde{\mathscr{D}})$ with $g > 0$ or on a closed leaf of the distribution) to prove non-existence and splitting theorems, related to results of Sections 2.3.3 and 2.3.4.

The next conditions allow us to simplify the presentation of results:

$$
\mathfrak{T}_X Y = 0 = \mathfrak{T}_Y X, \quad \mathfrak{T}_X^* Y = 0 = \mathfrak{T}_Y^* X \quad (X \in \mathscr{D},\ Y \in \widetilde{\mathscr{D}}), \tag{2.69}
$$
$$
\langle H_{\mathfrak{T}-\mathfrak{T}^*}, \tilde{H} \rangle = 0. \tag{2.70}
$$

For example, conditions (2.69) provide $H_{\mathfrak{T}}^{\perp} = 0$ and $\tilde{H}_{\mathfrak{T}}^{\top} = 0$; moreover, conditions (2.69)–(2.70) provide vanishing of $\bar{Q}$ in Lemma 2.70.

2.4.3.1 Harmonic Distributions

In the next two theorems we consider distributions with $\overline{S}_{\mathrm{mix}} \geq 0$ (for $S_{\mathrm{mix}} \geq 0$ on Riemannian manifolds, see also [SM] and [W4]).

Theorem 2.79. *Let $(M, g, \overline{\nabla})$ be a metric-affine space with complementary orthogonal integrable distributions $\mathscr{D}$ and $\widetilde{\mathscr{D}}$ satisfying (2.59), (2.69) and (2.70). Suppose that the leaves (M', g') of $\widetilde{\mathscr{D}}$ are closed manifolds and $\overline{S}_{\mathrm{mix}} \geq 0$. Then M splits.*

Proof. By (2.69), we get $H_{\mathfrak{T}^*-\mathfrak{T}}^{\perp} = 0$, then from (2.59) we find $H = 0$. By conditions and (2.61), we have
$$
\mathrm{div}^{\top} \xi = \overline{S}_{\mathrm{mix}} + \|\tilde{h}\|^2 + \|h\|^2, \tag{2.71}
$$
where $\xi = \tilde{H} + \frac{1}{2} \tilde{H}_{\mathfrak{T}-\mathfrak{T}^*}^{\top}$. Applying the Divergence Theorem to each leaf (a closed manifold), and since $\overline{S}_{\mathrm{mix}} \geq 0$, we get $\overline{S}_{\mathrm{mix}} = 0$, $h = 0$ and $\tilde{h} = 0$. By de Rham decomposition theorem, (M, g) splits. $\qquad\square$

Corollary 2.80. *Let $(M, g, \overline{\nabla})$ be a metric-affine space endowed with a distribution $\widetilde{\mathscr{D}}$ with integrable normal bundle satisfying (2.59), (2.69) and (2.70). Then $\widetilde{\mathscr{D}}$ has no closed leaves M' with $\overline{S}_{\mathrm{mix}\,|M'} > 0$.*

Proof. Let (M',g') be a closed leaf obeying the conditions. By conditions, As in the proof of Theorem 2.79, we have (2.71) with $\xi = \tilde{H} + \frac{1}{2}\tilde{H}^{\top}_{\mathfrak{T}-\mathfrak{T}^*}$ on M'. Applying the Divergence Theorem to this leaf, and since $\overline{S}_{\mathrm{mix}} > 0$, we get a contradiction. □

Corollary 2.81. *A codimension-one distribution $\widetilde{\mathscr{D}}$ of $(M,g,\overline{\nabla})$ with the Ricci curvature $\overline{\mathrm{Ric}} > 0$ and the properties (2.59), (2.69) and (2.70) has no closed leaves.*

Proof. For a codimension-one $\widetilde{\mathscr{D}}$, we have $T = 0$ and $\varepsilon_N \overline{\mathrm{Ric}}_{N,N} = \overline{S}_{\mathrm{mix}}$, where N is a unit (local) normal to the leaves. Thus, the claim follows from Corollary 2.80. □

Theorem 2.82. *Let $(M,g,\overline{\nabla})$ be a complete open metric-affine space endowed with complementary orthogonal harmonic foliations, and let conditions (2.69) and $\|\xi\|_g \in \mathrm{L}^1(M,g)$ with $2\xi = H^{\perp}_{\mathfrak{T}-\mathfrak{T}^*} + \tilde{H}^{\top}_{\mathfrak{T}-\mathfrak{T}^*}$, be satisfied. If $\overline{S}_{\mathrm{mix}} \geq 0$, then M splits.*

Proof. Under our assumptions $(T = H = \tilde{T} = \tilde{H} = 0)$, from (2.58) with $\bar{Q} = 0$ we get

$$\mathrm{div}\,\xi = \overline{S}_{\mathrm{mix}} + \|h\|^2 + \|\tilde{h}\|^2.$$

By Lemma 2.54 and since $\overline{S}_{\mathrm{mix}} \geq 0$, we get $\mathrm{div}\,\xi = 0$ and $\overline{S}_{\mathrm{mix}} = 0$. Thus, $h = 0 = \tilde{h}$. Hence, by de Rham decomposition theorem, (M,g) splits. □

2.4.3.2 Totally Umbilical Distributions

Here, we consider totally umbilical distributions with $\overline{S}_{\mathrm{mix}} \leq 0$.

Theorem 2.83. *Let $(M,g,\overline{\nabla})$ be a metric-affine space endowed with complementary orthogonal totally umbilical distributions $(\mathscr{D},\widetilde{\mathscr{D}})$ satisfying conditions (2.59), (2.69) and (2.70). Then $\widetilde{\mathscr{D}}$ has no closed leaves (M',g') with $\overline{S}_{\mathrm{mix}}|_{M'} < 0$.*

Proof. Let (M',g') be a closed totally umbilical leaf satisfying $\overline{S}_{\mathrm{mix}|M'} < 0$. By (2.69), we get $H^{\perp}_{\mathfrak{T}^*-\mathfrak{T}} = 0$, then from (2.59) we find $H = 0$. Thus, (2.61) reduces to

$$\mathrm{div}^{\top}\xi = \overline{S}_{\mathrm{mix}} - \|\tilde{T}\|^2 - \frac{p-1}{p}\|\tilde{H}\|^2 \tag{2.72}$$

on M', where $\xi = \tilde{H} + \frac{1}{2}\tilde{H}^{\top}_{\mathfrak{T}-\mathfrak{T}^*}$. Applying the Divergence Theorem to the leaf, and since $\overline{S}_{\mathrm{mix}|M'} < 0$, we get $\tilde{H} = 0$, $\tilde{T} = 0$ and $\overline{S}_{\mathrm{mix}} = 0$, a contradiction. □

Corollary 2.84. *A codimension-one distribution $\widetilde{\mathscr{D}}$ on $(M,g,\overline{\nabla})$ with $\overline{\mathrm{Ric}} < 0$ and conditions (2.59), (2.69) and (2.70) has no closed totally umbilical leaves.*

Theorem 2.85. *Let $(M,g,\overline{\nabla})$ be a complete open metric-affine space endowed with complementary orthogonal totally umbilical distributions $\mathscr{D}$ and $\widetilde{\mathscr{D}}$ defined on the complement to the "singularity set" Σ. Suppose that conditions (2.69),*

$$H_{\mathfrak{T}} = 0 = \tilde{H}_{\mathfrak{T}}, \qquad H_{\mathfrak{T}^*} = 0 = \tilde{H}_{\mathfrak{T}^*} \tag{2.73}$$

and $\|\xi\|_g \in \mathrm{L}^1(M,g)$, where $\xi = \tilde{H} + H$, are satisfied. If $\overline{S}_{\mathrm{mix}} \leq 0$, then M splits.

Proof. By assumptions, from (2.58) we get $\bar{Q} = 0$ and

$$\operatorname{div}\xi = \overline{S}_{\mathrm{mix}} - \|T\|^2 - \|\tilde{T}\|^2 - \frac{p-1}{p}\|\tilde{H}\|^2 - \frac{n-1}{n}\|H\|^2. \qquad (2.74)$$

From (2.74) and Lemma 2.54 and since $\overline{S}_{\mathrm{mix}} \leq 0$, we get $\operatorname{div}\xi = 0$. This yields vanishing of $T, \tilde{T}, H$ and $\tilde{H}$. By de Rham decomposition theorem, (M, g) splits. $\square$

Totally umbilical integrable distributions appear on doubly twisted products, see Section 1.3.2 and [PR, S2]. Conditions (2.69) are obviously satisfied for $B \times_{(v,u)} F$.

Corollary 2.86 (of Theorem 2.83). *Let* (2.59) *and* (2.70) *be satisfied along the fibres of a doubly twisted product* $M = B \times_{(v,u)} F$, *where* F *is a closed manifold. If* $\overline{S}_{\mathrm{mix}} \leq 0$, *then* M *is the direct product.*

Proof. By assumptions (and $\tilde{T} = 0 = H$), from (2.72) we get

$$\operatorname{div}^\top \xi = \overline{S}_{\mathrm{mix}} - \frac{p-1}{p}\|\tilde{H}\|^2,$$

where $\xi = \tilde{H} + \frac{1}{2}\tilde{H}_{\mathfrak{T}-\mathfrak{T}^*}^\top$ is tangent to fibres. Applying the Divergence Theorem to each fibre, and since $\overline{S}_{\mathrm{mix}} \leq 0$, we get $\overline{S}_{\mathrm{mix}} = 0$ and $\tilde{H} = 0$, i.e., $\nabla^\top u = 0 = \nabla^\perp v$. By the above, u and v are locally constant. By de Rham decomposition theorem, (M, g) is the direct product with factors $(B, c_1 g_B)$ and $(F, c_2 g_F)$ for some $c_1, c_2 > 0$. $\square$

Corollary 2.87 (of Theorem 2.85). *Let conditions* (2.73) *and* $\|\xi\|_g \in \mathrm{L}^1(M, g)$ *for the vector field* $\xi = \tilde{H} + H$ *be satisfied on a doubly twisted product* $M = B \times_{(v,u)} F$, *which is a complete open manifold. If* $\overline{S}_{\mathrm{mix}} \leq 0$, *then* M *is the direct product.*

Proof. Set $\mathscr{D} = \pi_*(TF)$. By conditions, and since the leaves and fibers are totally umbilical submanifolds, we have, see (2.74),

$$\operatorname{div}\xi = \overline{S}_{\mathrm{mix}} - \frac{p-1}{p}\|\tilde{H}\|^2 - \frac{n-1}{n}\|H\|^2.$$

Applying Lemma 2.54 to M, and using $\overline{S}_{\mathrm{mix}} \leq 0$, we get $\operatorname{div}\xi = 0$. Hence, $\overline{S}_{\mathrm{mix}} = 0$ and $H = 0 = \tilde{H}$, i.e., $\nabla^\top u = 0 = \nabla^\perp v$. By the above, $S_{\mathrm{mix}} = 0$; thus, u and v are constant. By de Rham decomposition theorem, (M, g) is the direct product with the factors $(B, c_1 g_B)$ and $(F, c_2 g_F)$ for some positive $c_1, c_2 \in \mathbb{R}$. $\square$

2.5 Codimension One Foliations of Finsler Spaces

Here, we study the following problem: *Find integral formulas for a closed codimension-one foliated manifold endowed with a set of linearly independent one-forms.* In [RW7], the problem was examined using approach of Randers norm. We continue this study, using (α, b)-norm of Section 1.2.2, determined by a Euclidean norm α, linearly independent 1-forms β_i, $(1 \le i \le p)$ and a function ϕ of p variables. We study a Riemannian structure, naturally arising from this norm and a codimension-one distribution $\ker \omega$ of 1-form $\omega \ne 0$, and the extrinsic geometry of $\ker \omega$ in terms of invariants of α, ω, β_i and ϕ. We apply the above to prove integral formulas for a closed Riemannian manifold endowed with a codimension-one distribution, which generalize the Reeb's formula (2.1) and its counterpart (2.2) for the second mean curvature. Using our norm, we get new estimates of the "non-umbilicity" of a codimension-one distribution and the energy of a vector field.

2.5.1 The Generalization of (α, β)-Norm

The following Minkowski norm [R9] (see also [JS] with homogeneous combinations of several Minkowski norms and one-forms) generalizes the (α, β)-norm.

Definition 2.88. Let $\phi : \Pi \to (0, \infty)$ be a smooth function on a parallelepiped $\Pi = \prod_{i=1}^{p} [-\delta_i, \delta_i]$ and $\boldsymbol{b} = (\beta_1, \ldots, \beta_p)$ a sequence of $p < n$ linearly independent 1-forms on $(\mathbb{R}^n, \alpha)$ of the norm $\alpha(\beta_i) < \delta_i$. Then, the $(\alpha, \boldsymbol{b})$-norm F on $\mathbb{R}^n$ is given by

$$F(y) = \alpha(y)\phi(s), \quad s = (s_1, \ldots, s_p), \quad s_i = \beta_i(y)/\alpha(y). \quad (2.75)$$

The indicatrix of F is a rotational hypersurface in $\mathbb{R}^n$ with the p-dimensional axis span$\{\beta_1^\sharp, \ldots, \beta_p^\sharp\}$, see Proposition 2.91 below. For $p = 1$, (2.75) defines the (α, β)-norm. By shifting the indicatrix of an (α, β)-norm, we obtain different Minkowski norms, called *navigation* (α, β)-*norms* [YZ]. The indicatrix of any such norm is still a rotational hypersurface, but the rotation axis does not pass through the origin in general. This is the particular case of $(\alpha, (\beta_1, \beta_2))$-norm.

Remark 2.89. Our norm (2.75) can be viewed as a canonical form of so called $(F_0, \boldsymbol{b})$-norm (for $F_0 = \alpha$), defined by $F = \sqrt{L(F_0, \beta_1, \ldots, \beta_p)}$, where $L : \mathbb{R}^{p+1} \to \mathbb{R}$ is a continuous function with the following properties: a) L is smooth and positive away from 0, b) L is positively homogeneous of degree 2, i.e., $L(\lambda y) = \lambda^2 L(y)$ for all $\lambda > 0$. Of course, set $\phi(s_1, \ldots, s_p) = \sqrt{L(1, s_1, \ldots, s_p)}$, then $F = F_0 \phi(\beta_1/F_0, \ldots, \beta_p/F_0)$. For $p = 1$, such combinations were named (F_0, β)-norms in [JS, p. 845], see also [RW6].

Obviously, a Minkowski norm F on $\mathbb{R}^n$ is the Euclidean one if and only F is $O(n)$-invariant. We will clarify the symmetry of indicatrices of $(\alpha, \boldsymbol{b})$-norms.

Definition 2.90. Let F be a Minkowski norm on $\mathbb{R}^n$ and G a subgroup of $GL(n, \mathbb{R})$. Then F is called G-*invariant* if the following is valid for some affine coordinates $(y^1, \ldots, y^n)$ of $\mathbb{R}^n$:

$$F(y^1, \ldots, y^n) = F(f(y^1, \ldots, y^n)), \quad y \in \mathbb{R}^n, \ f \in G. \quad (2.76)$$

Proposition 2.91. *A Minkowski norm F on $\mathbb{R}^n$ is G-invariant, where*

$$G = \left\{ \begin{pmatrix} C & 0 \\ 0 & \mathrm{id}_p \end{pmatrix}, \ C \in O(n-p,\mathbb{R}), \ 0 < p < n \right\},$$

if and only if there exist 1-forms β_i $(1 \le i \le p)$, for which F is $(\alpha,\boldsymbol{b})$-norm.

Proof. Let $F = \alpha \phi(\beta_1/\alpha, \ldots, \beta_p/\alpha)$ be the $(\alpha,\boldsymbol{b})$-norm. Let $\{e_1, \ldots, e_n\}$ be an $\langle \cdot, \cdot \rangle$-orthonormal frame such that $\bigcap_{i=1}^p \ker \beta_i = \mathrm{span}\{e_1, \ldots, e_{n-p}\}$. Then $\beta_i(y) = \sum_{j=n-p+1}^n \beta_i(e_j)y^j$, where

$$F(y) = \|y\| \phi \Big(\|y\|^{-1} \sum_{j=n-p+1}^n \beta_1(e_j)y^j, \ldots, \|y\|^{-1} \sum_{j=n-p+1}^n \beta_p(e_j)y^j \Big)$$

and $y = y^j e_j$, $\|y\|^2 = \sum_{j=1}^n (y^j)^2$. Hence, F is G-invariant.

Conversely, let F obey (2.76) for G in affine coordinates $y = (y^1, \ldots, y^n)$. If $p = n$, then for $G = \{\mathrm{id}_n\}$ one may take $\varphi = F$, $\beta_i = e_i^\flat$ and use 1-homogeneity (axiom M_2). Let $p \le n-1$. By restricting F on the $(n-p)$-dimensional linear subspace U given by p equations $y^{n-p+1} = \ldots = y^n = 0$, one obtains an $O(n-p)$-invariant Minkowski norm, which must be Euclidean. Thus, there exists $B > 0$, such that the norm $\alpha(y) = B\sqrt{(y^1)^2 + \ldots + (y^n)^2}$ on $\mathbb{R}^n$ obeys $\alpha|_U = F|_U$. Set

$$\tilde{\phi}(y) = F(y)/\alpha(y) \quad (y \ne 0).$$

Then $\tilde{\phi}$ is G-invariant, hence $\tilde{\phi}$ depends on p variables $y^{n-p+1}, \ldots, y^n$. Since $\tilde{\phi}$ is 0-homogeneous, we have $\tilde{\phi}(y) = \tilde{\phi}(By^{n-p+1}/\alpha(y), \ldots, By^n/\alpha(y))$, that is $\beta_i = Be_{n-p+i}^\flat$ for $i \in \{1, \ldots, p\}$. $\square$

Set $b_{ij} = \langle \beta_i, \beta_j \rangle = \langle \beta_i^\sharp, \beta_j^\sharp \rangle$. By the proof of Proposition 2.91, the $(\alpha,\boldsymbol{b})$-norm can be defined by certain function φ and mutually orthogonal 1-forms β_i of length $B > 0$. Thus, we can assume $b_{ij} = B^2 \delta_{ij}$. Define functions of variables $s = (s_1, \ldots, s_p)$,

$$\rho = \phi(\phi - \textstyle\sum_i s_i \dot\phi_i), \quad \rho_0^{ij} = \phi \ddot\phi_{ij} + \dot\phi_i \dot\phi_j, \quad \rho_1^i = \phi \dot\phi_i - \textstyle\sum_j s_j (\phi \ddot\phi_{ij} + \dot\phi_i \dot\phi_j),$$

where $\dot\phi_i = \frac{\partial \phi}{\partial s_i}$, $\ddot\phi_{ij} = \frac{\partial^2 \phi}{\partial s_i \partial s_j}$, etc. The following relations hold: $\partial_{s_i} \rho = \rho_1^i$ and $\partial^2_{s_i s_j} \rho = \partial_{s_j} \rho_1^i = -s_k \partial_{s_j} \rho_0^{ik}$. Assume throughout the section that $\rho > 0$, thus $\phi - \sum_i s_i \dot\phi_i > 0$, otherwise metric in (2.77) below is not positive definite for small s_i.

Proposition 2.92. *For the $(\alpha,\boldsymbol{b})$-norm, the bilinear form g_y $(y \ne 0)$ in (1.18) is*

$$\begin{aligned} g_y(u,v) = {} & \rho \langle u,v \rangle + \rho_0^{ij} \beta_i(u) \beta_j(v) + \rho_1^i (\beta_i(u)\langle y,v \rangle + \beta_i(v)\langle y,u \rangle)/\alpha(y) \\ & - \rho_1^i \beta_i(y)\langle y,u \rangle \langle y,v \rangle/\alpha^3(y). \end{aligned} \tag{2.77}$$

The Cartan torsion of the $(\alpha, \boldsymbol{b})$-norm is expressed by

$$2C_y(u,v,w) = \left(K_y(u,v)\langle \tilde{p}_y, w\rangle + K_y(v,w)\langle \tilde{p}_y, u\rangle + K_y(w,u)\langle \tilde{p}_y, v\rangle\right)/\alpha(y)$$

$$+ (1/\alpha(y)) \sum_{i,j,k} (\dot{\phi}_i \ddot{\phi}_{jk} + \dot{\phi}_j \ddot{\phi}_{ik} + \dot{\phi}_k \ddot{\phi}_{ij} + \phi \dddot{\phi}_{ijk}) \langle p_{yi}, u\rangle \langle p_{yj}, v\rangle \langle p_{yk}, w\rangle, \quad (2.78)$$

where $\tilde{p}_y = \rho_1^i p_{yi}$, $p_{yi} = (\beta_i^{\#} - s_i y)/\alpha(y)$ $(1 \le i \le p)$.

Proof. From (1.18) and (2.75) we find

$$g_y(u,v) = [F^2/2]_\alpha K_y(u,v)/\alpha(y) + [F^2/2]_{\alpha\alpha}\langle y, u\rangle\langle y, v\rangle/\alpha^2(y) \quad (2.79)$$

$$+ \sum_i ([F^2/2]_{\alpha\beta_i}/\alpha(y))(\langle y, u\rangle\beta_i(v) + \langle y, v\rangle\beta_i(u)) + \sum_{i,j} [F^2/2]_{\beta_i\beta_j}\beta_i(u)\beta_j(v).$$

Calculating derivatives of $\frac{1}{2}F^2 = \frac{1}{2}\alpha^2\phi^2(\beta_1/\alpha, \dots, \beta_p/\alpha)$,

$$[F^2/2]_\alpha = \alpha\rho, \quad [F^2/2]_{\beta_i} = \alpha\phi\dot{\phi}_i, \quad [F^2/2]_{\alpha\beta_i} = \rho_1^i, \quad [F^2/2]_{\beta_i\beta_j} = \rho_0^{ij},$$

$$[F^2/2]_{\alpha\alpha} = \rho + (\sum_i s_i \dot{\phi}_i)^2 + \phi \sum_{i,j} s_i s_j \ddot{\phi}_{ij} \quad (2.80)$$

and comparing (2.77) and (2.79), completes the proof of (2.77). Recall the classical *Euler Theorem*: if $H(z^1, \dots, z^q)$ is a positively homogeneous of degree r function then $\sum_k H_{z^k} z^k = rH$. The 0-homogeneity of $[\bar{F}^2/2]_{\mu\nu}$ in variables $F, \{\beta_k\}$ yields $[\bar{F}^2/2]_{F\mu\nu} F + \sum_k [\bar{F}^2/2]_{\beta_k\mu\nu} \beta_k = 0$ for $\mu, \nu \in \{F, \beta_1, \dots, \beta_p\}$; hence,

$$[\bar{F}^2/2]_{F\beta_i\beta_j} = -\sum_k (\beta_k/F) [\bar{F}^2/2]_{\beta_i\beta_j\beta_k},$$

$$[\bar{F}^2/2]_{FF\beta_i} = \sum_{j,k} (\beta_j/F)(\beta_k/F) [\bar{F}^2/2]_{\beta_i\beta_j\beta_k},$$

$$[\bar{F}^2/2]_{FFF} = -\sum_{i,j,k} (\beta_i/F)(\beta_j/F)(\beta_k/F) [\bar{F}^2/2]_{\beta_i\beta_j\beta_k}.$$

Using this, we calculate the Cartan torsion (1.19) of $(\alpha, \boldsymbol{b})$-norm as

$$2C_y(u,v,w) = \sum_i [F^2/2]_{\alpha\beta_i} \left(K_y(u,v)p_{yi}(w) + K_y(v,w)p_{yi}(u)\right.$$

$$\left. + K_y(w,u)p_{yi}(v)\right)/\alpha(y) + \sum_{i,j,k} [F^2/2]_{\beta_i\beta_j\beta_k} p_{yi}(u)p_{yj}(v)p_{yk}(w). \quad (2.81)$$

Then using equalities (2.80) and

$$[F^2/2]_{\beta_i\beta_j\beta_k} = (\dot{\phi}_i \ddot{\phi}_{jk} + \dot{\phi}_j \ddot{\phi}_{ik} + \dot{\phi}_k \ddot{\phi}_{ij} + \phi \dddot{\phi}_{ijk})/\alpha(y),$$

and comparing (2.81) and (2.78) completes the proof of (2.78). $\qquad \square$

One may show that in coordinate form, (2.77) reads as

$$g_{kl} = \rho a_{kl} + \rho_0^{ij}\beta_{ki}\beta_{lj} + \rho_1^i(\beta_{ki}\alpha_l + \beta_{li}\alpha_k) - s_i\rho_1^i\alpha_k\alpha_l,$$

where $\alpha_k = a_{km}y^m$, and for $p = 1$ this reduces to the (α, β)-metric (1.22).

Note that if $s_i = 0$ $(1 \le i \le p)$ then $\rho = 1$. By Proposition 2.92, g_y in (2.77) (for small s_i and $\rho > 0$) can be viewed as a perturbed scalar product $\langle \cdot, \cdot \rangle$.

Proposition 2.93. *For any smooth positive function ϕ defined in a neighborhood of the origin O of $\mathbb{R}^p$ there exists $\delta > 0$ such that any $p < n$ linear independent 1-forms $\beta_1, \ldots, \beta_p$ of α-norm less than δ determine by (2.75) the Minkowski norm F.*

Proof. The formula (2.77) shows that the inner products g_y for $\alpha(y) = 1$, depend uniformly of the 1-forms β_i $(i = 1, \ldots, p)$. For $\beta_1 = \ldots = \beta_p = 0$, $g_y = \rho \langle \cdot, \cdot \rangle$ is positive definite. Compactness of the α-unit sphere implies the statement. $\square$

Using [JS, Theorem 4.1], one can obtain a more accurate than Proposition 2.93 criterion for positive definiteness of g_y.
The following lemma can be used to examine regularity conditions of F in (2.75).

Lemma 2.94. *Given $c = (c_0, \ldots, c_p) \in \mathbb{R}^p$, linearly independent vectors $b^0, \ldots, b^p$ in $\mathbb{R}^n$ $(n > p)$ and reversible symmetric $n \times n$ matrix $a = (a_{ij})$, define $g_{ij} = a_{ij} + \sum_{k=0}^p c_k b_i^k b_j^k$. Then (see σ_λ in Section 1.5.3)*

$$\det g_{ij} = (\det a_{ij}) \sum_{|\lambda| \le p+1} \sigma_\lambda (a^{-1}(b^0) \otimes (b^0)^\flat, \ldots, a^{-1}(b^p) \otimes (b^p)^\flat) c^\lambda.$$

If vectors $b^1, \ldots, b^p$ are mutually a-orthogonal and $c_0 \|b^0\|_g^2 \ne -1$, then

$$\det g_{ij} = (\det a_{ij})\left(1 + c_0 \|b^0\|_a^2\right) \prod_{k=1}^p \left(1 + c_k\left(\|b^k\|_a^2 - \frac{c_0}{1 + c_0\|b^0\|_a^2} \langle b^0, b^k \rangle_a^2\right)\right).$$

Proof. Since rank of the matrix $\bar{g} - g = (\bar{g}_{ij} - g_{ij})$ does not exceed $p + 1$, the inequality $\mathrm{rank}(AB) \le \min\{\mathrm{rank}(A), \mathrm{rank}(B)\}$ (for ranks of arbitrary square matrices) shows that (1.83) reduces in our situation to the first equality of our Lemma. This way we get the first claim. If the vectors $b^0, \ldots, b^p$ are mutually g-orthogonal, then from Lemma 1.31 by induction we get

$$\det \bar{g}_{ij} = (\det g_{ij}) \prod_{k=0}^p (1 + c_k \|b^k\|_g^2).$$

The second claim follows from the above. $\square$

We restrict ourselves to regular (α, b)-norms alone, that is we assume that $\det g_y \ne 0$ $(y \ne 0)$. Let $\{e_1, \ldots, e_n\}$ be a basis of $\mathbb{R}^n$. A scalar product (metric) a on $\mathbb{R}^n$ and similarly, the metric g_y for any $y \ne 0$, define volume forms by $d\,\mathrm{vol}_a(e_1, \ldots, e_n) = \sqrt{\det b_{ij}}$ and $d\,\mathrm{vol}_{g_y}(e_1, \ldots, e_n) = \sqrt{\det g_y(e_i, e_j)}$. Then $d\,\mathrm{vol}_{g_y} = \mu_{g_y}(y)\,d\,\mathrm{vol}_a$ for some function $\mu_{g_y}(y) > 0$. Let $q_k = (q_k^1, \ldots, q_k^p) \in \mathbb{R}^p$ be unit eigenvectors with eigenvalues λ^k of the matrix $\{\rho_0^{ij} + \varepsilon^{-1}\rho_1^i\rho_1^j\}$. Thus, (q_k^i) is an orthogonal $p \times p$ matrix. Define vectors $\tilde{\beta}_k = q_k^i \beta_i$ $(1 \le k \le p)$. Set

$$\tilde{Y} = \varepsilon^{-1}\rho_1^i \beta_i - y^\flat/\alpha(y), \quad \varepsilon = s_j\rho_1^j. \tag{2.82}$$

Proposition 2.95. *Let* $\beta_1, \ldots, \beta_p$ *be a-orthogonal 1-forms and* $\varepsilon \|\tilde{Y}\|_{g_y}^2 \neq \rho$. *Then*

$$\mu_{g_y} = \rho^n (\rho - \varepsilon \|\tilde{Y}\|_a^2) \prod_{k=1}^{p} \left(1 + (\lambda^k / \rho)(\|\tilde{\beta}_k\|_a^2 + \frac{\varepsilon}{\rho - \varepsilon \|\tilde{Y}\|_a^2} \langle \tilde{Y}, \tilde{\beta}_k \rangle_a^2) \right).$$

Proof. The 1-forms $\tilde{\beta}_k$ are mutually a-orthogonal, and (2.77) takes the form

$$g_y(u, v) = \rho \langle u, v \rangle + \sum_{k=1}^{p} \lambda^k \tilde{\beta}_k(u) \tilde{\beta}_k(v) - \varepsilon \tilde{Y}(u) \tilde{Y}(v). \qquad (2.83)$$

From (2.83) and Lemma 2.94 with $c_0 = -\varepsilon/\rho$, $c_k = \lambda^k/\rho$ the claim follows. $\qquad \square$

Definition 2.96. Let (M^n, a) $(n \geq 2)$ be a Riemannian manifold. A *general* (α, b)-*metric F* on M (with pointwise linearly independent 1-forms β_i, $1 \leq i \leq p$) is a family of (α, b)-norms F_x in tangent spaces $T_x M$ depending smoothly on $x \in M$.

The reader might be interested to study the geometry (e.g., the bounds of curvature, and totally geodesic submanifolds) a sphere S^{m+1} endowed with a general (α, b)-metric and to compare with the construction of Berger's spheres.

2.5.2 The Modified Scalar Product

Let W be a hyperplane in a vector space V^{m+1} endowed with a Euclidean scalar product $\langle \cdot, \cdot \rangle$, and N a unit normal to W, i.e., $\langle N, v \rangle = 0$ $(v \in W)$ and $\langle N, N \rangle = 1$. Let $\boldsymbol{b} = \{\beta_1, \ldots, \beta_p\}$ be a set of p linearly independent 1-forms on V. If $W \neq \ker \beta_i$ $(1 \leq i \leq p)$, then $\beta_i^{\sharp \top} \neq 0$ (the projection of $\beta_i^{\sharp}$ onto W) and $|\beta_i(N)| < b_i$.

Remark 2.97. One may show that for any Minkowski norm on V, there are two normal directions to W, opposite when F is reversible, see Figure 2.1.

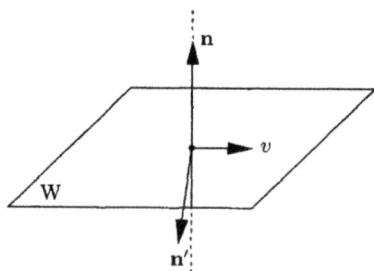

Fig. 2.1 Two directions **n** and **n**′ normal to W

By Remark 2.97, there exists a unique α-unit vector $n \in V$, which is g_n-orthogonal to W and lies in the same half-space as N:

$$g_n(n, v) = 0 \quad (v \in W), \quad \alpha(n) = 1, \quad \langle n, N \rangle > 0.$$

Below, in all expressions with s_i, ϕ and ρ's we assign the following values to s_i:

$$s_i = \beta_i(n), \quad 1 \le i \le p.$$

Observe that $v = F(n)^{-1} n$ is a g_n-unit normal to W, where $F(n) = \alpha \phi(s)$ with $s = (s_1, \dots, s_p)$. Put $g := g_n$. Thus, $g(n, n) = \phi^2(s)$ and, see (2.77) with $y = n$,

$$g(u, v) = \rho \langle u, v \rangle + \rho_0^{ij} \beta_i(u) \beta_j(v) + \rho_1^i (\beta_i(u) \langle n, v \rangle + \beta_i(v) \langle n, u \rangle) - \rho_1^i s_i \langle n, u \rangle \langle n, v \rangle. \tag{2.84}$$

Define the quantities (functions for a *general* (α, b)-*metric on* M)

$$\begin{aligned}
\gamma_1^i &= (\rho_1^i + \rho_0^{ij} s_j)/\rho = \dot{\phi}_i/(\phi - \textstyle\sum_j \dot{\phi}_j s_j) \quad (1 \le i \le p), \\
\gamma_2^{ij} &= \rho_0^{ij} - \gamma_1^i \rho_1^j - \gamma_1^j \rho_1^i - \gamma_1^i \gamma_1^j \rho_1^k s_k \quad (1 \le i, j \le p), \\
c_1 &= \gamma_1^i \beta_i(N) + (1 - \gamma_1^i \gamma_1^j b_{ij}^\top)^{1/2},
\end{aligned} \tag{2.85}$$

where $b_{ij}^\top := \langle \beta_i^\top, \beta_j^\top \rangle = b_{ij} - \beta_i(N) \beta_j(N)$. In what follows, we will assume that

$$b_{ij}^\top \gamma_1^i \gamma_1^j \le 1, \tag{2.86}$$

which becomes trivial for the Randers norm, see Example 2.101(i). By (2.86), c_1 is real. We will express n through the a-normal N to W and the quantities (2.85).

Lemma 2.98. *Let* (2.86) *be satisfied. Then*

$$n = c_1 N - \gamma_1^i \beta_i^\sharp, \tag{2.87}$$

$$g(u, v) = \rho \langle u, v \rangle + \gamma_2^{ij} \beta_i(u) \beta_j(v) \quad (u, v \in W); \tag{2.88}$$

moreover, the values $s_i = \beta_i(n)$ *can be found from the system*

$$s_i = c_1 \beta_i(N) - \gamma_1^j b_{ij} \quad (1 \le i \le p). \tag{2.89}$$

Proof. From (2.84) with $u = n$ and $v \in W$ and $g(n, v) = 0$ we find

$$\langle \rho n + \gamma_1^i \beta_i^\sharp, v \rangle = 0 \quad (v \in W). \tag{2.90}$$

From (2.90) and $\rho > 0$ we obtain $\rho n + \gamma_1^i \beta_i^{\sharp\top} = c_1 N$ for some real c_1. Using

$$1 = \langle n, n \rangle = c_1^2 - 2 c_1 \gamma_1^i \beta_i^\sharp + \gamma_1^i \gamma_1^j b_{ij}^\top,$$

and (2.86), we conclude that the above quadratic equation for c_1 has two real roots

$$(c_1)_\pm = \gamma_1^i \beta_i(N) \pm (1 - \gamma_1^i \gamma_1^j b_{ij}^\top)^{1/2}.$$

The value $c_1 := (c_1)_+$ provides inequality $\langle n, N \rangle > 0$, that proves (2.87). Thus, we get (2.89):

$$s_i = \beta_i(n) = \beta_i(c_1 N - \gamma_1^j \beta_j^\#) = c_1 \beta_i(N) - \gamma_1^j b_{ij} \quad (1 \le i \le p).$$

Finally, (2.88) follows from (2.84), (2.87) and $\langle n, u \rangle = -\gamma_1^j \beta_i(u)$ with $u \in W$. $\quad\square$

Remark 2.99. An interesting particular case appears when all vectors $\beta_i^\#$ belong to W, that is $\beta_i(N) = 0$ $(1 \le i \le p)$. Then, the system (2.89) is simplified to the form

$$\sum_i \dot{\phi}_i / \phi \, (b_{ij} - s_i s_j) = -s_j \quad (1 \le j \le p), \tag{2.91}$$

from which all $\dot{\phi}_i$ can be expressed through ϕ and $\{s_i\}$.

Let P be the matrix with elements $P_k^j = \gamma_2^{ij} b_{ik}^\top$. If γ_2^{ij} are "small" relative to $\rho > 0$, then the matrix $Q = \rho \, \mathrm{id} + P$ is non-singular, i.e.,

$$\det Q = \det[\rho \, \delta_k^j + P_k^j] \ne 0. \tag{2.92}$$

The condition (2.92) is trivial for Randers norm, see Example 2.101(i). Using the inverse matrix Q^{-1}, define the quantities:

$$\gamma_3^{ij} = -\gamma_2^{kj} (Q^{-1})_k^i \quad (1 \le i, j \le p). \tag{2.93}$$

Lemma 2.100. *Let* (2.86) *and* (2.92) *be satisfied, and the vectors u, U belong to W and obey*

$$g(u, v) = \langle U, v \rangle, \quad v \in W.$$

Then

$$\rho \, u = U + \gamma_3^{ij} \beta_i(U) \beta_j^{\#\top}. \tag{2.94}$$

Proof. By (2.88),

$$g(u, v) = \langle \rho \, u + \gamma_2^{ij} \beta_i(u) \beta_j^\#, v \rangle, \quad u, v \in W.$$

By conditions, and since $U, \beta_j^{\#\top} \in W$, we find

$$\rho \, u + \gamma_2^{ij} \beta_i(u) \beta_j^{\#\top} = U.$$

Applying β_k and using $\beta_k(\beta_j^{\#\top}) = b_{jk}^\top$ yields

$$(\rho \, \delta_k^j + P_k^j) \beta_j(u) = Q_k^j \beta_j(u) = \beta_k(U), \quad 1 \le k \le p,$$

and then (2.94). $\quad\square$

Example 2.101. Suppose that $p = 1$. For a hyperplane $W \subset V$ we have $s = \beta(n)$ and

$$c_1 = \gamma_1 \beta(N) + (1 - \gamma_1^2(b^2 - \beta(N)^2))^{1/2},$$

$$\gamma_1 = (\rho_1 + \rho_0 \beta(n))/\rho = \dot{\phi}/(\phi - s\dot{\phi}), \quad \gamma_3 = -\frac{\gamma_2}{\rho + (b^2 - \beta(N)^2)\gamma_2},$$

$$\gamma_2 = \rho_0 - \gamma_1 \rho_1(\beta(n)\gamma_1 + 2) = \phi(\phi^2\ddot{\phi} - \phi\dot{\phi}^2 + s\dot{\phi}^3)/(\phi - s\dot{\phi})^2,$$

where $b := \alpha(\beta) < 1$. Then (2.89) reduces to

$$\frac{\dot{\phi}}{\phi} = -\frac{s\sqrt{b^2 - s^2} + \beta(N)\sqrt{b^2 - \beta(N)^2}}{(b^2 - s^2 - \beta(N)^2)\sqrt{b^2 - s^2}},$$

which for $\beta^\sharp \in W$ reads $\dot{\phi}/\phi = -s/(b^2 - s^2)$, see (2.91) for $p = 1$.

We will illustrate the above metric g on V using some of (α, β)-norms.

(i) Consider *Randers norms*, i.e., $\phi(s) = 1 + s$, $|s| < b < 1$. For a hyperplane $W \subset V$ and $g = g_n$ we get

$$n = c_1 N - \beta^\sharp, \quad s = \beta(n) = cc_1 - 1, \quad \phi(s) = cc_1,$$

where $c_1 = c + \beta(N)$ and $c = \sqrt{1 - b^2 + \beta(N)^2} \in (0, 1]$. Then $\gamma_1 = 1$, $\gamma_2 = -cc_1$ and $\gamma_3 = c^{-2}$. Conditions (2.86) and (2.92) become trivial. Next, $\mu_g(n) = (cc_1)^{m+2}$ and

$$g(u, v) = (1 + s)\langle u, v\rangle - s\langle n, u\rangle\langle n, v\rangle + \beta(u)\langle n, v\rangle + \beta(v)\langle n, u\rangle + \beta(u)\beta(v).$$

(ii) Consider *generalized Kropina norms*, i.e., $\phi(s) = 1/s^l$ $(0 < s < b)$. For a hyperplane $W \neq \ker\beta$ in V and $g = g_n$ we get

$$c_1 = (b - 2\beta(N))/\sqrt{2b(b - \beta(N))}, \quad \beta(n) = s = \sqrt{b(b - \beta(N))/2},$$
$$\gamma_1 = -1/(2s) = -1/\sqrt{2b(b - \beta(N))}, \quad \gamma_2 = \gamma_3 = 0,$$

and $\mu_g(n) = \frac{4^{m+1}}{b^m(b - \beta(N))^{m+2}}$. Conditions (2.86) and (2.92) become trivial.

(iii) Consider *slope-norms*, i.e., $\phi(s) = \frac{1}{1-s}$ with $|s| < b < \delta_0 = \frac{1}{2}$. For a hyperplane $W \neq \ker\beta$ and $g = g_n$, by (2.89) we get $s = \beta(n)$ obeys 4th-degree equation

$$4s^4 - 4s^3 + (1 - 4b^2)s^2 + 2(b^2 + \beta(N)^2)s + b^4 - (b^2 + 1)\beta(N)^2 = 0,$$

and $s = \frac{1}{4}(1 - \sqrt{1 + 8b^2})$ if $\beta^\sharp \in W$, see (2.91). We find

$$\mu_g(n) = \frac{(1 - 2s)^{m-1}}{(1 - s)^{3m+3}}(2b^2 - 3s + 1), \quad c_1 = \frac{\beta(N) + \sqrt{(1 - 2s)^2 - b^2 + \beta(N)^2}}{1 - 2s},$$

$$\gamma_1 = \frac{1}{1 - 2s}, \quad \gamma_2 = \frac{1}{(1 - 2s)^2(1 - s)^3}, \quad \gamma_3 = \frac{1}{(1 - 2s)^3 + b^2 - \beta(N)^2}.$$

Thus, condition (2.92) becomes trivial and (2.86) reads as $(1-2s)^2 \geq b^2 - \beta(N)^2$.

Example 2.102. As for (i)–(iii), compute the quantities for the following norms.

(iv) A Minkowski norm is called a *polynomial* (α, β)-*norm* if $\phi(s) = \sum_{i=0}^{k} C_i s^i$, $C_0 = 1$, $C_k \neq 0$. Consider *quadratic Minkowski norms*, i.e., $\phi(s) = (1+s)^2$ with $|s| < b < \delta_0 = 1$. For a hyperplane $W \neq \ker \beta$ in V and $g = g_n$, from (2.89) we find that s obeys 4th-degree equation, see (2.91),

$$s^4 - 2s^3 + (1 - 4b^2 + 3\beta(N)^2)s^2 + 2(2b^2 - \beta(N)^2)s + 4b^4 - (4b^2+1)\beta(N)^2 = 0,$$

and $s = (1 - \sqrt{1 + 8b^2})/2$ if $\beta^\sharp \in W$. Then $c_1 = \frac{2\beta(N) + \sqrt{(1-s)^2 - 4(b^2 - \beta(N)^2)}}{1 - s}$ and

$$\gamma_1 = \frac{2}{1-s}, \quad \gamma_2 = \frac{2(3s-1)(1+s)^3}{(1-s)^2}, \quad \gamma_3 = \frac{2(3s-1)}{(1-s)^3 - 2(1-3s)^2(b^2 - \beta(N)^2)}$$

and $\mu_g(n) = (1+s)^{3m+3}(1-s)^{m-1}(2b^2 - 3s^2 + 1)$. The conditions (2.86) and (2.92) read as

$$(1-s)^2 \geq 4(b^2 - \beta(N)^2), \quad (1-s)^3 \neq 2(1-3s)(b^2 - \beta(N)^2).$$

(v) Consider *exponential norms*, i.e., $\phi(s) = e^{s/k}$, $|s| < b < \delta_0 := |k|$. For a hyperplane $W \neq \ker \beta$ in V and $g = g_n$, by (2.89), $s = \beta(n)$ obeys 4th-degree equation $s^4 - 2ks^3 + (k^2 - 2b^2 + \beta(N)^2)s^2 + 2b^2 ks + b^4 - (b^2 + k^2)\beta(N)^2 = 0$, and $s = (k - \sqrt{k^2 + 4b^2})/2$ if $\beta^\sharp$ is tangent to the foliation, see (2.91). Then we get

$$c_1 = \frac{\beta(N) + ((k-s)^2 - b^2 + \beta(N)^2)^{1/2}}{k-s},$$

$$\gamma_1 = \frac{1}{k-s}, \quad \gamma_2 = \frac{s\,e^{2s/k}}{k(k-s)^2}, \quad \gamma_3 = \frac{s}{(k-s)^3 + s(b^2 - \beta(N)^2)}.$$

and $\mu_g(n) = \frac{(k-s)^{m-1}}{k^{m+1}}(b^2 + k^2 - ks - s^2)\,e^{(2m+2)s/k}$. Conditions (2.86) and (2.92) read, respectively, $(k-s)^2 \geq b^2 - \beta(N)^2$ and $(k-s)^3 \neq -s(b^2 - \beta(N)^2)$. Figure 2.2 shows the dependence of $s = \beta(n)$ on $\beta(N) \in [-b, b]$ for above metrics (ii)–(v). For $\beta(N) = 0$ we get the values (a) 0.64, (b) -0.13, (c) -0.26, (d) -0.53 of s.

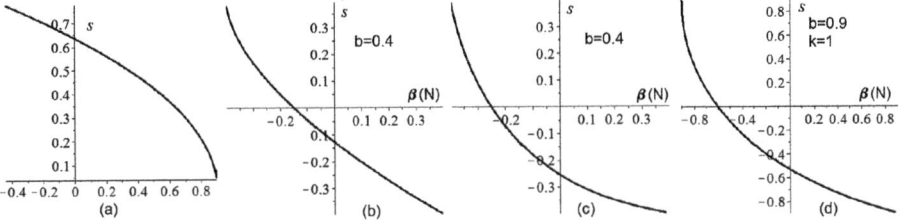

Fig. 2.2 Dependence of $s = \beta(n)$ on $\beta(N)$ for the norms: (**a**) Kropina, (**b**) Matsumoto, (**c**) quadratic, and (**d**) exponential

Example 2.103.

(i) Using Proposition 2.95 for $p = 2$, we find $\mu_g(y)$ and $\langle \beta_1, \beta_2 \rangle = 0$. By (2.83), using $\tilde{Y}$ of (2.82), $\tilde{\beta}_i = q_i^1 \beta_1 + q_i^2 \beta_2$ and $\varepsilon = \rho_1^1 s_1 + \rho_1^2 s_2$, we get

$$\mu_{g_y} = \rho^{m+1}\left(1 - \frac{\varepsilon}{\rho}|\tilde{Y}|_a^2\right)\left(1 + (\lambda^1/\rho)\left(|\tilde{\beta}_1|_a^2 + \frac{\varepsilon}{\rho - \varepsilon\|\tilde{Y}\|_a^2}\langle\tilde{Y}, \tilde{\beta}_1\rangle_a^2\right)\right)$$

$$\left(1 + (\lambda^2/\rho)\left(|\tilde{\beta}_2|_a^2 + \frac{\varepsilon}{\rho - \varepsilon\|\tilde{Y}\|_a^2}\langle\tilde{Y}, \tilde{\beta}_2\rangle_a^2\right)\right).$$

(ii) The shifted (α, β)-norms are a particular case of $(\alpha, (\beta_1, \beta_2))$-norms. For shifted Kropina norms $F = \alpha(1 + \frac{\alpha}{\beta_1} + \frac{\beta_2}{\alpha})$, hence $\phi = 1 + \frac{1}{s_1} + s_2$ for $s_1 > 0$, we have

$$\rho = (2+s_1)(1+s_1+s_1s_2)/s_1^2, \quad \rho_1^1 = -(4+3s_1+2s_1s_2)/s_1^3, \quad \rho_1^2 = (2+s_1)/s_1,$$
$$\rho_0^{11} = (3+2s_1+2s_1s_2)/s_1^4, \qquad \rho_0^{12} = \rho_0^{21} = -1/s_1^2, \qquad \rho_0^{22} = 1.$$

For a hyperplane $W \neq \ker \beta_i$ $(i = 1, 2)$ in V and the metric $g = g_n$ we get

$$c_1 = \frac{s_1^2 \beta_2(N) - \beta_1(N)}{s_1(2+s_1)}$$

$$+ \left(1 - \frac{b_{11} - \beta_1(N)^2}{s_1^2(2+s_1)^2} + \frac{2(b_{12} - \beta_1(N)\beta_2(N))}{(2+s_1)^2} - \frac{s_1^2(b_{22} - \beta_2(N)^2)}{(2+s_1)^2}\right)^{1/2},$$

$$\gamma_1^1 = -\frac{1}{s_1(2+s_1)}, \quad \gamma_1^2 = \frac{s_1}{2+s_1},$$

$$\gamma_2^{11} = -\frac{2 - s_1 - 10s_1^2 - 10s_1^3 - 3s_1^4 - s_1^2 s_2(2 - 2s_1^2 - s_1^3)}{s_1^4(2+s_1)},$$

$$\gamma_2^{12} = \frac{12 + 13s_1 + 3s_1^2 + s_1 s_2(2 - 2s_1 - s_1^2)}{s_1^2(2+s_1)}, \quad \gamma_2^{22} = \frac{4 + 3s_1 - s_1^2(1+s_2)}{s_1^2}.$$

If $\beta_i^\sharp \in W$ (and $\beta_1 \perp \beta_2$), then $s_1 = \beta_1(n)$ and $s_2 = \beta_2(n)$ satisfy

$$(1+2s_2)s_1^3 + b_{11} = 0, \quad (1+2s_1)s_1s_2^2 - b_{22}s_1^2 + b_{12} = 0.$$

Thus $s_1 = \left(\frac{b_{11}}{4b_{22}}(1+\sqrt{1+8b_{22}})\right)^{1/3}$ and $s_2 = \frac{1}{2}(b_{11}/s_1^3 - 1)$.

(iii) As in the case (i), for shifted slope-norms $F = \alpha\left(\frac{\alpha}{\alpha-\beta_1} + \frac{\beta_2}{\alpha}\right)$, hence $\phi = \frac{1}{1-s_1} + s_2$ with $\delta_i < 1$ we have

$$\rho = \frac{(1-2s_1)(1+s_2-s_1s_2)}{(1-s_1)^3}, \quad \rho_1^1 = \frac{1+2s_1(s_1s_2-s_2-2)}{(1-s_1)^4}, \quad \rho_1^2 = \frac{1-2s_1}{(1-s_1)^2},$$
$$\rho_0^{11} = (3-2s_1s_2+2s_2)/(1-s_1)^4, \quad \rho_0^{12} = \rho_0^{21} = 1/(1-s_1)^2, \quad \rho_0^{22} = 1.$$

For a hyperplane $W \neq \ker\beta_i$ $(1 \leq i \leq p)$ in V and the metric $g = g_n$ we get

$$c_1 = \frac{(1-s_1)^2\beta_2(N) + \beta_1(N)}{1-2s_1} + \Big(1 - \frac{(1-s_1)^4(b_{22} - \beta_2(N)^2)}{(1-2s_1)^2}$$
$$- \frac{2(1-s_1)^2(b_{12} - \beta_1(N)\beta_2(N))}{(1-2s_1)^2} - \frac{b_{11} - \beta_1(N)^2}{(1-2s_1)^2}\Big)^{1/2},$$

$$\gamma_1^1 = \frac{1}{1-2s_1}, \quad \gamma_1^2 = \frac{(1-s_1)^2}{1-2s_1}, \quad \gamma_2^{11} = \frac{1+2s_1+8s_1^2 + s_2(1+5s_1-6s_1^2)}{(1-s_1)^3(1-2s_1)},$$

$$\gamma_2^{22} = -\frac{1-3s_1+2s_1^2-4s_1^3+s_1^4+s_2(1-4s_1+3s_1^2)}{(1-s_1)^4},$$

$$\gamma_2^{12} = -\frac{1-5s_1+3s_1^2+4s_1^3+s_2(1-8s_1+17s_1^2-12s_1^3+2s_1^4)}{(1-2s_1)(1-s_1)^4}.$$

If $\beta_i^\sharp \in W$ (and $\beta_1 \perp \beta_2$), then s_1 and s_2 obey the system

$$b_{11} - 2s_1s_2(1-s_1)^2 = s_1, \quad (1-s_1)^2(b_{22} - 2s_2^2) = s_2.$$

Then $s_1 = -(2b_{11}s_2^2 - b_{11}b_{22})/b_{22}$, where s_2 is a root of a 6th-degree polynomial.

Similarly to graphs on Figure 2.2, one may graph surfaces in $\mathbb{R}^3$, showing dependence of $s_1 = \beta_1(n)$ and $s_2 = \beta_2(n)$ on $\beta_1(N)$ and $\beta_2(N)$ for the shifted (α, β)-metrics. For $\beta_i(N) = 0$ we get the following values: $s_1 \approx -0.79$ and $s_2 = -1.5$ for Kropina norm; and $s_1 \approx -0.42$ and $s_2 = s_1^3 - 2s_1^2 + s_1 \approx -0.84$ for slope-norm.

2.5.3 The Shape Operator

Let $(M^{m+1}, a = \langle\cdot,\cdot\rangle)$ $(m \geq 2)$ be a connected Riemannian manifold with the Levi-Civita connection $\bar\nabla$. Let N be a unit normal vector field to a codimension-one foliation $\mathscr{F}$ on (M,a). Due to Section 2.5.2, there exists a g_n-normal (to $\mathscr{F}$) vector field n such that $\langle n,N \rangle > 0$ and $\langle n,n \rangle = 1$. Define a new Riemannian metric $g := g_n$ on M, see (2.84), with the Levi-Civita connection ∇. Suppose that $\ker \beta_i \neq \mathscr{D}$ everywhere for all i, hence $|\beta_i(N)| < \sqrt{b_{ii}}$. By (2.89), $s_i = \beta_i(n)$ are smooth functions on M, and $v = n/\phi$ is a g-unit normal to the leaves.

The shape operators $\overline{A}$ and A^g of $\mathscr{D} = T\mathscr{F}$ and the curvature vectors of v- and N- curves for both metrics $\langle \cdot, \cdot \rangle$ and g are defined by

$$\overline{A}(u) = -\overline{\nabla}_u N, \quad A^g(u) = -\nabla_u v \quad (u \in \mathscr{D}), \tag{2.95}$$

$$Z = \nabla_v v, \qquad \overline{Z} = \overline{\nabla}_N N. \tag{2.96}$$

Let $\overline{T}^\sharp : \mathscr{D} \to \mathscr{D}$ be a linear operator a-adjoint to the integrability tensor $\overline{T}$ of $\mathscr{D}$,

$$2\overline{T}(u,v) = \langle [u,v], N \rangle \quad (u,v \in \mathscr{D}).$$

Let $\overline{\mathrm{Def}}_u$ be the deformation tensor of u, see (1.36). We express A^g through $\overline{A}$ and invariants of $\mathscr{D}$.

Proposition 2.104 (The Shape Operator). *Let $\beta_1, \ldots \beta_p$ be linearly independent 1-forms on (M^{m+1}, a) obeying (2.86) and (2.92), and let ϕ be a smooth function on $M \times \mathbb{R}^p$. Then*

$$\rho \, \phi A^g = -\mathscr{A} - \gamma_3^{ij} (\beta_i \circ \mathscr{A}) \otimes \beta_j^{\sharp \top}, \tag{2.97}$$

where γ_3^{ij} are defined in (2.93) and the linear operator $\mathscr{A} : \mathscr{D} \to \mathscr{D}$ is given by

$$\mathscr{A} = -\rho \, c_1 \overline{A} - \rho \gamma_1^i (\overline{\mathrm{Def}}_{\beta_i^\sharp})^\top + \frac{1}{2} n(\rho) \, \mathrm{id}^\top + (U^j \odot \beta_j^\top), \tag{2.98}$$

and the vector fields U^j are given by

$$U^j = \frac{1}{2} \left(n(\gamma_2^{ij}) \beta_i^{\sharp \top} + \gamma_2^{ij} (\overline{\nabla}_n \beta_i^{\sharp \top})^\top \right) - \rho \, (\overline{\nabla} \gamma_1^j)^\top$$
$$+ (\rho_0^{ij} - \gamma_1^j \rho_1^i) \left(\beta_i(N) (\overline{\nabla} c_1)^\top - (\gamma_1^k/2)(\overline{\nabla} b_{ik})^\top - b_{ik} (\overline{\nabla} \gamma_1^k)^\top \right)$$
$$+ (c_1 - \beta_k(N) \gamma_1^k) \left((\rho_0^{ij} - \gamma_1^j \rho_1^i) \beta_i(N) + c_1 \rho_1^j (1 + s_k \gamma_1^k) \right) \overline{Z}$$
$$+ \left(c_1 \rho_1^i (1 + s_k \gamma_1^k) \gamma_1^j - (\rho_0^{ij} - \gamma_1^j \rho_1^i)(c_1 - \beta_k(N) \gamma_1^k) \right) \overline{A}(\beta_i^{\sharp \top}). \tag{2.99}$$

Proof. By formula (1.9) for the Levi-Civita connection ∇ of g, we have

$$2g(\nabla_u n, v) = n(g(u,v)) + g([u,n], v) + g([v,n], u) - g([u,v], n) \tag{2.100}$$

for $u, v \in \mathscr{D}$. Assume $(\overline{\nabla}_X u)^\top = (\overline{\nabla}_X v)^\top = 0$ for $X \in T_x M$ at a given point $x \in M$. Using (2.84) and (2.88), we get

$$n(g(u,v)) = n(\rho \langle u,v \rangle) + n(\gamma_2^{ij} \beta_i(u) \beta_j(v))$$

$$= n(\rho)\langle u,v \rangle + \left[n(\gamma_2^{ij}) \beta_i(u) \beta_j(v) + \gamma_2^{ij} \left(\beta_i(u)(\overline{\nabla}_n(\beta_j^\top))(v) + \beta_i(v)(\overline{\nabla}_n(\beta_j^\top))(u) \right) \right],$$

$$g([u,v],n) = 2\rho\, c_1 \overline{T}(u,v),$$

$$g([u,n],v) = \rho \langle \overline{\nabla}_u n, v \rangle + \rho_0^{ij} \beta_i([u,n]) \beta_j(v) + \rho_1^i (\beta_i([u,n])\langle n,v \rangle + \beta_i(v)\langle n,[u,n] \rangle)$$

$$- \rho_1^i s_i \langle n,[u,n] \rangle \langle n,v \rangle,$$

where $u,v \in \mathscr{D}$. Using equalities $\langle n,v \rangle = -\gamma_1^i \beta_i(v)$ and

$$\langle \overline{\nabla}_u n, v \rangle = -\langle c_1 \overline{A}(u), v \rangle - \gamma_1^i \langle \overline{\nabla}_u \beta_i^\sharp, v \rangle - \beta_i(v) \langle \overline{\nabla} \gamma_1^i, u \rangle,$$

$$\beta_i([u,n]) = -\gamma_1^j \langle \overline{\nabla}_u \beta_j^\sharp, \beta_i^\sharp \rangle + \langle \beta_i(N) \overline{\nabla} c_1 - b_{ij} \overline{\nabla} \gamma_1^j$$

$$+ \beta_i(N) \left[(c_1 - \gamma_1^j \beta_j(N)) \overline{Z} + \gamma_1^j \overline{A}(\beta_j^{\sharp\top}) \right], u \rangle,$$

$$\langle n,[u,n] \rangle = \langle (c_1 - \gamma_1^j \beta_j(N)) \overline{\nabla} c_1 + (\gamma_1^i b_{ji} - c_1 \beta_j(N)) \overline{\nabla} \gamma_1^j$$

$$- c_1 \gamma_1^j \overline{\nabla}(\beta_j(N)) - c_1 \gamma_1^j \beta_j(N) \overline{Z}, u \rangle,$$

we then obtain

$$g([u,n],v) = -\rho\, c_1 \langle \overline{A}(u), v \rangle - \rho(\gamma_1^i \langle \overline{\nabla}_u \beta_i^\sharp, v \rangle + \beta_i(v) \langle \overline{\nabla} \gamma_1^i, u \rangle)$$

$$+ (\rho_0^{ij} - \rho_1^i \gamma_1^j)\langle \beta_i(N) \overline{\nabla} c_1 - (\tfrac{1}{2}\gamma_1^k \overline{\nabla} b_{ki} - b_{ik} \overline{\nabla} \gamma_1^k)$$

$$+ (c_1 - \beta_k(N)\gamma_1^k)(\beta_i(N)\overline{Z} - \overline{A}(\beta_i^{\sharp\top})), u \rangle \beta_j(v)$$

$$+ c_1 \rho_1^j (1 + s_k \gamma_1^k)\langle (c_1 - \beta_k(N)\gamma_1^k)\overline{Z} + \gamma_1^k \overline{A}(\beta_k^{\sharp\top}), u \rangle \beta_j(v),$$

where $u,v \in \mathscr{D}$. Formula for $g([v,n],u)$ is obtained from $g([u,n],v)$ after change $u \leftrightarrow v$. Substituting the above into (2.100), we find $g(\nabla_u n, v) = \langle \mathscr{A}(u), v \rangle$, where $\mathscr{A}$ is given in (2.98)–(2.99). In particular,

$$\langle 2\mathscr{A}(u), \beta_i^{\sharp\top} \rangle = -2\rho\, c_1 \langle \overline{A}(\beta_i^{\sharp\top}), u \rangle - 2\rho\, \gamma_1^j \langle \overline{\mathrm{Def}}_{\beta_i^\sharp}(\beta_j^\sharp), u \rangle$$

$$+ n(\rho)\beta_i(u) + \beta_j(u)\beta_i(U^j) + U^j(u)b_{ij}^\top.$$

By Lemma 2.100 and $g(\nabla_u n, v) = -\phi\, g(A^g(u), v)$, see (2.95), we get (2.97). $\qquad \square$

Corollary 2.105 (The Mean Curvature of $\mathscr{D}$). *Let conditions of Proposition 2.104 be satisfied. Then*

$$\rho\, \phi\, \sigma_1(A^g) = \rho\, c_1 \sigma_1(\overline{A}) - \frac{m}{2} n(\rho) + \rho\, \gamma_1^i (\overline{\mathrm{div}}\, \beta_i^\sharp - \beta_i(\overline{Z}) + N(\beta_i(N)))$$

$$- \beta_j(U^j) - \gamma_3^{ij} \langle \mathscr{A}(\beta_i^{\sharp\top}), \beta_j^\sharp \rangle, \tag{2.101a}$$

where U^j are given in (2.99) and

$$\langle \mathscr{A}(\beta_i^{\sharp\top}), \beta_j^\sharp \rangle = \rho\, c_1 \langle \overline{A}(\beta_i^{\sharp\top}), \beta_j^{\sharp\top} \rangle + \rho\, \gamma_1^k (\beta_k^{\sharp\top}(b_{ij}^\top)/2 - \beta_k(N)\langle \overline{A}(\beta_i^{\sharp\top}), \beta_j^\sharp \rangle)$$

$$- b_{ij}^\top (\tfrac{1}{2} n(\rho) + \beta_k(U^k)). \tag{2.101b}$$

Proof. Let $\{e_i\}$ be a local g-orthonormal frame of $\mathcal{D}$. We find, see (2.98)–(2.99),

$$\langle \overline{\mathrm{Def}}_{\beta_k^\sharp}(\beta_i^{\sharp\top}), \beta_j^{\sharp\top}\rangle = \frac{1}{2}\langle \overline{\nabla}b_{ij}^\top, \beta_k^{\sharp\top}\rangle - \beta_k(N)\langle \overline{A}(\beta_i^{\sharp\top}), \beta_j^\sharp\rangle.$$

Tracing (2.97) yields $\rho\phi\sigma_1(A^g) = -\sigma_1(\mathscr{A}) - \gamma_3^{jj}\langle \mathscr{A}(\beta_j^{\sharp\top}), \beta_i^\sharp\rangle$. Using Lemma 2.100, (2.101b) and $\mathrm{trace}(\overline{\mathrm{Def}}_{\beta_i^\sharp})^\top_{|T\mathscr{F}} = \overline{\mathrm{div}}\,\beta_i^\sharp - \beta_i(\overline{Z}) + N(\beta_i(N))$, we get (2.101a,b). □

In next corollary and proposition, we assume that $\mathcal{D}$ is integrable and $p = 1$.

Corollary 2.106 (The Second Mean Curvature). *If $p = 1$ and $\overline{\nabla}\beta^\sharp = 0$, then*

$$(\rho\phi)^2\sigma_2(A^g) = (\rho c_1)^2\sigma_2(\overline{A}) + \frac{1}{8}m(m-1)n(\rho)^2$$
$$-\frac{1}{2}(m-1)c_1\rho n(\rho)\sigma_1(\overline{A}) + \frac{1}{4}\beta(U)\langle 2\gamma_3\mathscr{A}(\beta^{\sharp\top}) + U, \beta^\sharp\rangle$$
$$-\frac{1}{4}(b^2 - \beta(N)^2)\langle 2\gamma_3\mathscr{A}(\beta^{\sharp\top}) + U, U\rangle + (\frac{m-1}{2}n(\rho)$$
$$-\rho c_1\sigma_1(\overline{A}))\langle\gamma_3\mathscr{A}(\beta^{\sharp\top}) + U, \beta^\sharp\rangle + \rho c_1\langle\gamma_3\mathscr{A}(\beta^{\sharp\top}) + U, \overline{A}(\beta^{\sharp\top})\rangle, \quad (2.102)$$

where $\mathscr{A} = -\rho c_1\overline{A} + (U\odot\beta^\top)$ and U is given in (2.99) with $j = 1$.

Proof. By our assumptions, $\overline{\mathrm{Def}}_{\beta^\sharp} = 0$. Thus, see Proposition 2.104,

$$\rho\phi A^g = \rho c_1\overline{A} - \frac{1}{2}n(\rho)\,\mathrm{id}^\top - A_1 - A_2,$$

where $A_1 = \frac{1}{2}U\otimes\beta^\top$ and $A_2 = (\frac{1}{2}U^b + \gamma_3(\beta\circ\mathscr{A}))\otimes\beta^{\sharp\top}$ are rank 1 matrices (thus $\sigma_2(A_i) = 0$) and $\mathscr{A} = -\rho c_1\overline{A} + \frac{1}{2}n(\rho)\,\mathrm{id}^\top + (U\odot\beta^\top)$ is symmetric. Applying the identity $\sigma_2(\sum_i P_i) = \sum_i\sigma_2(P_i) + \sum_{i<j}(\sigma_1(P_i)\sigma_1(P_j) - \sigma_1(P_iP_j))$, to matrices $P_1 = \rho c_1\overline{A}$, $P_2 = -\frac{1}{2}n(\rho)\,\mathrm{id}^\top$, $P_3 = -A_1$ and $P_4 = -A_2$, and using equalities $\langle(\beta\circ\mathscr{A})^\sharp, u\rangle = \langle\mathscr{A}(u^\top), \beta^\sharp\rangle$ and $\sigma_2(\mathrm{id}^\top) = m(m-1)/2$, we get

$$(\rho\phi)^2\sigma_2(A^g) = (\rho c_1)^2\sigma_2(\overline{A}) + m(m-1)n(\rho)^2/8$$
$$-\frac{1}{2}(m-1)c_1\rho n(\rho)\sigma_1(\overline{A}) + \sigma_1(A_1)\sigma_1(A_2) - \sigma_1(A_1A_2)$$
$$+((m-1)n(\rho)/2 - \rho c_1\sigma_1(\overline{A}))\sigma_1(A_1 + A_2) + \rho c_1\sigma_1(\overline{A}(A_1 + A_2)),$$

where $\sigma_1(A_1) = \beta(U)/2$, $\sigma_1(A_2) = \langle 2\gamma_3\mathscr{A}(\beta^{\sharp\top}) + U, \beta^\sharp\rangle/2$ and

$$\sigma_1(A_1A_2) = (b^2 - \beta(N)^2)\langle 2\gamma_3\mathscr{A}(\beta^\sharp) + U, U\rangle/4,$$
$$\sigma_1(\overline{A}(A_1 + A_2)) = \langle\gamma_3\mathscr{A}(\beta^{\sharp\top}) + U, \overline{A}(\beta^{\sharp\top})\rangle.$$

From the above (2.102) follows. □

Example 2.107. Let $\mathscr{F}$ be a codimension-one foliation of a Randers space. By Proposition 2.104, the shape operators of $\mathscr{F}$ satisfy the following, see calculations in [RW7, Proposition 1]:

$$c A^g = \overline{A} - \frac{1}{2} c^{-1} \hat{c}^{-2} (\hat{c} N - \beta^\sharp)(c \hat{c}) \, \mathrm{id}_m + \hat{c}^{-1} (\overline{\mathrm{Def}}_{\beta^\sharp})^\top_{|T\mathscr{F}}$$

$$+ \frac{1}{2} (U - \overline{A}(\beta^{\sharp\top})) \otimes \beta^\top + \frac{1}{2} c^{-2} \left(\overline{A}(\beta^{\sharp\top}) - \langle \overline{A}(\beta^{\sharp\top}), \beta^{\sharp\top} \rangle \beta^{\sharp\top} \right)$$

$$+ 2 \hat{c}^{-1} (\overline{\mathrm{Def}}_{\beta^\sharp} \beta^{\sharp\top})^\top + U + \beta(U) \beta^{\sharp\top} \big)^\flat \otimes \beta^{\sharp\top}, \tag{2.103}$$

where $U = \hat{c}^{-1} (\overline{\nabla}_{\hat{c} N - \beta^\sharp} \beta^{\sharp\top})^\top - c \overline{Z}$. Tracing (2.103), we obtain

$$c \, \sigma_1(A^g) = \sigma_1(\overline{A}) - \frac{m}{2} c^{-1} \hat{c}^{-2} (\hat{c} N - \beta^\sharp)(c \hat{c}) + \hat{c}^{-1} (\overline{\mathrm{div}} \, \beta^\sharp + \beta(\overline{Z}) - N(\beta(N)))$$

$$+ \frac{1}{2} (\beta(U) - \langle \overline{A}(\beta^{\sharp\top}), \beta^\sharp \rangle) + \frac{1}{2} c^{-2} (c^2 \langle \overline{A}(\beta^{\sharp\top}), \beta^\sharp \rangle$$

$$- 2 \hat{c}^{-1} (c \beta^{\sharp\top}(c) + \beta(N) \langle \overline{A}(\beta^{\sharp\top}), \beta^\sharp \rangle) + (2 - c^2) \beta(U))$$

$$= \sigma_1(\overline{A}) - \frac{m}{2} c^{-1} \hat{c}^{-2} (\hat{c} N - \beta^\sharp)(c \hat{c}) + \hat{c}^{-1} \overline{\mathrm{div}} \, \beta^\sharp - \hat{c}^{-1} N(\hat{c})$$

$$- (\hat{c} - c)(c \hat{c})^{-1} N(c) - (\hat{c} - c) c^{-2} \hat{c}^{-1} \langle \overline{A}(\beta^{\sharp\top}) + c \overline{Z}, \beta^\sharp \rangle.$$

Remark 2.108. The expression of Z through $\overline{Z}$, see (2.96), is long, we omit it here. For a codimension one foliation of a Randers space, see [RW7, Proposition 2],

$$Z = (c \hat{c})^{-1} \overline{Z} - c^{-1} \hat{c}^{-2} (\overline{\nabla} \hat{c})^\top + c^{-3} \hat{c}^{-1} \beta (\overline{Z} - \hat{c}^{-1} (\overline{\nabla} \hat{c})^\top) \beta^{\sharp\top}.$$

2.5.4 Around the Reeb Integral Formula and Its Counterpart

In this section we apply results of Sections 1.2.2 and 2.5.3 to prove an integral formula for a closed Riemannian manifold with a set of linearly independent 1-forms and a codimension-one distribution, which generalizes the integral formula (2.1).

Theorem 2.109. *Let g be a new Riemannian metric determined by a codimension-one distribution $\mathscr{D}$, 1-forms β_i $(1 \leq i \leq p)$ and a function $\phi(s)$, where $s = (s_1, \ldots, s_p)$, on a closed Riemannian manifold (M, a) with conditions (2.86) and (2.92). Then*

$$\int_M \mu_{g_n} (\rho \, \phi)^{-1} \{ \rho \, c_1 \sigma_1(\overline{A}) - (m/2) n(\rho) + \rho \, \gamma_1^i (\beta_i(\overline{Z}) - N(\beta_i(N))) + \beta_i(U^i)$$

$$- \gamma_3^{ij} \langle \mathscr{A}(\beta_i^{\sharp\top}), \beta_j^{\sharp\top} \rangle - \rho \, \phi (\beta_i^\sharp(\gamma_1^i \phi) + \gamma_1^i \phi \beta_i^\sharp(\log \mu_{g_n})) \} \, \mathrm{d} \, \mathrm{vol}_a = 0. \tag{2.104}$$

Proof. By (2.1) for metric g, we get $\int_M \mu_{g_n} \sigma_1(A^g) \, \mathrm{d} \, \mathrm{vol}_a = 0$. Corollary 2.105 and the equality $f^i \overline{\mathrm{div}} \, \beta_i^\sharp = \overline{\mathrm{div}} \, (f^i \beta_i^\sharp) - \beta_i^\sharp(f^i)$ with $f^i = \mu_{g_n} \gamma_1^i / \phi$, yield (2.104). □

The formula (2.104) is true when also all 1-forms are defined outside the "singularity set" Σ under convergence of some integrals, see Section 2.2.

For codimension-one totally geodesic and Riemannian foliations we get

Corollary 2.110. *In conditions of Theorem 2.109 for $p = 1$, let $b = \beta(n)$ and $\beta(N)$ be constant. (i) If $\overline{A} = 0$ and $q_2 \neq 0$, then either $\beta(\overline{Z}) \equiv 0$ or $\beta(\overline{Z})_x \cdot \beta(\overline{Z})_{x'} < 0$ for some points $x \neq x'$. (ii) If $\overline{Z} = 0$ and $q_1 \neq 0$, then either $\langle \overline{A}(\beta^{\sharp \top}), \beta^{\sharp} \rangle \equiv 0$ or $\langle \overline{A}(\beta^{\sharp \top}), \beta^{\sharp} \rangle_x \cdot \langle \overline{A}(\beta^{\sharp \top}), \beta^{\sharp} \rangle_{x'} < 0$ for some points $x \neq x'$.*

Example 2.111.

(i) For the Randers metric ($p = 1$ and $\phi = 1 + s$), by (2.104) we have

$$\int_M (c\,c_1)^{m+1} c^{-1} \Big((c\,c_1)\,\sigma_1(\overline{A}) - \frac{m+2}{2}\,(N + c_1^{-1}\beta^{\sharp})(c\,c_1) + c_1 N(c)$$
$$- (c_1 - c)\big[N(c) + \langle c^{-1}\overline{A}(\beta^{\sharp \top}) + \overline{Z}, \beta^{\sharp} \rangle \big] \Big)\,d\,\mathrm{vol}_a = 0, \tag{2.105}$$

which reduces to (2.1) when $\beta = 0$. If $\beta(N) = 0$, then (2.105) reads

$$\int_M c^{2m+1} \big(c^2 \sigma_1(\overline{A}) - (m+1)\,c\,N(c) - (m+2)\,\beta^{\sharp}(c) \big)\,d\,\mathrm{vol}_a = 0;$$

if b and $\beta(N) \neq 0$ are constant then (2.105) reads $\int_M \langle \overline{A}(\beta^{\sharp \top}) + c\overline{Z}, \beta^{\sharp} \rangle\,d\,\mathrm{vol}_a = 0$.

(ii) For a Kropina metric, if $\beta(N) = 0$ then $\mu_{g_n} = (2/b)^{2m+2}$, and

$$\gamma_1 = -\sqrt{2}/(2b), \quad \gamma_2 = 0, \quad c_1 = 1/\sqrt{2},$$
$$s = b/\sqrt{2}, \quad \rho = 4/b^2, \quad \rho_0 = 12/b^4, \quad \rho_1 = -8\sqrt{2}/b^3.$$

Hence, by Proposition 2.104 for $p = 1$,

$$\sigma_1(A^g) = \frac{1}{2} b\,\sigma_1(\overline{A}) - \frac{1}{2}\,\overline{\mathrm{div}}\,\beta^{\sharp} + \frac{m}{\sqrt{2}}\,n(b) + 12\,b\,\beta^{\sharp}(b),$$

and, we get integral formula

$$\int_M \Big(\frac{2}{b}\Big)^{2m+2} \Big\{ b\,\sigma_1(\overline{A}) + \sqrt{2}\,m\,n(b) - \frac{2m+1}{b}\,\beta^{\sharp}(b) \Big\}\,d\,\mathrm{vol}_a = 0,$$

which for $b = \mathrm{const}$ reduces to (2.1) for metric a.

Next, we assume that $\mathscr{D}$ is integrable and $p = 1$, and use (α, β)-metrics. The counterpart of the Reeb integral formula for the second mean curvature reads as (2.2),

$$\int_M (\sigma_2(\overline{A}) - (1/2)\overline{\mathrm{Ric}}_{N,N})\,d\,\mathrm{vol}_a = 0.$$

Here, $\overline{\mathrm{Ric}}_{N,N} = \mathrm{trace}_a(u \to R_{N,u}N)$ is the Ricci curvature of a in the N-direction. We will generalize (2.2) for general (α, β)-metrics on M. In this case, the volume form of g with μ_{g_n} given in (1.24) obeys

$$d\,\mathrm{vol}_g = \mu_{g_n}\,d\,\mathrm{vol}_a. \tag{2.106}$$

Let $\mathrm{Ric}^g_{v,v} = \mathrm{trace}_g(u \to R^g_{v,u} v)$ be the Ricci curvature of g in the v-direction, where $R^g_{u,v} = [\nabla_v, \nabla_u] - \nabla_{[v,u]}$ is the curvature tensor of the Levi-Civita connection of g.

Let D^v be the Chern connection, see (1.20), where $g(\nabla_u v, w)$ is given in (1.9). The *contorsion tensor* $\mathfrak{T} = D^v - \nabla$, see Section 1.2.1, is symmetric because both connections, ∇ and D^v, are torsion-free. By (1.20), $D^v_v v = \nabla_v v$ is valid; hence, $\mathfrak{T}_v v = 0$ (thus, v is geodesic for F if and only if it is geodesic for g). Comparing the curvature $R^D_{u,v} = [D^v_v, D^v_u] - D^v_{[v,u]}$ of D^v with $R^g_{u,v}$, we find

$$R^D_{v,u} - R^g_{v,u} = (\nabla_u \mathfrak{T})_v - (\nabla_v \mathfrak{T})_u - [\mathfrak{T}_v, \mathfrak{T}_u], \quad u \in TM. \tag{2.107}$$

In [CST], the Ricci curvature $\mathrm{Ric}^D_y = \mathrm{trace}_g(u \to R^D_{y,u} y)$ of (α, β)-metric is expressed through $\overline{\mathrm{Ric}}_y$ of α; in particular, $\overline{\nabla}\beta = 0$ provides $\mathrm{Ric}^D_y = \overline{\mathrm{Ric}}_y$ $(y \neq 0)$.

Let $C^\sharp_v$ be a $(1,1)$-tensor g-dual to the symmetric bilinear form $C_v(Z, \cdot, \cdot)$:

$$g(C^\sharp_v(u), v) = C_v(Z, u, v), \quad u, v \in TM,$$

where C is the Cartan torsion. Note that $A^g + C^\sharp_v$ is the shape operator of the leaves with respect to D^v, see [RW7]. By (1.20), we get

$$\mathfrak{T}_v = -C^\sharp_v \quad \text{and} \quad \mathrm{trace}\, \mathfrak{T}_v = -\sigma_1(C^\sharp_v) = -I_v(Z). \tag{2.108}$$

Unlike Theorem 2.109, the following theorem contains quantities dependent on C.

Theorem 2.112. *Let g be a new metric determined by a codimension-one foliation $\mathscr{F}$ $(T\mathscr{F} = \mathscr{D})$ and a 1-form β on (M, a) and a function $\phi(s)$ with conditions (2.86), (2.92) and $\overline{\nabla}\beta^\sharp = 0$. Then*

$$\int_M \Big\{ \Big[(c_1\rho)^2 (2\sigma_2(\overline{A}) - \overline{\mathrm{Ric}}_{N,N}) + \frac{1}{4}m(m-1)n(\rho)^2 - (m-1)c_1\rho\, n(\rho)\, \sigma_1(\overline{A})$$
$$+ \frac{1}{2}\beta(U)\langle 2\gamma_3 \mathscr{A}(\beta^{\sharp T}) + U, \beta^\sharp \rangle - \frac{1}{2}(b^2 - \beta(N)^2)\langle 2\gamma_3 \mathscr{A}(\beta^{\sharp T}) + U, U\rangle - (2\rho c_1\sigma_1(\overline{A})$$
$$- (m-1)n(\rho))\,\langle \gamma_3 \mathscr{A}(\beta^{\sharp T}) + U, \beta^\sharp \rangle + 2\rho\, c_1\langle \gamma_3 \mathscr{A}(\beta^{\sharp T}) + U, \overline{A}(\beta^{\sharp T})\rangle \Big](\rho\,\phi)^{-2}$$
$$- I_v((A^g + C^\sharp_v + \sigma_1(A^g)\,\mathrm{id})Z) - 2\sigma_1(A^g C^\sharp_v) - \sigma_1((C^\sharp_v)^2) \Big\}\mu_{g_n}\, d\,\mathrm{vol}_a = 0, \tag{2.109}$$

where A^g, $\mathscr{A}$ and U are given in Proposition 2.104, and μ_{g_n} is given in (1.24) with $y = n$, $s = \beta(n)$ and $I_v = \mathrm{trace}_{23} C_v$.

Proof. We will use the adjoint $(1,2)$-tensor $\mathfrak{T}^*$ defined in (1.15). Observe that $\mathfrak{T}^*_v v = 0$ and consider the vector field $\mathrm{trace}_g \mathfrak{T}^*$ (the trace of $\mathfrak{T}^*$ with respect to g). Let (e_i) be a g-orthonormal local frame on $\mathscr{D}$. One may assume $(\nabla_v e_i)^\top = 0$ and $(\nabla_{e_i} v)^\perp = 0$ at a point $x \in M$ and then calculate at x:

$$\sum_i g((\nabla_i \mathfrak{T})_v v, e_i) = 2\sum_i g(\mathfrak{T}^*_v e_i, A^g(e_i)) = 2\sigma_1(C^\sharp_v A^g),$$
$$\sum_i g((\nabla_v \mathfrak{T})_i v, e_i) = \mathrm{div}_g(\mathrm{trace}_g \mathfrak{T}^*), \quad \sum_i g([\mathfrak{T}_i, \mathfrak{T}_v]v, e_i) = -\sigma_1((C^\sharp_v)^2),$$

using the symmetry $\mathfrak{T}_i \, v = \mathfrak{T}_v \, e_i$. Then, applying (2.107) we get

$$\mathrm{Ric}^D_{v,v} - \mathrm{Ric}^g_{v,v} = \sum_i \left[g((\nabla_i \mathfrak{T})_v \, v, e_i) - g((\nabla_v \mathfrak{T})_i \, v, e_i) + g([\mathfrak{T}_i, \mathfrak{T}_v] \, v, e_i) \right]$$
$$= 2\sigma_1(C^\sharp_v A^g) - \sigma_1((C^\sharp_v)^2) - \mathrm{div}^\perp_g(\mathrm{trace}^\top \mathfrak{T}^*). \qquad (2.110)$$

From (2.110) and

$$\mathrm{div}^\perp_g(\mathrm{trace}_g \, \mathfrak{T}^*) = \mathrm{div}_g((\mathrm{trace}_g \, \mathfrak{T}^*)^\perp) - g(\mathrm{trace}_g \, \mathfrak{T}^*, \sigma_1(A^g) \, v - Z)$$

we obtain

$$\mathrm{div}_g((\mathrm{trace}_g \, \mathfrak{T}^*)^\perp) = \mathrm{Ric}^g_{v,v} - \mathrm{Ric}^D_{v,v}$$
$$+ g(\mathrm{trace}_g \, \mathfrak{T}^*, \sigma_1(A^g) \, v - Z) - 2\sigma_1(A^g C^\sharp_v) - \sigma_1((C^\sharp_v)^2). \qquad (2.111)$$

Then, using (1.20) and (2.108), we find

$$g(\mathrm{trace}_g \, \mathfrak{T}^*, v) = -\sum_i C_v(D^v_v \, v, e_i, e_i) = -\sigma_1(C^\sharp_v) = -I_v(Z),$$
$$g(\mathrm{trace}_g \, \mathfrak{T}^*, u) = -\sum_i C_v(D^v_u \, v, e_i, e_i) = I_v((A^g + C^\sharp_v)(u))$$

for $u \in \mathscr{D}$. By the above we obtain

$$g(\mathrm{trace}_g \, \mathfrak{T}^*, \sigma_1(A^g) \, v - Z) = -I_v((A^g + C^\sharp_v + \sigma_1(A^g) \, \mathrm{id})Z).$$

By our assumptions, $b = \mathrm{const}$ and $\bar{R}_{X,Y} \, \beta^\sharp = 0 \ (X, Y \in TM)$. Using

$$\mathrm{Ric}^D_{n,n} = \overline{\mathrm{Ric}}_{n,n} = c_1^2 \, \overline{\mathrm{Ric}}_{N,N} + \gamma_1^2 \, \overline{\mathrm{Ric}}_{\beta^\sharp, \beta^\sharp} - 2c_1 \gamma_1 \sum_i \langle \bar{R}_{N,e_i} \, \beta^\sharp, e_i \rangle$$

and $\mathrm{Ric}^D_{v,v} = \phi^{-2} \mathrm{Ric}^D_{n,n}$, we find $\mathrm{Ric}^D_{v,v} = (c_1/\phi)^2 \, \overline{\mathrm{Ric}}_{N,N}$. By the above, (2.2) and (2.106) for g, using (2.111) and Corollary 2.106, we get (2.109). $\qquad \square$

Recall that $F = \alpha + \beta$ is Berwald if and only if $\bar{\nabla} \beta^\sharp = 0$. In this case, the Finsler metric and the source metric have equal Riemann curvatures: $R_y = \bar{R}_y$ for $y \in TM \setminus 0$. Let $k_1 \leq k_2 \leq \ldots \leq k_m$ be the eigenvalues of A^g. One can consider the integral

$$U_{\mathscr{F}} = \int_M \sum_{i<j}(k_i - k_j)^2 \, d\mathrm{vol}_g,$$

which measures "how far from g-total umbilicity" is $\mathscr{F}$. Put $\mu_- = \inf_{y \in TM \setminus 0} \mu_{g_y}$.

Theorem 2.113. *Let g be a new Riemannian metric determined by a codimension-one foliation $\mathscr{F}$ and a 1-form β on a closed (M^{m+1}, a) and a function $\phi(s)$ with conditions (2.86), (2.92), $\bar{\nabla}\beta = 0$, $\beta(N) = \mathrm{const}$ and $\overline{\mathrm{Ric}}_{N,N} \leq -r < 0$. Then*

$$U_{\mathscr{F}} \geq m r (c_1/\phi)^2 \mu_- \mathrm{Vol}(M, a). \qquad (2.112)$$

In particular, if $c_1 \neq 0$ then $\mathscr{F}$ is nowhere g-totally umbilical.

Proof. One may show that

$$\sum_{i<j}(k_i - k_j)^2 = (m-1)\,\sigma_1^2(A^g) - 2m\,\sigma_2(A^g).$$

Hence, and by (2.2) for g we obtain

$$U_{\mathscr{F}} \geq -m\int_M 2\,\sigma_2(A^g)\,\mathrm{d}\,\mathrm{vol}_g = -m\int_M \mathrm{Ric}^g_{v,v}\mathrm{d}\,\mathrm{vol}_g.$$

By conditions, $\mathrm{Ric}^g_{v,v} = (c_1/\phi)^2\,\overline{\mathrm{Ric}}_{N,N}$ and $s,\rho,\rho_i,\gamma_i,c_1,\phi$ are constant. Thus,

$$U_{\mathscr{F}} \geq -m(c_1/\phi)^2\mu_-\int_M \overline{\mathrm{Ric}}_{N,N}\mathrm{d}\,\mathrm{vol}_a$$

that reduces to (2.112) by assumptions. $\square$

For Riemannian case, the *energy of a vector field* v is defined by

$$\mathscr{E}(v) = \frac{m+1}{2}\,\mathrm{Vol}(M,g) + \frac{1}{2}\int_M \|Dv\|_g^2\,\mathrm{d}\,\mathrm{vol}_g.$$

Definition of energy of vector fields can be found in many articles and books, this is a particular case of energy of maps between Riemannian manifolds. By (2.2) for g and the inequality $\|Dv\|_g^2 \geq \frac{2}{m}\,\sigma_2(A^g)$, see [BW], we get the following.

Theorem 2.114. *Let g be a new Riemannian metric determined by a codimension-one foliation $\mathscr{F}$ and a 1-form β on a closed (M^{m+1},a) and a function $\phi(s)$ with conditions (2.86), (2.92), $\overline{\nabla}\beta^\sharp = 0$ and $\beta(N) = \mathrm{const.}$ Then for a unit g-normal v,*

$$\mathscr{E}(v) \geq \mu_-\left(\frac{m+1}{2}\,\mathrm{Vol}(M,a) + \frac{c_1^2}{2m\,\phi^2}\int_M \overline{\mathrm{Ric}}_{N,N}\mathrm{d}\,\mathrm{vol}_a\right).$$

Chapter 3
Prescribing the Mean Curvature

This chapter provides conditions either necessary or sufficient for a given quantity (either a scalar function or a vector field) to be the mean curvature of a given foliation with respect to some Riemannian metric. The particular case of this quantity being identically zero (tautness) has been described separately. In the codimension-one case, the only obstructions for a scalar function to be the mean curvature of a foliation arise from Stokes' Theorem and a well known formula by H. Rummler relating mean curvature with the exterior derivative of the leaf volume form. The results obtained in [Os1, Os2, Os3] in general, and the second author [W2, W3] in particular cases, show that these obstructions depend on the topological structure of the foliation and are similar to those obtained in [KW] for the Gauss curvature of a closed oriented surface S: roughly speaking, a function $f : S \to \mathbb{R}$ can be realized as the Gauss curvature of S (with respect to some Riemannian structure) if and only if it attains at some points values of the same sign as the Euler characteristic of S. Also, geometric tautness has been shown [Ha, Su2] to be equivalent to some topological conditions imposed on foliations and their holonomy. Some partial results [SW] on prescribed mean curvature vectors of foliations of arbitrary codimension are described in a separate section.

3.1 Minimal Submanifolds

The mean curvature is one of the most important invariants of extrinsic geometry. Its vanishing defines either submanifolds or foliations called *minimal* since they appear to become critical points of the volume (or, area) functional. Originated in 18th century theory of minimal surfaces (or, hypersurfaces) of Euclidean spaces (or, other space forms or, other Riemannian submanifolds) has been widely developed (see [An, DHS], etc. and the bibliographies therein). Among other topics, the problem of existence of minimal immersions (or, embeddings) of a given surface (or, manifold) into the Euclidean space (or, another given Riemannian manifold) is of great

© Springer Nature Switzerland AG 2021
V. Rovenski, P. Walczak, *Extrinsic Geometry of Foliations*, Progress in Mathematics 339, https://doi.org/10.1007/978-3-030-70067-6_3

interest. Among the number of fascinating results concerning this problem, the most classical is the following one.

Proposition 3.1. *There exist no closed minimal surfaces in* $\mathbb{R}^3$.

Proof. Fix an origin o and find a point p of a closed surface Σ with maximal distance from o. Then, Σ lies on one side of the tangent space $T_p\Sigma$ and the Gaussian curvature $K = k_1 k_2$ of Σ at p (k_1 and k_2 being the principal curvatures at p) is nonnegative, in fact positive for a suitable choice of o. Therefore, k_1 and k_2 are either both positive or both negative and $k_1 + k_2 \neq 0$ at p. $\square$

Here, we consider a different problem: Given a foliation $\mathscr{F}$ of a closed manifold M, look for a Riemannian metric on M making all the leaves of $\mathscr{F}$ minimal. If this is possible, then such a foliation is called *geometrically taut*. A more general problem reads as: Given $\mathscr{F}$ as before and an, either scalar or vector, quantity (defined globally on M) look for such a Riemannian metric on M for which the given quantity becomes the mean curvature of (the leaves) of $\mathscr{F}$. Thinking about that we talk about the problem of *prescribing mean curvature* of $\mathscr{F}$. Note that the similar problems for single submanifolds are of little interest. For example, one has the following.

Proposition 3.2. *For any closed submanifold L of a manifold M there exists a Riemannian metric g on M for which L becomes minimal in (M, g).*

Remark 3.3. Let us construct g satisfying the condition of Proposition 3.2. Take a tubular neighborhood $\iota : L \times B(\varepsilon) \to M$, ε being small enough, put $U = \iota(L \times B(\varepsilon))$ and let g_U be the Riemannian structure on U obtained *via* ι from a product Riemannian structure on $L \times B(\varepsilon)$. Put $V = M \setminus L$ and take a parition of unity $\{f_U, f_V\}$ subordinated to the open covering $\{U, V\}$ of M. Take any Riemannian structure g_V on V and set $g = f_U g_U + f_V g_V$; g satisfies the required condition.

What is really interesting, it is the fact that the problems of geometric tautness and prescribing mean curvature for foliations are of topological nature: geometric tautness is equivalent to another condition called *topological tautness* and there exist only simple obstructions, related to the topological structure of $\mathscr{F}$, which do not allow given scalar functions (or, vector fields) become mean curvatures of $\mathscr{F}$ with respect to any Riemannian metric on the ambient space.

3.2 Tautness of Foliations

3.2.1 Rummler Formula

Here, we provide a simple formula, called by several authors *Rummler formula*, a tool in the study of geometric tautness of foliations. The formula appeared in,[1] where it has been applied for purposes similar to those here.

---

[1] H. Rummler, Quelques notions simples en géométie riemannienne et leurs applications aux feuilletages compact, Comment. Math. Helv. 54 (1979), 224–239.

Let $(M, \mathscr{F}, g)$ be a foliated Riemannian manifold, $\dim \mathscr{F} = p$, and H—the mean curvature vector of (the leaves of) $\mathscr{F}$: if (E_i) is a local orthonormal frame of $T\mathscr{F}$, then

$$H = \sum_{i=1}^{p} (\nabla_{E_i} E_i)^{\perp},$$

where $v^{\perp}$ denotes—as before—the $\mathscr{F}$-orthogonal component of a vector $v \in TM$. Also, let $\Omega_{\mathscr{F}}$ be the volume form of (the leaves of) $\mathscr{F}$:

$$\Omega_{\mathscr{F}}(X_1, \ldots, X_p) = \det[\langle X_i, E_j \rangle], \quad i, j = 1, \ldots, p \tag{3.1}$$

for all vector fields $X_1, \ldots, X_p$ on M.

Proposition 3.4. *For any vector field Z on M one has*

$$d\Omega_{\mathscr{F}}(Z, E_1, \ldots, E_p) = -\langle Z, H \rangle. \tag{3.2}$$

Proof. For Z tangent to $\mathscr{F}$, both sides are equal to 0, hence the equality holds. If Z is perpendicular to $\mathscr{F}$, then elementary calculations show that

$$
\begin{aligned}
d\Omega_{\mathscr{F}}(Z, E_1, \ldots, E_p) &= \sum_{i=1}^{p} (-1)^{i+1} \Omega_{\mathscr{F}}([Z, E_i], E_1, \ldots, \hat{E}_i, \ldots, E_p) \\
&= \sum_{i=1}^{p} \langle [Z, E_i], E_i \rangle = \sum_{i=1}^{p} \langle \nabla_{E_i} Z, E_i \rangle \\
&= -\langle \sum_{i=1}^{p} \nabla_{E_i} E_i, Z \rangle = -\langle Z, H \rangle,
\end{aligned}
$$

providing the proof. $\qquad \square$

If $\mathscr{F}$ is minimal on (M, g), then the volume form $\Omega_{\mathscr{F}}$ satisfies

$$d\Omega_{\mathscr{F}}(\cdot, X_1, \ldots, X_p) = 0$$

for all $X_1, \ldots, X_p$ tangent to $\mathscr{F}$. This motivates the following.

Definition 3.5. A k-form ω $(k \geq p)$ is said to be $\mathscr{F}$-*closed* whenever

$$d\omega(\cdot, \ldots, \cdot, X_1, \ldots, X_p) = 0$$

for all $X_1, \ldots, X_p$ tangent to $\mathscr{F}$.

Using this terminology, we may say that $\mathscr{F}$ is minimal on (M, g) if and only if its volume form $\Omega_{\mathscr{F}}$ is $\mathscr{F}$-closed.

Remark 3.6. Rummler formula holds also for any, integrable or not, plane field $\mathscr{D}$, the volume form defined by the formula analogous to (3.1) and the mean curvature vector $H = \operatorname{trace} A$ of $\mathscr{D}$, A being the shape operator of $\mathscr{D}$ defined in Section 1.3.

3.2.2 Foliation Currents

For any n-dimensional manifold M and $k \geq 0$, denote by $D^k = D^k(M)$ the space of (global) k-forms on M equipped with the C^∞-topology: D^k is a Frechet space with the Frechet structure defined by a family of semi-norms $\| \cdot \|_j$, $j = 0, 1, \ldots$, given by

$$\|\omega\|_j = \int_M \|\nabla^j \omega\| \, d\,\mathrm{vol},$$

∇ being, say, the Levi-Civita connection and d vol the volume form of a Riemannian structure g on M.

Definition 3.7. Elements of the space D_k dual to D^k, that is continuous linear functionals defined on D^k, are called *de Rham currents* or, shortly, *k-currents*.

Example 3.8.

(1) Any k-vector $v = v_1 \wedge \ldots \wedge v_k$, $v_i \in T_x M$, $x \in M$ defines a *Dirac current* z_v:

$$z_v(\omega) = \omega(v_1 \ldots, v_k), \quad \omega \in D^k.$$

Also, any linear combination of k-vectors defines a k-current in the obvious way.

(2) Any compact k-dimensional submanifold N of M defines a k-current z_N:

$$z_N(\omega) = \int_N \iota_N^* \omega, \quad \omega \in D^k,$$

$\iota_N : N \to M$ being the inclusion.

(3) If $p + q = n$ and ω is a q-form, then the formula

$$z_\omega(\eta) = \int_M \omega \wedge \eta, \quad \eta \in D^p,$$

defines a p-current z_ω. Thus, one has a natural continuous embedding $\iota : D^q \to D_p$. One can prove rigorously continuity of functionals z_v, z_N and z_ω.

Certainly, the set D_k of all k-currents on a given manifold M carries a natural structure of a linear space and can be equipped with a weak-$*$ topology:

$$z_n \to z \text{ in } D_k \text{ if and only if } z_n(\omega) \to z(\omega) \text{ for any } \omega \in D^k.$$

For this topology, one can construct the dual space D_k^* consisting, as usual, of all continuous linear functionals defined on D_k. Certainly, $D^k \subset D_k^*$. One of the most important results of the theory of currents tells us the following.

Theorem 3.9 (Theorem XIV in [Sc]). $D_k^* = D_k$, that is any continuous linear functional λ on the space D_k is given by

$$\lambda(z) = z(\omega), \quad z \in D_k,$$

for a k-form ω.

Moreover, the space $D = \bigoplus_k D_k$ of all the currents can be equipped with a linear operator $\partial : D \to D$, the dual of the exterior differential d:

$$\partial : D_k \to D_{k-1} \quad \text{and} \quad \partial(z)(\omega) = z(d\omega) \quad \text{for all} \quad z \in D_k \quad \text{and} \quad \omega \in D^{k-1}.$$

Certainly, $d^2 = 0$ implies $\partial^2 = 0$, therefore, one can consider the spaces $Z_k = \ker \partial \subset D_k$ of k-cycles and $B_k = \operatorname{im} \partial \in D_k$ of k-boundaries, and the corresponding homologies $H_k(M) = Z_k/B_k$.

Now, let $\mathscr{F}$ be a foliation of M, $\dim \mathscr{F} = p$. Assume that $\mathscr{F}$ is oriented and consider the closed convex cone $C_{\mathscr{F}}$ generated by all the Dirac currents z_v, where $v = v_1 \wedge \ldots \wedge v_p$ and $(v_1, \ldots, v_p)$ is a positive oriented frame of the space tangent to $\mathscr{F}$ at a point of M. In other words, $C_{\mathscr{F}}$ consists of all the limits of convergent sequences of linear combinations with positive coefficients of such Dirac currents:

$$z \in C_{\mathscr{F}} \iff z = \lim_{j \to \infty} \sum_{i=1}^{n_j} a_{ij} z_{v_{ij}},$$

where $n_j \in \mathbb{N}$, $a_{ij} \neq 0$, v_{ij} are positive frames of $T_{x_{ij}} \mathscr{F}$ and $x_{ij} \in M$.

Remark 3.10. Let us prove rigorously that for any closed leaf L of $\mathscr{F}$ the current z_L (defined in the same way as z_N in Example 3.8(2)) belongs to $C_{\mathscr{F}}$ and that $\partial z_L = 0$. Choose on L a positive volume form ω. For any n, cover L with finitely many pairwise disjoint sets A_{in} of ω-volumes a_{in} smaller than $1/n$, choose points $p_{in} \in A_{in}$ and such frames v_{in} of $T_{p_{in}} L$ that $\omega(v_{in}) = 1$. Then,

$$z_L = \lim_{n \to \infty} \sum_i a_{in} z_{v_{in}}.$$

Moreover, for any form η (of suitable rank), we have $\partial z_L(\eta) = z_L(d\eta) = \int_L d\eta = 0$, by the Stokes' Theorem.

By a *base* of a cone C contained in a topological vector space V we mean the set $C_1 = l^{-1}(1)$, where $l : V \to \mathbb{R}$ is a continuous linear functional positive on C (except of the origin). The cone C is said to be *compact* if its base is compact. Note that a subset K of the space D_k of k-currents is compact if it is closed and for any $\eta \in D^k$ the set $K(\eta) = \{z(\eta), z \in K\}$ is bounded in $\mathbb{R}$ (see, for instance, [Sc, p. 74]).

Proposition 3.11. *If $(M, \mathscr{F})$ is a compact foliated manifold, then the cone $C_{\mathscr{F}}$ of foliation currents is compact.*

Proof. Choose a Riemannian structure g on M and let $l : D_p \to \mathbb{R}$, $p = \dim \mathscr{F}$, be the functional defined by $\Omega_{\mathscr{F}}$, the volume form of $\mathscr{F}$ on (M, g). Certainly, l is positive on $C_{\mathscr{F}}$. If η is a p-form of norm c_0 and $z = \sum_i a_i z_{v_i}$, where v_i are g-unit p-vectors, $a_i \geq 0$ and $\sum_i a_i = 1$, then $|z(\eta)| \leq \sum_i a_i |\eta(v_i)| \leq c_0$. Therefore, $K(\eta) \subset [-c_0, c_0]$ for $K = l^{-1}(1)$ proving the compactness of the base K of $C_{\mathscr{F}}$. $\square$

3.2.3 Tautness

Here, we shall consider the following

Problem 3.12. Given $(M, \mathscr{F})$, when does there exist a Riemannian metric g on M for which the mean curvature of all the leaves of $\mathscr{F}$ vanishes identically?

For our convenience, we shall introduce the following notation: $B_{\mathscr{F}}$ denotes the closed linear subspace of D_p generated by the boundaries of all the Dirac currents of the form z_v with $v = w \wedge v_1 \wedge \ldots \wedge v_p$, where $w \in T_x M, v_i \in T_x \mathscr{F}$ and $x \in M$.

Definition 3.13. A p-dimensional foliation $\mathscr{F}$ of a manifold M is said to be

(1) *geometrically taut* whenever there exists a Riemannian metric g on M for which all the leaves of M become minimal submanifolds of (M, g);
(2) *topologically* (or, *homologically*) *taut* whenever the cone $C_{\mathscr{F}}$ of foliation currents intersects the space $B_{\mathscr{F}}$ trivially.

Our goal here is to show that these two types of tautness are equivalent, i.e. that geometric tautness is, in fact, a topological property of foliations. For this purpose, we need the following detail from linear algebra.

Let V be an n-dimensional vector space and W its subspace of dimension p. Assume that a p-form ω on V does not vanish on p-vectors of W. Following [Su2], define the projection $P_\omega : V \to W$ by

$$\iota_{P_\omega(v)}(\omega | \Lambda^p W) = \iota_{P_\omega(v)} \omega | \Lambda^{p-1} W = (\iota_v \omega) | \Lambda^{p-1} W, \quad v \in V,$$

and set

$$\tilde{\omega}(v_1, \ldots, v_p) = \omega(P_\omega(v_1), \ldots, P_\omega(v_p)), \quad v_1, \ldots, v_p \in V.$$

The form $\tilde{\omega}$ is pure (that is, $\tilde{\omega}(x)$ is a single wedge-product of p-covectors), therefore we call it the $\mathscr{F}$-*purification* of ω. From the above definition, it follows directly that

$$\tilde{\omega}(v, w_1, \ldots, w_{p-1}) = \omega(v, w_1, \ldots, w_{p-1}) \tag{3.3}$$

whenever $v \in V$, $w_j \in W$.

Remark 3.14.

(1) Let Ω be a nonzero element of $\Lambda^n \mathbb{R}^n$. According to equality $\dim \Lambda^{n-1} \mathbb{R}^n = n$, the map $v \mapsto \iota_v \Omega$ is an isomorphism between $\mathbb{R}^n$ and $\Lambda^{n-1} \mathbb{R}^n$. Choose $v \neq 0$ and a frame $e_1, \ldots, e_n$ of $\mathbb{R}^n$ for which $v = e_1$. Then $\Omega = a e_1 \wedge \ldots \wedge e_n$ and $\iota_v \Omega = a e_2 \wedge \ldots \wedge e_n$ is pure. We showed that any $(n-1)$-form ω on $\mathbb{R}^n$ is pure, therefore, its purification—with respect to any $(n-1)$-dimensional subspace on which ω is non-zero—coincides with ω.
(2) One may compute in $\mathbb{R}^4$ the purification $\tilde{\omega}$—with respect to the subspace generated by e_1 and e_2—of $\omega = e_1^* \wedge e_2^* + \sum_{i>1} a_i e_1^* \wedge e_i^* + b e_3^* \wedge e_4^*$, where (e_i^*) is the base dual to a base (e_i) of $\mathbb{R}^4$; and then express $\tilde{\omega}$ in the form of a product of two 1-forms. Of course, we have to assume that $1 + a_2 \neq 0$. Then, calculations

following the definition of the projection P_ω show that both vectors $P_\omega e_3$ and $P_\omega e_4$ are parallel to e_2. Obviously, $P_\omega e_i = e_i$ for $i = 1, 2$. Consequently, $\tilde{\omega}$ is the wedge-product of e_1^* and a suitable linear combination of e_i^*, $i = 2, 3, 4$.

Certainly, if $\mathscr{F}$ is a foliation of M and ω is a smooth p-form ($p = \dim \mathscr{F}$) on M non-vanishing on $\Lambda^p(T\mathscr{F})$, then the above procedure of purification can be applied pointwise on M providing a pure smooth p-form $\tilde{\omega}$.

Lemma 3.15. *If a p-form ω on M is $\mathscr{F}$-closed, then its purification $\tilde{\omega}$ is $\mathscr{F}$-closed, too.*

Proof. By (3.3) and integrability of $T\mathscr{F}$, if $Z, X_1, \ldots, X_p$ are arbitrary vector fields such that all the X_j's are tangent to $\mathscr{F}$, then

$$
\begin{aligned}
d\tilde{\omega}(Z, X_1, \ldots X_p) &= Z\tilde{\omega}(X_1, \ldots, X_p) + \sum_j (-1)^j X_j \tilde{\omega}(Z, X_1 \ldots, \hat{X}_j, \ldots X_p) \\
&+ \sum_j (-1)^{j+1} \tilde{\omega}([Z, X_j], X_1, \ldots, \hat{X}_j, \ldots, X_p) \\
&+ \sum_{i<j} (-1)^{i+j} \tilde{\omega}([X_i, X_j], Z, X_1 \ldots, \hat{X}_i,, \ldots, \hat{X}_j, \ldots, X_p) \\
&= Z\omega(X_1, \ldots, X_p) + \sum_j (-1)^j X_j \omega(Z, X_1 \ldots, \hat{X}_j, \ldots X_p) \\
&+ \sum_j (-1)^{j+1} \omega([Z, X_j], X_1, \ldots, \hat{X}_j, \ldots, X_p) \\
&+ \sum_{i<j} (-1)^{i+j} \omega([X_i, X_j], Z, X_1 \ldots, \hat{X}_i,, \ldots, \hat{X}_j, \ldots, X_p) \\
&= d\omega(Z, X_1, \ldots X_p) = 0,
\end{aligned}
$$

since in all the terms above at most one of the arguments is not tangent to $\mathscr{F}$. $\square$

In the proof of the main result of this section, we shall use also the following version of the Hahn-Banach Theorem (see, e.g. [KN]).

Theorem 3.16 (Hahn-Banach). *Let V be a Frechet space, W—a closed subspace of V and C—a compact convex cone in V (with vertex $0 \in V$). Any continuous linear functional $\rho : W \to \mathbb{R}$ positive on $W \cap (C \setminus 0)$ admits a continuous and positive on $C \setminus 0$ extension to the whole space V.*

Theorem 3.17. *A foliation $\mathscr{F}$ of a closed manifold M is geometrically taut if and only if it is topologically taut.*

Proof. First, assume that $\mathscr{F}$ is geometrically taut. Then, there exists a Riemannian metric g on M for which the volume form $\Omega_{\mathscr{F}}$ is $\mathscr{F}$-closed. If $z \in C_{\mathscr{F}}$ and $z \neq 0$, then $z(\Omega_{\mathscr{F}}) > 0$. On the other hand, if $z \in B_{\mathscr{F}}$, then $z(\Omega_{\mathscr{F}}) = 0$. Thus, $C_{\mathscr{F}} \cap B_{\mathscr{F}} = \{0\}$.

Now, assume that $\mathscr{F}$ is topologically taut, choose any Riemannian metric g on M and set $\omega = \Omega_{\mathscr{F}}$—the volume form of $\mathscr{F}$ on (M, g). Denote by W the subspace of D_p generated by all the elements of $C_{\mathscr{F}} \cup B_{\mathscr{F}}$. The formulae

$$
\lambda(z) = \begin{cases} z(\omega) & \text{when} \quad z \in C_{\mathscr{F}}, \\ 0 & \text{when} \quad z \in B_{\mathscr{F}} \end{cases}
$$

determine a linear functional $\lambda : W \to \mathbb{R}$, which is positive on $C_{\mathscr{F}}$. By the above Hahn-Banach Theorem, λ extends to $\hat{\lambda} : D_p \to \mathbb{R}$. By Theorem 3.9, $\hat{\lambda}$ is represented by a p-form $\hat{\omega}$. Certainly, $\hat{\omega}$ is $\mathscr{F}$-closed and positive on all positive oriented p-vectors $v_1 \wedge \ldots \wedge v_p$, $v_i \in T_x \mathscr{F}$, $x \in M$. Let $\tilde{\omega}$ be the $\mathscr{F}$-purification of $\hat{\omega}$. By Lemma 3.15 and (3.3), $\tilde{\omega}$ is also $\mathscr{F}$-closed and positive on $\mathscr{F}$ (in the same sense as $\hat{\omega}$). There exists a Riemannian metric $\tilde{g}$ on M for which $\tilde{\omega}$ coincides with $\tilde{\Omega}_{\mathscr{F}}$, the volume form of $\mathscr{F}$ on $(M, \tilde{g})$. By Rummler formula (3.2), the mean curvature vector $\tilde{H}$ of $\mathscr{F}$ with respect to $\tilde{g}$ vanishes identically. Thus, $\mathscr{F}$ is geometrically taut. $\square$

Example 3.18. Relatively simple examples of taut foliations are these in dimension one. Assuming orientability, a one-dimensional foliation consists of orbits of a nowhere vanishing vector field X. These orbits are minimal (in the sense considered here) if and only if they coincide with geodesics. Vector fields for which there exists a Riemannian metric making all the orbits geodesics can be called *geodesible*. Certainly, Theorem 3.17 applies to such vector fields. Also, geodesible vector fields with (say, isolated) singularities are of great interest. Geodesic vector fields (and totally geodesic foliations of dimensions higher than 1) on specific Riemannian manifolds (spheres, projective spaces and so on) have been studied intensively (see [Fe, Gl] and the bibliographies therein) and, in some cases, have been completely classified. Some of the algebraic tools used therein are rather far form the content of this book, so we refer interested readers to the original papers.

3.2.4 Tautness and Holonomy

Here, we intend to prove the following result of [Ha].

Theorem 3.19. *On compact manifolds, tautness of foliations depends only on their holonomy pseudogroups.*

Let us explain closer the meaning of Theorem 3.19. Given two compact foliated manifolds $(M_i, \mathscr{F}_i)$, $i = 0, 1$, denote by $\mathscr{H}_i$ pseudogroups of local diffeomorphisms of manifolds T_i representing holonomy pseudogroups of $\mathscr{F}_i$. Then, *if $\mathscr{H}_0$ is isomorphic to $\mathscr{H}_1$ (see Section 1.1.2), then $\mathscr{F}_0$ is taut if only if $\mathscr{F}_1$ is taut.*

Towards the proof of Theorem 3.19, let us consider first an arbitrary pseudogroup $\mathscr{G}$ of diffeomorphisms of a manifold T. Let $D_c^k = D^k(T_c)$ be the space of compactly supported k-forms on T and $D_c^k(T/\mathscr{G})$ be the quotient of D_c^k by the linear subspace generated by all the forms $\alpha - g^*\alpha$ with $\alpha \in D_c^k$ and $g \in \mathscr{G}$ such that $\mathrm{supp}\,\alpha$ is contained in the range of g and equipped with quotient topology obtained from the C^∞-topology in D_c^k. Since $d \circ g^* = g^* \circ d$ for all g, the exterior differential $d : D_c^k \to D_c^{k+1}$ induces a continuous differential $d : D_c^k(T/\mathscr{G}) \to D_c^{k+1}(T/\mathscr{G})$.

Proposition 3.20. *A morphism Φ between pseudogroups $\mathscr{G}_0$ and $\mathscr{G}_1$ on manifolds T_0 and T_1 induces continuous linear maps $\Phi^* : D_c^k(T_0/\mathscr{G}_0) \to D_c^k(T_1/\mathscr{G}_1)$, $k = 0, 1 \ldots$.*

Proof. Express a compactly supported k-form α on T_0 in the form

$$\alpha = \sum_{\phi \in \Phi} \alpha_\phi \tag{3.4}$$

with α_ϕ supported compactly in the domain of ϕ. The formula

$$\Phi^*(\alpha) = \sum_{\phi \in \Phi} (\phi^{-1})^* \alpha_\phi$$

defines a linear map $\Phi^* : D_c^k(T_0) \to D_c^k(T_1)$, which induces a required map of quotient spaces. Indeed, if $\alpha = \beta - g^*\beta$ for some β and $g \in \mathcal{G}_0$, then $\beta = \sum_{\phi \in \Phi} \beta_\phi$ for some β_ϕ supported in the domain of ϕ. Without loss of generality (using a suitable partition of unity), one may assume that $g^*\beta_\phi$ is supported in the domain of some $\psi_\phi \in \Phi$. Then, $\Phi^*\alpha = \sum_{\phi \in \Phi}((\phi_{-1})^*\beta_\phi - (\phi \circ g \circ \psi_\phi^{-1})^*(\phi_{-1})^*\beta_\phi)$ is of required form since $\phi \circ g \circ \psi_\phi^{-1} \in \mathcal{G}_1$.

It remains to prove that our definition is correct that is, independent of the decomposition (3.4). This is an elementary exercise left to the readers. □

Proposition 3.20 implies directly the following.

Corollary 3.21. *An isomorphism of Φ of pseudogroups $\mathcal{G}_0$ and $\mathcal{G}_1$ yields continuous linear isomorphisms $\Phi^* : D_c^k(T_0/\mathcal{G}_0) \to D_c^k(T_1/\mathcal{G}_1)$, $k = 0, 1 \ldots$.*

Let now $\mathcal{F}$ be a p-dimensional foliation of a manifold M and $\mathcal{H}$ the holonomy pseudogroup of $\mathcal{F}$ acting on a complete transversal T being the disjoint union of T_i's, transversals in foliated charts U_i belonging to a regular covering $\mathcal{U} = \{U_i\}$ of $(M, \mathcal{F})$. Recall that, given a submersion $\pi : M \to B$ ($\dim M = p + \dim B$) the integration $\int_\pi$ along the fibers of π maps $(p+k)$-forms on M into k-forms on B and is determined uniquely by the condition

$$\int_\pi \alpha \wedge \pi^*\beta(x) = \int_{\pi^{-1}(x)} \iota_x^*\alpha \cdot \beta(x)$$

for any p-form α on M, k-form β on B and $x \in B$, $\iota_x : \pi^{-1}(x) \to M$ being the inclusion map. The above integration along the fibres allows to define integration $\int_\mathcal{F}$ along the leaves of $\mathcal{F}$:

$$\int_\mathcal{F} : D_c^{p+k}(M) \to D_c^k(T/\mathcal{H}). \tag{3.5}$$

For any $(k+p)$-form α decomposed as $\alpha = \sum_i \alpha_i$ with α_i supported in U_i, $\int_\mathcal{F} \alpha$ is the member of $D_c^k(T/\mathcal{H})$ represented by the form $\sum_i \int_{\pi_i} \alpha_i$, where $\pi_i : U_i \to T_i \subset T$ is the submersion defining $\mathcal{F}$ in the chart U_i of $\mathcal{U}$.

Following the lines of the proof of Proposition 3.20, one may show that the above integration along the leaves is defined correctly.

We say that a $(p+k)$-form α is $\mathcal{F}$-*trivial* whenever $\alpha(v_1, \ldots v_{p+k}) = 0$ for all vectors $v_1, \ldots, v_{p+k} \in T_xM$ such that $v_i \in T_x\mathcal{F}$ for $i = 1, \ldots, p$. Therefore, a form is $\mathcal{F}$-closed (see Definition 3.5) iff its differential is $\mathcal{F}$-trivial. With this terminology we have the following.

Proposition 3.22. *The kernel of $\int_{\mathscr{F}}$ in (3.5) coincides with the linear space gener-ated by all $\mathscr{F}$-trivial $(p+k)$-forms and the differentials of all $\mathscr{F}$-trivial $(p+k-1)$-forms.*

Proof. First, if $U = \mathbb{R}^p \times \mathbb{R}^q$ $(p+q=n)$ is a foliated chart and ω is a com-pactly supported $(p+k)$-form on U, then $\omega = \alpha + \beta$, where β is $\mathscr{F}$-trivial, $\alpha = \sum_I h_I dx^I \wedge dx^{q+1} \wedge \ldots \wedge dx^{p+q}$, $I = (i_1, \ldots, i_k)$, $1 \le i_1 < \ldots < i_k \le q$ and $dx^I = dx^{i_1} \wedge \ldots \wedge dx^{i_k}$. Then, $\omega \in \ker \int_{\mathscr{F}}$ if and only if for each I as above, $\int_{\mathbb{R}^p} h_I dx^{q+1} \wedge \ldots \wedge dx^{p+q} = 0$, therefore $h_I dx^{q+1} \wedge \ldots \wedge dx^{p+q} = d\gamma_I$ for some $(p-1)$-forms γ_I. Put $\gamma = (-1)^k \sum_I dx^I \wedge \gamma_I$. Certainly, γ is $\mathscr{F}$-trivial and $d\gamma = \alpha + \eta$ for an $\mathscr{F}$-trivial η. Consequently, $\omega = d\gamma - \eta + \beta$ with β, γ and η being $\mathscr{F}$-trivial.

Passing to the general case, take, as in the above construction of $\int_{\mathscr{F}}$, a regular covering $\mathscr{U} = \{U_i\}$ by foliated charts and submersions $\pi_i : U_i \to \mathbb{R}^q$ defining $\mathscr{F}$. Certainly, if $\omega = d\gamma + \beta$, γ and β being $\mathscr{F}$-trivial, then $\omega = \sum_i \omega_i$, $\omega_i = d\gamma_i + \beta_i$, where β_i and γ_i are supported in U_i and $\mathscr{F}$-trivial. Consequently, $\int_{\pi_i} \beta_i = 0$, $\int_{\pi_1} d\gamma_i = 0$ and, finally, $\int_{\mathscr{F}} \omega = 0$.

Assume conversely that $\omega = \sum_i \omega_i$ as before and that $\int_{\mathscr{F}} \omega = 0$, i.e., there exist k-forms β_{ji} on T_i which are compactly supported and satisfy

$$\int_{\pi_i} \omega_i = \sum_j (h_{ji}^* \beta_{ij} - \beta_{ji})$$

for each i; hereafter, h_{ij}'s are elementary holonomy transformations from open sub-domains of T_j's into T_i's. Choose $(p+k)$-forms α_{ji} supported compactly in $U_i \cap U_j$ and such that $\int_{\pi_i} \alpha_{ji} = \beta_{ji}$. Since $\pi_j = h_{ji} \circ \pi_i$ on $U_i \cap U_j$, $\int_{\pi_i} \alpha_{ij} = h_{ji}^* \beta_{ij}$. There-fore, if $\eta_i = \omega_i - \sum_j (\alpha_{ij} - \alpha_{ji})$, then $\int_{\pi_i} \eta_i = 0$ and, by the first step of our proof, $\eta_i = \beta_i + d\gamma_i$, β_i and γ_i being $\mathscr{F}$-trivial and supported in U_i. Since $\omega = \sum_i \eta_i$, $\omega = \beta + d\gamma$ for some $\mathscr{F}$-trivial forms β and γ. □

The above Proposition implies the following.

Proposition 3.23. *A compactly supported p-form ω_0 along the leaves of $\mathscr{F}$ (that is, a section of the p-th exterior power of the cotangent bundle of $\mathscr{F}$) coincides with the restriction of an $\mathscr{F}$-closed form ω iff $d \int_{\mathscr{F}} \omega_0 = 0$ in $D_c^1(T/\mathscr{H})$.*

Proof. Implication "⇒" is obvious. Assume that $d \int_{\mathscr{F}} \omega_0$ and ω_0 is the restriction of a p-form η on M. Then $\int_{\mathscr{F}} \eta = \int_{\mathscr{F}} \omega_0$ and $d \int_{\mathscr{F}} \eta = d \int_{\mathscr{F}} \omega_0 = 0$. Therefore, there exists an $\mathscr{F}$-trivial form α for which $d(\eta - \alpha)$ is $\mathscr{F}$-trivial. The form $\omega = \eta - \alpha$ is $\mathscr{F}$-closed and restricts to ω_0 along the leaves of $\mathscr{F}$. □

The final step leading towards Theorem 3.19 is the following.

Proposition 3.24. *A foliation $\mathscr{F}$ of a closed manifold M is geometrically taut if and only if, for a representation $\mathscr{H}$ of the holonomy pseudogroup acting on a transversal T, there exists a smooth, compactly supported, non-negative function $f : T \to \mathbb{R}$, which is strictly positive on a set intersecting all the $\mathscr{H}$-orbits and satisfies $df = 0$ in $D_c^1(T/\mathscr{H})$.*

Proof. Again, implication "⇒" is obvious: $\int_{\mathscr{F}} \omega_0$, ω_0 being the Riemannian volume form of $\mathscr{F}$ arising from the Riemannian metric g_0 making all the leaves minimal is equivalent in $D_c^0(T/\mathscr{H})$ to a function f satisfying the condition of our Proposition.

To get the converse let us show first that existence of f satisfying our condition is independent of holonomy representations. Assume that such f exists for T and $\mathscr{H}$ and let $\mathscr{H}'$ be a pesudogroup on T' isomorphic, *via* $\boldsymbol{\Phi} : \mathscr{H}' \to \mathscr{H}$, to $\mathscr{H}$. Let $K' \subset T'$ be a compact set meeting all the orbits of $\mathscr{H}'$. Cover K' by domains $U_1, \ldots, U_r$ of members ϕ_i of $\boldsymbol{\Phi}$ and choose other members $\phi_{r+i} : U_{r+i} \to T$, $i = 1, \ldots, s$, of $\boldsymbol{\Phi}$ such that $\operatorname{supp} f \subset \bigcup_{i=1}^{s} \phi_{r+i}(U_{r+i})$. Then, choose on T a partition of unity $(h_i, i = 1, \ldots, r+s)$ subordinated to the covering of T by the complement of $\operatorname{supp} f$ and all the sets $\phi_i(U_i)$, $i = 1, \ldots, r+s$, and put $f' = \sum_{i=1}^{r+s}(h_i f) \circ \phi_i$. This $f' : T' \to \mathbb{R}$ satisfies our condition.

Now, let $\mathscr{U} = \{U_i\}$ be a finite regular covering of M by foliated charts and let $\pi_i : U_i \to T_i$ be the corresponding projections. Let $V_i \subset \overline{V}_i \subset U_i$ be a bit smaller open and relatively compact sets covering M. For each i, choose a closed p-form α_i which, for any plaque P of U_i, satisfies $\int_P \alpha_i = 1$, is strictly positive on $P \cap V_i$ and has a compact support there. Let f be a function on T, the union of T_i's, satisfying our condition, and $f_i = f|T_i$. Then, $\omega = \sum_i (h_i \circ \pi_i) \cdot \alpha_i$ is positive on all the leaves and its integral $\int_{\mathscr{F}} \omega$ is equivalent to f. $\qquad\square$

Now, the reader should be able to extract from this section arguments providing the proof of Theorem 3.19.

3.3 Prescribing Mean Curvature in Codimension One

Throughout this section, $(M, \mathscr{F})$ is a closed foliated manifold, $(\dim M = n$ and $\operatorname{codim} \mathscr{F} = 1)$, both, M and $\mathscr{F}$ being oriented; therefore, $\mathscr{F}$ is also transversally oriented. The orientation convention used here says that $(N, X_1, \ldots, X_{n-1})$ is a positively oriented frame on M when N is a positive normal and $(X_1, \ldots, X_{n-1})$ is a positively oriented frame for $\mathscr{F}$. The section is devoted to the following (posed and solved partially in [W2], and solved completely in [Os3]) problem.

Problem 3.25. Given $(M, \mathscr{F})$ as above, describe the set $\operatorname{Mean}(\mathscr{F})$ of all smooth functions $h : M \to \mathbb{R}$, which can be realized as mean curvature functions of $\mathscr{F}$ with respect to some Riemannian metric g on M.

Certainly, this problem is closely related to Problem 3.12 which can be reformulated in the following way: *When $0 \in \operatorname{Mean}(\mathscr{F})$?* Therefore, the methods used in this section will be similar to those of Section 3.2.3. Moreover, we shall need some results about topology of codimension-one foliations, which will be discussed briefly in the subsection below.

3.3.1 Novikov Components

Given $(M, \mathcal{F})$ as above and two points x and y of M, we shall say that x *precedes* y and write $x \prec y$ whenever there exists a positive oriented and transverse to $\mathcal{F}$ curve $\gamma: [0,1] \to M$ with $\gamma(0) = x$ and $\gamma(1) = y$. It is not too difficult to show that the relation $\prec$ is transitive ($x \prec y \prec z \Rightarrow x \prec z$ for $x, y, z \in M$) and that it is possible that $x \prec y \prec x$: this happens when x and y lie on a loop transverse to $\mathcal{F}$. Moreover, if x, x' belong to the same leaf L and $x \prec y$, then $x' \prec y$. Indeed, taking a curve c on L connecting x and x', deforming it slightly to make it transverse to $\mathcal{F}$ and joining such a c with a positive oriented and transverse to $\mathcal{F}$ curve γ connecting x to y, one can obtain a positive oriented and transverse to $\mathcal{F}$ curve γ' connecting x' to y. In the same way, one can show that $x \prec y \Rightarrow x \prec y'$ when y and y' belong to the same leaf. These observations show that one can define the relation $\prec$ of precedence in the space of leaves:

$$L_1 \prec L_2 \iff x_1 \prec x_2 \quad \text{for some (equiv., any)} \quad x_1 \in L_1, x_2 \in L_2.$$

Again, the precedence of leaves is transitive and it may happen that $L_1 \prec L_2 \prec L_1$.

Using the above relation $\prec$ one can define the relation $\equiv$ of equivalence of points of M (and similarly, of leaves of $\mathcal{F}$):

$$x \equiv y \iff \quad \text{either} \quad x = y \quad \text{or} \quad x \prec y \prec x.$$

Definition 3.26. Equivalence classes of the above relation $\equiv$ are called *Novikov components* of $\mathcal{F}$.

Our observations above show that Novikov components are *saturated* by the leaves of $\mathcal{F}$: if L is a leaf intersecting a Novikov's component N, then $L \subset N$. Results of [No] provide the following.

Proposition 3.27 ([No]). *Let N be a Novikov component of $(M, \mathcal{F})$. Then either*

(i) $N = M$, *or*
(ii) $N = L$, *a single closed leaf, or*
(iii) N *is an open saturated set bounded by finitely many closed leaves.*

The following example provides a fact, which can be used as a hint towards the proof of Proposition 3.27 above.

Example 3.28. One can show that any non-compact leaf L on a closed foliated manifold $(M, \mathcal{F})$ is intersected by a closed loop transverse to $\mathcal{F}$. Indeed, let us cover M with finite family $\{U_i\}$ of foliated charts. Since L is noncompact, L intersects one of these charts, say U_1, infinitely many times. Let p_1 and p_2 be two points of the intersection of L with the transversal $T_1 \subset U_1$. Since L is connected (therefore, path-connected), we can find a (regular) curve γ on L joining p_1 to p_2. Modifying γ, step by step in the charts U_i containing the curve, is such a way that the new curve (denoted still by γ) is transverse to $\mathcal{F}$ and joining the end points q_1 and q_2 of the new γ in U_1 by an arc transverse to $\mathscr{F}$ we get a loop which intersects L and is transverse to $\mathscr{F}$.

The relation $\prec$ defined above induces that for Novikov components: given two such components N_1 and N_2, the relation $N_1 \prec N_2$ holds whenever there exists a positive oriented and transverse to $\mathscr{F}$ curve with origin in N_1 and end point in N_2. The family $\mathscr{N}$ of all the Novikov components of $\mathscr{F}$ is partially ordered by $\prec$ and any ordered family of Novikov components admits maximal elements. Therefore, Zorn Lemma implies the existence of maximal (and, minimal) components of $\mathscr{F}$. Certainly, if $N = M$ is the only Novikov component of $\mathscr{F}$ then it is maximal and minimal at the same time. A component $N = L$ consisting of a single closed leaf cannot be neither maximal nor minimal: indeed, there exists a positive oriented and transverse to $\mathscr{F}$ arc $\gamma : [0,1] \to M$ intersecting L at a single point and then the Novikov components N_0 and N_1 containing, respectively, the points $\gamma(0)$ and $\gamma(1)$ precede and follow N: $N_0 \prec N \prec N_1$. In the same way, if N is of type (iii) in Proposition 3.27 and the boundary of N contains two leaves L_0 and L_1 such that the positively oriented normals of $\mathscr{F}$ point inwards along L_0 and outwards along L_2, then N cannot be neither minimal nor maximal: indeed if γ_0 and γ_1 are positive oriented and transverse to $\mathscr{F}$ arcs intersecting L_0 and L_1 at single points, then the component N_0 containing the origin of γ_0 together with the component N_1 containing the end point of γ_1 satisfy $N_0 \prec N \prec N_1$. This discussion implies the following.

Corollary 3.29. *If $(M, \mathscr{F})$ contains more than one Novikov component, then maximal (resp., minimal) components N are open saturated sets bounded by finitely many closed leaves such that the positively oriented normals of $\mathscr{F}$ point inwards (resp., outwards) everywhere along the boundary ∂N of N.*

Oshikiri [Os3] calls an open saturated domain U of $(M, \mathscr{F})$ *positive* (resp., *negative*) whenever the positive oriented normals point outwards (resp., inwards) everywhere along the boundary ∂U. Therefore, maximal Novikov components are negative while minimal Novikov components are positive in this sense.

Example 3.30.

(i) If $\mathscr{F}$ admits a dense leaf, then M is the only Novikov component of $\mathscr{F}$.

(iii) The Reeb foliation $\mathscr{R}$ of the three-dimensional sphere S^3 consists of three Novikov components: two of them coincide with the Reeb components of $\mathscr{R}$ (and are of type (iii)) while the third one coincides with the two dimensional torus, the only closed leaf of $\mathscr{R}$ separating its Reeb components (and is of type (ii) of Proposition 3.27).

Remark 3.31.

(i) Given an odd number $n > 3$, one may construct a foliation $\mathscr{F}$ of S^3 which consists of exactly n Novikov components. Begin with the standard Reeb foliation of S^3. For $n = 5$, choose a closed transversal Γ contained in one of the Reeb components $\mathscr{F}$ and perform turbulization (compare Section 2.2) along Γ. We get a foliation with two closed leaves and three open Novikow components. For $n > 5$, perform this procedure more times.

(ii) One may construct on S^3 a foliation with infinitely many Novikov components. Begin again with the standard Reeb foliation. Thicken T^2, the unique toral leaf to get a "thick torus" $T^2 \times [-\varepsilon, \varepsilon]$, choose an infinite sequence $0 < t_1 < t_2 < t_3 \ldots < \varepsilon$, take tori $T_i = T^2 \times \{t_i\}$, $i = 1, 2, 3, \ldots$, as the leaves of the foliation under construction and fill each zone $Z_i = T^2 \times (t_i, t_{i+1})$ with suitable surfaces approaching asymptotically both boundary tori, T_i and T_{i+1}.

3.3.2 Consequences of Rummler Formula

Let g be a Riemannian metric on M. Since $\mathrm{codim}\,\mathscr{F} = 1$ and $\mathscr{F}$ is transversely oriented, the (scalar) mean curvature function σ_1 of $\mathscr{F}$ can be defined as $\sigma_1 = \langle H, N \rangle$, N and H being, respectively, the positive oriented unit normal and the mean curvature vector of $\mathscr{F}$. In this case, Rummler formula (3.2) reads as

$$d\Omega_{\mathscr{F}} = -\sigma_1 \cdot \Omega_M, \tag{3.6}$$

$\Omega_{\mathscr{F}}$ and Ω_M being, respectively, the Riemannian volume forms of $\mathscr{F}$ and M. The above and the Stokes' Theorem implies immediately that

$$\int_M \sigma_1 = 0 \quad \text{and} \quad \int_U \sigma_1 = \sum_{i=1}^k \varepsilon_i \mathrm{vol}(L_i) \tag{3.7}$$

when U is an open saturated domain bounded by the leaves $L_1, \ldots, L_k$ and $\varepsilon_i = 1$ (resp., $\varepsilon_i = -1$) whenever N points inwards (resp., outwards) U along L_i.

Equation (3.7) implies directly the following.

Proposition 3.32.

(i) *If $f \in \mathrm{Mean}(\mathscr{F})$, then either $f \equiv 0$ or $f(x) \cdot f(y) < 0$ for some points x and y of M.*

(ii) *If $f \in \mathrm{Mean}(\mathscr{F})$ and $N \neq M$ is a maximal (resp., minimal) Novikov component of $\mathscr{F}$, then $f(x) > 0$ (resp., $f(x) < 0$) at some points x of N.*

3.3.3 Characterization of Mean Curvature Functions

Here, we shall show that the "sign conditions" of Proposition 3.32 are not only necessary but also sufficient for belonging f to $\mathrm{Mean}(\mathscr{F})$. This is similar to the following result of [KW]: A function $f : S \to \mathbb{R}$ can be realized as the Gauss curvature of a closed surface S if and only if (1) S contains points x, where the sign of $f(x)$ is the same as that of the *Euler characteristic* $\chi(S)$ when $\chi(S) \neq 0$ and (2) either $f \equiv 0$ on S or S contains points x and y such that $f(x) < 0 < f(y)$ when $\chi(S) = 0$. More precisely, we have the following.

Theorem 3.33 ([Os3]). *Let $\mathcal{F}$ be a transversely oriented codimension-one foliation of a closed oriented manifold M.*

(i) *If the only Novikov component N of $\mathcal{F}$ coincides with M, then* $\mathrm{Mean}(\mathcal{F})$ *consists of 0 and all the functions f such that $f(x) \cdot f(y) < 0$ for some points x and y of M.*

(ii) *Otherwise,* $\mathrm{Mean}(\mathcal{F})$ *consists of all the functions f positive at some points of any maximal Novikov component of $\mathcal{F}$ and negative at some points of any minimal component of $\mathcal{F}$.*

Proof. The theorem follows directly from Lemmas 3.36–3.38 below. $\square$

This theorem together with Theorem 3.17 implies directly another description of topological tautness for codimension-one foliations.

Corollary 3.34 ([Su2]). *A transversely oriented codimension-one foliation $\mathcal{F}$ of a closed oriented manifold M is topologically taut if and only if*

(i) *any closed leaf of $\mathcal{F}$ meets a loop transverse to $\mathcal{F}$,*
    equivalently if and only if
(ii) *there exists a loop transverse to $\mathcal{F}$ which intersects all the leaves of $\mathcal{F}$.*

Remark 3.35. One may prove equivalence of conditions (i) and (ii) of Corollary 3.34. Implication (ii) $\Rightarrow$ (i) is obvious.

(i) $\Rightarrow$ (ii). By Example 3.28, every leaf meets a transverse loop. Therefore, we can find a finite number of transverse loops Γ_i, $i = 1, \ldots, m$ such that their saturations U_i cover M. For any i and j such that $U_i \cap U_j \neq \emptyset$ choose a leaf $L_{ij} \subset U_i \cap U_j$. This leaf meets Γ_i and Γ_j at some points p_i and p_j. Connect these points by a segment a curve in L_{ij} and modify it slightly to get a segment Σ_{ij} which connects some points of Γ_i and Γ_j and is transverse to the foliation. A suitable join of Γ_i's and Σ_{ij}'s provides a loop which satisfies our condition.

Lemma 3.36 ([Os3]). *If $f : M \to \mathbb{R}$ satisfies conditions (i) and (ii) of Proposition 3.32, then there exists a volume form Ω on M such that*

$$\int_M f \cdot \Omega = 0 \quad and \quad \int_D f \cdot \Omega < 0 \tag{3.8}$$

for any positive domain $D \subset M$.

Proof. Denote by $N_1, \ldots, N_k$ (resp., by $N_{k+1}, \ldots, N_{k+l}$) all the minimal (resp., maximal) Novikov components of $\mathcal{F}$. For any $i \leq k$ (resp., $j \leq l$) denote by a_i (resp., b_j) the number of all the maximal components which can be reached from N_i by a path which is transverse to $\mathcal{F}$ and positive oriented (resp., the number of all the minimal components from which N_{k+j} can be reached by a path like that). From the definition of Novikov components and properties of the relation $\prec$ of Section 3.3.1 it follows easily that each N_{k+j} can be reached form some N_i by such a path and that for each $i \leq k$ there exists $j \leq l$ such that N_{k+j} can be reached from N_i in such a way. This implies that

$$\sum_{i=1}^{k} a_i - \sum_{j=1}^{l} b_j = 0. \tag{3.9}$$

Now, take a Riemannian metric g_0 on M and denote by ω_0 its volume form. Take also a function f satisfying our conditions and deform ω_0 locally, in neighborhoods of some interior points of Novikov components, to get a volume form Ω such that

$$c = \int_{M \setminus (\cup_i N_i \cup \cup_j N_{k+j})} f\Omega < 1/2, \quad \int_{N_1} f\Omega = c - a_1 < 0,$$
$$\int_{N_i} f\Omega = -a_i \quad \text{and} \quad \int_{N_j} f\Omega = b_j \tag{3.10}$$

for all $i \in \{2, \dots, k\}$ and $j \in \{1, \dots, l\}$.

From (3.9) and (3.10), the equality $\int_M f\Omega = 0$ follows directly.

To complete the proof take any positive saturated domain D and define the sets $I \subset \{1, \dots, k\}$ and $J \subset \{1, \dots, l\}$, respectively, by the conditions $i \in I \Leftrightarrow N_i \subset D$ and $j \in J \Leftrightarrow N_{k+j} \subset D$. Since D is positive,

$$\sum_{i \in I} a_i - \sum_{j \in J} b_j \geq 1. \tag{3.11}$$

Indeed, if $N_i \subset D$ can be connected to some N_{k+j} by a path which is transverse to $\mathscr{F}$ and positive oriented, then $N_{k+j} \subset D$. Also, there exists $N_i \subset D$ which can be connected by such a path to a maximal Novikov component lying outside of D. Inequality (3.11) implies that

$$\int_D f\Omega \leq -\sum_{i \in I} a_i + \sum_{j \in J} b_j + 2c \leq -1 + 2c < 0.$$

(One may split the proof of the above into two cases: $N_1 \subset D$ and $N_1 \subset M \setminus D$.) $\square$

Lemma 3.37 ([Os2]). *If a volume form Ω satisfies (3.8), then*

$$z(f\Omega) < 0 \tag{3.12}$$

for any $z \in \partial^{-1}(C_{\mathscr{F}} \cap B_{n-1})$.

Proof. We sketch the idea of the proof. Its details depend on several results in foliation theory which are too far from the main topic of our book.

Assume that Ω satisfies (3.8) and take a current z such that $\partial z \in C_{\mathscr{F}}$. Then, by [HL, Theorem 7.2], the support of ∂z coincides with the union of a family $\mathscr{C}$ of compact leaves. If $\mathscr{C} = \{L_1 \cdots, L_k\}$ is finite, then $\partial z = \sum_{i=1}^{k} a_i z_{L_i}$ is a linear combination of integration currents (see Example 3.8(2)) over L_i's with real positive coefficients a_i's. Since ∂z represents zero in real homology of M, the family $\mathscr{C}$ splits in a number of subfamilies $\mathscr{C}_j$, $j = 1, \dots, m$, for which the corresponding parts ξ_j of ∂z represent zero in integral homology. Using some arguments of [CG, Lemma 1.1] one can show that each of such parts can be represented as ∂z_{D_j} for a positive saturated domain D_j. Therefore, $\partial z = \sum_{j=1}^{m} b_j \partial z_{D_j}$ for some positive constants b_j. Consequently, $z = \sum_{j=1}^{m} b_j z_{D_j} + a \cdot z_M$ for some $a \in \mathbb{R}$ and, by (3.8),

$$z(f\Omega) = \sum_{j=1}^{m} b_j \int_{D_j} f\Omega + a \int_M f\Omega < 0.$$

Assume that $\mathscr{C}$ is infinite. Note that: (i) the union $C(\mathscr{F})$ of all compact leaves is compact ([CC], vol. I, Theorem 6.1.1), (ii) a leaf L contained in the interior of $C(\mathscr{F})$ belongs to a foliated bundle $L \times [0,1]$ (ibid., Theorem 5.3.4), (iii) the complement of $C(\mathscr{F})$ contains at most finitely many components which are not foliated bundles as above (ibid., Corollary 5.2.9) and (iv) the boundary of any component of $M \setminus C(\mathscr{F})$ is a finite union of leaves (ibid., Lemma 5.2.5). Therefore, all but finitely many leaves of $\mathscr{C}$ are contained in foliated bundles.

Now, choose a compact leaf L which belongs to a foliated bundle $E = L \times [0,1]$ which intersects non-trivially the support of our current z. The restriction to E of the boundary ∂z is of the form $z_L \times \mu$ for a Radon measure μ on the interval $[0,1]$. That is, $\partial z(\eta) = \int_0^1 (\int_L \omega) d\mu$ for any $(n-1)$-form η supported in E. Denote by J the set of all $t \in [0,1]$ corresponding to compact leaves of $\mathscr{F}$ in E. There exists a positive domain D such that $L \times \{0\} \subset \partial D$ (compare [HL, Thm. 7.7]). Set $D_t = D \cup (L \times [0,t])$ for $t \in J$ and choose $t_0 \in J$ for which

$$\int_0^1 \left(\int_{D_t} f\Omega \right) d\mu - \mu([0,1]) \cdot \int_{D_{t_0}} f\Omega \leq 0.$$

(This is possible, otherwise we would have the inequality

$$\int_0^1 \left(\int_{D_t} f\Omega \right) d\mu > \mu([0,1]) \cdot \int_{D_{t_0}} f\Omega$$

for all $t \in J$, which after integrating over $[0,1]$ with respect to μ leads—since μ is supported in J—towards a contradiction of the form $0 < 0$.) Define a current z_0 by

$$z_0(\eta) = \int_0^1 \left(\int_{D_t} \eta \right) d\mu - \mu([0,1]) \cdot \int_{D_{t_0}} \eta, \quad \eta \in D^n.$$

Then, $\partial z_0 = \partial(z|E) - a \cdot z_{L_0}$ with $a = \mu([0,1])$ and $L_0 \subset E$ being the leaf corresponding to $t_0 \in J$.

Since the support of ∂z is compact, proceeding as above finitely many times we can find compact leaves $L_1, \ldots, L_k$, positive constants $a_1, \ldots, a_k$ and a current $\zeta \in D_{n-1}$ such that $\partial \zeta = \partial z - \sum_i a_i z_{L_i}$ and $\zeta(f\omega) \leq 0$. This reduces the proof to the previous case ($\mathscr{C}$ finite) and completes the proof of the lemma. □

Lemma 3.38 ([Os1]). *If there exists a volume form Ω satisfying (3.12), then $f \in$* Mean$(\mathscr{F})$.

Proof. Since $\int_M f\Omega = 0$, there exists an $(n-1)$-form η such that $d\eta = -f\Omega$. The restriction to B_{n-1} of the linear functional $\eta : D_{n-1} \to \mathbb{R}$ determined by this form does not depend on the choice of such an η. By the second property of Ω in (3.12), applying the Hahn-Banach Theorem (Thm. 3.16) to $\eta|B_{n-1}$ we get a linear functional $\lambda \in D_{n-1}^*$ positive at all non-zero elements of $C_{\mathscr{F}}$ and equal to η on B_{n-1}.

By Theorem 3.9, such λ is represented by an $(n-1)$-form χ. Since $\chi = \eta$ on B_{n-1}, $d\chi = -\Omega$. Now, one can find a Riemannian metric g_0 on M for which the orthogonal complement of $T\mathscr{F}$ coincides with $\ker \chi$, the volume form $\Omega_{\mathscr{F}}$ of $\mathscr{F}$ coincides with χ and the volume form of (M, g_0) coincides with Ω. By Rummler formula (3.6), the mean curvature function of $\mathscr{F}$ with respect to g_0 coincides with f. $\qquad\square$

Example 3.39. Already in 20'th of the last century, Kneser [Kn] classified one-dimensional C^2-foliations of the two-dimensional torus T^2. These are either minimal or consist of countably many annuli A_k foliated either by (i) circles, or (ii) lines spiraling from one boundary components towards the other component with the same orientation, or (ii) lines spiraling from one boundary components towards the other component with the opposite orientation. Annuli of type (iii) are called *Reeb* or *Poincaré components*. The number of such components is always finite and a foliation like that is transversely oriented if and only if this number is even. Such a foliation is taut if and only if Reeb components do not exist. In this case, T^2 itself is the only Novikov component of a foliation. Otherwise, maximal and minimal Novikov components correspond precisely to suitably oriented Reeb components. Theorem 3.33 allows to classify easily all the mean curvature functions (with respect to all Riemannian structures on T^2) of a given foliation.

Remark 3.40. One may classify one-dimensional foliations $\mathscr{F}$ of the two-dimensional torus T^2 such that there exists a Riemannian metric on T^2 for which all the leaves of $\mathscr{F}$ have constant curvature. Existence of Reeb (Poincaré) components may be an obstruction. For example, if the whole T^2 is the union of two such components with common boundaries, then the curvature should be positive at some points of one of the components, negative—at some points of the other one. If the leaf curvatures would be constant, then the circles which form the common boundaries should have positive and negative curvature at the same time. This is just impossible. To solve the question in general, one should discuss different situations analogous to this above.

3.4 Prescribing Mean Curvature in Higher Codimension

Throughout this section, M is a closed m-dimensional manifold equipped with a foliation $\mathscr{F}$ of arbitrary dimension p and X is a vector field on M transverse to $\mathscr{F}$ at all the points $x \in M$ at which $X(x) \neq 0$. We study the following problem.

Problem 3.41. Does there exist a Riemannian metric g on M for which X becomes the mean curvature vector of $\mathscr{F}$?

We provide partial solutions of this problem in two cases: far from singular points of X (Section 3.4.2) and around singular points of X (Section 3.4.3).

3.4.1 Notation

For any $k \in \{1,\ldots,m\}$ and any closed set A of M denote by $D_k(A)$ the space of k-currents supported in A. By definition, $D_k = D_k(M)$ is the dual of D^k, the space of differentiable k-forms on M with the C^∞ topology.

Also let X be a smooth vector field on M and let Σ be the set of all the singular points of X. Given A, a closed subset of M contained in $M \setminus \Sigma$, consider the following subsets of D_p determined by $\mathscr{F}$ and X:

- $C_\mathscr{F}(A)$, the closed convex cone generated by all the Dirac currents of the form $e_1 \wedge \cdots \wedge e_p$, where $(e_1,\ldots,e_p)$ is a positively oriented frame of $T_x\mathscr{F}$, the space tangent to the leaf through x, $x \in A$;
- $\tilde{C}_{\mathscr{F},X}(A)$, the closed convex cone generated by all the boundaries of the form $\partial(-X(x) \wedge e_1 \wedge \cdots \wedge e_p)$, where $x \in A$ and $e_1,\ldots,e_p$ are as above;
- $C_{\mathscr{F},X}(A)$, the closed convex cone generated by the union $C_\mathscr{F}(A) \cup \tilde{C}_{\mathscr{F},X}(A)$;
- $P_{\mathscr{F},X}(A)$, the closed linear space generated by all the Dirac currents $X(x) \wedge v_1 \cdots \wedge v_{p-1}$, where $v_1,\ldots,v_{p-1} \in T_x\mathscr{F}$ and $x \in A$.

We write $C_\mathscr{F}$ instead of $C_\mathscr{F}(M)$ and, if $\Sigma = \emptyset$, $\tilde{C}_{\mathscr{F},X}$, $C_{\mathscr{F},X}$ and $P_{\mathscr{F},X}$ instead of $\tilde{C}_{\mathscr{F},X}(M), C_{\mathscr{F},X}(M)$ and $P_{X,\mathscr{F}}(M)$, respectively. In [Su1], it was proved that the cone $C_\mathscr{F}$ has a compact base, that is, that there exists a continuous linear functional $\lambda : D_k \to \mathbb{R}$ positive on $C_\mathscr{F} \setminus 0$ and such that the set $\lambda^{-1}(1) \cap C_\mathscr{F}$ is compact. Obviously, the same holds for all the cones $C_\mathscr{F}(A)$ with A closed in M. However, Sullivan's argument cannot be applied to the cones $\tilde{C}_{\mathscr{F},X}(A)$ (and, *a fortiori*, to $C_{\mathscr{F},X}(A)$). In fact, in some situations these cones do not have compact bases.

Example 3.42. Let M be a closed oriented three-dimensional manifold equipped with a two-dimensional foliation $\mathscr{F}$ and a nowhere vanishing vector field X transverse to $\mathscr{F}$. If $(M,\mathscr{F})$ contains a Reeb component R bounded by a two-dimensional torus T, then both currents $\int_T$ and $-\int_T$ belong to $\tilde{C}_{\mathscr{F},X} \subset C_{\mathscr{F},X}$. Indeed, $\int_T = \partial \int_R$, $-\int_T = \partial \int_{M \setminus R}$ and the integrals $\int_R$ and $\int_{M \setminus R}$ can be expressed as limits of convex combinations of Dirac currents of the form $X(x) \wedge e_1 \wedge e_2$, where $e_i \in T_x\mathscr{F}$ and $x \in M$. Since the cones $\tilde{C}_{\mathscr{F},X}$ and $C_{\mathscr{F},X}$ contain two non-trivial elements $\pm c, c \in D_2$, they cannot have compact bases. The same argument applies to any codimension-one foliation $\mathscr{F}$ which admits an either positive or negative domain U: Again, the currents $\pm \int_{\partial U}$ belong to $\tilde{C}_{\mathscr{F},X}$.

Now assume that $X = H_{\mathscr{F},g}$ is the mean curvature vector of $\mathscr{F}$ for some Riemannian metric g. Let $\Omega = \Omega_{\mathscr{F},g}$ be the volume form of $\mathscr{F}$ with respect to g: Ω is the differential p-form which gives the p-volume on the leaves and vanishes if any argument is a vector orthogonal to the leaf. Rummler formula (3.2) implies that

$$\Omega | \tilde{C}_{\mathscr{F},X}(A) \setminus 0 > 0. \tag{3.13}$$

Since obviously $\Omega | C_\mathscr{F} \setminus 0 > 0$, condition (3.13) implies that

$$\Omega | C_{\mathscr{F},X}(A) \setminus 0 > 0. \tag{3.14}$$

Moreover, in this case we have the following.

Lemma 3.43. *If $X = H_{\mathscr{F},g}$ and $A \subset M \setminus \Sigma$ is closed in M, then the cone $C_{\mathscr{F},X}(A)$ has a compact base.*

Proof. Set $B = C_{\mathscr{F},X}(A) \cap \Omega^{-1}(1)$. We have to show that $B \subset D_p$ is compact, i.e. that any continuous linear functional $\lambda : D_p \to \mathbb{R}$ is bounded on B.

According to Schwartz's Theorem 3.9, any such λ can be represented by a differential p-form ω. Take any current

$$c = \sum_i t_i e_1^i \wedge \cdots \wedge e_p^i - \sum_j t_j \partial (X(x_j) \wedge e_1^j \wedge \cdots \wedge e_p^j),$$

where $x_j \in A$, the $(e_1^k, \dots, e_p^k)$'s are positive oriented orthonormal frames of $T\mathscr{F}$ at certain points of A and all the coefficients t_i and t_j are positive. Note that

$$\Omega(\partial(X(x_j) \wedge e_1^j \wedge \cdots \wedge e_p^j)) = d\Omega(X(x_j) \wedge e_1^j \wedge \cdots \wedge e_p^j) = -\|X(x_j)\|^2$$

by (3.2). Hence if $c \in B$, then

$$1 = \Omega(c) = \sum_i t_i + \sum_j \|X(x_j)\|^2 t_j \geq \alpha \left(\sum_i t_i + \sum_j t_j \right),$$

where

$$\alpha = \min\{1, \min\{\|X(x)\|^2 : x \in A\}\} > 0.$$

Therefore,

$$|\omega(c)| \leq \sum_i t_i |\omega(e_1^i \wedge \cdots \wedge e_p^i)|$$
$$+ \sum_j t_j |d\omega(X(x_j) \wedge e_1^j \wedge \cdots \wedge e_p^j)| \leq \alpha^{-1} \cdot (\|\omega\| + \|d\omega\| \cdot \|X\|),$$

where $\|X\| = \max\{\|X(x)\|; x \in M\}$,

$$\|\omega\| = \max\{|\omega(\xi)|; \xi \in \Lambda^p T_x M, \|\xi\| = 1, x \in M\},$$

and so on. □

Finally, it is obvious that if $X = H_{\mathscr{F},g}$, then

$$\Omega_{\mathscr{F}} | P_{\mathscr{F},X}(A) \equiv 0 \tag{3.15}$$

for any closed set A contained in $M \setminus \Sigma$. Conditions (3.14) and (3.15) imply that in this case the cone $C_{\mathscr{F},X}(A)$ intersects the space $P_{\mathscr{F},X}(A)$ trivially, i.e.,

$$C_{\mathscr{F},X}(A) \cap P_{\mathscr{F},X}(A) = \{0\}.$$

3.4.2 Away from Singularities

Here, we consider the problem of prescribing mean curvature assuming that it is already solved in a neighborhood of the singular set Σ of a given vector field X. More precisely, given a closed oriented manifold M with an oriented foliation $\mathscr{F}$ and a vector field X on M, which vanishes precisely on Σ, we prove the following.

Theorem 3.44. *Suppose that g_0 is a Riemannian metric on a neighborhood U of Σ on which $X = H_{\mathscr{F},g_0}$. Let B be a smooth compact m-dimensional submanifold with boundary for which*

$$\Sigma \subset \mathrm{Int}(B) \subset B \subset U.$$

Let A denote the closure of $M \setminus B$, so that $\partial A = \partial B = A \cap B$. Then there exists a Riemannian metric g on M extending $g_0|_B$ and such that $X = H_{\mathscr{F},g}$ on M if and only if the cone $C_{\mathscr{F},X}(A)$ has a compact base and intersects the space $P_{\mathscr{F},X}(A)$ trivially.

Proof. In the previous section it was shown that if the metric g exists, then the cone $C_{\mathscr{F},X}(A)$ has a compact base and intersects the space $P_{\mathscr{F},X}(A)$ in 0. To prove the converse, let us impose these conditions on the cone $C_{\mathscr{F},X}(A)$.

In order to obtain the desired metric g, we shall first construct a differential p-form Ω and then construct g that makes Ω the volume form along the leaves of $\mathscr{F}$. Let Ω_0 be the volume form of $\mathscr{F}$ on U with respect to g_0.

Claim. *Under our hypotheses, the sum $\ker(\Omega_0|B) + P_{\mathscr{F},X}(A)$ also intersects the cone $C_{\mathscr{F},X}(A)$ trivially.*

To prove the Claim, take a small normal neighborhood $N \subset U$ of $A \cap B$ such that each point x in N has a unique shortest g_0-geodesic γ_x joining x to a point of $A \cap B$ and so that γ_x is entirely contained in N. If we denote the endpoint of γ_x by $\mathrm{pr}(x)$, then the resulting projection $\mathrm{pr} : N \to A \cap B$ becomes a smooth map. Also equip the bundle $T\mathscr{F}|U$ with a linear connection ∇ (for example, the one induced by the Levi-Città connection on (U, g_0)) and let τ_x denote parallel transport along γ_x. Finally choose a smooth function $f : M \to \mathbb{R}$ supported in N and equal to 1 at all the points of $A \cap B$. With all these tools in hand, define a projection map $\pi : W \to D_p(A \cap B)$, W being the closure of the linear subspace of $D_p(A)$ generated by the union $C_{\mathscr{F},X}(A) \cup P_{\mathscr{F},X}(A) \cup D_p(A \cap B)$, in the following way. First, for generators of $C_{\mathscr{F},X}(A)$ and $P_{\mathscr{F},X}(A)$ put

$$\pi(e_1 \wedge \cdots \wedge e_p) = f(x) \cdot \tau_x(e_1) \wedge \cdots \wedge \tau_x(e_p),$$
$$\pi(-\partial(X(x) \wedge e_1 \wedge \cdots \wedge e_p)) = -f(x) \cdot \partial(X(\mathrm{pr}(x)) \wedge \tau_x e_1 \wedge \cdots \wedge \tau_x e_p),$$
$$\pi(X(x) \wedge v_1 \wedge \cdots \wedge v_{p-1}) = f(x) \cdot X(\mathrm{pr}(x)) \wedge \tau_x v_1 \wedge \cdots \wedge \tau_x v_{p-1},$$

whenever $e_1 \wedge \cdots \wedge e_p$ is a positively oriented frame of $T_x\mathscr{F}$, $v_j \in T_x\mathscr{F}$ and $x \in N$. Then π is well defined, since the generators of the cone (in the first two formulas) are linearly independent, as can easily be shown by evaluating a linear combination of them on appropriate differential p-forms, and this cone meets the subspace $P_{\mathscr{F},X}(A)$

(whose generators appear in the third formula) trivially. If $x \notin N$, then set π equal to 0 in all these cases. Next, put $\pi(c) = c$ for all $c \in D_p(A \cap B)$. From the above formulas it is clear that π is continuous where defined. Therefore, it can be extended over W by linearity and continuity. This projection $\pi : W \to D_p(A \cap B)$ maps the cone $C_{\mathscr{F},X}(A)$ into the cone $C_{\mathscr{F},X}(A \cap B)$ and the subspace $P_{\mathscr{F},X}(A)$ into the subspace $P_{\mathscr{F},X}(A \cap B)$.

Now, suppose that a non-zero element c of $C_{\mathscr{F},X}(A)$ decomposes as $c = c_1 + c_2$ with $c_1 \in D_p(B)$ and $c_2 \in P_{\mathscr{F},X}(A)$. Since $c_1 = c - c_2$ is supported in $A \cap B$, $\pi(c_1) = c_1$. Therefore, $\pi(c) \neq 0$ since otherwise $c_1 = -\pi(c_2)$ would be supported in $A \cap B$ while c itself would lie in the intersection $C_{\mathscr{F},X}(A) \cap P_{\mathscr{F},X}(A) = \{0\}$. Consequently, $\Omega_0(c_1) = \Omega_0(\pi(c)) \neq 0$, that is $c_1 \notin \ker \Omega_0$. This proves our Claim.

Next, Ω_0 on $D_p(B)$ extends to a continuous linear functional $\lambda_0 : D_p(B) + P_{\mathscr{F},X}(A) \to \mathbb{R}$ by setting $\lambda_0 = 0$ on $P_{\mathscr{F},X}(A)$, for Ω_0 vanishes on the generators of $P_{\mathscr{F},X}(A)$. Furthermore, since Ω_0 is positive on currents in $C_{\mathscr{F},X}(A) \setminus 0$ where it is defined, the same holds for λ_0, so we can apply the Hahn-Banach Theorem 3.16 to extend λ_0 to a continuous linear functional $\lambda : D_p \to \mathbb{R}$ that is positive on $C_{\mathscr{F},X}(A) \setminus 0$. Clearly $\lambda = \Omega_0$ on $D_p(B)$ and $\lambda \equiv 0$ on $P_{\mathscr{F},X}(A)$. As before, λ is represented by a unique globally defined p-form ω.

Let Ω denote the purification of ω with respect to $T\mathscr{F}$ ([Su2] and Section 3.2.3 here). Recall that this means that if $x \in M$ and $P_\omega : T_x M \to T_x \mathscr{F}$ is the projection defined by the condition

$$\iota_{P_\omega(v)}(\omega | \Lambda^p T_x \mathscr{F}) = (\iota_{P_\omega(v)} \omega) | \Lambda^{p-1} T_x \mathscr{F} = (\iota_v \omega) | \Lambda^{p-1} T_x \mathscr{F}, \quad v \in T_x M,$$

then

$$\Omega(v_1, \ldots v_p) = \omega(P_\omega(v_1), \ldots, P_\omega(v_p))$$

for all $v_1, \ldots, v_p \in T_x M$, $x \in M$, and that from the above it follows directly that

$$\Omega(w, v_1, \ldots, v_{p-1}) = \omega(w, v_1, \ldots, v_{p-1}), \tag{3.16}$$

whenever $w \in T_x M$, $v_j \in T_x \mathscr{F}$, $x \in M$. Since the bundle $T\mathscr{F}$ is integrable, for any $w \in T_{x_0} M$, $x_0 \in M$ one can find a $(p+1)$-dimensional submanifold N such that $w \in TN$ and $T_x \mathscr{F} \subset T_x N$ for all $x \in N$. If $\iota_N : N \to M$ is the canonical inclusion, then $\iota_N^* \Omega = \iota_N^* \omega$ and consequently $\iota_N^* d\Omega = \iota_N^* d\omega$. In particular,

$$d\Omega(w, v_1, \ldots v_p) = d\omega(w, v_1, \ldots, v_p) \tag{3.17}$$

for any $v_1, \ldots, v_p \in T_{x_0} \mathscr{F}$. Equalities (3.16) and (3.17) show that Ω is strictly positive on $C_{\mathscr{F},X}(A) \setminus 0$ and vanishes identically on $P_{\mathscr{F},X}(A)$. Also, since $\omega = \Omega_0$ on $D_p(B)$ and Ω_0 is pure, the forms Ω and Ω_0 coincide on B.

Now decompose the tangent bundle TM over A into the direct sum

$$TM = T\mathscr{F} \oplus \ker \Omega = T\mathscr{F} \oplus \mathrm{Span}(X) \oplus E, \tag{3.18}$$

where E is the $(m - p - 1)$-dimensional subbundle of $\ker \Omega$ defined by the equation

$$\iota_w d\Omega | \Lambda^p T_x \mathscr{F} = 0.$$

Note that $\ker \Omega$ coincides with the g_0-orthogonal complement of $T\mathscr{F}$ at all points of B. Also, Rummler's formula (3.2) implies that E is g_0-orthogonal to X on B. Choose a Riemannian metric $g = \langle \cdot, \cdot \rangle$ on M making all components of the decomposition in (3.18) orthogonal and such that the volume form $\Omega_{\mathscr{F},g}$ equals Ω, while

$$\langle X, Z \rangle = -d\Omega(Z, E_1, \ldots, E_p), \tag{3.19}$$

where $E_1, \ldots, E_p$ is any g-orthonormal positive oriented local frame of $T\mathscr{F}$. Note that such a metric exists since $d\Omega(X, E_1, \ldots E_p) < 0$ on $M \setminus \Sigma$. Also, without loss of generality, we may assume that $g|T\mathscr{F} \otimes T\mathscr{F} = g_0|T\mathscr{F} \otimes T\mathscr{F}$ and $g|E \otimes E = g_0|E \otimes E$ at all the points of B. Then the Riemannian metrics g and g_0 will coincide on B. Comparing (3.2) and (3.19) yields the equality $X = H_{\mathscr{F},g}$. $\qquad\square$

The above theorem (applied to the case $\Sigma = U = \emptyset$, $A = M$), yields the following.

Corollary 3.45. *A nowhere vanishing vector field X on a closed foliated manifold $(M, \mathscr{F})$ can be the mean curvature of $\mathscr{F}$ (for some Riemannian metric g on M) if and only if the cone $C_{\mathscr{F},X}$ has a compact base and intersects the subspace $P_{\mathscr{F},X}$ trivially.*

Let us now discuss some consequences of the conditions in Theorem 3.44 and Corollary 3.45. We have already seen that if X is nonsingular and transverse to a codimension one foliation $\mathscr{F}$ and the cone $C_{\mathscr{F},X}$ has a compact base, then $(M, \mathscr{F})$ contains no positive/negative domains. Triviality of the intersection of $C_{\mathscr{F},X}$ and space $P_{\mathscr{F},X}$ implies transversality of X and $\mathscr{F}$ along $M \setminus \Sigma$. Indeed, if $X(x_0) \neq 0$ belongs to $T_{x_0}\mathscr{F}$ and $v_1, \ldots v_{p-1}$ are such that $(X(x_0), v_1, \ldots, v_{p-1})$ is a positive oriented frame of $T_{x_0}\mathscr{F}$, then $X(x_0) \wedge v_1 \wedge \cdots \wedge v_{p-1}$ belongs to the intersection $C_{\mathscr{F}} \cap P_{\mathscr{F},X}(A) \subset C_{\mathscr{F},X}(A) \cap P_{\mathscr{F},X}(A)$ for any closed set $A \subset M \setminus \Sigma$, which contains x_0. Another consequence of triviality of this intersection is described in the following.

Example 3.46. Let N be a compact $(p+1)$-dimensional submanifold (with boundary and corners) of $(M, \mathscr{F})$ such that TN is spanned by X and $T\mathscr{F}$ with the orientation given by $(X(x), v_1, \ldots, v_p)$, where $(v_1, \ldots, v_p)$ is a positive oriented frame of $T_x\mathscr{F}$, and suppose that the boundary ∂N decomposes into $\partial^\pitchfork N$, a submanifold tangent to X, and a union $\partial_+^\top N \cup \partial_-^\top N$ of pieces of leaves, with X pointing outwards along $\partial_+^\top N$ and inwards along $\partial_-^\top N$ (Figure 3.1). Then the p-current $\int_{\partial N}$, which

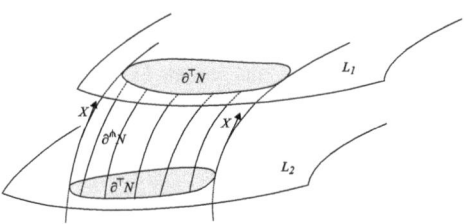

Fig. 3.1 A submanifold spanned by X and $\mathscr{F}$

takes a p-form α to the integral $\int_{\partial N} \alpha$ (where ∂N has the induced orientation) has a decomposition

$$\int_{\partial N} = \int_{\partial_+^\top N} - \int_{\partial_-^\top N} + \int_{\partial^\pitchfork N}$$

with $-\int_{\partial N} \in \tilde{C}_{\mathscr{F},X}$, $\int_{\partial_+^\top N}$ and $\int_{\partial_-^\top N} \in C_{\mathscr{F}}$, and $\int_{\partial^\pitchfork N} \in P_{\mathscr{F},X}$.

Now suppose that the flow ϕ_t of $-X$ (or $-fX$, for some positive function f) maps the closed bounded domain $\partial_+^\top N$ on a leaf into itself, i.e., $\partial_-^\top N = \phi_{t_0}(\partial_+^\top N) \subset \partial_+^\top N$ for some $t_0 > 0$ (Figure 3.2). Then $\int_{\partial_+^\top N} - \int_{\partial_-^\top N} = \int_B \in C_{\mathscr{F}}$, where $B = \partial_+^\top N \setminus \partial_-^\top N$, and consequently,

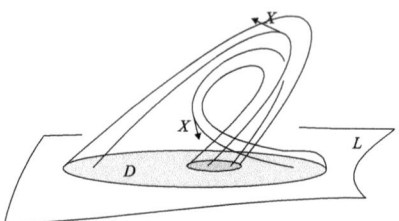

Fig. 3.2 A vector field, which cannot be mean curvature

$$-\int_{\partial^\pitchfork N} \in C_{\mathscr{F},X} \cap P_{\mathscr{F},X},$$

so X cannot be the mean curvature of $\mathscr{F}$ for any Riemannian metric on M. In fact, the mean curvature $H = H_{\mathscr{F},g}$ is the negative gradient of the leaf volume, thus the volumes of pieces of leaves have to increase when deformed by the flow of $-H$ (or $-fH$, $f > 0$).

The previous example can be generalized to a construction with a submanifold N of dimension $p+k+1$ in the presence of a suitable invariant measure, so that a family of pieces of leaves flows backwards into itself without preserving single leaves.

Example 3.47. Let T and D be compact manifolds (possibly with boundary) of dimensions k and p, respectively, and suppose that an embedding $h_0 : T \times D \to M$ is given such that the induced foliation $h_0^*(\mathscr{F})$ coincides with the vertical foliation $\mathscr{V}$ of $T \times D$ whose leaves are $\{y\} \times D$, $y \in T$. Suppose that the flow (ϕ_t) of $-X$ (or of $-fX$, for some positive function f) generates an immersion $h : T \times D \times I \to M$, defined by setting $h(x,y,t) = \phi_t(h_0(x,y))$, where I is an interval $[0,t_0]$, such that the induced foliation $h^*(\mathscr{F})$ on $T \times D \times I$ has leaves $\{y\} \times D \times \{t\}$, $(y,t) \in T \times I$. Also suppose that ϕ_{t_0} takes $N_0 = h_0(T \times D)$ into itself, and that this is the only identification of points under h. It follows that there exist a smooth embedding $\phi : T \to T$ and uniquely determined subsets $D_y \subset D$ for each $y \in \phi(T)$ such that

$$h(\{y\} \times D \times \{t_0\}) = \phi_{t_0} \circ h_0(\{y\} \times D) = h_0(\{\phi(y)\} \times D_{\phi(y)}). \qquad (3.20)$$

Suppose that D and $\mathscr{F}$ are oriented and h takes the orientation of D into the orientation of the leaves of $\mathscr{F}$. In this case, we can extend the argument of Example 3.46 to show that the vector field X cannot be the mean curvature of $\mathscr{F}$ for any Riemannian metric on M. To this end, take any positive ϕ-invariant Borel measure μ on T. Such a measure exists by the Krylov–Bogolyubov Theorem ([Wal, Cor. 6.9.1]). The boundary of $N = h(T \times D \times I)$ admits a decomposition $\partial N = \partial^\top N \cup \partial^\pitchfork N$, where $\partial^\top N$ is the closure of $N_0 \setminus \phi_{t_0}(N_0)$. For every value of the parameter $y \in T$ we have

$$\int_{\{y\}\times\partial D\times I} = -\partial \int_{\{y\}\times D\times I} + \int_{\{y\}\times D\times 0} - \int_{\{y\}\times D\times\{t_0\}},$$

where these integrals are interpreted as p-currents operating on p-forms on $T \times D \times I$. Integration over T yields

$$\int_T \int_{\{y\}\times\partial D\times I} d\mu = -\int_T \partial \int_{\{y\}\times D\times I} d\mu + \int_T \int_{\{y\}\times D\times\{0\}} d\mu - \int_T \int_{\{y\}\times D\times\{t_0\}} d\mu. \tag{3.21}$$

Then h_* will take the left hand side of (3.21) into $P_{\mathscr{F},X}$ and the first term on the right hand side into $\tilde{C}_{\mathscr{F},X}$. Next, we shall show that h_* takes the sum of the two remaining terms on the right hand side of (3.21) into $C_{\mathscr{F}}$. From (3.20) and the invariance of the measure μ under ϕ we get

$$h_* \int_T \int_{\{y\}\times D\times\{t_0\}} d\mu = h_* \int_T \int_{\{\phi(y)\}\times D_{\phi(y)}\times\{0\}} d\mu = h_* \int_{\phi(T)} \int_{\{y\}\times D_y\times\{0\}} d\mu.$$

Therefore, setting $D_y = \emptyset$ for $y \in T \setminus \phi(T)$, we get

$$h_* \int_T \int_{\{y\}\times D\times\{0\}} d\mu - h_* \int_T \int_{\{y\}\times D\times\{t_0\}} d\mu = h_* \int_T \int_{\{y\}\times(D\setminus D_y)\times\{0\}} d\mu,$$

which belongs to $C_{\mathscr{F}}$, as desired. Consequently

$$h_* \int_T \int_{\{y\}\times\partial D\times I} d\mu \in C_{\mathscr{F},X} \cap P_{\mathscr{F},X},$$

and X cannot be the mean curvature vector field of $\mathscr{F}$ for any metric.

To give a concrete example of this situation, let $M = S^1 \times S^p \times \mathbb{R}/\sim$, where $\sim$ is the equivalence relation generated by setting $(y,x,t+1) \sim (\phi(y), f_y(x), t)$ for diffeomorphisms $\phi : S^1 \to S^1$ and $f_y : S^p \to S^p$ for every $y \in S^1$, with the property that $\psi(y,x) = (\phi(y), f_y(x))$ is a diffeomorphism of $S^1 \times S^p$ onto itself. (Thus, M is the suspension of ψ.) Let $\mathscr{F}$ be the foliation with the slices $\{x\} \times S^p \times \{t\}$ as leaves, oriented by the standard orientation of S^p. Letting $T = S^1$ and fixing a p-ball $D \subset S^p$, we suppose that for every y, $f_y(D) \subset D$. (For example, ϕ could be a rotation and every f_y could be the identity on S^p, or f_y could contract D into a smaller concentric ball.) Let $h_0 : S^1 \times D \equiv S^1 \times D \times \{0\} \to M$ be the inclusion and set $X = -\partial/\partial t$. Then the flow (ϕ_t) of $-X$ on M preserves $\mathscr{F}$ and at time $t_0 = 1$ satisfies

$$\phi_1 \circ h_0(T \times D) \subset h_0(T \times D),$$

so the preceding argument shows that there is no Riemannian metric, for which X is the mean curvature.

3.4.3 At Singular Sets

The behavior of vector fields in neighborhoods of singular points can be very complicated. In the case of mean curvature vectors of foliations, one can get immediate obstructions. For instance, such vector fields take values in the orthogonal complements of tangent bundles of foliations. Let us look at this situation more closely.

So, again let $(M, \mathscr{F})$ be a closed foliated manifold equipped with a Riemannian structure g. Let $X = H_{\mathscr{F},g}$ be the mean curvature vector and $E = T^{\perp}\mathscr{F}$, the orthogonal complement of $T\mathscr{F}$, the tangent bundle of $\mathscr{F}$. Then X takes values in E and $\Omega = \Omega_{\mathscr{F},g}$, the volume form of $\mathscr{F}$ with respect to g, vanishes on $P_{\mathscr{F},E}$, the closed subspace of D_p generated by all the Dirac currents of the form $w \wedge v_1 \cdots \wedge v_{p-1}$, where $w \in E_x$, $v_i \in T_x\mathscr{F}$ and $x \in M$. If Σ is, as before, the set of all the singular points of X, then by (3.2) Ω also vanishes on $B_{\mathscr{F}}(\Sigma)$, the closed subspace of D_p generated by all the boundaries of the form $\partial(w \wedge v_1 \wedge \cdots \wedge v_p)$ with $w \in T_xM$, $v_i \in T_x\mathscr{F}$ and $x \in \Sigma$. Consequently, $\Omega \equiv 0$ on the sum $P_{\mathscr{F},E} + B_{\mathscr{F}}(\Sigma)$. Since Ω is positive on the cone $C_{\mathscr{F}} \setminus 0$, it follows that

$$C_{\mathscr{F}} \cap (P_{\mathscr{F},E} + B_{\mathscr{F}}(\Sigma)) = \{0\}. \tag{3.22}$$

On the other hand, if K is an arbitrary compact subset of M satisfying the condition

$$C_{\mathscr{F}} \cap (P_{\mathscr{F},E} + B_{\mathscr{F}}(K)) = \{0\} \tag{3.23}$$

analogous to (3.22), and $(U_k)_{k=1}^{\infty}$ is a decreasing nested family of open neighborhoods of K such that $\bigcap_{k=1}^{\infty} \overline{U_k} = K$, then $B_{\mathscr{F}}(K) = \bigcap_{k=1}^{\infty} B_{\mathscr{F}}(\overline{U_k})$ and, in view of the existence of a compact base of $C_{\mathscr{F}}$, we get

$$C_{\mathscr{F}} \cap (P_{\mathscr{F},E} + B_{\mathscr{F}}(\overline{U_k})) = \{0\}$$

for k large enough. By arguments similar to those of [Su2], we get the following.

Proposition 3.48. *If $K \subset M$ is compact, $E \subset TM$ is a subbundle complementary to $T\mathscr{F}$ and condition (3.23) is satisfied, then there exists a Riemannian metric g on M such that E is orthogonal to $\mathscr{F}$ and $H_{\mathscr{F},g} \equiv 0$ in a neighborhood of K. Furthermore, given any Riemanian metric g_0 on M, g can be chosen so that $g(v,w)$ coincides with $g_0(v,w)$ whenever v and w belong to E.*

The following elementary fact can be obtained by easy calculation (or found in the literature, for example, in [Os1] or [W2]).

Lemma 3.49. *If g and $g' = e^{2\phi} g$ are conformally equivalent Riemannian metrics on a foliated manifold $(M, \mathscr{F})$, $\dim \mathscr{F} = p$, then the mean curvature vectors H and H' of $\mathscr{F}$ with respect to g and g' are related by the formula*

$$H' = e^{-2\phi}\left(H - p(\nabla\phi)^{\perp}\right), \tag{3.24}$$

where $(\nabla\phi)^{\perp}$ denotes the component of the g-gradient of ϕ orthogonal to $\mathscr{F}$.

From (3.24) it follows that if $H = 0$ on an open subset U of M and $X = (\nabla f)^{\perp}$ for some g and $f \in C^{\infty}(M)$, then $H' = X$ on U when $g' = e^{2\phi} g$ with

$$\phi = \frac{1}{2} \log \frac{p}{2(f+c)},$$

where c is a positive constant greater than $\max_{x \in M}|f(x)|$. Also, if X is a gradient field (i.e., $X = \nabla^0 f$ is the gradient of a function f with respect to some Riemannian metric g_0), X takes values in E, a subbundle of TM complementary to $T\mathscr{F}$, and the conditions of Proposition 3.48 are satisfied, then there exists a Riemannian metric g on U for which E is orthogonal to $\mathscr{F}$, the volume form $\Omega_{\mathscr{F}}$ of $\mathscr{F}$ (with respect to g) is $\mathscr{F}$-closed (i.e., vanishes on $B_{\mathscr{F}}(\overline{U})$) and $g(v, w)$ coincides with $g_0(v, w)$ whenever v and $w \in E$. For this g, $H_{\mathscr{F},g} = 0$ and $(\nabla f)^{\perp} = X$ on U. Therefore, Proposition 3.48 and Theorem 3.44 imply the following.

Theorem 3.50. *If X is a vector field on $(M, \mathscr{F})$, which takes values in a subbundle $E \subset TM$ complementary to $T\mathscr{F}$, E and $\Sigma = \{x \in M : X(x) = 0\}$ satisfy (3.22), $X|_U$ is a gradient field for some open neighborhood U of Σ, and the cone $C_{\mathscr{F},X}(A)$ has a compact base and intersects the subspace $P_{\mathscr{F},X}(A)$ trivially for every closed set $A \subset M \setminus \Sigma$, then there exists a Riemannian metric g on M for which $X = H_{\mathscr{F},g}$, the mean curvature of the foliation $\mathscr{F}$ on (M, g).*

We complete this section with some remarks on gradient vector fields. These fields play an important role in the theory of smooth dynamical systems. For instance (see [PM]), Morse–Smale gradient fields are open and dense in the space of all gradients. They have a number of simple properties: they admit no closed orbits and the limit sets of their orbits consist of singularities (infinitely many if more than one). These properties can be expressed in terms of currents in the following way.

Let X be a vector field and Σ, as before, the set of all the singular points of X. Consider the closed convex cone $C_X \subset D_1$ generated by all the Dirac currents $X(x)$, $x \in M$, and the closed linear subspace $P_\Sigma \subset D_1$ generated by all the 1-currents c for which there exist currents z supported in Σ such that $\partial c = \partial z$.

If $X = \nabla f$ for some $f \in C^{\infty}(M)$ and some Riemannian metric g, then

$$C_X \cap P_\Sigma \subset D_1(\Sigma). \tag{3.25}$$

Indeed, if $c = \lim_{n \to \infty} c_n$ belongs to P_Σ, where $c_n = \sum_i t_{n,i} X(x_{n,i})$ for some $t_{n,i} > 0$ and $x_{n,i} \in M$, then setting $X_{n,i} = X(x_{n,i})$ we have

$$\|X_{n,i}\|^2 = g(X_{n,i}, X_{n,i}) = g(X_{n,i}, \nabla f) = \langle X_{n,i}, df \rangle,$$

and therefore,

$$\lim_{n\to\infty} \sum_i t_{n,i} \|X_{n,i}\|^2 = \langle c, df \rangle = \langle \partial c, f \rangle = \langle \partial z, f \rangle = \langle z, df \rangle = 0$$

for some $z \in D_1$ supported in Σ. This implies condition (3.25).

Figure 3.3 below shows some situations when (3.25) is not satisfied.

Unfortunately, the cone C_X need not have a compact base even if X is a gradient field. For example, if $M = S^1$ is a standard unit circle parametrized by the angle θ, $f : S^1 \to [0,1]$ is a function which has exactly two critical points, x_0 with $f(x_0) = 0$ and x_1 with $f(x_1) = 1$, and is strictly increasing on oriented arcs I_0 and I_1 with initial point x_0 and end point x_1, then the Dirac currents corresponding to the vectors $\pm(\partial/\partial\theta)(x_0)$ and $\pm(\partial/\partial\theta)(x_1)$ belong to $C_{\nabla f}$, and, therefore, $C_{\nabla f}$ has no compact base. A slight modification of this example provides a gradient field Y, for which C_Y

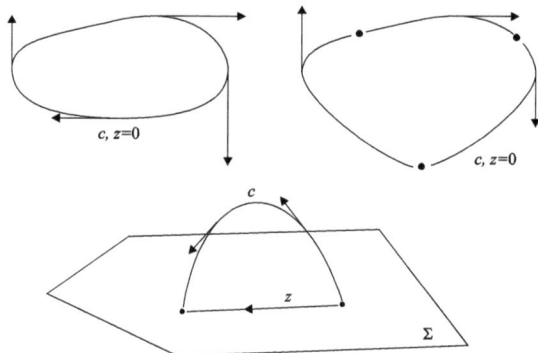

Fig. 3.3 Some non-gradient phenomena

has a compact base. In fact, take $f : S^1 \to [0,1]$ such that $f|J_0 \equiv 0$ and $f|J_1 \equiv 1$ for some disjoint non-trivial arcs J_0 and J_1 contained in S^1, while f is strictly increasing on the complementary arcs I_0 and I_1 whose endpoints coincide with those of J_0 and J_1. Then there exists a 1-form η on S^1 which coincides with $d\theta$ on I_0 and $-d\theta$ on I_1. In this case, η is positive on $C_Y \setminus 0$ and the set $\eta^{-1}(1) \cap C_Y$ is compact. If (3.25) is satisfied, the cone C_X is compact, and U is an open subset of M for which $\overline{U} \cap \Sigma = \emptyset$, then, again by the Hahn-Banach Theorem 3.16, there exists a 1-form ω such that $\omega > 0$ on $C_X(\overline{U}) \setminus 0$ and $\omega \equiv 0$ on P_Σ. Since P_Σ contains all the 1-cycles, ω is exact, i.e. $\omega = df$ for some function f. Since ω is positive on $C_X(\overline{U}) \setminus 0$, $df(X(x)) > 0$, whenever $x \in \overline{U}$. Therefore, one can find a Riemannian metric g on U for which X is orthogonal to $\ker df$ and $\|X\|^2 = df(X)$. For this g, $X = \nabla f$ on U.

The problem of realization of a vector field as a gradient in a neighborhood of its singular set seems to be much more delicate. In [Os4], mean curvature vectors of codimension-one foliations has been studied. For example, one has the following

Proposition 3.51. *For any vector field Z transverse to a codimension-one foliation $\mathscr{F}$ of a closed oriented manifold M, there exists a smooth function f on M with $\mathrm{supp}(f) = M$ such that $fZ = H_{\mathscr{F},g}$ for some Riemannian metric g on M.*

Proof. First, we solve the following, rather easy, "sub-question": for any nonsingular vector field X on a closed manifold M, there exists a smooth function f on M for which $\mathrm{supp}\, Xf = M$. Next, we take an arbitrary Riemannian structure g on M and let N be the g-unit normal (to $\mathscr{F}$) vector field, transversely oriented (with respect to $\mathscr{F}$) in the same direction as Z. Then, $Z = \psi N + X$ for a positive ϕ and X tangent to $\mathscr{F}$. Let a Riemannian metric $\tilde{g}$ coincide with g on vectors tangent to $\mathscr{F}$ and make Z $\tilde{g}$-unit and $\tilde{g}$-orthogonal to $\mathscr{F}$. It is not too hard to prove that there exists a smooth function ϕ such that $\mathrm{supp}(\tilde{h} - aZ(\phi)) = M$ for some $a > 0$. (Here, $\tilde{h}$ is, certainly, the mean curvature of $\mathscr{F}$ with respect to $\tilde{g}$.) Put $\hat{g} = e^{2\phi} \cdot \tilde{g}$. The mean curvature vector $\hat{H}$ of $\mathscr{F}$ with respect to $\hat{g}$ equals to $\hat{H} = e^{-2\phi}(\tilde{h} - aZ(\phi))Z$. $\qquad\square$

Chapter 4
Variational Formulae

Variational formulae for geometric quantities are important in studying critical points of functionals on a manifold and flows of metrics. The fundamental question (similar to the question on existence of canonical metrics on a manifold) reads as: *What are the optimal metrics (in a specified sense) on a smooth foliated manifold?* Our goal here is to examine the actions on a manifold for different types of variations. Apart from varying among all metrics, we also deal with the case when the varying metric remains fixed along the distribution, and the complementary case when metric varies only along the distribution—preserving its orthogonal complement and the metric on it. This approach applies to finding an optimal extension of a metric that is defined only along a distribution—which is a problem of the relationship between sub-Riemannian and Riemannian geometry. When viewed as a functional of the metric, the total S_{mix} can be considered an analogue of the Einstein–Hilbert action, and the "best" metrics (for the above question) are proposed to be among critical metrics of the action. We study Einstein–Hilbert type actions and Godbillon-Vey type numbers, and characterize critical metrics in certain distinguished classes of almost-product structures. Of course, such Riemannian structures may not exist, but if they do, they usually have an interesting geometry.

4.1 "Optimally Placed" Distributions

The problem of minimizing geometric quantities has been studied for a long time: recall, for example, isoperimetric inequalities and estimates of total curvature of submanifolds. In the context of foliations, Gluck–Ziller [GZ] initiated studies of the problem of minimizing functions like volume and total bending defined for k-plane fields on manifolds. In all the cases mentioned above, the authors consider a fixed Riemannian manifold and look for objects (e.g. submanifolds, foliations) minimizing geometric quantities defined usually as total curvatures of different types.

© Springer Nature Switzerland AG 2021
V. Rovenski, P. Walczak, *Extrinsic Geometry of Foliations*, Progress
in Mathematics 339, https://doi.org/10.1007/978-3-030-70067-6_4

Here, we are interested in the problem that concerns "optimally placed" distributions or foliations, i.e., ones minimizing the total mixed scalar curvature.

Given p_1-dimensional plane $\mathscr{P}_1$ in $\mathbb{R}^n$, denote by $U(\mathscr{P}_1)$ the set of all p_1-dimensional planes of $\mathbb{R}^n$ uniquely projecting onto $\mathscr{P}_1$. Taking orthonormal basis e_i $(1 \leq i \leq p_1)$ of $\mathscr{P}_1$, and extending it to orthonormal basis e_i $(1 \leq i \leq n)$ of $\mathbb{R}^n$, we represent any $\widetilde{\mathscr{P}} \in U(\mathscr{P}_1)$ as a linear graph over $\mathscr{P}_1$ with values in orthogonal complement, i.e., by the system $x_j = \sum_{i=1}^{p_1} a_{ji} x_i$ $(p_1 < j \leq n)$. The $p_1 p_2$ elements of the matrix $A = (a_{ji})$ can be chosen as local coordinates on real Grassmannian $G_{p_1}(\mathbb{R}^n)$ in a domain $U(\mathscr{P}_1)$, in particular, $\dim G_{p_1}(\mathbb{R}^n) = p_1 p_2$. To any one-parameter variation $\mathscr{P}_1(t)$ of $\mathscr{P}_1$ there corresponds the matrix-function $A(t) = (a_{ji}(t))$. Assuming $A(t)$ of a class C^2 with $A_1 = (dA/dt)(0)$ and $A_2 = (d^2A/dt^2)(0)$, we have $A(t) = tA_1 + \frac{1}{2}t^2 A_2 + o(t^2)$. For $\widetilde{\mathscr{P}} \in U(\mathscr{P}_1)$, the stationary values $0 \leq \alpha_1 \leq \ldots \leq \alpha_{p_1} \leq \pi/2$ of angle between unit vectors in $\widetilde{\mathscr{P}}$ and $\mathscr{P}_1$ are called the *angles* between these planes. It is known that there exist orthonormal bases e_i $(1 \leq i \leq p_1)$ of $\mathscr{P}_1$ and $\tilde{e}_j$ $(1 \leq j \leq p_1)$ of $\widetilde{\mathscr{P}}$ such that $\langle e_i, \tilde{e}_j \rangle = \cos \alpha_i \delta_{ij}$ (the property can be taken as the definition of angles α_i). The angles α_i $(1 \leq i \leq p_1)$ determine the relative position of two p_1-dimensional planes in $\mathbb{R}^n$, and the distance is given by

$$\rho(\widetilde{\mathscr{P}}, \mathscr{P}_1) = (\textstyle\sum_{i=1}^{p_1} \alpha_i^2)^{1/2}.$$

Choosing the basis, one may represent a variation $\mathscr{P}_1(t)$ of $\mathscr{P}_1$ using the matrix $A(t)$, where the diagonal matrix A_1 consists of $a_{ji}^{(1)} = \delta_{ji} \cos \alpha_i$ $(i \leq p_1, \, p_1 < j \leq n)$.

We use similar notations as in Section 1.4.2: $\mathscr{D}_1$ and $\mathscr{D}_2$ with $\dim \mathscr{D}_i = p_i \geq 2$ $(i = 1, 2)$, $p_1 + p_2 = n$ are smooth complementary orthogonal distributions on a Riemannian manifold (M^n, g). Here, we call the *partial Ricci curvature related to* $\mathscr{D}_1$ the following symmetric bilinear form on the tangent bundle TM:

$$\mathrm{Ric}_1(X, Y) = \textstyle\sum_{\alpha > p_1} g(R(e_\alpha, X)Y, e_\alpha).$$

The definition of Ric_2 (related to $\mathscr{D}_2$) is similar. Note that Ric_i differs from $\mathrm{Ric}^\perp$ or $\mathrm{Ric}^\top$, and we have $\mathrm{Ric} = \mathrm{Ric}_1 + \mathrm{Ric}_2$. Note that $\mathrm{S}_{\mathrm{mix}} = \mathrm{trace}_g \, \mathrm{Ric}_i |_{\mathscr{D}_i}$ for $i = 1, 2$.

Definition 4.1. For any point $x \in M$ and orthonormal bases e_i $(1 \leq i \leq p_1)$ of $\mathscr{D}_1(x)$ and ε_j $(1 \leq j \leq p_2)$ of $\mathscr{D}_2(x)$, consider the bilinear form $\mathscr{I}_x(\vec{\mu}, \vec{\mu}) = \sum_{i,j=1}^{p_1} \Phi_{ij} \mu_i \mu_j$ with $\vec{\mu} = (\mu_1, \ldots, \mu_{p_1})$ and the coefficients

$$\Phi_{ij} = \big((\mathrm{Ric}_1 - \mathrm{Ric}_2)(\varepsilon_j, \varepsilon_j) - (\mathrm{Ric}_1 - \mathrm{Ric}_2)(e_i, e_i)\big) \delta_{ij} \\ + 2\big(\langle R(e_i, e_j)\varepsilon_i, \varepsilon_j \rangle + \langle R(e_i, \varepsilon_j)\varepsilon_i, e_j \rangle\big). \tag{4.1}$$

We say that $\mathscr{I}$ is *quasi-positive* if $\mathscr{I}_x$ is positive definite for arbitrary adapted orthonormal frame $\{e_i, \varepsilon_j\}$ at any $x \in M \setminus \Sigma$, where Σ is a set of zero volume.

In the next theorem, using the partial Ricci curvature, we calculate the variations of the *total mixed scalar curvature* $I_K : \mathscr{D}_1 \to \int_M \mathrm{S}_{\mathrm{mix}} \, d \mathrm{vol}_g$.

Theorem 4.2. *A distribution $\mathcal{D}_1$ on a closed Riemannian manifold (M, g) is a critical point for the functional I_K if and only if*

$$(\text{Ric}_1 - \text{Ric}_2)(y_1, y_2) = 0, \quad y_i \in \mathcal{D}_i. \tag{4.2}$$

It is a point of local minimum if the form $\mathscr{I}$ (of Definition 4.1) is quasi-positive.

Proof. Let $\varphi : M \to \mathbb{R}$ be a smooth nonnegative function with a local support $V \subset M$. Without loss of generality, assume that $p_1 \leq p_2$. Due to discussion above, any variation $\mathcal{D}_1(t)$ ($|t| < 1$) on V can be represented by orthogonal vector fields $e_i(t) = e_i \cos(t \mu_i \varphi) + \varepsilon_i \sin(t \mu_i \varphi)$, where e_i ($i \leq p_1$) is an orthonormal frame of $\mathcal{D}_{1|V}$, ε_j ($j \leq p_2$) an orthonormal frame in $\mathcal{D}_2$ on V, and $\mu_i \geq 0$ ($i \leq p_1$) are real numbers. Set $\varepsilon_i(t) = \varepsilon_i \cos(t \mu_i \varphi) - e_i \sin(t \mu_i \varphi)$ for $1 \leq i \leq p_1$ and $\varepsilon_j(t) = \varepsilon_j$ for $p_1 < j \leq p_2$. Consequently, $\varepsilon_j(t)$ ($j \leq p_2$) span $\mathcal{D}_2(t)$ on V. We have $\dot{e}_i = \varphi \mu_i \varepsilon_i$ and $\dot{\varepsilon}_i = -\varphi \mu_i e_i$ on V for $i \leq p_1$. We extend distributions outside of V as $\mathcal{D}_1(t) = \mathcal{D}_1$ and define $I(t) = I_K(\mathcal{D}_1(t))$ for all t. The mixed scalar curvature of $\mathcal{D}_1(t)$ over V, see (1.34), is

$$\mathrm{S}_{\mathrm{mix}}(t) := \mathrm{S}_{\mathrm{mix}}(\mathcal{D}_1(t)) = \sum_{i=1}^{p_1} \sum_{j=1}^{p_2} \langle R(e_i(t), \varepsilon_j(t)) \varepsilon_j(t), e_i(t) \rangle.$$

Derivation by t at $t = 0$ (using symmetries of curvature tensor) yields on V

$$\dot{\mathrm{S}}_{\mathrm{mix}} = 2 \varphi \sum_{i,j=1}^{p_1} \left(\langle R(e_i, \varepsilon_j) \varepsilon_j, \mu_i \varepsilon_i \rangle + \langle R(e_i, \varepsilon_j)(-\mu_j e_j), e_i \rangle \right)$$

$$= 2 \varphi \sum_{i=1}^{p_1} \mu_i (\text{Ric}_1 - \text{Ric}_2)(e_i, \varepsilon_i).$$

In view of $\mathrm{S}_{\mathrm{mix}}(t) = \mathrm{S}_{\mathrm{mix}}(0)$ outside of V, we have

$$\frac{d}{dt} I(t)|_{t=0} = 2 \int_V \varphi \sum_{i=1}^{p_1} \mu_i (\text{Ric}_1 - \text{Ric}_2)(e_i, \varepsilon_i) \, d\,\text{vol}.$$

Since $e_i, \varepsilon_i, \mu_i$ and φ are arbitrary, $\mathcal{D}_1$ is critical for I_K if and only if (4.2) is satisfied.

Now, let $\mathcal{D}_1$ be a critical point for I_K. Consider an arbitrary variation $\mathcal{D}_1(t)$ ($|t| < 1$) of $\mathcal{D}_1$ with a local support $V \subset M$. Then,

$$\ddot{e}_i = -\varphi^2 \mu_i^2 e_i, \quad \ddot{\varepsilon}_i = -\varphi^2 \mu_i^2 \varepsilon_i$$

on V for $i \leq p_1$, where $\varphi : M \to \mathbb{R}$ a smooth nonnegative function supported in V. The second derivative of $\mathrm{S}_{\mathrm{mix}}(t)$ at $t = 0$ on V is

$$(1/2) \ddot{\mathrm{S}}_{\mathrm{mix}} = \varphi^2 \Big\{ \sum_{j=1}^{p_1} \mu_j^2 (\text{Ric}_1 - \text{Ric}_2)(\varepsilon_j, \varepsilon_j) - \sum_{i=1}^{p_1} \mu_i^2 (\text{Ric}_1 - \text{Ric}_2)(e_i, e_i)$$

$$+ 2 \sum_{i,j=1}^{p_1} \mu_i \mu_j \big(\langle R(e_i, e_j) \varepsilon_i, \varepsilon_j \rangle + \langle R(e_i, \varepsilon_j) \varepsilon_i, e_j \rangle \big) \Big\}.$$

Thus, $\frac{d^2}{dt^2} I(t)|_{t=0} = 2 \int_V \varphi^2 \sum_{i,j=1}^{p_1} \Phi_{ij}(x) \mu_i \mu_j d\,\text{vol}$, where $\Phi_{ij}(x)$ is defined in (4.1) and $\mu_i \in \mathbb{R}$ are arbitrary numbers. From above it follows that $\frac{d^2}{dt^2} I(t)|_{t=0} > 0$ when the form $\mathscr{I}$ is quasi-positive on M. $\qquad\square$

Suppose that dim $\mathscr{D}_2 = 1$, and let $N \in \mathscr{D}_2$ be a unit vector field on a domain $V \subset M$. Then $\mathrm{Ric}_1(\mathscr{D}_1, \mathscr{D}_2) = 0$ and (4.2) is reduced to

$$\mathrm{Ric}(N, X) = 0, \quad X \in \mathscr{D}_1. \tag{4.3}$$

Only one term of $\mathscr{I}$ is presented: $\Phi_{11} = \mathrm{Ric}(\varepsilon_1, \varepsilon_1) - \mathrm{Ric}(N, N)$. Hence, $\mathscr{I}$ is positive definite at $x \in M$ if and only if $\mathrm{Ric}(X, X) - \mathrm{Ric}(N, N)\|X\|^2 > 0$ for all non-zero $X \perp N$ in $T_x M$. With this remark, we have the following.

Corollary 4.3. *A unit vector field N orthogonal to a codimension-one distribution $\mathscr{D}_1$ on a compact Riemannian manifold (M, g) is a critical point for the functional $I_2 : N \to \int_M \mathrm{Ric}(N, N)\, d\,\mathrm{vol}$, if and only if (4.3) is valid. It is a point of local minimum if the bilinear form*

$$\mathscr{I}_{2, N}(X, X) = \mathrm{Ric}(X, X) - \mathrm{Ric}(N, N)\, g(X, X)$$

is positive definite on the space of sections of $\mathscr{D}_1$.

Example 4.4.

(a) If $\mathscr{D}_1$ and $\mathscr{D}_2$ are curvature-invariant, then $\mathrm{Ric}_i(\mathscr{D}_1, \mathscr{D}_2) = 0$ $(i = 1, 2)$, hence $\mathscr{D}_1$ is critical for I_K. A distribution $\mathscr{D}_1$ is called *curvature-invariant* if $R(X, Y)Z \in \mathscr{D}_1$ for all $X, Y, Z \in \mathscr{D}_1$.

(b) For an Einstein manifold M^4 of non-constant sectional curvature, any "optimally placed" two-dimensional distribution consists of planes with maximum or minimum sectional curvature. To see this, we use the following characteristic property of Einstein 4-manifolds among Riemannian manifolds (M^4, g): "the sectional curvature $K(Q) = K(Q^\perp)$ for any 2-plane Q", see [Bes, Corollary 1.129]. Hence, e.g., the product $S^2(1) \times \mathbb{H}^2(-1)$ is not an Einstein manifold.

Extreme values of I_K can be used to estimate the total energy and bending of a distribution or a vector field, see [RW4].

4.2 Adapted Variations of Metric

Recall two general points of view on canonical Riemannian metrics, e.g., [CM]:

- **Curvature conditions**, when we impose the constancy of Riemannian tensor, or of its traces, namely, the Ricci and the scalar curvatures, and obtain space forms, Einstein manifolds and Yamabe metrics. Using the kernel of a first-order linear differential operator, we get locally symmetric metrics, metrics with parallel Ricci tensor and harmonic curvature metrics. Adding potential, we get Ricci solitons, etc.
- **Critical metrics** of suitable functionals are, by definition, tensorial solutions of associated Euler–Lagrange equations: the Einstein–Hilbert action (most famous geometrically meaningful functional), also quadratic curvature functionals, etc.

One can define and study new functionals that have suitable classical canonical metrics as a subset of their critical points. The purpose of this section is variational formulae for the quantities of the extrinsic geometry of a pseudo-Riemannian almost-product (e.g. foliated) manifold. We use *adapted variations* of metrics to study functionals defined as the geometric part of the Einstein–Hilbert action, with the difference that the scalar curvature is replaced by S_{mix} or S_{ex} of a non-degenerate distribution. We find the Euler–Lagrange equations and characterize critical metrics in several classes of almost-product and foliated manifolds.

4.2.1 Variational Formulae

The next approach to problems of codimension-one foliations was introduced in [RW2]: Given a foliation and a property Q of a hypersurface, which depends on its principal curvatures, study Riemannian metrics that minimize the integral of Q (of the leaves) in the class of variations of metrics for which the unit normal to leaves is the same for all metrics of the family of variations. Adapted variations, which we apply to a foliation of any codimension, allow us to generalize the approach described above: the metric changes in such a way as to preserve the normal distribution.

Let M^{n+p} be a smooth manifold with a pseudo-Riemannian metric g and an almost-product structure $(\widetilde{\mathscr{D}}, \mathscr{D})$ with $\dim \widetilde{\mathscr{D}} = n$ and $\dim \mathscr{D} = p$, see Section 1.3.1. Denote by $\perp$ and $\top$ the g-orthogonal projections onto $\mathscr{D}$ and $\widetilde{\mathscr{D}}$, respectively.

Definition 4.5. A tensor $B \in \mathrm{Sym}^2(M)$ is said to be *adapted* if $B(X^\top, Y^\perp) = 0$ for any $X, Y \in \mathfrak{X}_M$. Let $\mathfrak{M} \equiv \mathfrak{M}(\widetilde{\mathscr{D}}, \mathscr{D})$ be the space of all adapted symmetric tensors on $(M, \widetilde{\mathscr{D}}, \mathscr{D})$, and let $\mathrm{Riem}(M, \widetilde{\mathscr{D}}, \mathscr{D}) = \mathrm{Riem}(M) \cap \mathfrak{M}$, where $\mathrm{Riem}(M) \subset \mathrm{Sym}^2(M)$, be the subspace of pseudo-Riemannian metrics of a given signature.

We study pseudo-Riemannian structures on M, critical for the functional

$$J_{mix} : g \to \int_M S_{mix}(g) \, d\,\mathrm{vol}_g \tag{4.4}$$

with respect to smooth variations $\{g_t \in \mathrm{Riem}(M, \widetilde{\mathscr{D}}, \mathscr{D}) : |t| < \varepsilon\}$ of $g_0 = g$, preserving orthogonality of $\widetilde{\mathscr{D}}$ and $\mathscr{D}$. Let the induced infinitesimal variations, presented by a symmetric $(0,2)$-tensor $B_t \equiv \partial g_t / \partial t \in \mathfrak{M}$, be supported in a relatively compact domain $\Omega \subset M$ (and $\Omega = M$ when M is closed), i.e., $g_t = g$ outside Ω for $|t| < \varepsilon$. Let $\mathfrak{M}_{\widetilde{\mathscr{D}}}$ and $\mathfrak{M}_{\mathscr{D}}$ be the spaces of symmetric $(0,2)$-tensors obeying $B(X^\perp, Y) = 0$, respectively, $B(X^\top, Y) = 0$, for any $X, Y \in \mathfrak{X}_M$. Then the decomposition

$$\mathfrak{M} = \mathfrak{M}_{\widetilde{\mathscr{D}}} \oplus \mathfrak{M}_{\mathscr{D}} \tag{4.5}$$

is orthogonal with respect to the inner product induced on $\mathfrak{M}$ by $g \in \mathrm{Riem}(M, \widetilde{\mathscr{D}}, \mathscr{D})$.

For each $(0,2)$-tensor B (and its dual $(1,1)$-tensor $B^\sharp$) tangent to M we define its components $B^\top, B^\perp \in \Gamma(T^*M \otimes T^*M)$ by setting $B^\top(X,Y) = B(X^\top, Y^\top)$ and $B^\perp(X,Y) = B(X^\perp, Y^\perp)$. If $B \in \mathrm{Sym}^2(M)$, then $B \in \mathfrak{M} \Longleftrightarrow B = B^\top + B^\perp$, see (4.5). In particular, $g = g^\perp + g^\top$ for any $g \in \mathrm{Riem}(M, \widetilde{\mathscr{D}}, \mathscr{D})$. Note that if $B \in \mathfrak{M}$ then $\widetilde{\mathscr{D}}$ and $\mathscr{D}$ are $B^\sharp$-invariant. Our purpose is to compute the directional derivatives

$$D_g J_{\mathrm{mix}} : T_g \mathrm{Riem}(M, \widetilde{\mathscr{D}}, \mathscr{D}) \equiv \mathfrak{M} \to \mathbb{R}$$

for any $g \in \mathrm{Riem}(M, \widetilde{\mathscr{D}}, \mathscr{D})$ and study the geometry of almost-product or foliated manifolds $(M, \mathscr{D}, \widetilde{\mathscr{D}})$, where g is a critical point of J_{mix} with respect to all adapted variations supported in domain Ω. Taking into account (4.5), we can restrict ourselves to the cases $D_g J_{\mathrm{mix}} : \mathfrak{M}_\mathscr{D} \to \mathbb{R}$ and $D_g J_{\mathrm{mix}} : \mathfrak{M}_{\widetilde{\mathscr{D}}} \to \mathbb{R}$, when g is critical either for $g^\perp$-variations or for $g^\top$-variations. Moreover, all variational formulas for $B_t \in \mathfrak{M}_{\widetilde{\mathscr{D}}}$ have the same form as those obtained for variations (4.6), only with the roles of distributions $\widetilde{\mathscr{D}}$ and $\mathscr{D}$ interchanged. Thus, it is sufficient to work with special curves $\{g_t\}_{|t|<\varepsilon}$ starting at $g \in \mathrm{Riem}(M, \widetilde{\mathscr{D}}, \mathscr{D})$ called $g^\perp$-*variations*:

$$\{g_t = g_t^\perp + g^\top : |t| < \varepsilon\}, \tag{4.6}$$

as the associated infinitesimal variations B_t lie in $\mathfrak{M}_\mathscr{D}$. We adopt the notations

$$\partial_t \equiv \partial/\partial t, \quad B = \dot{g} := \partial_t g_t|_{t=0}. \tag{4.7}$$

For any variations g_t it is known how the Levi-Civita connection changes [AH],

$$2g_t(\partial_t(\nabla_X^t Y), Z) = (\nabla_X^t B)(Y,Z) + (\nabla_Y^t B)(X,Z) - (\nabla_Z^t B)(X,Y), \quad X, Y, Z \in \mathfrak{X}_M. \tag{4.8}$$

Lemma 4.6. *Let* $\{E_a, \mathscr{E}_i\}$ *be a local* $(\widetilde{\mathscr{D}}, \mathscr{D})$-*adapted and orthonormal for* $t = 0$ *frame, that evolves by adapted variations* g_t *according to*

$$\partial_t E_a = -(1/2) B_t^\sharp(E_a), \qquad \partial_t \mathscr{E}_i = -(1/2) B_t^\sharp(\mathscr{E}_i).$$

Then, for all t, $\{E_a, \mathscr{E}_i\}$ *is a* g_t-*orthonormal frame adapted to* $(\widetilde{\mathscr{D}}, \mathscr{D})$.

Proof. For $\{E_a\}$ (and similarly for $\{\mathscr{E}_i\}$) we have

$$\partial_t(g_t(E_a, E_b)) = g_t(\partial_t E_a, E_b) + g_t(E_a, \partial_t E_b) + (\partial_t g_t)(E_a, E_b)$$
$$= B_t(E_a, E_b) - \tfrac{1}{2} g_t(B_t^\sharp(E_a), E_b) - \tfrac{1}{2} g_t(E_a, B_t^\sharp(E_b)) = 0. \qquad \square$$

Lemma 4.7. *For* $g^\perp$-*variations* g_t *of metric and* $X, Y \in \mathfrak{X}_\mathscr{D}, Z \in \mathfrak{X}_{\widetilde{\mathscr{D}}}$ *we have*

$$2\langle \partial_t \tilde{h}(X,Y), Z \rangle = \langle (\tilde{h} - \tilde{T})(B^\sharp(X), Y) + (\tilde{h} + \tilde{T})(X, B^\sharp(Y)), Z \rangle - (\nabla_Z B)(X,Y), \tag{4.9}$$
$$2\partial_t \tilde{H} = -\nabla^\top(\mathrm{trace}\, B^\sharp), \qquad \partial_t h = -B^\sharp \circ h, \qquad \partial_t H = -B^\sharp(H), \tag{4.10}$$

where the tensors h, H, T *and* $\tilde{h}, \tilde{H}, \tilde{T}$ *are given in Section 1.3.1.*

Proof. Recall that $\tilde{C}_N = \tilde{A}_N + \tilde{T}_N^\sharp$ is the conullity operator of $\mathscr{D}$ relative to the unit vector $N \in \tilde{\mathscr{D}}$. Let $B = \partial_t g_t$ correspond to our $g^\perp$-variation g_t. Using (4.8), the symmetry of B, and the property $B(\cdot, \tilde{\mathscr{D}}) = 0$, we obtain

$$
\begin{aligned}
2g_t(\partial_t \tilde{h}(X,Y), N) &= g_t(\partial_t(\nabla_X^t Y + \nabla_Y^t X), N) \\
&= (\nabla_X^t B)(Y,N) + (\nabla_Y^t B)(X,N) - (\nabla_N^t B)(X,Y) \\
&= B(\tilde{C}_N(X),Y) + B(\tilde{C}_N(Y),X) - (\nabla_N^t B)(X,Y)
\end{aligned}
$$

for all $X, Y \in \mathscr{D}$ and a unit vector $N \in \tilde{\mathscr{D}}$. The above and

$$
B(\tilde{C}_N(X),Y) = g((\tilde{A}_N + \tilde{T}_N^\sharp)(X), B^\sharp(Y)) = g((\tilde{h}+\tilde{T})(X, B^\sharp(Y)), N)
$$

yield (4.9). Next, tracing (4.9) and using skew-symmetry of $\tilde{T}$, we deduce (4.10)$_1$,

$$
\begin{aligned}
2g(\partial_t \tilde{H}, X) &= 2\sum_i g(\partial_t(\tilde{h}(\mathscr{E}_i, \mathscr{E}_i)), X) = 2\sum_i g(\partial_t \tilde{h}(\mathscr{E}_i, \mathscr{E}_i) + 2\tilde{h}(\partial_t \mathscr{E}_i, \mathscr{E}_i), X) \\
&= -\sum_i (\nabla_X B)(\mathscr{E}_i, \mathscr{E}_i) = -X(\operatorname{trace} B^\sharp).
\end{aligned}
$$

By (4.8), for any $X \in \mathscr{D}$ and $E_a, E_b \in \mathfrak{X}_{\tilde{\mathscr{D}}}$,

$$
\begin{aligned}
2g_t(\partial_t h(E_a, E_b), X) &= g_t(\partial_t(\nabla_a^t E_b + \nabla_b^t E_a), X) \\
&= (\nabla_a^t B)(X, E_b) + (\nabla_b^t B)(X, E_a) - (\nabla_X^t B)(E_a, E_b) \\
&= -B(\nabla_a^t E_b, X) - B(\nabla_b^t E_a, X) = -2B(h(E_a, E_b), X).
\end{aligned}
$$

This proves (4.10)$_2$. Next (by $B(E_a, \cdot) = 0$ and the equality $H = \sum_a h(E_a, E_a)$ for a local orthonormal frame (E_a) of $\tilde{\mathscr{D}}$) we may derive (4.10)$_3$

$$
\begin{aligned}
g_t(\partial_t H, X) &= \sum_a g_t(\partial_t(\nabla_a E_a), X) = (\nabla_a B)(E_a, X) - (1/2)(\nabla_X B)(E_a, E_a) \\
&= -\sum_a B(\nabla_a E_a, X) = -B(H, X) = -g(B^\sharp(H), X). \qquad \square
\end{aligned}
$$

By (4.10)$_{2,3}$, $g^\perp$-variations preserve the properties of $\tilde{\mathscr{D}}$: to be totally umbilical and harmonic distribution.

For any $(0,2)$-tensors P, Q (and S) on TM, we define a $(0,2)$-tensor $\Upsilon_{P,Q}$ by

$$
\langle \Upsilon_{P,Q}, S \rangle = \sum_{\lambda,\mu} \varepsilon_\lambda \varepsilon_\mu \left(S(P(e_\lambda, e_\mu), Q(e_\lambda, e_\mu)) + S(Q(e_\lambda, e_\mu), P(e_\lambda, e_\mu)) \right),
$$

where $\{e_\lambda\}$ is an orthonormal basis of TM and $\varepsilon_\lambda = \langle e_\lambda, e_\lambda \rangle \in \{-1, 1\}$. Note that

$$
\Upsilon_{P,Q} = \Upsilon_{Q,P} \quad \text{and} \quad \Upsilon_{P,Q_1+Q_2} = \Upsilon_{P,Q_1} + \Upsilon_{P,Q_2}
$$

for all $(0,2)$-tensors P, Q, Q_i. We clarify the geometrical sense of $\Upsilon_{h,h}$ and $\Upsilon_{T,T}$. Let g be positive definite on $\tilde{\mathscr{D}}$, then $\Upsilon_{h,h} = 0$ if and only if $h = 0$. For any $X \in \mathscr{D}$ we have

$$
\langle \Upsilon_{h,h}, X^\flat \otimes X^\flat \rangle = 2\sum_{a,b} \varepsilon_a \varepsilon_b \langle X, h(E_a, E_b) \rangle^2.
$$

Since all ε_a are of the same sign, the above sum is equal to zero if and only if every summand vanishes. This yields $h = 0$. Therefore, $\Upsilon_{h,h}$ can be viewed as a "measure of non-total geodesy" of $\mathscr{D}$. Similarly, if $\Upsilon_{T,T} = 0$ then for any $X \in \mathscr{D}$ we have

$$\langle \Upsilon_{T,T}, X^\flat \otimes X^\flat \rangle = 2 \sum\nolimits_{a,b} \varepsilon_a \varepsilon_b \langle X, T(E_a, E_b) \rangle^2.$$

Hence, if g is definite on $\widetilde{\mathscr{D}}$ ($\varepsilon_a = \varepsilon_b$) then the equality $\Upsilon_{T,T} = 0$ is equivalent to $T = 0$. Therefore, $\Upsilon_{T,T}$ can be viewed as a "measure of non-integrability" of $\mathscr{D}$.

Set $A_i := A_{E_i}$, $T_i^\sharp = T_{E_i}^\sharp$ (dual to h, T), etc. Define a self-adjoint $(1,1)$-tensor

$$\mathscr{K} = \sum\nolimits_i \varepsilon_i [T_i^\sharp, A_i] = \sum\nolimits_i \varepsilon_i (T_i^\sharp A_i - A_i T_i^\sharp)$$

(with zero trace), which vanishes when $\mathscr{D}$ is either integrable or totally umbilical. Indeed, if $\mathscr{D}$ is integrable then $T_i^\sharp = 0$ ($i = 1, \ldots, p$), hence $\mathscr{K} = 0$. If $\mathscr{D}$ is totally umbilical, then every operator A_i is a multiple of identity and $\mathscr{K}$ vanishes as well.

Proposition 4.8. *For $g^\perp$-variations of g, i.e., $B \in \mathfrak{M}_{\mathscr{D}}$, we have*

$$\partial_t \|h\|^2 = -(1/2)\langle \Upsilon_{h,h}, B \rangle, \tag{4.11a}$$

$$\partial_t \|H\|^2 = -B(H, H), \tag{4.11b}$$

$$\partial_t \|\tilde{h}\|^2 = \langle \operatorname{div} \tilde{h} + \mathscr{K}^\flat, B \rangle - \operatorname{div}\langle \tilde{h}, B \rangle, \tag{4.11c}$$

$$\partial_t \|\tilde{H}\|^2 = \langle (\operatorname{div} \tilde{H})g^\perp, B \rangle - \operatorname{div}((\operatorname{trace} B^\sharp)\tilde{H}), \tag{4.11d}$$

$$\partial_t \|T\|^2 = (1/2)\langle \Upsilon_{T,T}, B \rangle, \tag{4.11e}$$

$$\partial_t \|\tilde{T}\|^2 = 2\langle \widetilde{\mathscr{T}}^\flat, B \rangle. \tag{4.11f}$$

Proof. Take an adapted orthonormal frame and assume that $\nabla_a \mathscr{E}_i \in \widetilde{\mathscr{D}}_x$ at a point $x \in M$. In the calculations below we use (4.8), Lemma 4.6 and vanishing of B on $\widetilde{\mathscr{D}}$. We also use skew-symmetry of $T, \tilde{T}$ and symmetry of $h, \tilde{h}$ and B, hence, for example, $\operatorname{trace}(\tilde{T}_a^\sharp \tilde{A}_a B^\sharp) = -\operatorname{trace}(\tilde{A}_a \tilde{T}_a^\sharp B^\sharp)$.

First, we obtain (4.11f) with Casorati type operator $\widetilde{\mathscr{T}}$, see Section 1.3.2:

$$\partial_t \|\tilde{T}\|^2 = 2 \sum\nolimits_{i,j,a} \varepsilon_i \varepsilon_j \varepsilon_a \langle \tilde{T}(\mathscr{E}_i, \mathscr{E}_j), E_a \rangle \langle \tilde{T}(\partial_t \mathscr{E}_i, \mathscr{E}_j) + \tilde{T}(\mathscr{E}_i, \partial_t \mathscr{E}_j), E_a \rangle$$
$$= \sum\nolimits_{i,a} \varepsilon_i \varepsilon_a \langle ((\tilde{T}_a^\sharp)^2 B^\sharp + \tilde{T}_a^\sharp B^\sharp \tilde{T}_a^\sharp)(\mathscr{E}_i), \mathscr{E}_i \rangle = 2 \operatorname{trace}(B^\sharp \sum\nolimits_a (\tilde{T}_a^\sharp)^2).$$

Next, we obtain (4.11c)

$$\partial_t \|\tilde{h}\|^2 = 2 \sum\nolimits_{i,j,a} \varepsilon_i \varepsilon_j \varepsilon_a \langle \tilde{h}(\mathscr{E}_i, \mathscr{E}_j), E_a \rangle \langle \partial_t (\tilde{h}(\mathscr{E}_i, \mathscr{E}_j)), E_a \rangle$$
$$= \sum\nolimits_{i,a} \varepsilon_i \varepsilon_a \big(B(\tilde{T}_a^\sharp(\mathscr{E}_i), \tilde{A}_a(\mathscr{E}_i)) + B(\mathscr{E}_i, \tilde{T}_a^\sharp \tilde{A}_a(\mathscr{E}_i)) \big)$$
$$+ \langle \widetilde{\operatorname{div}} \tilde{h} - \langle \tilde{h}, \tilde{H} \rangle, B \rangle - \operatorname{div}\langle \tilde{h}, B \rangle = \langle \operatorname{div} \tilde{h} + \mathscr{K}^\flat, B \rangle - \operatorname{div}\langle \tilde{h}, B \rangle.$$

To get (4.11d), we apply the equality

$$\langle \nabla(\text{trace}\, B^{\sharp}), \tilde{H} \rangle = \text{div}((\text{trace}\, B^{\sharp})\tilde{H}) - (\text{div}\,\tilde{H})\,\text{trace}\, B^{\sharp}$$

for $\partial_t \|\tilde{H}\|^2 = 2\langle \partial_t \tilde{H}, \tilde{H}\rangle = -\langle \nabla(\text{trace}\, B^{\sharp}), \tilde{H}\rangle$. Next, we prove (4.11a,b),

$$\partial_t \|H\|^2 = B(H,H) + 2\langle \partial_t H, H\rangle = B(H,H) - 2\langle B^{\sharp}(H), H\rangle = -B(H,H),$$
$$\partial_t \|h\|^2 = \partial_t \sum_{i,a,b} \varepsilon_i\, \varepsilon_a\, \varepsilon_b\, \langle h(E_a, E_b), \mathscr{E}_i\rangle^2 = -\sum_{a,b} \varepsilon_a\, \varepsilon_b\, B(h(E_a, E_b), h(E_a, E_b)).$$

Finally, we have (4.11e):

$$\partial_t \|T\|^2 = \partial_t \sum_{i,a,b} \varepsilon_i\, \varepsilon_a\, \varepsilon_b \langle T(E_a, E_b), \mathscr{E}_i\rangle^2 = \sum_{a,b} \varepsilon_a\, \varepsilon_b\, B(T(E_a, E_b), T(E_a, E_b)). \quad \square$$

Corollary 4.9. *For $g^{\perp}$-variations of metric we have*

$$\partial_t\, \mathrm{S}_{\mathrm{ex}} = \langle \frac{1}{2}\Upsilon_{h,h} - H^{\flat} \otimes H^{\flat},\, B\rangle, \tag{4.12a}$$

$$\partial_t\, \tilde{\mathrm{S}}_{\mathrm{ex}} = \langle (\text{div}\,\tilde{H})\, g^{\perp} - \text{div}\,\tilde{h} - \widetilde{\mathscr{K}}^{\flat},\, B\rangle + \text{div}(\langle \tilde{h}, B\rangle - (\text{trace}_g B)\tilde{H}). \tag{4.12b}$$

Proof. By (1.45), (4.11a,b) yield (4.12a), and (4.11c,d) yield (4.12b). $\qquad\qquad\square$

4.2.2 Euler–Lagrange Equations for the Total $\mathrm{S}_{\mathrm{mix}}$

If one examines the Einstein–Hilbert action for all compactly supported variations in $\dim M \neq 2$, then only Ricci-flat metrics are critical. Instead, variations of metric are required to keep the volume of the considered relatively compact domain Ω, and in consequence yield a more interesting, broader class of *Einstein metrics* as the critical points for the action. In analogy to this approach, we consider $g^{\perp}$-variations of the action (4.4) that preserve the volume of Ω. To get them, we conformally scale arbitrary $g^{\perp}$-variations along the distribution $\mathscr{D}$, by a function ϕ that—for our convenience — depends only on the variation parameter t (i.e., it is constant on Ω for any given t). While these volume preserving variations are not any more compactly supported inside Ω (they do not vanish on the boundary of Ω for this particular choice of function ϕ), they accomplish our goal, leading to "inhomogeneous" versions of the Euler–Lagrange equations obtained for $g^{\perp}$-variations.

For arbitrary $f \in L^1(\Omega, d\,\mathrm{vol}_g)$ denote the mean value of f on Ω by

$$f(\Omega, g) = \mathrm{Vol}^{-1}(\Omega, g) \int_{\Omega} f\, d\,\mathrm{vol}_g. \tag{4.13}$$

Together with a family g_t of (4.6), consider on Ω the metrics

$$\bar{g}_t = \phi_t g_t^{\perp} + g^{\top},\quad \phi_t := \big(\mathrm{Vol}(\Omega, g_t)/\mathrm{Vol}(\Omega, g)\big)^{-2/p},\quad |t| < \varepsilon. \tag{4.14}$$

Note that the following formula for variation of the volume form, see [AH], is valid for any variation g_t (of a metric) with $B = \partial_t g_{t|t=0}$:

$$\partial_t (d\,\mathrm{vol}_g)|_{t=0} = \frac{1}{2}(\mathrm{trace}_g\, B)\,d\,\mathrm{vol}_g. \tag{4.15}$$

As $\bar{g}_t$ are $\mathscr{D}$-conformal to g_t with constant scale ϕ_t, their volume forms are related as

$$d\,\mathrm{vol}_{\bar{g}_t} = \phi_t^{p/2}\,d\,\mathrm{vol}_{g_t}. \tag{4.16}$$

Remark 4.10. (i) One may show that $\mathrm{Vol}(\Omega, \bar{g}_t) = \mathrm{Vol}(\Omega, g)$ for all t. As $\bar{g}_t$ are $\mathscr{D}$-conformal to g_t with constant scale ϕ_t, their volume forms are related by (4.16), hence, $\mathrm{Vol}(\Omega, \bar{g}_t) = \int_\Omega d\,\mathrm{vol}_{\bar{g}_t} = \mathrm{Vol}(\Omega, g)$. Then, we differentiate (4.16) to obtain

$$\partial_t (d\,\mathrm{vol}_{\bar{g}_t}) = \frac{1}{2}\big(\mathrm{trace}\, B_t^\sharp - (\mathrm{trace}_{g_t}\, B_t)(\Omega, g_t)\big)\,d\,\mathrm{vol}_{\bar{g}_t}.$$

Finally, we use (4.15) and the fact that $\phi_0 = 1$ and $\phi_t' = -\frac{\phi_t}{p}(\mathrm{trace}_{g_t}\, B_t)(\Omega, g_t)$.
(ii) For $\bar{g}_t$ in (4.14), using Proposition 4.8 with $B = sg^\perp$ for a certain function s on M, we obtain

$$H_{\bar{g}} = \phi^{-1}H, \quad \tilde{H}_{\bar{g}} = \tilde{H}, \quad h_{\bar{g}} = \phi^{-1}h, \quad \tilde{h}_{\bar{g}} = \phi\,\tilde{h},$$
$$(\|T\|_{\bar{g}})^2 = \phi\,\|T\|_g^2, \quad (\|h_{\bar{g}}\|_{\bar{g}})^2 = \phi^{-1}\|h\|_g^2, \quad (\|\tilde{h}_{\bar{g}}\|_{\bar{g}})^2 = \|\tilde{h}\|_g^2,$$
$$(\|\tilde{T}\|_{\bar{g}})^2 = \phi^{-2}\|\tilde{T}\|_g^2, \quad (\|H_{\bar{g}}\|_{\bar{g}})^2 = \phi^{-1}\|H\|_g^2, \quad (\|\tilde{H}_{\bar{g}}\|_{\bar{g}})^2 = \|\tilde{H}\|_g^2, \tag{4.17}$$

where subscript $\bar{g}$ corresponds to quantities calculated with respect to $\bar{g}$.

Next we give two technical lemmas.

Lemma 4.11. *For all $g^\perp$-variations of metric and all $g^\perp$-variations preserving the volume of Ω the evolution of div of a t-dependent vector field X is given by*

$$\partial_t(\mathrm{div}\, X) = \mathrm{div}(\partial_t\, X) + (1/2)\,X(\mathrm{trace}\, B^\sharp). \tag{4.18}$$

Proof. First, consider arbitrary $g^\perp$-variation g_t. Differentiating the formula $\mathrm{div}\, X \cdot d\,\mathrm{vol}_g = \mathscr{L}_X(d\,\mathrm{vol}_g)$, see (1.77), we obtain (4.18). $\qquad\square$

Lemma 4.12. *For any $g^\perp$-variation g_t or $\bar{g}_t$ of (4.14), supported in $\Omega \subset M$ we get*

$$\frac{d}{dt}\int_\Omega \mathrm{div}(H+\tilde{H})\,d\,\mathrm{vol}_g = \begin{cases} 0 & \text{for } g_t, \\ \frac{1}{2}\,\mathrm{div}\left(\frac{2-p}{p}H - \tilde{H}\right)(\Omega, g)\int_\Omega(\mathrm{trace}_g\, B)\,d\,\mathrm{vol}_g & \text{for } \bar{g}_t. \end{cases}$$

Proof. Using t-derivatives of mean curvatures and the volume form, equalities (4.16) and (4.17), we get

$$\frac{d}{dt}\int_\Omega \mathrm{div}(H+\tilde{H})\,d\,\mathrm{vol}_g = \int_\Omega \partial_t\,(\mathrm{div}(H+\tilde{H}))\,d\,\mathrm{vol}_g + \int_\Omega \mathrm{div}(H+\tilde{H})\,\partial_t\,(d\,\mathrm{vol}_g)$$
$$= \int_\Omega \left(-\mathrm{div}\,\nabla^\top(\mathrm{trace}\, B^\sharp) - \mathrm{div}(B^\sharp(H)) + \frac{1}{2}\,\mathrm{div}\left((\mathrm{trace}\, B^\sharp)(H+\tilde{H})\right)\right)d\,\mathrm{vol}_g = 0,$$

since all the above terms are integrals of divergences of vector fields supported in Ω. For $g^{\perp}$-variations preserving the volume of Ω, all the following derivatives with respect to t will be calculated at $t = 0$. By Lemma 4.11 and

$$\phi'_t = -\frac{\phi_t}{p}\,(\mathrm{trace}_{g_t}\,B_t)(\Omega, g_t), \tag{4.19}$$

we have

$$\partial_t\,(\mathrm{div}\,H_{\bar{g}}) = \partial_t\,(\mathrm{div}\,H) + \frac{1}{p}\,(\mathrm{div}\,H)\cdot(\mathrm{trace}_g\,B)(\Omega, g),$$

while $\tilde{H}_{\bar{g}} = \tilde{H}$, and hence $\partial_t\,(\mathrm{div}\,\tilde{H}_{\bar{g}}) = \partial_t\,(\mathrm{div}\,\tilde{H})$. We also have

$$\partial_t\,(\mathrm{d\,vol}_{\bar{g}}) = \partial_t\,(\mathrm{d\,vol}_g) - \frac{1}{2}\,(\mathrm{trace}_g\,B)(\Omega, g)\,\mathrm{d\,vol}_{\bar{g}}.$$

Thus,

$$\frac{\mathrm{d}}{\mathrm{d}t}\int_{\Omega}\mathrm{div}(H_{\bar{g}} + \tilde{H}_{\bar{g}})\,\mathrm{d\,vol}_{\bar{g}} = \int_{\Omega}\partial_t\,\mathrm{div}(H + \tilde{H})\,\mathrm{d\,vol}_g + \int_{\Omega}\mathrm{div}(H + \tilde{H})\,\partial_t\mathrm{d\,vol}_g$$
$$= (\mathrm{trace}_g\,B)(\Omega, g)\int_{\Omega}\mathrm{div}\left(\frac{2-p}{2p}H - \frac{1}{2}\tilde{H}\right)\mathrm{d\,vol}_g. \qquad \square$$

Next we compare adapted variations of (4.4) associated with metrics $\bar{g}_t$ and g_t.

Proposition 4.13. *The $g^{\perp}$-variations of a metric $g \in \mathrm{Riem}(M, \widetilde{\mathscr{D}}, \mathscr{D})$ for the action (4.4) associated with $\bar{g}_t$ and g_t are related by*

$$\frac{\mathrm{d}}{\mathrm{d}t}J_{\mathrm{mix}}(\bar{g}_t)\big|_{t=0} = \frac{\mathrm{d}}{\mathrm{d}t}J_{\mathrm{mix}}(g_t)\big|_{t=0} - \frac{1}{2}\,\mathrm{S}^*_{\mathrm{mix}}(\Omega, g)\int_{\Omega}(\mathrm{trace}_g\,B)\,\mathrm{d\,vol}_g, \tag{4.20}$$

where

$$\mathrm{S}^*_{\mathrm{mix}} = \mathrm{S}_{\mathrm{mix}} - \frac{2}{p}\,(\mathrm{S}_{\mathrm{ex}} + 2\|\tilde{T}\|^2 - \|T\|^2 + \mathrm{div}\,H). \tag{4.21}$$

Proof. Let us fix a $g^{\perp}$-variation g_t, see $(4.6)_1$. By (1.46) and Lemma 4.12, we have

$$\frac{\mathrm{d}}{\mathrm{d}t}J_{\mathrm{mix}}(g_t) = \frac{\mathrm{d}}{\mathrm{d}t}\int_{\Omega}Q(g_t)\,\mathrm{d\,vol}_{g_t},$$

where $Q(g) := \mathrm{S}_{\mathrm{mix}} - \mathrm{div}(H + \tilde{H})$ is represented using (1.46) as

$$Q(g) = \mathrm{S}_{\mathrm{ex}}(g) + \widetilde{\mathrm{S}}_{\mathrm{ex}}(g) + \|\tilde{T}\|^2_g + \|T\|^2_g \tag{4.22}$$

(with S_{ex} and $\widetilde{\mathrm{S}}_{\mathrm{ex}}$ defined in Section 1.3.2). Hence, and by (4.17),

$$Q(\bar{g}_t) = Q(g_t) + (\phi_t^{-1} - 1)\,\mathrm{S}_{\mathrm{ex}}(g_t) + (\phi_t^{-2} - 1)\,\|\tilde{T}\|^2_{g_t} + (\phi_t - 1)\,\|T\|^2_{g_t}.$$

Differentiating the above at $t = 0$ and using $\phi_0 = 1$, we get

$$\partial_t Q(\bar{g}_t)_{|t=0} = \partial_t Q(g_t)_{|t=0} - \phi_0'\left(S_{ex}(g) + 2\|\tilde{T}\|_g^2 - \|T\|_g^2\right),$$

where $\phi_0' = -\frac{1}{p}(\mathrm{trace}_g B)(\Omega, g)$, see (4.19). Using Lemma 4.12, we obtain

$$\frac{d}{dt} J_{mix}(g_t)_{|t=0} = \int_\Omega \left\{\partial_t Q(g_t)_{|t=0} + \frac{1}{2} Q(g)\,\mathrm{trace}_g B\right\} d\,\mathrm{vol}_g,$$

$$\frac{d}{dt} J_{mix}(\bar{g}_t)_{|t=0} = \int_\Omega \left\{\partial_t Q(\bar{g}_t)_{|t=0} + \frac{1}{2} Q(g)(\mathrm{trace}_g B + p\,\phi_0')\right\} d\,\mathrm{vol}_g$$

$$+ \frac{d}{dt}\int_\Omega \mathrm{div}(H_{\bar{g}_t} + \tilde{H}_{\bar{g}_t})\,d\,\mathrm{vol}_{\bar{g}_t\,|t=0}. \tag{4.23}$$

Hence,

$$\frac{d}{dt} J_{mix}(\bar{g}_t)_{|t=0} = \int_\Omega \left\{\partial_t Q(\bar{g}_t)_{|t=0} + \frac{1}{2} Q(g)\left(\mathrm{trace}_g B - (\mathrm{trace}_g B)(\Omega, g)\right)\right\} d\,\mathrm{vol}_g$$

$$+ \frac{d}{dt}\int_\Omega \mathrm{div}(H_{\bar{g}_t} + \tilde{H}_{\bar{g}_t})\,d\,\mathrm{vol}_{\bar{g}_t\,|t=0} = \int_\Omega \partial_t Q(g_t)_{|t=0}\,d\,\mathrm{vol}_g + \frac{1}{2}\int_\Omega Q(g)(\mathrm{trace}_g B)\,d\,\mathrm{vol}_g$$

$$+ \frac{1}{2}\left(\frac{2}{p}(S_{ex} + 2\|\tilde{T}\|_g^2 - \|T\|_g^2 + \mathrm{div}\,H) - Q(g) - \mathrm{div}(H + \tilde{H})\right)(\Omega, g)\int_\Omega(\mathrm{trace}_g B)\,d\,\mathrm{vol}_g.$$

Using definition of $Q(g)$ and (4.21) we get (4.20). $\qquad\square$

It should be stressed that we work with two types of variations of metric, (4.6) and (4.14); the second of which preserves the volume of Ω. Formulas containing S_{mix}^* correspond to (4.14). To obtain similar formulas, corresponding to 1-parameter variations of the form (4.6), one should merely delete the mean value terms $S_{mix}^*(\Omega, g)$ in the previous identities. Considering a closed manifold M instead of Ω, we obtain S_{mix}^* of (4.20)–(4.21) without $\mathrm{div}\,H$-term.

Next theorem provides the Euler–Lagrange equations of the variational principle $\delta J_{mix}(g) = 0$ on a relatively compact domain Ω of a manifold M with an almost-product structure. These have a form $P = \lambda\,g^\top$ (on $\widetilde{\mathcal{D}}$) and $P = \lambda\,g^\perp$ (on $\mathcal{D}$) for certain tensors P and functions λ on M. In addition to tensors $\mathrm{Ric}^\perp$, $\widetilde{\mathcal{A}}$, $\widetilde{\mathcal{T}}$ and $\mathrm{Def}_{\mathcal{D}}$ defined in Section 1.3.2, define the symmetric $(0, 2)$-tensor Ψ by the identity

$$\Psi(X, Y) = \mathrm{trace}(A_Y A_X + T_Y^\sharp T_X^\sharp), \quad X, Y \in \mathfrak{X}_{\mathcal{D}}. \tag{4.24}$$

Theorem 4.14 (Euler–Lagrange Equations). *A metric* $g \in \mathrm{Riem}(M, \widetilde{\mathcal{D}}, \mathcal{D})$ *is critical for the action (4.4) with respect to* $g^\perp$*-variations of metric if and only if*

$$\mathrm{Ric}^\perp - \langle\tilde{h}, \tilde{H}\rangle + \widetilde{\mathcal{A}}^\flat - \widetilde{\mathcal{T}}^\flat + \widetilde{\mathcal{K}}^\flat + H^\flat \otimes H^\flat - \tfrac{1}{2}\Upsilon_{h,h} - \tfrac{1}{2}\Upsilon_{T,T} + \Psi - \mathrm{Def}_{\mathcal{D}}\,H$$

$$= \tfrac{1}{2}\left(S_{mix} - S_{mix}^*(\Omega, g) + \mathrm{div}(\tilde{H} - H)\right)g^\perp. \tag{4.25}$$

Proof. Applying Corollary 4.9 to (4.22), using (1.32) and removing integrals of divergences of vector fields compactly supported in Ω, we get

$$\int_M \dot{Q}\,d\,\mathrm{vol}_g = \int_M \langle\mathrm{div}(\tilde{H}\,g^\perp - \tilde{h}) + 2\widetilde{\mathcal{T}}^\flat - \widetilde{\mathcal{K}}^\flat - H^\flat \otimes H^\flat + \tfrac{1}{2}\Upsilon_{h,h} + \tfrac{1}{2}\Upsilon_{T,T}, B\rangle d\,\mathrm{vol}_g,$$

where $B = \{\partial_t g_t\}_{|t=0} \in \mathfrak{M}_{\mathscr{D}}$. Notice that $\mathrm{trace}_g B = \langle B, g^\perp \rangle$. By (4.23) we have

$$\frac{d}{dt} J_{\mathrm{mix}}(g_t)_{|t=0} = \int_M \Big\langle \mathrm{div}(\tilde{H} g^\perp - \tilde{h}) + 2\widetilde{\mathscr{F}}^\flat - \widetilde{\mathscr{K}}^\flat - H^\flat \otimes H^\flat + \frac{1}{2}\Upsilon_{h,h} + \frac{1}{2}\Upsilon_{T,T}$$
$$+ \frac{1}{2}\big(\mathrm{S}_{\mathrm{mix}} - \mathrm{div}(H + \tilde{H})\big) g^\perp, \ B \Big\rangle \mathrm{d\,vol}_g . \tag{4.26}$$

By (4.26) and Proposition 4.13 we obtain

$$\frac{d}{dt} J_{\mathrm{mix}}(\bar{g}_t)_{|t=0} = \int_M \Big\langle \mathrm{div}(\tilde{H} g^\perp - \tilde{h}) + 2\widetilde{\mathscr{F}}^\flat - \widetilde{\mathscr{K}}^\flat - H^\flat \otimes H^\flat + \frac{1}{2}\Upsilon_{h,h} + \frac{1}{2}\Upsilon_{T,T}$$
$$+ \frac{1}{2}\big(\mathrm{S}_{\mathrm{mix}} - \mathrm{S}^*_{\mathrm{mix}}(\Omega, g) - \mathrm{div}(\tilde{H} + H)\big) g^\perp, \ B \Big\rangle \mathrm{d\,vol}_g . \tag{4.27}$$

If g is critical for the action J_{mix} with respect to $g^\perp$-variations, then the integral in (4.27) is zero for arbitrary symmetric tensor $B \in \mathfrak{M}$ vanishing on $\widetilde{\mathscr{D}}$. That yields

$$\mathrm{div}\,\tilde{h} - 2\widetilde{\mathscr{F}}^\flat + \widetilde{\mathscr{K}}^\flat + H^\flat \otimes H^\flat - \frac{1}{2}\Upsilon_{h,h} - \frac{1}{2}\Upsilon_{T,T} = \frac{1}{2}\big(\mathrm{S}_{\mathrm{mix}} - \mathrm{S}^*_{\mathrm{mix}}(\Omega, g) + \mathrm{div}(\tilde{H} - H)\big) g^\perp.$$
$$\tag{4.28}$$

Using the partial Ricci tensor, see Proposition 1.36, and replacing $\mathrm{div}\,\tilde{h}$ in (4.28) according to (1.41), we rewrite (4.28) as (4.25). $\qquad\square$

Differentiating (4.14) with respect to t, we see that variations $\partial_t \bar{g}$ preserving the volume of Ω are not compactly supported in Ω, and hence they do not form a subclass of adapted variations. Comparing (4.28) for general adapted variations and those given by (4.14), one can see that metrics critical for all adapted variations supported inside a domain Ω remain critical for variations (4.14) preserving the volume of this set if and only if the equality $(p - 2)\int_\Omega \mathrm{div}(\tilde{H} - H)\,\mathrm{d\,vol}_g = 0$ is valid. The above equality is satisfied, in particular, when we consider as Ω the entire, closed manifold M. Also note that for general adapted variations g_t the Euler–Lagrange equations are supposed to hold everywhere on M, since the set Ω containing the support of a variation is assumed to be arbitrary. On the other hand, metrics critical with respect to variations preserving the volume of a fixed domain Ω satisfy the Euler–Lagrange equation only at the points of Ω (and since Ω explicitly appears in that equation, it cannot be assumed free).

Example 4.15. Let both distributions be totally geodesic. Then (4.25) reads

$$\mathrm{Ric}^\perp - \widetilde{\mathscr{F}}^\flat - \frac{1}{2}\Upsilon_{T,T} + \Psi = \frac{1}{2}\big(\mathrm{S}_{\mathrm{mix}} - \mathrm{S}^*_{\mathrm{mix}}(\Omega, g)\big) g^\perp,$$

where $\Psi(X, Y) = \mathrm{trace}_g(T_Y^\sharp T_X^\sharp)$, $X, Y \in \mathfrak{X}_{\mathscr{D}}$. Also

$$\mathrm{S}_{\mathrm{mix}} = \|T\|^2 + \|\tilde{T}\|^2, \qquad \mathrm{S}^*_{\mathrm{mix}} = \frac{p-4}{p}\|\tilde{T}\|^2 + \frac{p+2}{p}\|T\|^2,$$

see (4.21). This is the case of Hopf fibrations, when $\mathscr{D}$ is a non-integrable, totally geodesic distribution with integrable orthogonal complement.

The variational formulas of Section 4.2.1 can be applied to different functionals which depend on extrinsic geometry of distributions; in particular, the *total extrinsic scalar curvature*, i.e., integrals of extrinsic scalar curvatures $\widetilde{S}_{ex}$ and S_{ex}. Since the variational formulas for both these quantities are similar, we shall examine only $\widetilde{S}_{ex}$. We consider adapted variations of the functional

$$J_{\widetilde{ex}}(g) : g \to \int_M \widetilde{S}_{ex}(g) \, d\mathrm{vol}_g . \tag{4.29}$$

Note that for $p = 1$ we have $\widetilde{S}_{ex} = 0$ for any metric.

Proposition 4.16. *A pseudo-Riemannian metric* $g \in \mathrm{Riem}(M, \widetilde{\mathscr{D}}, \mathscr{D})$ *with* $\dim \mathscr{D} > 1$ *is critical for the action (4.29) with respect to all adapted variations if and only if*

$$\mathrm{div}\, \tilde{h} + \widetilde{\mathscr{H}^{b}} = -\frac{1}{2(p-1)} (\widetilde{S}_{ex} - \widetilde{S}_{ex}^*(\Omega, g)) g^{\perp}, \quad (g^{\perp}\text{-variations}), \tag{4.30a}$$

$$\tilde{H}^b \otimes \tilde{H}^b - \frac{1}{2}\Upsilon_{\tilde{h},\tilde{h}} = \frac{1}{n}\widetilde{S}_{ex}g^{\top}, \quad \widetilde{S}_{ex} = \widetilde{S}_{ex}^*(\Omega, g) \text{ for } n \neq 2 \ (g^{\top}\text{-variations}), \tag{4.30b}$$

where $\widetilde{S}_{ex}^* = \widetilde{S}_{ex}$ *for variations* $\bar{g}_t$, *and* $\widetilde{S}_{ex}^* = 0$ *for variations* g_t.

Proof. Using equation for $\widetilde{S}_{ex}$ dual to (4.12a), we write the $g^{\top}$-variation of $\widetilde{S}_{ex}$ as

$$\partial_t \widetilde{S}_{ex} = \langle \frac{1}{2}\Upsilon_{\tilde{h},\tilde{h}} - \tilde{H}^b \otimes \tilde{H}^b, B \rangle, \tag{4.31}$$

interchanging the roles of $\mathscr{D}$ and $\widetilde{\mathscr{D}}$. Using (4.12b), (4.15), (4.31), and removing divergences of compactly supported vector fields, we obtain for $g^{\perp}$-variations:

$$\frac{d}{dt} J_{\widetilde{ex}}(g_t)_{|t=0} = \int_M \langle (\mathrm{div}\, \tilde{H}) g^{\perp} - \mathrm{div}\, \tilde{h} - \widetilde{\mathscr{H}^b} + \frac{1}{2}\widetilde{S}_{ex}g^{\perp}, B \rangle \, d\mathrm{vol}_g,$$

and for $g^{\top}$-variations:

$$\frac{d}{dt} J_{\widetilde{ex}}(g_t)_{|t=0} = \int_M \langle \frac{1}{2}\Upsilon_{\tilde{h},\tilde{h}} - \tilde{H}^b \otimes \tilde{H}^b + \frac{1}{2}\widetilde{S}_{ex}g^{\top}, B \rangle \, d\mathrm{vol}_g .$$

In the case of variations preserving the volume of Ω, we get for $g^{\perp}$-variations

$$\frac{d}{dt} J_{\widetilde{ex}}(\bar{g}_t)_{|t=0} = \frac{d}{dt} J_{\widetilde{ex}}(g_t)_{|t=0} - \frac{1}{2}\widetilde{S}_{ex}(\Omega, g) \int_M (\mathrm{trace}_g B) \, d\mathrm{vol}_g,$$

and for $g^{\top}$-variations:

$$\frac{d}{dt} J_{\widetilde{ex}}(\bar{g}_t)_{|t=0} = \frac{d}{dt} J_{\widetilde{ex}}(g_t)_{|t=0} - \frac{n-2}{2n}\widetilde{S}_{ex}(\Omega, g) \int_M (\mathrm{trace}_g B) \, d\mathrm{vol}_g .$$

Therefore, we obtain the following Euler–Lagrange equations for the action (4.29) (terms $\widetilde{S}^*_{ex}$ appear only in case of $\bar{g}_t$—variations preserving the volume of Ω):

$$\operatorname{div}\tilde{h} + \widetilde{\mathscr{H}^\flat} = \frac{1}{2}\left(2\operatorname{div}\tilde{H} + \widetilde{S}_{ex} - \widetilde{S}^*_{ex}(\Omega,g)\right)g^\perp \quad \text{for } g^\perp\text{-variations,} \quad (4.32a)$$

$$\tilde{H}^\flat \otimes \tilde{H}^\flat - \frac{1}{2}\Upsilon_{\tilde{h},\tilde{h}} = \frac{1}{2}\left(\widetilde{S}_{ex} - \frac{n-2}{n}\widetilde{S}^*_{ex}(\Omega,g)\right)g^\top \quad \text{for } g^\top\text{-variations.} \quad (4.32b)$$

Taking traces of (4.32a,b) yields

$$\operatorname{div}\tilde{H} = \frac{p}{2(1-p)}\left(\widetilde{S}_{ex} - \widetilde{S}^*_{ex}(\Omega,g)\right) \quad \text{for } g^\perp\text{-variations,} \quad (4.33a)$$

$$(n-2)\left(\frac{1}{2}\widetilde{S}_{ex} - \widetilde{S}^*_{ex}(\Omega,g)\right) = 0 \quad \text{for } g^\top\text{-variations.} \quad (4.33b)$$

Using (4.33a) in (4.32a) and (4.33b) in (4.32b), yield Euler–Lagrange equations (4.30a,b) and completes the proof. □

4.2.3 Particular Cases

Assume that a pseudo-Riemannian manifold (M^{n+p}, g) is endowed with an n-dimensional foliation $\mathscr{F}$. Since $\widetilde{\mathscr{D}} = T\mathscr{F}$, we obtain dual to (1.41) equations

$$\operatorname{Ric}^\top = \operatorname{div}h + \langle h, H\rangle - \mathscr{A}^\flat - \widetilde{\Psi} + \operatorname{Def}_{\mathscr{F}}\tilde{H}, \qquad d_{\mathscr{F}}\tilde{H} = 0.$$

Note that $\widetilde{\Psi}(X,Y) = \operatorname{trace}_g(\tilde{A}_Y\tilde{A}_X)$, see (4.24), and definition (4.21) takes the form

$$S^*_{mix} = S_{mix} - \begin{cases} \frac{2}{p}\left(S_{ex} + 2\|\tilde{T}\|^2 + \operatorname{div}H\right) & \text{for } g^\perp\text{-variations,} \\ \frac{2}{n}\left(\widetilde{S}_{ex} - \|\tilde{T}\|^2 + \operatorname{div}\tilde{H}\right) & \text{for } g^\top\text{-variations.} \end{cases} \quad (4.34)$$

The system of (4.25) and its dual equation

$$\operatorname{Ric}^\top - \langle h, H\rangle + \mathscr{A}^\flat + \tilde{H}^\flat \otimes \tilde{H}^\flat - \frac{1}{2}\Upsilon_{\tilde{h},\tilde{h}} - \frac{1}{2}\Upsilon_{\tilde{T},\tilde{T}} + \widetilde{\Psi} - \operatorname{Def}_{\mathscr{F}}\tilde{H}$$
$$= \frac{1}{2}\left(S_{mix} - S^*_{mix}(\Omega,g) + \operatorname{div}(H - \tilde{H})\right)g^\top \quad \text{for } g^\top\text{-variations,} \quad (4.35)$$

admits many solutions (e.g., for twisted products), see discussion in [BDRS, RZ1].

Let $\mathscr{D}$ and $\widetilde{\mathscr{D}}$ determine totally umbilical foliations. One may show (replacing h and $\tilde{h}$ due to definition of total umbilicity) that

$$\Upsilon_{h,h} = \frac{2}{n}H^\flat \otimes H^\flat, \quad \mathscr{A}^\flat = \frac{1}{n^2}\|H\|^2 g^\top, \quad \Psi = \frac{1}{n}H^\flat \otimes H^\flat, \quad S_{ex} = \frac{n-1}{n}\|H\|^2,$$

for $\mathscr{D}$ and similarly for another distribution. Moreover, the fundamental equation (1.41) and the Euler–Lagrange equation (4.25) read as

$$\mathrm{Ric}^{\perp} + \frac{1}{n} H^{\flat} \otimes H^{\flat} - \mathrm{Def}_{\mathscr{D}} H = \frac{1}{p}\left(\frac{p-1}{p}\|\tilde{H}\|^2 + \mathrm{div}\,\tilde{H}\right) g^{\perp}, \qquad (4.36)$$

$$\mathrm{Ric}^{\perp} + H^{\flat} \otimes H^{\flat} - \mathrm{Def}_{\mathscr{D}} H$$
$$= \frac{1}{2}\left(\mathrm{S}_{\mathrm{mix}} - \mathrm{S}_{\mathrm{mix}}^{*}(\Omega,g) + \frac{2(p-1)}{p^2}\|\tilde{H}\|^2 + \mathrm{div}(\tilde{H}-H)\right) g^{\perp}. \qquad (4.37)$$

4.2.3.1 Critical Adapted Metrics on Foliations

We will characterize critical metrics for the action (4.4) in some distinguished classes of foliations.

Theorem 4.17. *Let complementary orthogonal distributions $\mathscr{D}^p$ and $\widetilde{\mathscr{D}}^n$ with $n, p > 1$ determine totally umbilical foliations of a pseudo-Riemannian manifold (M,g). Then g is critical for the action (4.4) with respect to $g^{\perp}$-variations if and only if the leaves of $\widetilde{\mathscr{D}}$ are totally geodesic and*

$$\mathrm{Ric}^{\perp} = (\mathrm{S}_{\mathrm{mix}}/p)\,g^{\perp} \quad \text{with} \quad \begin{cases} (\mathrm{div}\,\tilde{H})(\Omega,g) = 0 & \text{if } p \neq 2, \\ \mathrm{S}_{\mathrm{mix}} = \mathrm{const} & \text{if } p = 2. \end{cases} \qquad (4.38)$$

Proof. By (4.36),

$$\mathrm{S}_{\mathrm{mix}} = \frac{n-1}{n}\|H\|^2 + \frac{p-1}{p}\|\tilde{H}\|^2 + \mathrm{div}(H+\tilde{H}).$$

Let g be critical for $g^{\perp}$-variations. The difference of (4.37) and (4.36) provides

$$\frac{n-1}{n} H^{\flat} \otimes H^{\flat} = \frac{1}{2}\left(\frac{n-1}{n}\|H\|^2 + \frac{p-1}{p}\|\tilde{H}\|^2 - \mathrm{S}_{\mathrm{mix}}^{*}(\Omega,g) + \frac{2(p-1)}{p}\,\mathrm{div}\,\tilde{H}\right) g^{\perp}.$$

As the tensor $H^{\flat} \otimes H^{\flat}$ has rank ≤ 1 and $g^{\perp}$ has rank $p > 1$, we obtain $H = 0$; hence, the leaves of $\mathscr{D}$ are totally geodesic. By (4.36), the tensor $\mathrm{Ric}^{\perp}$ is conformal on $\widetilde{\mathscr{D}}$ (i.e., $\widetilde{\mathscr{D}}$-conformal). We also get $\mathrm{S}_{\mathrm{mix}}^{*} = \mathrm{S}_{\mathrm{mix}}$ and $\mathrm{S}_{\mathrm{mix}} + \frac{p-2}{p}\,\mathrm{div}\,\tilde{H} = \mathrm{S}_{\mathrm{mix}}^{*}(\Omega,g)$. Thus, $\int_{\Omega}(\mathrm{div}\,\tilde{H})\,d\mathrm{vol} = 0$ for $p \neq 2$. The proof of converse statement is similar. $\square$

Example 4.18. For a doubly twisted product $M_1 \times_{(f_1,f_2)} M_2$ we get

$$A_Y = -Y(f_1)\,\widetilde{\mathrm{id}}, \qquad h = -(\nabla^{\perp} f_1)\,g^{\top}, \qquad H = -n\nabla^{\perp} f_1,$$

(and similarly for $\tilde{A}_X, \tilde{h}, \tilde{H}$) where $X \in \widetilde{\mathscr{D}}$ and $Y \in \mathscr{D}$ are unit vectors. In this case,

$$\mathrm{div}\,H = -n\Delta^{\perp} f_1 - n^2\|\nabla^{\perp} f_1\|^2, \qquad \mathrm{div}\,\tilde{H} = -p\Delta^{\top} f_2 - p^2\|\nabla^{\top} f_2\|^2,$$

see (1.33). By (1.46),

$$\mathrm{S}_{\mathrm{mix}} = \mathrm{div}(H+\tilde{H}) + \frac{n-1}{n}\|H\|^2 + \frac{p-1}{p}\|\tilde{H}\|^2.$$

Let g be critical for the action (4.4) with respect to $g^\perp$-variations. By Theorem 4.17, the leaves (of $\widetilde{\mathscr{D}}$) are totally geodesic, and (4.38) hold. Note that

$$\Delta^\top e^{pf_2} = e^{pf_2}\left(p\Delta^\top f_2 + p^2 \|\nabla^\top f_2\|^2\right).$$

We conclude that a pseudo-Riemannian doubly twisted product metric g is critical for (4.4) with respect to $g^\perp$-variations if and only if

(i) $\mathrm{Ric}^\perp$ is $\mathscr{D}$-conformal;
(ii) $\Delta^\top e^{pf_2} = 0$; hence, e^{pf_2} is $\widetilde{\mathscr{D}}$-harmonic when $g^\perp$ is definite;
(iii) $(M_1, e^{f_1}g_1)\times_{f_2}(M_2,g_2)$ is a twisted product and f_1 doesn't depend on M_2.

There are no nonconstant positive harmonic functions on a complete manifold (M,g) with $\mathrm{Ric} \geq 0$, but such functions exist on (M,g) of nonnegative curvature outside a compact set. Bounded harmonic functions on open manifolds with non-negative curvature were described in [LT]; positive harmonic functions correspond to 'large ends' of the manifold (e.g., the fiber of the submersion in Section 4.2.3.3). The next theorem continues Example 4.15: one of distributions is integrable.

Theorem 4.19. *Let a distribution $\mathscr{D}$ be nowhere integrable and $\widetilde{\mathscr{D}}$ tangent to a to-tally geodesic Riemannian foliation of a pseudo-Riemannian manifold (M,g). Then g is critical for the action (4.4) with respect to $g^\perp$-variations if and only if*

$$\text{(a) } \mathrm{Ric}^\perp = (S_{\mathrm{mix}}/p)\,g^\perp \quad \text{and} \quad \text{(b) } S_{\mathrm{mix}} = \text{const}, \quad \text{when } p \neq 4.$$

Proof.

(a) By conditions, $p > 1$, $h = 0$, $\tilde{h} = 0$ and $T = 0$. Thus, (1.41) reads as

$$\mathrm{Ric}^\perp = -\widetilde{\mathscr{T}}^\flat. \tag{4.39}$$

Tracing (4.39), we find $S_{\mathrm{mix}} = \|\tilde{T}\|^2$. From (4.25) we obtain

$$\mathrm{Ric}^\perp - \widetilde{\mathscr{T}}^\flat = \frac{1}{2}\left(S_{\mathrm{mix}} - S^*_{\mathrm{mix}}(\Omega,g)\right)g^\perp \quad \text{for } g^\perp\text{-variations}, \tag{4.40}$$

where $S^*_{\mathrm{mix}} = \frac{p-4}{p}\|\tilde{T}\|^2$, see (4.34). Adding (4.39) and (4.40), we obtain

$$\mathrm{Ric}^\perp = \frac{1}{4}\left(S_{\mathrm{mix}} - S^*_{\mathrm{mix}}(\Omega,g)\right)g^\perp. \tag{4.41}$$

(b) Tracing (4.41) and using trace $\mathrm{Ric}^\perp = S_{\mathrm{mix}}$, we get $(p-4)S_{\mathrm{mix}} = p\,S^*_{\mathrm{mix}}(\Omega,g)$, hence, $S_{\mathrm{mix}} = \frac{p}{p-4}S^*_{\mathrm{mix}}(\Omega,g)$ when $p \neq 4$. This and (4.41) complete the proof. $\square$

Theorem 4.20. *Let $\mathscr{F}$ be a totally geodesic foliation of a pseudo-Riemannian mani-fold (M,g) with integrable normal bundle $\mathscr{D}$. Then g is critical for J_{mix} with respect to $g^\perp$-variations if and only if*

$$\mathrm{div}\left(\tilde{h} - \frac{1}{p}\tilde{H}g^{\perp}\right) = 0, \tag{4.42a}$$

and g is critical for the action J_{mix} with respect to $g^{\top}$-variations if and only if

$$\tilde{H}^{\flat} \otimes \tilde{H}^{\flat} - \frac{1}{2}\Upsilon_{\tilde{h},\tilde{h}} = \frac{1}{n}\,\tilde{S}_{\mathrm{ex}}\,g^{\top} \quad \text{and} \quad S_{\mathrm{ex}} = \text{const} \quad \text{when } n \neq 2. \tag{4.42b}$$

Proof. Using (4.28) and its dual with $\tilde{T} = 0$, rewrite (4.25) and (4.35) as

$$\mathrm{div}(\tilde{h} - \tilde{H}g^{\perp}) + H^{\flat} \otimes H^{\flat} - \frac{1}{2}\Upsilon_{h,h} = \frac{1}{2}\left(S_{\mathrm{ex}} + \tilde{S}_{\mathrm{ex}} - S^{*}_{\mathrm{mix}}(\Omega, g)\right)g^{\perp} \quad (g^{\perp}\text{-var-s}), \tag{4.43a}$$

$$\mathrm{div}(h - Hg^{\top}) + \tilde{H}^{\flat} \otimes \tilde{H}^{\flat} - \frac{1}{2}\Upsilon_{\tilde{h},\tilde{h}} = \frac{1}{2}\left(S_{\mathrm{ex}} + \tilde{S}_{\mathrm{ex}} - S^{*}_{\mathrm{mix}}(\Omega, g)\right)g^{\top} \quad (g^{\top}\text{-var-s}). \tag{4.43b}$$

To show (4.42b) $\Rightarrow$ (4.42a), observe that $h = 0$; hence, (4.43a) reads:

$$\mathrm{div}\left(\tilde{h} - \tilde{H}g^{\perp}\right) = \frac{1}{2}\left(\tilde{S}_{\mathrm{ex}} - S^{*}_{\mathrm{mix}}(\Omega, g)\right)g^{\perp}, \tag{4.44}$$

where $S^{*}_{\mathrm{mix}} = S_{\mathrm{mix}}$. Taking trace of (4.44) yields

$$(1 - p)\,\mathrm{div}\,\tilde{H} = \frac{p}{2}\left(\tilde{S}_{\mathrm{ex}} - S^{*}_{\mathrm{mix}}(\Omega, g)\right). \tag{4.45}$$

From (4.44) and (4.45) we obtain (4.42a).

To show (4.42a) $\Rightarrow$ (4.42b), from (4.43b) with $h = 0$ we obtain for $g^{\top}$-variations,

$$\tilde{H}^{\flat} \otimes \tilde{H}^{\flat} - \frac{1}{2}\Upsilon_{\tilde{h},\tilde{h}} = \frac{1}{2}\left(\tilde{S}_{\mathrm{ex}} - S^{*}_{\mathrm{mix}}(\Omega, g)\right)g^{\top}, \tag{4.46}$$

where $S^{*}_{\mathrm{mix}} = S_{\mathrm{mix}} - \frac{2}{n}(\tilde{S}_{\mathrm{ex}} + \mathrm{div}\,\tilde{H})$. Tracing (4.46) yields

$$\tilde{S}_{\mathrm{ex}} = \frac{n}{2}\left(\tilde{S}_{\mathrm{ex}} - S^{*}_{\mathrm{mix}}(\Omega, g)\right). \tag{4.47}$$

From (4.46) and (4.47) we get the first equality of (4.42b). Next, (4.47) yields the second equality of (4.42b) for $n \neq 2$. The proof of converse statements is similar. $\square$

4.2.3.2 Critical Adapted Metrics on Flows

Let the distribution $\tilde{\mathscr{D}}$ be spanned by a nonsingular vector field N. Then N defines a flow (a one-dimensional foliation), see Example 1.38. Assume that $g(N,N) = \varepsilon_N \in \{-1, 1\}$. The action (4.4) reduces itself to

$$J_{\mathrm{mix}}(g) = \varepsilon_N \int_M \mathrm{Ric}_{N,N}\,\mathrm{d}\,\mathrm{vol}_g. \tag{4.48}$$

Thus, (4.34) takes the form

$$S^*_{mix} = \varepsilon_N \, \text{Ric}_{N,N} - 2 \begin{cases} \frac{2}{p} \|\tilde{T}\|^2 + \frac{1}{p} \, \text{div} \, H & \text{for } g^\perp\text{-variations,} \\ \varepsilon_N(N(\tilde{\tau}_1) - \tilde{\tau}_2) - \|\tilde{T}\|^2 & \text{for } g^\top\text{-variations.} \end{cases}$$

From Theorem 4.14 we obtain the following.

Corollary 4.21 (Euler–Lagrange Equations). *Metric $g \in \text{Riem}(M^{p+1}, \widetilde{\mathscr{D}}, \mathscr{D})$ is critical for the action (4.48) with respect to all adapted variations if and only if*

$$\varepsilon_N \left(R_N + (\tilde{A}_N)^2 - (\tilde{T}_N^\sharp)^2 + [\tilde{T}_N^\sharp, \tilde{A}_N] \right)^\flat - \tilde{\tau}_1 \tilde{h}_{sc} + H^\flat \otimes H^\flat - \text{Def}_{\mathscr{D}} H$$

$$= \tfrac{1}{2} \left(\varepsilon_N \text{Ric}_{N,N} - S^*_{mix}(\Omega, g) + \text{div}(\varepsilon_N \tilde{\tau}_1 N - H) \right) g^\perp \quad (g^\perp\text{-variations}), \quad (4.49a)$$

$$\varepsilon_N \text{Ric}_{N,N} + S^*_{mix}(\Omega, g) - 4\|\tilde{T}\|^2 - \text{div}(\varepsilon_N \tilde{\tau}_1 N + H) = 0 \quad (g^\top\text{-variations}). \quad (4.49b)$$

Proof. An easy computation shows that

$$\begin{aligned}
&\widetilde{\mathscr{A}} = \varepsilon_N(\tilde{A}_N)^2, \quad \langle \tilde{h}_{sc} N, \tilde{H} \rangle = \tilde{\tau}_1 \tilde{h}_{sc}, \quad \Psi = H^\flat \otimes H^\flat, \quad \widetilde{\Psi} = (\varepsilon_N \tilde{\tau}_2 - \|\tilde{T}\|^2) g^\top, \\
&\mathscr{A} = \|H\|^2 \, \text{id}^\top, \quad \mathscr{T} = 0, \quad \langle h, H \rangle = \|H\|^2 g^\top, \\
&H = \varepsilon_N \nabla_N N, \quad h = H g^\top, \quad \|h\|^2 = \|H\|^2, \\
&\tilde{H} = \varepsilon_N \tilde{\tau}_1 N, \quad \tilde{\tau}_1 = \varepsilon_N \, \text{trace}_g \, \tilde{h}_{sc}, \quad \|\tilde{h}\|^2 = \varepsilon_N \tilde{\tau}_2, \quad \text{Def}_{\widetilde{\mathscr{D}}} \tilde{H} = \varepsilon_N N(\tilde{\tau}_1) g^\top. \quad (4.50)
\end{aligned}$$

Notice that $(H^\flat \otimes H^\flat)(X, Y) = g(H, X) g(H, Y)$. Substituting (4.50) and

$$H^\flat \otimes H^\flat - \frac{1}{2} \Upsilon_{h,h} = 0 = S_{ex}, \quad \widetilde{S}_{ex} = \varepsilon_N(\tilde{\tau}_1^2 - \tilde{\tau}_2), \quad \widetilde{\mathscr{T}} = \varepsilon_N(\tilde{T}_N^\sharp)^2$$

into (4.25) yields (4.49a). Substituting (4.50) and

$$h = H g^\top, \quad \tilde{H}^\flat \otimes \tilde{H}^\flat - (1/2)\Upsilon_{\tilde{h}, \tilde{h}} = \varepsilon_N(\tilde{\tau}_1^2 - \tilde{\tau}_2) g^\top, \quad \Upsilon_{\tilde{T}, \tilde{T}} = 2\|\tilde{T}\|^2 g^\top$$

into equation dual to (4.25) yields (4.49b). $\qquad \square$

By (1.32), we have $\text{div} \, \tilde{h} = N(\tilde{h}_{sc}) - \tilde{\tau}_1 \tilde{h}_{sc}$ and $\text{div} \, h = (\text{div} \, H) g^\top$.

Corollary 4.22 (of Theorem 4.19). *Let a unit vector field N generate a geodesic Riemannian flow (see Section 1.2.1) on a pseudo-Riemannian manifold (M^{p+1}, g). Then g is critical for the action (4.48) with respect to $g^\perp$-variations if and only if*

$$R_N = (1/p) \text{Ric}_{N,N} \, \text{id}^\perp \quad \text{and} \quad \text{Ric}_{N,N} = \text{const} \quad \text{when} \quad p \neq 4. \quad (4.51)$$

Moreover, if p is odd then $\text{Ric}_{N,N} = 0$ and M splits, and if $\text{Ric}_{N,N} \neq 0$ then p is even.

Proof. By Theorem 4.19, we have (4.51), and (4.39) reads $R_N = -(\tilde{T}_N^\sharp)^2$. Tracing this, we obtain $\varepsilon_N \text{Ric}_{N,N} = \|\tilde{T}\|^2$. In our case, (4.21) reads

$$S^*_{mix} = \begin{cases} \frac{p-4}{p} \|\tilde{T}\|^2 & \text{for } g^\perp\text{-variations,} \\ 3 \|\tilde{T}\|^2 & \text{for } g^\top\text{-variations.} \end{cases}$$

For a geodesic Riemannian N-flow, (4.49a-b) reduce to

$$\varepsilon_N \big(R_N - (\tilde{T}_N^{\sharp})^2 \big)^{\flat} = (1/2) \big(\varepsilon_N \mathrm{Ric}_{N,N} - \mathrm{S}_{\mathrm{mix}}^{*}(\Omega, g) \big) g^{\perp} \quad \text{for } g^{\perp}\text{-variations},$$
$$\varepsilon_N \mathrm{Ric}_{N,N} = -\mathrm{S}_{\mathrm{mix}}^{*}(\Omega, g) + 4 \|\tilde{T}\|^2 \quad \text{for } g^{\top}\text{-variations}.$$

For p odd, the skew-symmetric operator $\tilde{T}_N^{\sharp}$ has zero eigenvalue; hence, $R_N = 0 = \tilde{T}$; and by de Rham Decomposition Theorem, (M, g) splits. $\square$

Finally, observe that we can examine codimension-one foliations and distributions with critical metrics for other actions with respect to adapted variations, for example, (4.29). Since the case of $p = 1$ is trivial for this action, we consider $n = 1$ instead. Next result provides applications to foliations whose leaves have constant second mean curvature, see [RW2, Section 1.1.1].

Proposition 4.23. *Let (M, g) be a pseudo-Riemannian manifold, and a distribution $\widetilde{\mathscr{D}}$ spanned by a complete in Ω unit vector field N (i.e., all the trajectories of N passing through Ω extend and stay in Ω all the time). If g is critical for the action (4.29) with respect to all adapted variations, then*

$$\tilde{\tau}_1 = 0, \quad \tilde{\tau}_2 = \mathrm{const}. \tag{4.52}$$

Proof. From (4.30a) we obtain

$$\nabla_N \tilde{h}_{sc} - \tilde{\tau}_1 \tilde{h}_{sc} + \varepsilon_N [\tilde{T}_N^{\sharp}, \tilde{A}_N]^{\flat} = 0.$$

Tracing the above yields $N(\tilde{\tau}_1) = \tilde{\tau}_1^2$, and in view of completeness in Ω of the flow of N, the only solution is $\tilde{\tau}_1 = 0$, hence $(4.52)_1$. From (4.30b) with $n = 1$ and $\tilde{H}^{\flat} \otimes \tilde{H}^{\flat} - \frac{1}{2}\Upsilon_{\tilde{h},\tilde{h}} = \varepsilon_N(\tilde{\tau}_1^2 - \tilde{\tau}_2) g^{\top}$ we obtain

$$2\tilde{\sigma}_2 = \varepsilon_N \widetilde{\mathrm{S}}_{\mathrm{ex}}, \quad \widetilde{\mathrm{S}}_{\mathrm{ex}} = \widetilde{\mathrm{S}}_{\mathrm{ex}}(\Omega, g),$$

which together with $\widetilde{\mathrm{S}}_{\mathrm{ex}} = \widetilde{\mathrm{S}}_{\mathrm{ex}}^{*}$ and $\tilde{\tau}_1 = 0$ yields $\tilde{\tau}_2 = -\varepsilon_N \widetilde{\mathrm{S}}_{\mathrm{ex}}^{*}(\Omega, g)$. Hence critical metrics of (4.29) with respect to all adapted variations are those with constant $\tilde{\tau}_2$. $\square$

From the proof of Proposition 4.23 it follows that for $n = 1$ the critical metrics of the action (4.29) with respect to all adapted variations also satisfy the equation

$$\nabla_N \tilde{h}_{sc} + \varepsilon_N [\tilde{T}_N^{\sharp}, \tilde{A}_N]^{\flat} = 0.$$

4.2.3.3 Conformal Submersions

Conformal (and Riemannian) submersions form an important class of mappings, which were investigated also in relation with Einstein equations, see survey in [FIP].

Definition 4.24. A differentiable mapping $\pi : (M, g) \to (\hat{M}, \hat{g})$ of smooth pseudo-Riemannian manifolds is called a *(horizontally) conformal submersion* if

1. π is a submersion, i.e., it is surjective and has maximal rank,
2. π_* restricted to the distribution orthogonal to the fibers of π is a conformal mapping (i.e., there exists function $f \in C^1(M)$, called *dilation* of the submersion, such that $g(X,Y) = e^{2f}\hat{g}(\pi_*X, \pi_*Y)$ for all vectors X,Y orthogonal to fibers of π).

For $p = 1$ and positive definite metrics any submersion is conformal. Denote by $\widetilde{\mathcal{D}}$ the (integrable) distribution tangent to the fibers of conformal submersion, and assume that $\widetilde{\mathcal{D}}$ and its orthogonal complement $\mathcal{D}$ are non-degenerate. For a conformal submersion, $\mathcal{D}$ is totally umbilical whose second fundamental form obeys

$$\tilde{h} = -(\nabla^\top f)\, g^\perp. \tag{4.53}$$

Of course, for $g_{new} = g^\top \oplus e^{-2f} g^\perp$, our π becomes a Riemannian submersion, thus, $\tilde{h}_{new} = 0$. Then we use $\tilde{h} = e^{2f}(\tilde{h}_{new} - (\nabla^\top f)\, g^\perp$, see also [RWo1, Lemma 2.2].

Among conformal submersions, those with totally umbilical fibers provide an example of particularly interesting geometry. While the adapted variations g_t preserve the orthogonality of two distributions, we can consider their particular class which preserves the structure of conformal submersion with totally umbilical fibers.

Definition 4.25. A tensor $B \in \mathfrak{M}_{\mathcal{D}}$ is $\widetilde{\mathcal{D}}$-*conformal* if $B = s\, g^\perp$ for some $s \in C^\infty(M)$. An adapted variation g_t is *biconformal* if there exist $s_1, s_2 \in C^\infty(M, \mathbb{R})$ such that

$$\partial_t g_t^\perp = s_1 g_0^\perp, \quad \partial_t g_t^\top = s_2 g_0^\top.$$

Given $g \in \mathrm{Riem}(M, \widetilde{\mathcal{D}}, \mathcal{D})$, the subspace of $\mathfrak{M}$, consisting of *biconformal* adapted tensors, splits into the direct sum of $\mathcal{D}$- and $\widetilde{\mathcal{D}}$-conformal components.

Proposition 4.26. *Let* $\pi : (M^{n+p}, g) \to (\hat{M}^p, \hat{g})$ *be a conformal submersion with totally umbilical fibers, and* g_t *an adapted variation of* g. *Then all mappings* $\pi : (M, g_t) \to (\hat{M}, \hat{g})$ *are conformal submersions with totally umbilical fibers if and only if variation* g_t *is* $\mathcal{D}$-*conformal and*

$$\nabla\left(B - (\mathrm{trace}\, B^\sharp / n)\, g^\top\right) = 0. \tag{4.54}$$

Proof. If all the mappings $\pi : (M, g_t) \to (\hat{M}, \hat{g})$ are conformal submersions, then $e^{-2f_t} g_t^\perp = \pi^*(\hat{g})$ for some $f_t \in C^\infty(M)$. Differentiating the above we get

$$e^{-2f_t} \partial_t g_t^\perp - 2\partial_t f_t\, e^{-2f_t} g_t^\perp = 0.$$

Hence, $\partial_t g_t^\perp = s\, g_0^\perp$ for $s = 2\partial_t f_t$, and variation g_t is a $\mathcal{D}$-conformal. If $\widetilde{\mathcal{D}}$ is g_t-totally umbilical for all t, then $h = \frac{1}{n} H g_t^\top$, and by Lemma 4.7 for $\tilde{h}$, we get

$$\frac{2}{n}\left(B(X,Y)H + \langle X,Y\rangle \partial_t H\right) = \frac{2}{n} B(X,Y)H - \nabla B(X,Y)$$

for all $X, Y \in \widetilde{\mathcal{D}}$. Using Lemma 4.7 for $\tilde{H}$ yields $\frac{1}{n}\langle X,Y\rangle \nabla(\mathrm{trace}\, B^\sharp) = \nabla B(X,Y)$.

On the other hand, if (4.54) is satisfied and the variation is $\mathscr{D}$-conformal, then from the uniqueness of the solution of ODE, $h = \frac{1}{n}Hg_t^\top$ and $e^{-2f_t}g_t^\perp = \pi^*\hat{g}$ for all t; hence, all $\pi : (M, g_t) \to (\hat{M}, \hat{g})$ are conformal submersions with umbilical fibers. $\square$

Note that (4.54) is satisfied, in particular, by biconformal variations. Together with a family g_t of $\{g_t^\perp + g^\top : |t| < \varepsilon\}$, one may consider on Ω the metrics (4.14) and examine the metrics critical for (4.4) with respect to $\mathscr{D}$-conformal variations.

Remark 4.27. Using (4.13) and (4.27) with $B = sg^\perp$ for some $s \in C^\infty(M)$, we find the Euler–Lagrange equation for the action (4.4) and variations (4.14) of metrics,

$$(p-1)\operatorname{div}\tilde{H} + \frac{p-2}{2}\left(S_{ex} + \|\tilde{T}\|^2\right) + \frac{p}{2}\left(\tilde{S}_{ex} + \|T\|^2 - S^*_{mix}(\Omega, g)\right) = 0, \quad (4.55)$$

where $S^*_{mix} = S_{mix} - \frac{2}{p}\left(S_{ex} + 2\|\tilde{T}\|^2 - \|T\|^2\right)$.

The mixed scalar curvature is an important tool in investigation of conformal submersions with totally umbilical fibers. The following formula (used in [Za] to obtain integral formulas and existence conditions for such mappings) is just a particular case of (1.46), expressed in terms of f and H:

$$S_{mix} = -p\Delta^\top f - p\|\nabla^\top f\|^2 + \|\tilde{T}\|^2 + \operatorname{div}H + \frac{n-1}{n}\|H\|^2. \quad (4.56)$$

Next, we will present the Euler–Lagrange equations for biconformal variations on the domains of conformal submersions with totally umbilical fibers.

Proposition 4.28 (Euler–Lagrange Equations). *Let* $\pi : (M^{n+p}, g) \to (\hat{M}^p, \hat{g})$, *where* $p > 1$, *be a conformal submersion with totally umbilical fibers. Then g is critical for the action (4.4) with respect to biconformal variations if and only if*

$$-2p(p-1)\Delta^\top f - p^2(p-1)\|\nabla^\top f\|^2 + \frac{(p-2)(n-1)}{n}\|H\|^2$$
$$+ (p-2)\|\tilde{T}\|^2 = p\,S^*_{mix}(\Omega, g) \quad \text{for } \mathscr{D}\text{-conformal variations,} \quad (4.57a)$$
$$p(p-1)(n-2)\|\nabla^\top f\|^2 + 2(n-1)\operatorname{div}H + (n-1)\|H\|^2$$
$$+ n\|\tilde{T}\|^2 = n\,\tilde{S}^*_{mix}(\Omega, g) \quad \text{for } \tilde{\mathscr{D}}\text{-conformal variations} \quad (4.57b)$$

(one of conditions for one type of conformality and both for biconformality), where

$$S^*_{mix} = -p\left(\Delta^\top f + \|\nabla^\top f\|^2\right) + \frac{p-4}{p}\|\tilde{T}\|^2 + \frac{(n-1)(p-2)}{np}\|H\|^2 + \frac{p-2}{p}\operatorname{div}H,$$
$$\tilde{S}^*_{mix} = -p\frac{n-2}{n}\left(\Delta^\top f + \|\nabla^\top f\|^2\right) + \frac{n+2}{n}\|\tilde{T}\|^2 + \operatorname{div}H + \frac{n-1}{n}\|H\|^2. \quad (4.58)$$

Proof. For conformal submersions with totally umbilical fibers we have

$$T = 0, \quad S_{ex} = \frac{n-1}{n}\|H\|^2, \quad \tilde{S}_{ex} = \frac{p-1}{p}\|\tilde{H}\|^2,$$

and from (4.53) we get $\tilde{H} = -p\nabla^\top f$. Using this, rewrite (4.55) as (4.57a). For $\widetilde{\mathcal{D}}$-conformal variations of metrics on the domain of conformal submersion with umbilical fibers, a formula analogous to (4.55) yields (4.57b). Using (4.56) in (4.21), we get remaining formulas (4.58). $\qquad\square$

Corollary 4.29. *Let* $\pi : (M^{n+p}, g) \to (\hat{M}^p, \hat{g})$, *where* $p > 1$ *and* $g^\perp > 0$, *be a conformal submersion with complete totally geodesic fibers. Then* g *is critical for the action* (4.4) *with respect to biconformal variations if and only if* $e^{\lambda f}$, *where* $\lambda = \frac{1}{2n}(pn + (p-2)(n-2)) > 0$, *is a fiberwise harmonic function.*

Proof. From (4.57a,b) we obtain

$$p(p-1)\big(\Delta^\top f + \lambda \|\nabla^\top f\|^2\big) = \frac{p-2}{2}\,\widetilde{S}^*_{\mathrm{mix}}(\Omega, g) - \frac{p}{2}\,S^*_{\mathrm{mix}}(\Omega, g).$$

Using the identity $\Delta^\top f + \lambda \|\nabla^\top f\|^2 = \frac{1}{\lambda} e^{-\lambda f}\Delta^\top e^{\lambda f}$ in the above yields

$$\Delta^\top e^{\lambda f} = G\lambda e^{\lambda f}, \tag{4.59}$$

where $G = \frac{1}{2p(p-1)}\big((p-2)\widetilde{S}^*_{\mathrm{mix}} - p\,S^*_{\mathrm{mix}}\big)(\Omega, g)$. Equation (4.59) is an eigenvalue problem of operator $\Delta^\top$ on every fiber, and $e^{\lambda f}$ is its positive solution; hence, $G = 0$ and $e^{\lambda f}$ is fiber wise harmonic. For closed fibers, (4.59) admits only fiber wise constant solutions f. If we allow our variations not to preserve the volume of Ω, then again $G = 0$ and (4.59) becomes the fiberwise Laplace equation for $e^{\lambda f}$. $\qquad\square$

4.3 General Variations of Metric

Here, we provide variational formulae for the quantities of extrinsic geometry of almost-product (e.g. foliated) pseudo-Riemannian manifolds for general variations of metrics, and their application to study the Einstein–Hilbert type actions. Given a pair $(M, \widetilde{\mathcal{D}})$ of a manifold and a distribution, we study pseudo-Riemannian structures g non-degenerate on $\widetilde{\mathcal{D}}$ and critical for the functional

$$J_{\mathrm{mix},\widetilde{\mathcal{D}}} : g \to \int_M S_{\mathrm{mix}}(\widetilde{\mathcal{D}}, g)\, d\mathrm{vol}_g, \tag{4.60}$$

and Ω is a relatively compact domain of M (and $\Omega = M$ if M is closed), containing supports of variations of the metric. Together with general variations of metric, we consider also variations that preserve the volume of the manifold or partially preserve the metric (e.g. on the distribution). For each of those cases, we express the Euler–Lagrange equation for (4.60) in terms of extrinsic geometry of a pair $(\widetilde{\mathcal{D}}, \mathcal{D})$. Given examples of critical metrics are related to contact and 3-Sasakian manifolds, geodesic Riemannian flows, codimension-one foliations and distributions of interesting geometry (e.g. totally umbilical and minimal). We also write the Euler–Lagrange equations of the action in the form of Einstein field equations, which involve a new kind of Ricci curvature and might be useful in physics.

4.3.1 Variational Formulae

We consider smooth 1-parameter variations $\{g_t \in \mathrm{Riem}(M): |t| < \varepsilon\}$ of the metric $g_0 = g$ on $(M, \widetilde{\mathscr{D}}, g)$. We adopt notations (4.7), but we shall also write B instead of B_t to make formulas easier to read, wherever it does not lead to confusion.

Definition 4.30. A family g_t $(|t| < \varepsilon)$ will be called a $g^\pitchfork$-*variation* for $(\widetilde{\mathscr{D}}, \mathscr{D})$ if only the metric on $\widetilde{\mathscr{D}}$ is preserved, i.e.,

$$g_t(X,Y) = g(X,Y) \quad (X,Y \in \mathfrak{X}_{\widetilde{\mathscr{D}}},\ |t| < \varepsilon).$$

If $\widetilde{\mathscr{D}}$ and $\mathscr{D}$ remain orthogonal, then g_t is an *adapted variation*, see Section 4.2; moreover, if only the metric on $\widetilde{\mathscr{D}}$ changes, i.e., $g_t(X,Y) = g(X,Y)$ $(X,Y \in \mathfrak{X}_{\mathscr{D}})$, then we obtain a $g^\top$-*variation*.

Here, we allow the g_t-orthogonal complement $\mathscr{D}(t)$ of a distribution $\widetilde{\mathscr{D}}$ to vary, this enables us to consider arbitrary variations of the metric. Indeed, one can show that variational formulae for general variations g_t can be obtained by summing of the corresponding formulas from Proposition 4.33 given below. This follows from the fact that every variation of g is decomposed into the sum of $g^\pitchfork$- and $g^\top$-variations. Such decomposition is not possible using only adapted variations of metrics.

Let $\mathscr{D} = \mathscr{D}(0)$ be the g-orthogonal complement of $\widetilde{\mathscr{D}}$. While the distributions $\widetilde{\mathscr{D}}$ and $\mathscr{D}$ may not be g_t-orthogonal for $t > 0$, we can assume that they span the tangent bundle. For any $X \in TM$, let $X_{\widetilde{\mathscr{D}}}$ denote the g_0-orthogonal projection of X onto $\widetilde{\mathscr{D}}$ and let $X_{\mathscr{D}}$ denote the g_0-orthogonal projection of X onto $\mathscr{D}$.

Let V be the linear subspace of $TM \times TM$ spanned by $(\mathscr{D} \times \widetilde{\mathscr{D}}) \cup (\widetilde{\mathscr{D}} \times \mathscr{D})$. Thus, the product $TM \times TM$ is the sum of three subbundles, $\widetilde{\mathscr{D}} \times \widetilde{\mathscr{D}}$, $\mathscr{D} \times \mathscr{D}$ and V. For any $g_t \in \mathrm{Riem}(M)$, we have $g_t = g_{t\,|\,\mathscr{D} \times \mathscr{D}} + g_{t\,|\,\mathscr{D} \times \widetilde{\mathscr{D}}} + g_{t\,|\,\widetilde{\mathscr{D}} \times \mathscr{D}} + g_{t\,|\,\widetilde{\mathscr{D}} \times \widetilde{\mathscr{D}}}$, where

$$g_{t\,|\,\widetilde{\mathscr{D}} \times \widetilde{\mathscr{D}}}(X,Y) = g_t(X_{\widetilde{\mathscr{D}}}, Y_{\widetilde{\mathscr{D}}}), \quad g_{t\,|\,\mathscr{D} \times \mathscr{D}}(X,Y) = g_t(X_{\mathscr{D}}, Y_{\mathscr{D}}),$$
$$g_{t\,|\,\mathscr{D} \times \widetilde{\mathscr{D}}}(X,Y) = g_t(X_{\mathscr{D}}, Y_{\widetilde{\mathscr{D}}}), \quad g_{t\,|\,\widetilde{\mathscr{D}} \times \mathscr{D}}(X,Y) = g_t(X_{\widetilde{\mathscr{D}}}, Y_{\mathscr{D}});$$

thus, $g_{t\,|\,\mathrm{V}}(X,Y) = g_t(X_{\mathscr{D}}, Y_{\widetilde{\mathscr{D}}}) + g_t(X_{\widetilde{\mathscr{D}}}, Y_{\mathscr{D}})$, and we can present g_t in the form

$$g_t = \begin{pmatrix} g_{t\,|\,\mathscr{D} \times \mathscr{D}} & g_{t\,|\,\mathscr{D} \times \widetilde{\mathscr{D}}} \\ g_{t\,|\,\widetilde{\mathscr{D}} \times \mathscr{D}} & g_{t\,|\,\widetilde{\mathscr{D}} \times \widetilde{\mathscr{D}}} \end{pmatrix}.$$

Similarly, $B_t = B_t^\perp + B_t^{\mathrm{V}} + \widetilde{B}_t$, where $B_t^\perp = \partial_t g_{t\,|\,\mathscr{D} \times \mathscr{D}}$, $\widetilde{B}_t = \partial_t g_{t\,|\,\widetilde{\mathscr{D}} \times \widetilde{\mathscr{D}}}$ and $B_t^{\mathrm{V}} = \partial_t g_{t\,|\,\mathrm{V}}$. For $g^\pitchfork$-variations $g_t = g_{t\,|\,\mathscr{D} \times \mathscr{D}} + g_{t\,|\,\mathrm{V}} + g_{0\,|\,\widetilde{\mathscr{D}} \times \widetilde{\mathscr{D}}}$ and for $g^\top$-variations $g_t = g_{0\,|\,\mathscr{D} \times \mathscr{D}} + g_{t\,|\,\widetilde{\mathscr{D}} \times \widetilde{\mathscr{D}}}$ (as $g_{t\,|\,\mathrm{V}} = g_{0\,|\,\mathrm{V}} = 0$) we have, respectively,

$$B_t = B_t^\perp + B_t^{\mathrm{V}} = \begin{pmatrix} B_{t\,|\,\mathscr{D} \times \mathscr{D}}^\perp & B_{t\,|\,\mathscr{D} \times \widetilde{\mathscr{D}}}^{\mathrm{V}} \\ B_{t\,|\,\widetilde{\mathscr{D}} \times \mathscr{D}}^{\mathrm{V}} & 0 \end{pmatrix}, \quad B_t = \widetilde{B}_t = \begin{pmatrix} 0 & 0 \\ 0 & \widetilde{B}_{t\,|\,\widetilde{\mathscr{D}} \times \widetilde{\mathscr{D}}} \end{pmatrix}.$$

By the above, the derivative B_t of any variation g_t can be decomposed into sum of derivatives of $g^\pitchfork$- and $g^\top$-variations. Denote by $^\top$ and $^\perp$ the g_t-orthogonal (t-dependent) projections onto $\widetilde{\mathscr{D}}$ and $\mathscr{D}(t)$, respectively.

Lemma 4.31. *Let g_t be a $g^\pitchfork$-variation of g with $B_t = \partial_t g_t$ and let $\{E_a, \mathscr{E}_i\}$ be a local $(\widetilde{\mathscr{D}}, \mathscr{D})$-adapted and orthonormal for $t = 0$ frame, that evolves according to*

$$\partial_t E_a = 0, \qquad \partial_t \mathscr{E}_i = -(1/2)\,(B_t^\sharp(\mathscr{E}_i))^\perp - (B_t^\sharp(\mathscr{E}_i))^\top. \tag{4.61}$$

Then, for all t, $\{E_a(t) \equiv E_a, \mathscr{E}_i(t)\}$ is a g_t-orthonormal frame adapted to $(\widetilde{\mathscr{D}}, \mathscr{D}(t))$.

Proof. By $E_a(t) \equiv E_a \in \widetilde{\mathscr{D}}$, we get for $g^\pitchfork$-variation $\partial_t(g_t(E_a, E_b)) = 0$. Also,

$$\partial_t(g_t(E_a, \mathscr{E}_i(t))) = (\partial_t g_t)(E_a, \mathscr{E}_i(t)) + g_t(E_a, \partial_t \mathscr{E}_i(t))$$
$$= B_t(E_a, \mathscr{E}_i(t)) - (1/2)\,g_t((B_t^\sharp(\mathscr{E}_i(t)))^\perp, E_a) - g_t(E_a, B_t^\sharp(\mathscr{E}_i(t))^\top) = 0.$$

Since $g_t(E_a, \mathscr{E}_i(t)) = 0$ then $\mathscr{E}_i(t) \in \mathscr{D}(t)$, and for any X we have $g_t(\mathscr{E}_i(t), X^\top) = 0$. We complete the proof by computing

$$\partial_t(g_t(\mathscr{E}_i(t), \mathscr{E}_j(t))) = (\partial_t g_t)(\mathscr{E}_i(t), \mathscr{E}_j(t)) + g_t(\partial_t \mathscr{E}_i(t), \mathscr{E}_j(t)) + g_t(\mathscr{E}_i(t), \partial_t \mathscr{E}_j(t))$$
$$= B_t(\mathscr{E}_i(t), \mathscr{E}_j(t)) - (1/2)g_t((B_t^\sharp(\mathscr{E}_i(t)))^\perp, \mathscr{E}_j(t)) - (1/2)g_t(\mathscr{E}_i(t), (B_t^\sharp \mathscr{E}_j(t))^\perp) = 0. \ \square$$

Evolution of $\mathscr{D}(t)$ gives rise to the evolution of both $\widetilde{\mathscr{D}}$- and $\mathscr{D}(t)$-components of any vector X on M.

Lemma 4.32. *Let g_t be a $g^\pitchfork$-variation of g. Then for any vector X_t on M, we have*

$$\partial_t(X^\top) = (\partial_t X)^\top + (B^\sharp(X^\perp))^\top, \qquad \partial_t(X^\perp) = (\partial_t X)^\perp - (B^\sharp(X^\perp))^\top.$$

Proof. Using the frame from Lemma 4.31, we can write

$$X^\top = \sum_a \varepsilon_a g_t(X_t, E_a)E_a, \quad X^\perp = \sum_i \varepsilon_i g_t(X_t, \mathscr{E}_i(t))\mathscr{E}_i(t). \tag{4.62}$$

Then

$$B_t(E_a, E_b) = (\partial_t g_t)(E_a, E_b) = \partial_t(g_t(E_a, E_b)) - g_t(\partial_t E_a, E_b) - g_t(E_a, \partial_t E_b) = 0,$$

hence, $(B^\sharp(X^\top))^\top = 0$, which implies

$$(B^\sharp(X))^\top = (B^\sharp(X^\perp))^\top. \tag{4.63}$$

The proof follows from differentiating (4.62) and using (4.61) and (4.63). $\square$

Define the $(1,2)$-tensors for all $X, Y, Z \in \mathfrak{X}_M$:

$$\alpha(X,Y) = \big(A_{X^\perp}(Y^\top) + A_{Y^\perp}(X^\top)\big)/2, \quad \theta(X,Y) = \big(T_{X^\perp}^\sharp(Y^\top) + T_{Y^\perp}^\sharp(X^\top)\big)/2,$$
$$\tilde{\delta}_Z(X,Y) = \big(\langle \nabla_{X^\top} Z, Y^\perp \rangle + \langle \nabla_{Y^\top} Z, X^\perp \rangle\big)/2.$$

Similar tensors for $\widetilde{\mathscr{D}}$ will be denoted using $\tilde{}$ notation. Note that α, θ and $\tilde{\delta}_Z$ are symmetric and vanish for $(X,Y) \in (\widetilde{\mathscr{D}} \times \widetilde{\mathscr{D}}) \cup (\mathscr{D} \times \mathscr{D})$. If $X^\top = 0$ and $Y^\perp = 0$, then $\theta(X,Y) = \frac{1}{2} T_X^\sharp(Y)$. The key (and technical) result of this section is the following.

Proposition 4.33. *Let g_t be a $g^\pitchfork$-variation of g. Then*

$$\partial_t \|\tilde{h}\|^2 = \langle \operatorname{div} \tilde{h} - 4\Upsilon_{\tilde{\alpha},\theta} + \widetilde{\mathscr{K}}^\flat, B \rangle - \operatorname{div}\langle \tilde{h}, B \rangle,$$

$$\partial_t \|\tilde{H}\|^2 = \langle (\operatorname{div}\tilde{H}) g^\perp + 4\langle \theta, \tilde{H} \rangle, B \rangle - \operatorname{div}((\operatorname{trace}_{\mathscr{D}} B^\sharp)\tilde{H}),$$

$$\partial_t \|h\|^2 = 2\operatorname{div}\langle \alpha, B \rangle - 2\langle (\operatorname{div}\alpha)_{|V} + \Upsilon_{\alpha,\tilde{\alpha}+\tilde{\theta}} + (1/2)\Upsilon_{h,h}, B \rangle,$$

$$\partial_t \|H\|^2 = 2\langle \langle \tilde{\theta} - \tilde{\alpha}, H \rangle + H^\flat \odot \tilde{H}^\flat - \tilde{\delta}_H, B \rangle - B(H,H) + 2\operatorname{div}((B^\sharp(H))^\top),$$

$$\partial_t \|\tilde{T}\|^2 = 2\langle \widetilde{\mathscr{T}}^\flat + \Upsilon_{\tilde{\theta},\theta-\alpha} - (\operatorname{div}\tilde{\theta})_{|V}, B \rangle + 2\operatorname{div}\langle \tilde{\theta}, B \rangle,$$

$$\partial_t \|T\|^2 = (1/2)\langle \Upsilon_{T,T}, B \rangle.$$

Tensors $\Upsilon_{P,Q}, \mathscr{K}, \mathscr{T}$, etc. were defined in Sections 4.2.1 and 1.3.2. The proof of Proposition 4.33 is similar to the proof of Proposition 4.8 but rather long, see [RZ2], and we omit it. Using (1.45) and Proposition 4.33, for $g^\pitchfork$-variations we obtain

$$\partial_t S_{ex} = \langle 2\langle \tilde{\theta} - \tilde{\alpha}, H \rangle - H^\flat \otimes H^\flat + (1/2)\Upsilon_{h,h} + 2H^\flat \odot \tilde{H}^\flat - 2\tilde{\delta}_H$$
$$+ 2(\operatorname{div}\alpha)_{|V} + 2\Upsilon_{\alpha,\tilde{\alpha}+\tilde{\theta}}, B \rangle + 2\operatorname{div}\left((B^\sharp(H))^\top - \langle \alpha, B \rangle\right),$$

$$\partial_t \widetilde{S}_{ex} = \langle (\operatorname{div}\tilde{H})g^\perp + 4\langle \theta, \tilde{H} \rangle - \operatorname{div}\tilde{h} + 4\Upsilon_{\tilde{\alpha},\theta} - \widetilde{\mathscr{K}}^\flat, B \rangle + \operatorname{div}(\langle \tilde{h}, B \rangle - (\operatorname{trace}_{\mathscr{D}} B)\tilde{H}).$$

4.3.2 Euler–Lagrange Equations for the Total S_{mix}

Here, we present the Euler–Lagrange equation for (4.60) with different kinds of variations of metric. For arbitrary variations of the metric, the Euler–Lagrange equation is the condition that the gradient of the functional $\delta J_{mix,\widetilde{\mathscr{D}}}(g)$ vanishes, where

$$\frac{d}{dt} J_{mix,\widetilde{\mathscr{D}}}(g_t)_{|t=0} = \int_M \langle \delta J_{mix,\widetilde{\mathscr{D}}}, B \rangle \, d\operatorname{vol}_g$$

for any variation g_t with $B = \partial_t g_{t|t=0}$. One can also consider variations that preserve the volume of Ω. In this case, using (4.15), we have

$$0 = \partial_t \int_M d\operatorname{vol}_g = \int_M \partial_t(d\operatorname{vol}_g) = \int_M \left(\frac{1}{2}\operatorname{trace} B\right) d\operatorname{vol}_g = \frac{1}{2}\int_M \langle g, B \rangle \, d\operatorname{vol}_g.$$

Hence, a metric g is critical for the volume-preserving variations if and only if $\int_M \langle \delta J_{mix,\widetilde{\mathscr{D}}}, B \rangle \, d\operatorname{vol}_g = 0$ is valid for all B satisfying $\int_M \langle g, B \rangle \, d\operatorname{vol}_g = 0$. The corresponding Euler–Lagrange equation is now

$$\delta J_{mix,\widetilde{\mathscr{D}}} = \lambda g, \tag{4.65}$$

where $\lambda \in \mathbb{R}$ is an arbitrary constant. Note that the Euler–Lagrange equation for arbitrary variations is a special case of (4.65), with $\lambda = 0$.

We also consider volume-preserving $g^\top$- and $g^\pitchfork$-variations. For $g^\top$-variations B is restricted to $\widetilde{\mathscr{D}} \times \widetilde{\mathscr{D}}$, and for $g^\pitchfork$-variations B vanishes on $\widetilde{\mathscr{D}} \times \widetilde{\mathscr{D}}$. Hence, the Euler–Lagrange equation is still (4.65), only either restricted to $\widetilde{\mathscr{D}} \times \widetilde{\mathscr{D}}$ for $g^\top$-variations, or considered on $V \cup (\mathscr{D} \times \mathscr{D})$ for $g^\pitchfork$-variations.

Theorem 4.34 (Euler–Lagrange Equation). *A metric* $g \in \mathrm{Riem}(M, \widetilde{\mathscr{D}}, \mathscr{D})$ *is critical for* (4.60) *with respect to volume-preserving* $g^\pitchfork$-*variations if and only if*

$$\mathrm{Ric}^\perp - \langle \tilde{h}, \tilde{H} \rangle + \widetilde{\mathscr{A}^\flat} - \widetilde{\mathscr{T}^\flat} + \widetilde{\mathscr{K}^\flat} + H^\flat \otimes H^\flat - \frac{1}{2}\Upsilon_{h,h} - \frac{1}{2}\Upsilon_{T,T} + \Psi - \mathrm{Def}_\mathscr{D} H$$

$$= \frac{1}{2}\left(\mathrm{S}_{\mathrm{mix}} + \mathrm{div}(\tilde{H} - H) + 2\lambda \right) g^\perp, \tag{4.66a}$$

$$2\langle \theta, \tilde{H} \rangle + (\mathrm{div}(\alpha - \tilde{\theta}))_{|V} + \langle \tilde{\theta} - \tilde{\alpha}, H \rangle + H^\flat \odot \tilde{H}^\flat - \tilde{\delta}_H + 2\Upsilon_{\tilde{\alpha}, \theta} + \Upsilon_{\alpha, \tilde{\alpha}} + \Upsilon_{\theta, \tilde{\theta}} = 0. \tag{4.66b}$$

A metric $g \in \mathrm{Riem}(M, \widetilde{\mathscr{D}}, \mathscr{D})$ *is critical for the action* (4.60) *with respect to volume-preserving* $g^\top$-*variations if and only if the dual to* (4.66a) *is valid:*

$$\mathrm{Ric}^\top - \langle h, H \rangle + \mathscr{A}^\flat - \mathscr{T}^\flat + \mathscr{K}^\flat + \tilde{H}^\flat \otimes \tilde{H}^\flat - \frac{1}{2}\Upsilon_{\tilde{h},\tilde{h}} - \frac{1}{2}\Upsilon_{\tilde{T},\tilde{T}} + \widetilde{\Psi} - \mathrm{Def}_{\widetilde{\mathscr{D}}} \tilde{H}$$

$$= \frac{1}{2}\left(\widetilde{\mathrm{S}}_{\mathrm{mix}} + \mathrm{div}(H - \tilde{H}) + 2\lambda \right) g^\top. \tag{4.66c}$$

The (4.66a) *is equivalent to the following* (*no use of* $\mathrm{Ric}^\perp$, *similarly for* (4.66c)):

$$\mathrm{div}\,\tilde{h} - 2\widetilde{\mathscr{T}^\flat} + \widetilde{\mathscr{K}^\flat} + H^\flat \otimes H^\flat - \frac{1}{2}\Upsilon_{h,h} - \frac{1}{2}\Upsilon_{T,T} = \frac{1}{2}\left(\mathrm{S}_{\mathrm{mix}} + \mathrm{div}(\tilde{H} - H) + 2\lambda \right) g^\perp. \tag{4.67}$$

Proof. Let g_t be a $g^\pitchfork$-variation and let $Q(g) := \mathrm{S}_{\mathrm{mix}} - \mathrm{div}(H + \tilde{H})$. Then

$$\frac{\mathrm{d}}{\mathrm{d}t} J_{\mathrm{mix},\widetilde{\mathscr{D}}}(g_t)_{|t=0} = \frac{\mathrm{d}}{\mathrm{d}t} \int_\Omega Q(g_t)\,\mathrm{d\,vol}_{g_t\,|t=0} + \frac{\mathrm{d}}{\mathrm{d}t} \int_\Omega \mathrm{div}(H + \tilde{H})\,\mathrm{d\,vol}_{g_t\,|t=0}.$$

Differentiating $\mathrm{div}\,X \cdot \mathrm{d\,vol}_g = \mathscr{L}_X(\mathrm{d\,vol}_g)$ and using (4.15), we obtain $\partial_t(\mathrm{div}\,X) = \mathrm{div}(\partial_t X) + (1/2)X(\mathrm{trace}\,B^\sharp)$ for any t-dependent vector field X. In particular,

$$\frac{\mathrm{d}}{\mathrm{d}t} \int_\Omega \mathrm{div}(H + \tilde{H})\,\mathrm{d\,vol}_{g_t} = \int_M \partial_t(\mathrm{div}(H + \tilde{H}))\,\mathrm{d\,vol}_{g_t} + \int_M \mathrm{div}(H + \tilde{H})\,\partial_t(\mathrm{d\,vol}_{g_t})$$

$$= \int_M \mathrm{div}(\partial_t(H + \tilde{H}))\,\mathrm{d\,vol}_{g_t} + \int_M (1/2)\left((H + \tilde{H})\,\mathrm{trace}\,B^\sharp\right)\mathrm{d\,vol}_{g_t}$$

$$+ \int_M (1/2)(\mathrm{trace}\,B^\sharp)\,\mathrm{div}(H + \tilde{H})\,\mathrm{d\,vol}_{g_t}$$

$$= \int_M \mathrm{div}(\partial_t(H + \tilde{H}))\,\mathrm{d\,vol}_{g_t} + (1/2)\int_M \mathrm{div}((H + \tilde{H})\,\mathrm{trace}\,B^\sharp)\,\mathrm{d\,vol}_{g_t},$$

where $B = \{\partial_t g_t\}_{|t=0}$. For $g^{\pitchfork}$-variations supported in Ω, both vector fields $\partial_t(H + \tilde{H})$ and $(\mathrm{trace}\, B^{\sharp})(H + \tilde{H})$ vanish on $\partial\Omega$, hence, by the Divergence Theorem and (1.13), $\frac{\mathrm{d}}{\mathrm{dt}} \int_{\Omega} \mathrm{div}(H + \tilde{H})\, \mathrm{d}\,\mathrm{vol}_g = 0$. We have therefore

$$\frac{\mathrm{d}}{\mathrm{dt}} J_{\mathrm{mix},\widetilde{\mathscr{D}}}(g_t)_{|t=0} = \frac{\mathrm{d}}{\mathrm{dt}} \int_{\Omega} Q(g_t)\, \mathrm{d}\,\mathrm{vol}_{g_t\,|t=0},$$

and $Q(g)$ can be presented using (1.46) as (4.22). Applying Proposition 4.33 to (4.22), using (1.32) and removing integrals of divergences of vector fields compactly supported in Ω, we get

$$\int_M \partial_t Q(g_t)_{|t=0}\, \mathrm{d}\,\mathrm{vol}_g = \int_M \Big\langle 2\widetilde{\mathscr{T}}^{\flat} - \widetilde{\mathscr{K}}^{\flat} - \mathrm{div}\,\tilde{h} - H^{\flat} \otimes H^{\flat} + \frac{1}{2}\Upsilon_{h,h} + \frac{1}{2}\Upsilon_{T,T}$$
$$+ 4\Upsilon_{\tilde{\alpha},\theta} + 4\langle \theta, \tilde{H}\rangle + (\mathrm{div}\,\tilde{\alpha})g^{\perp} + 2(\mathrm{div}\,\alpha)_{|V} + 2\Upsilon_{\alpha,\tilde{\alpha}} + 2\langle \tilde{\theta} - \tilde{\alpha}, H\rangle$$
$$+ 2H^{\flat} \odot \tilde{H}^{\flat} - 2\tilde{\delta}_H + 2\Upsilon_{\tilde{\theta},\theta} - 2(\mathrm{div}\,\tilde{\theta})_{|V},\ B\Big\rangle \mathrm{d}\,\mathrm{vol}_g. \tag{4.68}$$

Since

$$\frac{\mathrm{d}}{\mathrm{dt}} J_{\mathrm{mix},\widetilde{\mathscr{D}}}(g_t)_{|t=0} = \int_M \partial_t Q(g_t)_{|t=0}\, \mathrm{d}\,\mathrm{vol}_g + \int_M Q(g)\, (\partial_t \mathrm{d}\,\mathrm{vol}_{g_t\,|t=0}),$$

by (4.68) and (4.15), we have

$$\frac{\mathrm{d}}{\mathrm{dt}} J_{\mathrm{mix},\widetilde{\mathscr{D}}}(g_t)_{|t=0} = \int_M \Big\langle 4\Upsilon_{\tilde{\alpha},\theta} - \mathrm{div}\,\tilde{h} + 2\widetilde{\mathscr{T}}^{\flat} - \widetilde{\mathscr{K}}^{\flat} - H^{\flat} \otimes H^{\flat} + \frac{1}{2}\Upsilon_{h,h} + \frac{1}{2}\Upsilon_{T,T}$$
$$+ 4\langle \theta, \tilde{H}\rangle + 2(\mathrm{div}(\alpha - \tilde{\theta}))_{|V} + 2\Upsilon_{\alpha,\tilde{\alpha}} + 2\langle \tilde{\theta} - \tilde{\alpha}, H\rangle + 2H^{\flat} \odot \tilde{H}^{\flat}$$
$$- 2\tilde{\delta}_H + 2\Upsilon_{\tilde{\theta},\theta} + \frac{1}{2}\big(\,\mathrm{S}_{\mathrm{mix}} + \mathrm{div}(\tilde{H} - H)\big)g^{\perp},\ B\Big\rangle \mathrm{d}\,\mathrm{vol}_g. \tag{4.69}$$

If g is critical for $J_{\mathrm{mix},\widetilde{\mathscr{D}}}$ with respect to $g^{\pitchfork}$-variations, then the integral in (4.69) is zero for arbitrary symmetric $(0,2)$-tensor B vanishing on $\widetilde{\mathscr{D}} \times \widetilde{\mathscr{D}}$. This yields the Euler–Lagrange equation, which we can decompose into two independent parts: its V and $\mathscr{D} \times \mathscr{D}$ components, obtaining (4.66b) and

$$\mathrm{div}\,\tilde{h} - 2\widetilde{\mathscr{T}}^{\flat} + \widetilde{\mathscr{K}}^{\flat} + H^{\flat} \otimes H^{\flat} - \frac{1}{2}\Upsilon_{h,h} - \frac{1}{2}\Upsilon_{T,T} = \frac{1}{2}\big(\mathrm{S}_{\mathrm{mix}} + \mathrm{div}(\tilde{H} - H)\big)g^{\perp}. \tag{4.70}$$

For volume-preserving $g^{\pitchfork}$-variations, the Euler–Lagrange equation (4.70) reduces to (4.67). Using (1.41) (with the tensor $\mathrm{Ric}^{\perp}$) and replacing $\mathrm{div}\,\tilde{h}$ in (4.70) according to (1.41), we rewrite (4.66a) as (4.67). Finally, since all variational formulae for $g^{\top}$-variations are dual to the $\mathscr{D} \times \mathscr{D}$ components of the variational formulae for $g^{\pitchfork}$-variations, we can take the dual equation to (4.67) to obtain the following Euler–Lagrange equation for volume-preserving $g^{\top}$-variations:

$$\mathrm{div}\,h - 2\mathscr{T}^{\flat} + \mathscr{K}^{\flat} + \tilde{H}^{\flat} \otimes \tilde{H}^{\flat} - \frac{1}{2}\Upsilon_{\tilde{h},\tilde{h}} - \frac{1}{2}\Upsilon_{\tilde{T},\tilde{T}} = \frac{1}{2}\big(\mathrm{S}_{\mathrm{mix}} + \mathrm{div}(H - \tilde{H}) + 2\lambda\big)g^{\top}.$$

Using the dual of Lemma 1.36 yields (4.66c).   □

Remark 4.35.

(i) Equations (4.66a,c) coincide with the equations for adapted variations of metric. However, (4.66b) corresponds to the variation of the orthogonal complement of $\widetilde{\mathscr{D}}$, and cannot be obtained by means of adapted variations. We can compare the Euler–Lagrange equation for different types of variations. To get the Euler–Lagrange equation for arbitrary $g^\pitchfork$-variations (respectively, $g^\top$-variations) not necessarily preserving volume of (M,g), one should merely set $\lambda = 0$ in the Euler–Lagrange equation obtained for volume-preserving $g^\pitchfork$-variations (respectively, $g^\top$-variations). To obtain the Euler–Lagrange equation for arbitrary volume preserving variations g_t one should consider both Euler–Lagrange equations for volume-preserving $g^\pitchfork$- and $g^\top$-variations, with the same, arbitrary $\lambda \in \mathbb{R}$.

(ii) If $\mathscr{D}$ is integrable, then for $X \in \widetilde{\mathscr{D}}$ and $N \in \mathscr{D}$ we have

$$\langle \theta, \tilde{H} \rangle (X,N) = \frac{1}{2} \langle \tilde{H}, T_N^\sharp(X) \rangle, \quad \Upsilon_{\tilde{\alpha},\theta}(X,N) = -\frac{1}{2} \langle \tilde{H}, T_N^\sharp(X) \rangle.$$

It follows that the Euler–Lagrange equation (4.66b) does not depend on the integrability tensor of $\widetilde{\mathscr{D}}$, but depends only on its extrinsic geometry.

Generally speaking, it is difficult to find critical points of (4.60) for arbitrary variations of the metric. A trivial example of such a metric is the metric product of manifolds, i.e., with integrable and totally geodesic both $\widetilde{\mathscr{D}}$ and $\mathscr{D}$. There exist many interesting examples of metrics that are critical with respect to volume-preserving variations, or volume-preserving $g^\pitchfork$- and $g^\top$-variations considered separately—we shall present some of them below. The volume-preserving $g^\pitchfork$- and $g^\top$-variations generalize other variations considered in literature, e.g., variations among associated metrics on a contact manifold [Bl], discussed in Section 4.3.3.1.

4.3.3 Particular Cases

Here, we study the Euler–Lagrange equations (4.25), assuming a certain (co)dimension of the distribution $\widetilde{\mathscr{D}}$ or the existence of an additional (contact, 3-Sasakian) structure on the manifold M. In these special conditions, we get examples of metrics that are critical for the action (4.60) with respect to previously discussed variations.

4.3.3.1 Contact Metric Structure

Let $\widetilde{\mathscr{D}}$ be tangent to the foliation by flowlines of a vector field $N \neq 0$ on M, see Example 1.38. Based on the Theorem 4.34, we obtain the following.

Theorem 4.36 (Euler–Lagrange Equation). *Suppose that $g(N,N) = \varepsilon_N$ with respect to $g \in \mathrm{Riem}(M, \widetilde{\mathscr{D}}, \mathscr{D})$. Then g is critical for (4.60) with respect to all volume-preserving $g^{\pitchfork}$-variations if and only if*

$$\varepsilon_N \left(R_N + (\tilde{A}_N)^2 - (\tilde{T}_N^{\sharp})^2 + [\tilde{T}_N^{\sharp}, \tilde{A}_N] \right)^{\flat} - \tilde{\tau}_1 \tilde{h}_{sc} + H^{\flat} \otimes H^{\flat} - \mathrm{Def}_{\mathscr{D}} H$$

$$= \frac{1}{2} \left(\varepsilon_N \mathrm{Ric}_{N,N} + \mathrm{div}(\varepsilon_N \tilde{\tau}_1 N - H) + 2\tilde{\lambda} \right) g^{\perp}, \qquad (4.71\text{a})$$

$$\mathrm{div}^{\perp} \tilde{T}_N^{\sharp}|_{\mathscr{D}} + 2\, (\tilde{T}_N^{\sharp}(H))^{\flat} = 0; \qquad (4.71\text{b})$$

and g is critical with respect to volume-preserving $g^{\top}$-variations if and only if

$$\varepsilon_N \mathrm{Ric}_{N,N} - 4\|\tilde{T}\|^2 - \mathrm{div}(\varepsilon_N \tilde{\tau}_1 N + H) = 2\lambda. \qquad (4.71\text{c})$$

Moreover, the metric g is critical for (4.60) for all volume-preserving variations if and only if equations (4.71a-c) hold with the same constant, $\tilde{\lambda} = \lambda$.

Proof. The proof of (4.71a,c) is similar to the proof of (4.49a,b). Let X be orthogonal to N with $\nabla_Z X \in \widetilde{\mathscr{D}}$ for all $Z \in TM$. We have $\theta = 0$. Since

$$2\,(\mathrm{div}\,\alpha)(X,N) = \langle \nabla_N H - \tilde{\tau}_1 H, X \rangle,$$
$$2\,\langle \tilde{\theta} - \tilde{\alpha}, H \rangle(X,N) = -\langle \tilde{T}_N^{\sharp}(H) + \tilde{A}_N(H), X \rangle,$$
$$2\,(H^{\flat} \otimes H^{\flat})(X,N) = \langle \tilde{\tau}_1 H, X \rangle,$$
$$2\,\tilde{\delta}_H(X,N) = \langle \nabla_N H, X \rangle, \quad 2\,\Upsilon_{\alpha,\tilde{\alpha}}(X,N) = \langle \tilde{A}_N(H), X \rangle,$$

the Euler–Lagrange equation (4.66b) reduces to

$$(\mathrm{div}\,\tilde{\theta})_{|V} = \langle \tilde{\theta}, H \rangle. \qquad (4.72)$$

For $X \in \mathscr{D}$ such that $\nabla_Z X \in \widetilde{\mathscr{D}}$ for all $Z \in TM$ we have

$$2\,\mathrm{div}\,\tilde{\theta}(X,N) = \sum_i \varepsilon_i \langle (\nabla_i \tilde{T}_N^{\sharp})(X), \mathscr{E}_i \rangle + \varepsilon_N \langle \nabla_N (\tilde{T}_N^{\sharp}(X)), N \rangle$$
$$= (\mathrm{div}^{\perp} \tilde{T}_N^{\sharp})(X) + \langle \tilde{T}_N^{\sharp}(H), X \rangle.$$

Hence, (4.72) can be written as (4.71b). $\qquad \square$

Restricting to the case of a geodesic Riemannian flow, we obtain the following.

Corollary 4.37. *Let N generate a geodesic Riemannian flow on (M^{p+1}, g). Then*

(i) *the metric g is critical for the action (4.60) with respect to volume-preserving $g^{\pitchfork}$-variations if and only if the following system of equations hold:*

$$R_N = (1/p)\,\mathrm{Ric}_{N,N}\,\mathrm{id}^{\perp}, \quad \mathrm{Ric}_{X,N} = 0 \quad (X \in \mathscr{D}), \quad \mathrm{Ric}_{N,N} = \mathrm{const}. \qquad (4.73)$$

(ii) *the metric g is critical for the action (4.60) with respect to volume-preserving $g^{\top}$-variations if and only if $(4.73)_3$ is valid.*

Proof. (i) By $(1.49)_1$ we get $R_N = -(\tilde{T}_N^\sharp)^2$ and (4.71a) takes the form

$$4\varepsilon_N(R_N)^\flat = (\varepsilon_N \operatorname{Ric}_{N,N} + 2\lambda)g^\perp,$$

which together with $\operatorname{Ric}_{N,N} = \operatorname{trace} R_N$ yields $(4.73)_{1,3}$. A Riemannian geodesic flow locally gives rise to a Riemannian submersion with totally geodesic fibers. Such mappings can be described by the tensor $\mathcal{O}$, see (1.25), which is antisymmetric with respect to g and satisfies $\mathcal{O}_X Y = \tilde{T}(X,Y)$ for $X,Y \in \mathcal{D}$. Hence, for $X,Y \in \mathcal{D}$ we have

$$\langle \tilde{T}_N^\sharp(X), Y \rangle = \langle \tilde{T}(X,Y), N \rangle = \langle \mathcal{O}_X Y, N \rangle = -\langle \mathcal{O}_X N, Y \rangle,$$

and then we obtain $\tilde{T}_N^\sharp(X) = -\mathcal{O}_X N$. For a geodesic Riemannian N-flow, (4.71b) reduces to equation

$$\operatorname{div}^\perp \tilde{T}_N^\sharp(X) = 0, \quad X \in \mathcal{D},$$

that we shall now examine. Suppose that $X \in \mathcal{D}$ and $\nabla_Z X \in \widetilde{\mathcal{D}}$ for all $Z \in TM$. Using an adapted frame with $\mathcal{E}_i \in \mathcal{D}$ at a point, the condition $\nabla_N N = 0$, and the antisymmetry of $\nabla_Z \mathcal{O}$ for all $Z \in TM$, we obtain

$$(\operatorname{div}^\perp \tilde{T}_N^\sharp)(X) = \sum_i \langle \nabla_{\mathcal{E}_i} \tilde{T}_N^\sharp(X), \mathcal{E}_i \rangle = -\sum_i \langle \nabla_{\mathcal{E}_i} \mathcal{O}_X N, \mathcal{E}_i \rangle$$
$$= -\sum_i \langle (\nabla_{\mathcal{E}_i} \mathcal{O})_X N, \mathcal{E}_i \rangle = \sum_i \langle (\nabla_{\mathcal{E}_i} \mathcal{O})_X \mathcal{E}_i, N \rangle.$$

By Codazzi equation (1.38), adjusted to our definitions of R and Ric, we get

$$(\operatorname{div}^\perp \tilde{T}_N^\sharp)(X) = -\sum_i \langle R(\mathcal{E}_i, X)\mathcal{E}_i, N \rangle = -\operatorname{Ric}_{X,N},$$

thus $(4.73)_2$. (ii) For volume-preserving $g^\top$-variations we get the Euler-Lagrange equation (4.71c), which for geodesic Riemannian flows is equal to $\varepsilon_N \operatorname{Ric}_{N,N} = -(2/3)\lambda$. $\qquad\square$

Next, we show that for associated metric on a contact manifold the Euler–Lagrange equation (4.66b) is valid. We will use this fact later to show that Sasakian 3-structures are a natural source of metrics critical for the action (4.60).

Proposition 4.38. *Let* (ϕ, ξ, η, g) *be a contact metric structure on* M. *Then the Euler–Lagrange equation (4.71b) is satisfied for* $N = \xi$

Proof. For all X,Y such that $\langle X, \xi \rangle = \langle Y, \xi \rangle = 0$ we have

$$d\eta(X,Y) = -\frac{1}{2}\eta([X,Y]) = -\frac{1}{2}\langle [X,Y], \xi \rangle = -\langle \tilde{T}_\xi^\sharp(X), Y \rangle = \langle X, \tilde{T}_\xi^\sharp(Y) \rangle.$$

Hence, it follows from (1.52), that $\langle X, \phi(Y) \rangle = \langle X, \tilde{T}_\xi^\sharp(Y) \rangle$. Since $\phi(\xi) = 0$, we get $\tilde{T}_\xi^\sharp = \phi$. By Remark 1.43, (4.71b) reduces to $\operatorname{div}^\perp \tilde{T}_\xi^\sharp = 0$, which takes the form

$$(\operatorname{div}^\perp \phi)(Y) = 0 \quad (Y \in \mathcal{D}). \tag{4.74}$$

For $Y \in \mathscr{D}$, the formula in [Bl, Corollary 6.1],

$$2\langle(\nabla_X\phi)Y,Z\rangle = \langle N^{(1)}(Y,Z),\phi X\rangle + 2d\eta(\phi Y,X)\eta(Z) - 2d\eta(\phi Z,X)\eta(Y),$$

where $N^{(1)}(Y,Z) = N_\phi(Y,Z) + 2d\eta(Y,Z)\xi$ and N_ϕ is the Nijenhuis torsion of ϕ, yields $2\langle(\nabla_i\phi)(Y),\mathscr{E}_i\rangle = \langle N_\phi(Y,\mathscr{E}_i),\phi(\mathscr{E}_i)\rangle$. Considering an orthonormal ϕ-basis, i.e., assuming that $\mathscr{E}_{i+p/2} = \phi(\mathscr{E}_i)$ for $i = 1,\dots,p/2$, we then obtain

$$\sum_{i=1}^{p}\langle N_\phi(Y,\mathscr{E}_i),\phi(\mathscr{E}_i)\rangle = -\sum_{i=1}^{p}\langle N_\phi(Y,\phi(\mathscr{E}_i)),\phi^2(\mathscr{E}_i)\rangle.$$

Hence, $(\mathrm{div}^\perp\phi)(Y) = 0$ and (4.71b) is satisfied.   □

In [Bl], the action (4.60), which reduces to

$$J_{\mathrm{mix},\widetilde{\mathscr{D}}} : g \to \int_M \mathrm{Ric}_{N,N}(g)\,\mathrm{d}\,\mathrm{vol}_g, \tag{4.75}$$

has been studied on the metrics associated to a given contact form.

Proposition 4.39. *Any K-contact metric g, i.e., N is a Killing field, is critical for the action* (4.75) *with respect to both volume-preserving $g^\pitchfork$- and $g^\top$-variations.*

Proof. Integral curves of ξ are geodesics for the contact metric structure, see Remark 1.43. On the other hand, a nonsingular Killing vector field defines a Riemannian flow. Thus, in case of a K-contact structure, we can use Corollary 4.22. If (M,g) is a K-contact manifold, then by $R(X,\xi)\xi = X - \eta(X)\xi$, see [Bl, Theorem 7.2], (4.73)$_{1,3}$ are satisfied with $\mathrm{Ric}_{N,N} = p$. By Proposition 4.38, (4.73)$_2$ is valid.   □

Definition 4.40 ([Bl], p. 24). A contact structure is *regular* if ξ is regular as a vector field, that is, every point of the manifold has a neighborhood such that any integral curve of ξ passing through the neighborhood passes through it only once.

Theorem 4.41 (Theorem 10.12 in [Bl]). *An associated metric g on a compact regular contact manifold (M,η) is critical for the action* (4.75) *considered on the set of metrics associated to η if and only if it is K-contact.*

We have $\langle\xi,\xi\rangle = 1$ for any associated metric and the volume form of associated metric on a contact manifold can be expressed only in terms of η and $d\eta$. Therefore, variations of the metric restricted to the set of all associated metrics form a subclass of the volume preserving $g^\pitchfork$-variations. Hence, on compact regular contact manifolds Proposition 4.39 and Theorem 4.41 together give the following characterization of some critical metrics—for a larger space of variations.

Corollary 4.42. *Let (M,η) be a compact regular contact manifold and let g be an associated metric. Then g is critical for the action* (4.75) *for volume-preserving $g^\pitchfork$-variations if and only if g is K-contact.*

Flowlines of Reeb fields on contact manifolds are often described as having "maximally non-integrable" orthogonal distributions. We illustrate this notion by showing that contact metric structures are critical points of the action:

$$J_{\tilde{T}} : g \rightarrow \int_M \|\tilde{T}\|^2 \, d\,\mathrm{vol}_g . \tag{4.76}$$

This is the total norm of the integrability tensor of the varying orthogonal complement of a fixed distribution $\widetilde{\mathscr{D}}$.

Theorem 4.43. *A metric* $g \in \mathrm{Riem}(M, \widetilde{\mathscr{D}}, \mathscr{D})$ *is critical for the action* (4.76) *with respect to volume-preserving $g^{\pitchfork}$-variations if and only if*

$$2\widetilde{\mathscr{T}}^{\flat} = -\left(\frac{1}{2}\|\tilde{T}\|^2 - \lambda\right) g^{\perp}, \quad \Upsilon_{\tilde{\theta}, \theta - \alpha} = (\mathrm{div}\,\tilde{\theta})_{|V}. \tag{4.77}$$

Proof. Let g_t be a $g^{\pitchfork}$-variation. Using Proposition 4.33, we obtain

$$\frac{d}{dt} J_{\tilde{T}}(g_t)_{|t=0} = \int_M \langle 2\Upsilon_{\tilde{\theta}, \theta - \alpha} - 2\,(\mathrm{div}\,\tilde{\theta})_{|V} + 2\widetilde{\mathscr{T}}^{\flat} + \frac{1}{2}\|\tilde{T}\|^2 g^{\perp}, \ B\rangle \, d\,\mathrm{vol}_g .$$

Decomposing the resulting Euler–Lagrange equation into parts defined on $\mathscr{D} \times \mathscr{D}$ and V, yields (4.77). □

As expected, distributions with integrable orthogonal complement are critical for (4.76). Using the results for the contact metric structure, we get the following.

Corollary 4.44. *Let* (ϕ, ξ, η, g) *be a contact metric structure on M and let* $\widetilde{\mathscr{D}}$ *be spanned by* ξ. *Then g is critical for the action* (4.76) *with respect to volume-preserving $g^{\pitchfork}$-variations.*

Proof. Using results from the proof of Proposition 4.38, we compute

$$\|\tilde{T}\|^2 = \sum_{i,j}\langle \tilde{T}_{\xi}^{\sharp}(\mathscr{E}_i), \mathscr{E}_j\rangle^2 = \sum_i \langle \phi(\mathscr{E}_i), \phi(\mathscr{E}_i)\rangle^2 = p.$$

We also have $\widetilde{\mathscr{T}} = (\tilde{T}_{\xi}^{\sharp})^2 = -\mathrm{id}^{\perp}$; hence, $\widetilde{\mathscr{T}}^{\flat} = -g^{\perp}$. By the above, $(4.77)_1$ is satisfied; $(4.77)_2$ reduces to $(\mathrm{div}\,\tilde{\theta})_{|V} = 0$ and is valid by Proposition 4.38. □

If each of contact 3-structures is Sasakian, then we obtain a *Sasakian 3-structure*. According to [Ka], *any contact 3-structure is necessarily a Sasakian 3-structure*.

Proposition 4.45. *The metric of a Sasakian 3-structure on M is critical for the action* (4.60) ($\widetilde{\mathscr{D}}$ *is spanned by the characteristic vector fields*), *with respect to both volume-preserving $g^{\pitchfork}$-variations and $g^{\top}$-variations.*

Proof. Since every ξ_i defines a Sasakian structure, we can use the following formulas for any unit vectors X, Y orthogonal to ξ_i (so we can also have $X = \xi_j$):

$$R(X, Y)\xi_i = \eta_i(Y)X - \eta_i(X)Y, \quad R(X, \xi_i)Y = \eta_i(Y)X - \langle X, Y\rangle\xi_i.$$

The above formulas are consistent with their counterparts for ξ_j and ξ_k, and yield the following: $\mathrm{Ric}^\perp = 3g^\perp$ and $\mathrm{Ric}^\top = pg^\top$. We also have

$$-\frac{1}{2}\Upsilon_{\tilde{T},\tilde{T}}(\xi_i,\xi_j) = \tilde{\Psi}(\xi_i,\xi_j) = -p\langle\xi_a,\xi_b\rangle,$$

$$\tilde{\mathscr{T}}^b(X,Y) = -3\langle X,Y\rangle \quad (X,Y \in \mathscr{D})$$

and $\|\tilde{T}\|^2 = 3p$. It follows that (4.66a,c) are satisfied, but never with the same $\lambda \in \mathbb{R}$. The remaining Euler–Lagrange equation (4.66b) reduces to $(\mathrm{div}\,\tilde{\theta})_{|V} = 0$. Since g is K-contact for ξ_a, we have $\nabla_Y\xi_a = -\phi_a(Y)$ for all $Y \in TM$, see [Bl], and $\tilde{T}_i^\sharp(Y) = \phi_i(Y)$, and it follows that $(\mathrm{div}\,\tilde{\theta})(Y,\xi_i) = (\mathrm{div}\,\phi_i)(Y)$. For any contact metric structure, we have $\nabla_{\xi_i}\xi_i = 0$ and $\phi_i(\xi_i) = 0$. Then

$$\begin{aligned}
(\mathrm{div}\,\tilde{\theta})(X,\xi_i) &= \sum_l \langle(\nabla_l\phi_i)(X),\mathscr{E}_l\rangle + \langle(\nabla_{\xi_j}\phi_i)(X),\xi_j\rangle \\
&\quad + \langle(\nabla_{\xi_k}\phi_i)(X),\xi_k\rangle + \langle(\nabla_{\xi_i}\phi_i)(X),\xi_i\rangle \\
&= \sum_l \langle(\nabla_l\phi_i)(X),\mathscr{E}_l\rangle + \langle(\nabla_{\xi_j}\phi_i)(X),\xi_j\rangle + \langle(\nabla_{\xi_k}\phi_i)(X),\xi_k\rangle,
\end{aligned}$$

and similarly for ξ_j,ξ_k. The formula we obtained above, when considered for a contact metric structure (ϕ_i,ξ_i,η_i,g) on M, is precisely $(\mathrm{div}^\perp\phi_i)(X)$. In the proof of Proposition 4.38 we showed that (4.74) is valid; hence, (4.66b) is satisfied. $\qquad\square$

4.3.3.2 Non-integrable Distributions

Here, we examine the action (4.60) for a fixed, non-integrable distribution $\widetilde{\mathscr{D}}$ on (M,g) and concentrate on particular examples of critical metrics. For this purpose, we set as $\widetilde{\mathscr{D}}$ the distribution orthogonal to the Reeb fields on contact and 3-Sasakian manifolds. In this setting, dual to the one considered in Section 4.3.3.1, K-contact and 3-Sasakian metrics are critical for the action (4.60) for volume-preserving $g^\pitchfork$- and $g^\top$-variations. We show how a K-contact metric can be slightly modified and still remain critical. Finally, we consider $g^\pitchfork$-variations of a codimension-one distribution, and give an example of contact metric structure (that is, not K-contact) critical with respect to them.

Proposition 4.46. *Let $\widetilde{\mathscr{D}}$ be a totally geodesic distribution with a totally geodesic, integrable orthogonal complement $\mathscr{D}$ on (M,g). Then*

(i) g is critical for the action (4.60) with respect to volume-preserving $g^\pitchfork$-variations if and only if

$$-\frac{1}{2}\Upsilon_{T,T} = \left(\frac{1}{2}\|T\|^2 + \lambda\right)g^\perp; \tag{4.78}$$

(ii) g is critical for (4.60) for volume-preserving $g^\top$-variations if and only if

$$-2\mathscr{T}^b = \left(\frac{1}{2}\|T\|^2 + \lambda\right)g^\top$$

for some $\lambda \in \mathbb{R}$.

Proof. The Euler–Lagrange equation (4.66b) is always satisfied, since by the assumptions all its terms vanish. Equation (4.67) becomes (4.78).   □

Formulas in Proposition 4.46 cannot hold together with the same constant λ. Thus, in the setting of Proposition 4.46, one can find metrics critical for volume-preserving $g^{\pitchfork}$- and $g^{\top}$-variations, but only considered separately (each with different constant λ in certain Euler–Lagrange equation). Note that (4.78) yields that the mapping $\mathscr{D} \ni X \mapsto T_X^{\sharp} \in T^*M \otimes TM$ is conformal with respect to the metric g and the metric induced by g on $T^*M \otimes TM$. Since

$$\frac{1}{2}\Upsilon_{T,T}(X,Y) = \sum_{a,b} \varepsilon_a \varepsilon_b \langle T(E_a,E_b),X \rangle \langle T(E_a,E_b),Y \rangle = -\operatorname{trace}(T_Y^{\sharp}T_X^{\sharp})$$

can be related to the Killing form on $SO(n)$, it is natural to look for examples of critical metrics on manifolds with the action of this group.

Example 4.47.

(i) Let (M,g) be a 3-Sasakian manifold and let $\widetilde{\mathscr{D}}$ be the distribution orthogonal to all integral manifolds of the Reeb fields ξ_1, ξ_2, ξ_3. Then g is critical for the action (4.60) with respect to volume-preserving $g^{\pitchfork}$- and $g^{\top}$-variations. Indeed, we have $\Upsilon_{T,T} = 2ng^{\perp}$ and $\mathscr{T}^{\flat} = -3g^{\top}$, see Proposition 4.46.

(ii) Let (M,g) be a K-contact manifold and let $\widetilde{\mathscr{D}}$ be the orthogonal distribution of the Reeb field. Assume that $\overline{g} = \phi g^{\top} + \psi g^{\perp}$, with positive functions $\phi, \psi \in C^{\infty}(M)$, then the relations between geometric quantities for $\overline{g}$ and g are

$$\|T\|_{\overline{g}}^2 = \psi \phi^{-2} \|T\|_g^2, \quad \mathscr{T}_{\overline{g}}^{\flat} = \psi \phi^{-1} \mathscr{T}^{\flat}.$$

We also have $\tau_1 = -\frac{n}{2} \psi^{-1/2} \phi^{-1} N(\phi)$ on $(M,\overline{g})$. It follows that for all positive functions $\phi, \psi \in C^{\infty}(M)$ satisfying $N(\phi) = 0$ and $\psi^{-1/2}\phi = \text{const}$ the metric $\overline{g}$ is critical for the action (4.60) with respect to volume-preserving $g^{\pitchfork}$- and $g^{\top}$-variations.

If $\widetilde{\mathscr{D}}$ is a codimension-one distribution, then every $g^{\pitchfork}$-variation corresponds to a family of foliations (by curves g_t-orthogonal to $\widetilde{\mathscr{D}}$) that share the same transversal geometry. For example, if $\widetilde{\mathscr{D}}$ is totally geodesic or umbilical, all foliations corresponding to metrics g_t are either Riemannian or conformal. Thus, $g^{\pitchfork}$-variation can provide a tool to find the locally 'best' (e.g., minimizing a functional) metrics for foliations of some fixed transverse property. Using equations dual to the ones formulated in Theorem 4.36 and some easy computations, we obtain the following.

Proposition 4.48. *Let $\widetilde{\mathscr{D}}$ be a codimension-one distribution on a manifold M^{n+1} with unit normal field N. A metric g on M is critical for the action (4.60) with respect to volume-preserving $g^{\pitchfork}$-variations if and only if*

$$\varepsilon_N \operatorname{Ric}_{N,N} - 4\|T\|^2 - \operatorname{div}(\varepsilon_N \tau_1 N + \tilde{H}) = 2\lambda, \tag{4.79a}$$

$$(\operatorname{div}_{\widetilde{\mathscr{D}}} A_N)^{\sharp} = \nabla^{\top} \tau_1 \tag{4.79b}$$

for some $\lambda \in \mathbb{R}$. *A metric g on M is critical for the action* (4.60) *with respect to volume-preserving* $g^\top$-*variations if and only if:*

$$\varepsilon_N \left(R_N + A_N^2 - (T_N^\sharp)^2 + [T_N^\sharp, A_N] \right)^\flat - \tau_1 h_{sc} + \tilde{H}^\flat \otimes \tilde{H}^\flat - \mathrm{Def}_{\widetilde{\mathscr{D}}} \tilde{H}$$

$$-\frac{1}{2} \left(\varepsilon_N \mathrm{Ric}_{N,N} + \mathrm{div}(\varepsilon_N \tau_1 N - \tilde{H}) \right) g^\top = \lambda g^\top \qquad (4.79c)$$

for some $\lambda \in \mathbb{R}$. *A metric g on M is critical for the action* (4.60) *with respect to all volume-preserving variations if and only if* (4.79a–c) *hold, with the same* λ.

Example 4.49. The following contact metric structure on $\mathbb{R}^3$, see Example 1.67,

$$\eta = \frac{1}{2} (dz - y\, dx), \quad g = \frac{1}{4} \begin{pmatrix} 1 + y^2 + z^2 & z & -y \\ z & 1 & 0 \\ -y & 0 & 1 \end{pmatrix},$$

is not K-contact. Using an adapted orthonormal frame: $E_1 = 2(\frac{\partial}{\partial x} - z\frac{\partial}{\partial y} + y\frac{\partial}{\partial z})$, $E_2 = 2\frac{\partial}{\partial y}$, $E_3 = N = 2\frac{\partial}{\partial z}$, one can show that in $\{E_1, E_2\}$ basis of $\mathscr{D}$ we have $\tilde{A}_N = \begin{pmatrix} 0 & -1 \\ -1 & 0 \end{pmatrix}$, $\tilde{T}_N^\sharp = \begin{pmatrix} 0 & 1 \\ -1 & 0 \end{pmatrix}$. Thus, $\mathrm{Ric}_{N,N} = 0$, $\tilde{H} = 0 = H$, $\tau_1 = 0$, and $R_N + A_N^2 - (T_N^\sharp)^2 + [T_N^\sharp, A_N]$ is not conformal, hence (4.79c) is not satisfied, but (4.79a) is valid. Computations show that $\nabla_{E_1} E_2 = -2E_3$ and $\nabla_{E_1} E_3 = 2E_2$ are the only non-vanishing derivatives of vector fields (E_a) from the frame. We get $\widetilde{\mathrm{div}} A_N = 0$, hence also (4.79b) is satisfied. Thus, g is critical for the action (4.60) with respect to volume-preserving $g^\pitchfork$-variations.

4.4 Einstein–Hilbert Type Action

From a mathematical point of view, a *space-time* of general relativity is a d-dimensional time-oriented (that is with a given timelike vector field) Lorentzian manifold, see [BEE]. A space-time admits a global time function (i.e., increasing function along each future directed nonspacelike curve) if and only if it is stable causal; in particular, a globally hyperbolic space-time is naturally endowed with a codimension-one foliation (the level hypersurfaces of a time-function), see [BS]. Most of known canonical metrics (e.g., Einstein metrics, $\mathrm{Ric} = \lambda g$ where $\lambda \in \mathbb{R}$) are obtained by variational principle, their classification is a deep problem [Bes, CM].

Inspired by the Einstein-Cartan theory of gravity (allowing space-time to have torsion, in addition to Ricci curvature, and relating torsion to spin of matter), we investigate the Einstein–Hilbert type action in the framework of metric-affine geometry. The *mixed Einstein–Hilbert action* on a pseudo-Riemannian manifold (M, g) endowed with a linear connection $\bar{\nabla} = \nabla + \mathfrak{T}$ and a distribution $\widetilde{\mathscr{D}}$, is given by

$$\bar{J}_{\widetilde{\mathscr{D}}} : (g,\mathfrak{T}) \mapsto \int_M \left\{ \frac{1}{2\mathfrak{a}} \left(\overline{S}_{\mathrm{mix}}(\widetilde{\mathscr{D}};g,\mathfrak{T}) - 2\Lambda \right) + \mathscr{L}(g,\mathfrak{T}) \right\} \mathrm{d}\,\mathrm{vol}_g, \tag{4.80}$$

see [RZ3, RZ4], where Λ is a constant (analogous to the "cosmological constant"), $\mathscr{L}$—Lagrangian describing the "matter contents", and $\mathfrak{a}$—the coupling constant. This is an analog of the Einstein–Hilbert action, where the scalar curvature is replaced by the mixed scalar curvature $\overline{S}_{\mathrm{mix}}$ with respect to connection $\bar{\nabla}$, see (1.50). The geometrical part of (4.80), i.e., $\Lambda = \mathscr{L} = 0$ and $\mathfrak{a} = 1$, is the total mixed scalar curvature on $(M,g,\bar{\nabla})$ with a distribution or a foliation, and we regard it as a functional of the pseudo-Riemannian metric g and the contorsion tensor $\mathfrak{T}$,

$$\bar{J}_{\mathrm{mix},\widetilde{\mathscr{D}}} : (g,\mathfrak{T}) \mapsto \int_M \overline{S}_{\mathrm{mix}}(\widetilde{\mathscr{D}};g,\mathfrak{T}) \,\mathrm{d}\,\mathrm{vol}_g. \tag{4.81}$$

The action (4.81) with $\mathfrak{T} = 0$ is (4.60). The action (4.80) with $\mathfrak{T} = 0$,

$$J_{\widetilde{\mathscr{D}}} : g \to \int_M \left\{ \frac{1}{2\mathfrak{a}} \left(S_{\mathrm{mix}}(\widetilde{\mathscr{D}};g) - 2\Lambda \right) + \mathscr{L}(g) \right\} \mathrm{d}\,\mathrm{vol}_g, \tag{4.82}$$

has been introduced in [BDRS] for globally hyperbolic space-times (and in [R7] for distributions of any dimension) as analog of the Einstein–Hilbert action with the scalar curvature replaced by S_{mix}. The main goal of Sections 4.4.1 and 4.4.2 is the Euler–Lagrange equation for (4.82), presented in the form of Einstein field equation,

$$\mathrm{Ric}_{\widetilde{\mathscr{D}}} - (1/2)\,\mathrm{Scal}_{\widetilde{\mathscr{D}}} \cdot g + \Lambda\,g = \mathfrak{a}\Theta, \tag{4.83}$$

where $\Theta_{\mu\nu} = -2\partial\mathscr{L}/\partial g^{\mu\nu} + g_{\mu\nu}\mathscr{L}$ is the stress-energy type tensor, while Ricci and scalar curvatures are replaced by a Ricci type curvature $\mathrm{Ric}_{\widetilde{\mathscr{D}}}$ (called the *mixed Ricci curvature*) and its trace $\mathrm{Scal}_{\widetilde{\mathscr{D}}}$. Consequently, (4.83) contains the *mixed Einstein tensor* $\mathrm{G}_{\widetilde{\mathscr{D}}} := \mathrm{Ric}_{\widetilde{\mathscr{D}}} - (1/2)\,\mathrm{Scal}_{\widetilde{\mathscr{D}}} \cdot g$, which can be used for studying perturbations $S + \varepsilon\, S_{\mathrm{mix}}$ ($\varepsilon \in \mathbb{R}$) related to equations of the theory of relativity.

4.4.1 Variable Metric

Here, we use variational formulae for extrinsic geometry (see Sections 4.2.1 and 4.3.1), to derive Euler–Lagrange equations of (4.82) and apply them to build the *mixed Ricci tensor* $\mathrm{Ric}_{\widetilde{\mathscr{D}}}$ obeying (4.83).

For $B_t = \partial_t g_t$, we can choose a nice evolution of a local orthonormal frame.

Lemma 4.50. *Let a $(\mathscr{D}, \widetilde{\mathscr{D}})$-adapted and orthonormal for $t = 0$ frame $\{E_a(t), \mathscr{E}_i(t)\}$ be evolved by $\{g_t\}_{|t|<\varepsilon}$ according to*

$$\partial_t E_a = -(1/2)\,B_t^\sharp(E_a)^\top, \qquad \partial_t \mathscr{E}_i = -(1/2)\,B_t^\sharp(\mathscr{E}_i)^\perp - B_t^\sharp(\mathscr{E}_i)^\top.$$

Then $\{E_a(t), \mathscr{E}_i(t)\}$ is a g_t-orthonormal frame such that $\{E_a(t)\} \subset \mathscr{D}$ and $\{\mathscr{E}_i(t)\} \subset \widetilde{\mathscr{D}}_t$ for all $|t| < \varepsilon$.

Proof. By the general Leibniz rule and given differential equations, we have

$$\partial_t(g_t(E_a, E_b)) = B(E_a, E_b) + g_t(\partial_t E_a, E_b) + g_t(E_a, \partial_t E_b)$$
$$= B_t(E_a, E_b) - \frac{1}{2}g_t(B_t^\sharp(E_a), E_b) - \frac{1}{2}g_t(E_a, B_t^\sharp(E_b))$$
$$= B_t(E_a, E_b) - B_t(E_a, E_b) = 0.$$

The above and $E_a(0) \in \mathscr{D}$ provide that $\{E_a(t)\}$ is a g_t-orthogonal frame of $\mathscr{D}$. Also,

$$\partial_t(g_t(E_a, \mathscr{E}_i)) = (\partial_t g_t)(E_a(t), \mathscr{E}_i(t)) + g_t(\partial_t E_a(t), \mathscr{E}_i(t)) + g_t(E_a(t), \partial_t \mathscr{E}_i(t))$$
$$= B_t(E_a(t), \mathscr{E}_i(t)) - g_t(E_a(t), B_t^\sharp(\mathscr{E}_i(t))) = 0.$$

Since $\mathscr{E}_i(0) \in \mathscr{D}^\perp$, we get $\mathscr{E}_i(t) \in \mathscr{D}_t^\perp$ (i.e., g_t-orthogonal to $\mathscr{D}$). Finally, computing

$$\partial_t(g_t(\mathscr{E}_i, \mathscr{E}_j)) = g_t(\partial_t \mathscr{E}_i(t), \mathscr{E}_j(t)) + g_t(\mathscr{E}_i(t), \partial_t \mathscr{E}_j(t)) + (\partial_t g_t)(\mathscr{E}_i(t), \mathscr{E}_j(t))$$
$$= B_t(\mathscr{E}_i(t), \mathscr{E}_j(t)) - \frac{1}{2}g_t(B_t^\sharp(\mathscr{E}_i(t))^\perp, \mathscr{E}_j(t)) - \frac{1}{2}g_t(\mathscr{E}_i(t), (B_t^\sharp \mathscr{E}_j(t))^\perp) = 0,$$

we show the orthonormality of $\{\mathscr{E}_i(t)\}$ for all t.   □

Using (4.8), Lemma 4.50 and decomposition of tensors w.r.t. $g = g^\top + g^\perp$, one can obtain variational formulae for extrinsic geometry (see proof in [RZ2]).

Remark 4.51. The Lagrangian of (4.82) is a sum of two terms $\mathscr{L}_{gr} + \mathscr{L}$, where $\mathscr{L}_{gr} = \frac{1}{2a}(\mathrm{S}_{\mathrm{mix}} - 2\Lambda)$ is the gravity Lagrangian, see (4.82), and the matter Lagrangian $\mathscr{L}$ depends only on g and not on its derivatives. Set

$$J_m : g \to \int_M \mathscr{L}(g)\,\mathrm{d}\,\mathrm{vol}_g.$$

Variation of metric g with $B = \partial_t g_t|_{t=0}$ produces the tensor Θ such that

$$\frac{\mathrm{d}}{\mathrm{d}t}J_m(g_t)|_{t=0} = \frac{1}{2}\int_M \langle \Theta, B \rangle\,\mathrm{d}\,\mathrm{vol}_g.$$

Definition 4.52. Define the symmetric $(0,2)$-tensor $\mathrm{Ric}_{\widetilde{\mathscr{D}}}$ (referred to as the *mixed Ricci curvature*) by its restrictions on 3 complementary subbundles of $TM \times TM$,

$$\mathrm{Ric}_{\widetilde{\mathscr{D}}|\mathscr{D}\times\mathscr{D}} = \mathrm{Ric}^\perp - \langle \tilde{h}, \tilde{H} \rangle + \widetilde{\mathscr{A}}^\flat - \widetilde{\mathscr{T}}^\flat + \mathscr{K}^\flat + H^\flat \otimes H^\flat$$
$$- \frac{1}{2}\Upsilon_{h,h} - \frac{1}{2}\Upsilon_{T,T} + \Psi - \mathrm{Def}_{\mathscr{D}} H + \mu_1 g^\perp,$$
$$\mathrm{Ric}_{\widetilde{\mathscr{D}}|V} = -4\langle \theta, \tilde{H} \rangle - 2(\mathrm{div}(\alpha - \tilde{\theta}))|_V - 2\langle \tilde{\theta} - \tilde{\alpha}, H \rangle$$
$$- 2H^\flat \odot \tilde{H}^\flat + 2\tilde{\delta}_H - 4\Upsilon_{\tilde{\alpha},\theta} - 2\Upsilon_{\alpha,\tilde{\alpha}} - 2\Upsilon_{\theta,\tilde{\theta}},$$
$$\mathrm{Ric}_{\widetilde{\mathscr{D}}|\widetilde{\mathscr{D}}\times\widetilde{\mathscr{D}}} = \mathrm{Ric}^\top - \langle h, H \rangle + \mathscr{A}^\flat - \mathscr{T}^\flat + \mathscr{K}^\flat + \tilde{H}^\flat \otimes \tilde{H}^\flat$$
$$- \frac{1}{2}\Upsilon_{\tilde{h},\tilde{h}} - \frac{1}{2}\Upsilon_{\tilde{T},\tilde{T}} + \widetilde{\Psi} - \mathrm{Def}_{\widetilde{\mathscr{D}}} \tilde{H} + \mu_2 g^\top,$$
$$\tag{4.84}$$

where $(4.84)_3$ is dual to $(4.84)_1$, $\mu_1 = \mu_2 = 0$ if $n = p = 1$, and for $n + p > 2$ we have

$$\mu_1 = -\frac{n-1}{p+n-2} \operatorname{div}(\tilde{H} - H), \quad \mu_2 = \frac{p-1}{p+n-2} \operatorname{div}(\tilde{H} - H). \tag{4.85}$$

Applying (1.41), we find that (4.84) is equivalent to

$$\begin{aligned}
\operatorname{Ric}_{\widetilde{\mathscr{D}}\,|\,\mathscr{D}\times\mathscr{D}} &= \operatorname{div}\tilde{h} + \widetilde{\mathscr{H}}^\flat - 2\widetilde{\mathscr{T}}^\flat + H^\flat \otimes H^\flat - \tfrac{1}{2}\Upsilon_{h,h} - \tfrac{1}{2}\Upsilon_{T,T} + \mu_1 g^\perp, \\
\operatorname{Ric}_{\widetilde{\mathscr{D}}\,|\,V} &= 2\tilde{\delta}_H + 2\operatorname{div}(\tilde{\theta} - \alpha) - 4\langle\theta,\tilde{H}\rangle - 2\langle\tilde{\theta} - \tilde{\alpha}, H\rangle \\
&\quad - 4\Upsilon_{\tilde{\alpha},\theta} - 2\Upsilon_{\alpha,\tilde{\alpha}} - 2\Upsilon_{\tilde{\theta},\theta} - 2H^\flat \odot \tilde{H}^\flat, \\
\operatorname{Ric}_{\mathscr{D}\,|\,\widetilde{\mathscr{D}}\times\widetilde{\mathscr{D}}} &= \operatorname{div}h + \mathscr{H}^\flat - 2\mathscr{T}^\flat + \tilde{H}^\flat \otimes \tilde{H}^\flat - \tfrac{1}{2}\Upsilon_{\tilde{h},\tilde{h}} - \tfrac{1}{2}\Upsilon_{\tilde{T},\tilde{T}} + \mu_2 g^\top.
\end{aligned} \tag{4.86}$$

The trace of $\operatorname{Ric}_{\widetilde{\mathscr{D}}}$ is

$$\operatorname{Scal}_{\widetilde{\mathscr{D}}} = \operatorname{S}_{\mathrm{mix}} + \frac{p-n}{n+p-2} \operatorname{div}(\tilde{H} - H). \tag{4.87}$$

Theorem 4.53 (Euler–Lagrange Equations). *A metric g on M with non-degenerate $\widetilde{\mathscr{D}}$ is critical for (4.82) if and only if g satisfies (4.83), where $\operatorname{Ric}_{\widetilde{\mathscr{D}}}$ is given by (4.84).*

Proof. The Euler–Lagrange equations (4.66a-c) for (4.60) consist of V-, $\mathscr{D} \times \mathscr{D}$- and $\widetilde{\mathscr{D}} \times \widetilde{\mathscr{D}}$- components. Thus, for the action (4.82) we obtain (4.86). Substituting (4.86) with arbitrary μ_1 and μ_2 into (4.83) and comparing with (4.86), we find (μ_1, μ_2) in (4.85) as solution of a linear system

$$(p-2)\mu_1 + n\mu_2 = \operatorname{div}(\tilde{H} - H), \quad (n-2)\mu_2 + p\mu_1 = -\operatorname{div}(\tilde{H} - H).$$

This completes the proof. □

Corollary 4.54. *A metric g on M with non-degenerate $\widetilde{\mathscr{D}}$ is critical for the action (4.82) with respect to all adapted variations of g if and only if g solves (4.83), where $\operatorname{Ric}_{\widetilde{\mathscr{D}}}$ is built from $(4.86)_{1,3}$, that is $\operatorname{Ric}_{\widetilde{\mathscr{D}}}\,|_V = 0$.*

Components $(4.86)_{1,3}$ can be obtained using adapted variations of metric, see Section 4.2. The component $(4.86)_2$ is not symmetric under the change $\widetilde{\mathscr{D}} \leftrightarrow \mathscr{D}$, because $\operatorname{S}_{\mathrm{mix}}(g_t)$ (when $\widetilde{\mathscr{D}}$ is fixed) differs from the mixed scalar curvature for g_t when $\mathscr{D}$ is fixed while $\widetilde{\mathscr{D}}_t$ changes. We will show that the replacement of $\operatorname{S}_{\mathrm{mix}}$ by $\operatorname{Scal}_{\widetilde{\mathscr{D}}}$ in (4.82) leads to the same Euler–Lagrange equations (4.83).

Proposition 4.55. *The Euler–Lagrange equations for $J_{\widetilde{\mathscr{D}}}$ are the same as for*

$$\hat{J}_{\widetilde{\mathscr{D}}}: g \to \int_M \left\{ \frac{1}{2\mathfrak{a}} \left(\operatorname{Scal}_{\widetilde{\mathscr{D}}}(g) - 2\Lambda \right) + \mathscr{L}(g) \right\} d\operatorname{vol}_g.$$

Proof. Consider variations of metric with compact support contained in $\Omega \subset M$. By the Divergence Theorem, for any vector fields X_t and metrics g_t we have

$$\int_\Omega \operatorname{div} X_t \, d\operatorname{vol}_{g_t} = \int_{\partial\Omega} \langle X_t, \nu \rangle \, dA_{g_t},$$

v being outward-pointing normal unit vector field to $\partial\Omega$. If $\partial_t g$ and $\partial_t X$ are supported in Ω, then the RHS integral does not depend on t, and we have

$$\frac{d}{dt}\int_\Omega \operatorname{div} X_t \, d\operatorname{vol}_{g_t} = 0.$$

Thus, for $X_t = \frac{p-n}{n+p-2}(\tilde{H}-H)$, see (4.87), replacing S_{mix} by $\operatorname{Scal}_{\widetilde{\mathscr{D}}}$ in (4.82), we get the same Euler–Lagrange equations.   □

4.4.2 The Mixed Field Equations for Space-Times

Here, we consider the tensor $\operatorname{Ric}_{\widetilde{\mathscr{D}}}$ for space-times, see (4.88). For physically relevant applications, one should take $\varepsilon_N = -1$ (and $n = 1$), while $g_{|\mathscr{D}} > 0$.

Proposition 4.56. *For $\widetilde{\mathscr{D}}$ spanned by N, the tensor $\operatorname{Ric}_{\widetilde{\mathscr{D}}}$ in (4.84) is given by*

$$\operatorname{Ric}_{\widetilde{\mathscr{D}}\,|\,\mathscr{D}\times\mathscr{D}} = \varepsilon_N(R_N + (\tilde{A}_N)^2 - (\tilde{T}_N^\sharp)^2 + [\tilde{T}_N^\sharp,\tilde{A}_N])^\flat + H^\flat \otimes H^\flat - \tilde{\tau}_1 \tilde{h}_{sc} - \operatorname{Def}_{\mathscr{D}} H,$$

$$\operatorname{Ric}_{\widetilde{\mathscr{D}}}(\cdot,N)_{|\,\mathscr{D}} = \operatorname{div}^\perp \tilde{T}_N^\sharp{}_{|\mathscr{D}} + 2(\tilde{T}_N^\sharp(H))^\flat, \tag{4.88}$$

$$\operatorname{Ric}_{\widetilde{\mathscr{D}}}(N,N) = \varepsilon_N \operatorname{Ric}_{N,N} - 2\|\tilde{T}\|^2 - \operatorname{div} H.$$

Proof. By (4.85), we have

$$\mu_1 = 0, \quad \mu_2 = \operatorname{div}(\tilde{H}-H) = \varepsilon_N(N(\tilde{\tau}_1) - \tilde{\tau}_1^2) - \operatorname{div} H.$$

Substituting the values of (1.47), (1.48) and

$$\Upsilon_{h,h} = 2H^\flat \otimes H^\flat, \quad \widetilde{\mathscr{T}} = \varepsilon_N(\tilde{T}_N^\sharp)^2, \quad \Upsilon_{T,T} = 0$$

into $(4.86)_1$ yields $(4.88)_1$. Substituting the values of (1.47), (1.48) and

$$h = Hg^\top, \quad \tilde{H}^\flat \otimes \tilde{H}^\flat = \varepsilon_N \tilde{\tau}_1^2 g^\top, \quad \mathscr{K} = 0 = \mathscr{T},$$
$$\Upsilon_{\tilde{h},\tilde{h}} = 2\varepsilon_N \tilde{\tau}_2 g^\top, \quad \Upsilon_{\tilde{T},\tilde{T}} = 2\|\tilde{T}\|^2 g^\top$$

into $(4.86)_3$ yields $(4.88)_3$. The proof of $(4.88)_2$ is the same as for (4.71b).   □

Using (1.49), (4.86) and the arguments in the proof of Proposition 4.56, we find the following equivalent form of $\operatorname{Ric}_{\widetilde{\mathscr{D}}}$ in Proposition 4.56, which does not explicitly include the curvature tensor:

$$\operatorname{Ric}_{\widetilde{\mathscr{D}}\,|\,\mathscr{D}\times\mathscr{D}} = \nabla_N \tilde{h}_{sc} - \tilde{\tau}_1 \tilde{h}_{sc} + \varepsilon_N([\tilde{T}_N^\sharp,\tilde{A}_N] - 2(\tilde{T}_N^\sharp)^2)^\flat,$$

$$\operatorname{Ric}_{\widetilde{\mathscr{D}}}(\cdot,N)_{|\,\mathscr{D}} = \operatorname{div}^\perp \tilde{T}_N^\sharp{}_{|\mathscr{D}} + 2(\tilde{T}_N^\sharp(H))^\flat,$$

$$\operatorname{Ric}_{\widetilde{\mathscr{D}}}(N,N) = \varepsilon_N(N(\tilde{\tau}_1) - \tilde{\tau}_2) - \|\tilde{T}\|^2, \tag{4.89}$$

In this case, $\operatorname{Scal}_{\widetilde{\mathscr{D}}} = \varepsilon_N \operatorname{Ric}_{N,N} + \operatorname{div}(\varepsilon_N \tilde{\tau}_1 N - H)$.

Theorem 4.53 and Proposition 4.56 (see also Theorem 4.36) yield the following.

Corollary 4.57 (The Mixed Gravitational Field Equations). *A metric g on M with $\mathscr{D}$ spanned by a unit vector field N is critical for (4.82) if and only if g is a solution of (4.83) with* $\mathrm{Ric}_{\widetilde{\mathscr{D}}}$ *given in (4.88).*

Remark 4.58. Substituting (4.89) into (4.83), we find that (4.83) for $\mathrm{Ric}_{\widetilde{\mathscr{D}}}$ given in (4.89) splits into the system

$$\nabla_N \tilde{h}_{sc} - \tilde{\tau}_1 \tilde{h}_{sc} - \varepsilon_N \big(2(\tilde{T}_N^\sharp)^2 + [\tilde{A}_N, \tilde{T}_N^\sharp] \big)^\flat + \|\tilde{T}\|^2$$
$$- (1/2) \big(\varepsilon_N (2N(\tilde{\tau}_1) - \tilde{\tau}_1^2 - \tilde{\tau}_2) - 2\Lambda \big) g^\perp = \mathfrak{a}\Theta_{|\mathscr{D}\times\mathscr{D}},$$
$$\mathrm{div}^\perp \tilde{T}_N^\sharp |_{\mathscr{D}} + 2\big(\tilde{T}_N^\sharp(H)\big)^\flat = \mathfrak{a}\Theta_{\cdot, N|\mathscr{D}}, \tag{4.90}$$
$$(1/2)\varepsilon_N(\tilde{\tau}_1^2 - \tilde{\tau}_2) - (3/2)\|\tilde{T}\|^2 + \Lambda = \mathfrak{a}\Theta_{N,N}.$$

Corollary 4.59. *A metric g on M with a non-degenerate distribution* $\widetilde{\mathscr{D}} = \mathrm{Span}(N)$ *and integrable $\mathscr{D}$ (e.g., for a globally hyperbolic space-time, see [BDRS]) is critical for (4.82) if and only if the non-zero components of* $\mathrm{Ric}_{\widetilde{\mathscr{D}}}$ *are given by*

$$\mathrm{Ric}_{\widetilde{\mathscr{D}}|\mathscr{D}\times\mathscr{D}} = \nabla_N \tilde{h}_{sc} - \tilde{\tau}_1 \tilde{h}_{sc}, \quad \mathrm{Ric}_{\widetilde{\mathscr{D}}}(N,N) = \varepsilon_N(N(\tilde{\tau}_1) - \tilde{\tau}_2),$$

see (4.89) with $\tilde{T} = 0$. In this case, (4.83) splits into the system

$$\nabla_N \tilde{h}_{sc} - \tilde{\tau}_1 \tilde{h}_{sc} - \frac{1}{2}\big(\varepsilon_N(2N(\tilde{\tau}_1) - \tilde{\tau}_1^2 - \tilde{\tau}_2) - 2\Lambda\big)g^\perp = \mathfrak{a}\Theta_{|\mathscr{D}\times\mathscr{D}},$$
$$\frac{1}{2}\varepsilon_N(\tilde{\tau}_1^2 - \tilde{\tau}_2) + \Lambda = \mathfrak{a}\Theta_{N,N}.$$

Proof. Since $(4.90)_2$ is valid for an integrable $\mathscr{D}$, variations g_t, which are constant on $\widetilde{\mathscr{D}}$, give the same Euler–Lagrange equations as adapted $g^\perp$-variations. □

Example 4.60. Let a vector field N generate a geodesic Riemannian flow on a pseudo-Riemannian manifold (M^{p+1}, g). Then $\tilde{h} = 0 = h$ and

$$\mathrm{Ric}_{\widetilde{\mathscr{D}}}(X,N) = -\mathrm{Ric}_{X,N}.$$

Consider the Hopf fibration $S^{2m+1} \to \mathbb{C}P^m$, with N tangent to the fibers $\{S^1\}$. We have $\tilde{T}(X,Y) = \langle JX, Y\rangle N$ for the standard almost complex structure J on $\mathbb{R}^{2m+2}$ and metric $g > 0$, and hence $\tilde{T}_N^\sharp(X) = J(X)$ and $\|\tilde{T}\|^2 = 2m$. Then

$$(\nabla_Z \tilde{T}_N^\sharp)(X) = (\nabla_Z J)(X) = 0,$$

and $(4.90)_2$ with $\Theta_{|\widetilde{\mathscr{D}}\times\mathscr{D}} = 0$ is valid. Moreover, $(4.90)_{1,3}$ are satisfied with

$$\mathfrak{a}\Theta_{N,N} = \Lambda - 3, \quad \mathfrak{a}\Theta_{|\mathscr{D}\times\mathscr{D}} = (2 - m + \Lambda)g^\perp.$$

The reader can find more examples in [RZ2, RZ3] (and in [BDRS] for space-times).

4.4.3 Variable Connection

Let variations of g and $\mathfrak{T}$ be supported in a relatively compact domain Ω. The Euler–Lagrange equations for (4.80) with varying $\mathfrak{T}$ (which are analogous to the spin-connection equation in the Einstein-Cartan theory of gravity) are divided into 8 sets according to the decomposition $TM = \widetilde{\mathscr{D}} \oplus \mathscr{D}$.

Theorem 4.61. *The Euler–Lagrange equation for (4.80) with fixed g and all variations of $\mathfrak{T}$ is the following algebraic system with spin tensor $s_{\mu\nu}^c = 2\partial\mathscr{L}/\partial\mathfrak{T}_{\mu\nu}^c$:*

$$\langle \tilde{H}_{\mathfrak{T}^*} - \tilde{H}, Z\rangle\langle X,Y\rangle + \langle \tilde{H}_{\mathfrak{T}} + \tilde{H}, Y\rangle\langle X,Z\rangle = -(\mathfrak{a}/2)\,\langle s(X,Y),Z\rangle,$$

$$\langle H_{\mathfrak{T}^*} - H, W\rangle\langle U,V\rangle + \langle H_{\mathfrak{T}} + H, V\rangle\langle U,W\rangle = (\mathfrak{a}/2)\,\langle s(U,V),W\rangle,$$

$$\langle \tilde{H}_{\mathfrak{T}^*} + H, U\rangle\langle X,Y\rangle - \langle (A_U - T_U^\sharp + \mathfrak{T}_U)X, Y\rangle = (\mathfrak{a}/2)\,\langle s(X,Y),U\rangle,$$

$$\langle H_{\mathfrak{T}^*} + \tilde{H}, X\rangle\langle U,V\rangle - \langle (\tilde{A}_X - \tilde{T}_X^\sharp + \mathfrak{T}_X)U, V\rangle = (\mathfrak{a}/2)\,\langle s(U,V),X\rangle,$$

$$\langle \tilde{H}_{\mathfrak{T}} - H, U\rangle\langle X,Y\rangle + \langle (A_U + T_U^\sharp - \mathfrak{T}_U)Y, X\rangle = (\mathfrak{a}/2)\,\langle s(X,U),Y\rangle,$$

$$\langle H_{\mathfrak{T}} - \tilde{H}, X\rangle\langle U,V\rangle + \langle (\tilde{A}_X + \tilde{T}_X^\sharp - \mathfrak{T}_X)V, U\rangle = (\mathfrak{a}/2)\,\langle s(U,X),V\rangle,$$

$$2\langle \tilde{T}_X^\sharp U, V\rangle + \langle \mathfrak{T}_U V + \mathfrak{T}_V^* U, X\rangle = (\mathfrak{a}/2)\,\langle s(X,U),V\rangle,$$

$$2\langle T_U^\sharp X, Y\rangle + \langle \mathfrak{T}_X Y + \mathfrak{T}_Y^* X, U\rangle = (\mathfrak{a}/2)\,\langle s(U,X),Y\rangle, \quad (4.91)$$

for all $X,Y,Z \in \widetilde{\mathscr{D}}$ and $U,V,W \in \mathscr{D}$. Here, $(4.91)_{2,4,6,8}$ are dual to $(4.91)_{1,3,5,7}$.

Proof. Set $S = \partial_t \mathfrak{T}^t|_{t=0}$ for a one-parameter family $\mathfrak{T}^t$ $(|t| < \varepsilon)$ of $(1,2)$-tensors. Using Lemma 2.70 and removing integrals of divergences of compactly supported (in a domain Ω) vector fields and factors ε_μ, we get

$$\frac{\mathrm{d}}{\mathrm{d}t} \int_\Omega \bar{S}_{\mathrm{mix}}(\mathfrak{T}^t)\,\mathrm{d}\,\mathrm{vol}_g\,|_{t=0}$$

$$= \frac{1}{2} \int_M \sum \big\{ \langle S_a E_b, E_c\rangle\big(\langle \tilde{H}_{\mathfrak{T}^*} - \tilde{H}, E_c\rangle\langle E_a, E_b\rangle + \langle \tilde{H}_{\mathfrak{T}} + \tilde{H}, E_b\rangle < E_a, E_c\rangle\big)$$

$$+ \langle S_a E_b, \mathscr{E}_i\rangle\big(\langle \tilde{H}_{\mathfrak{T}^*} + H, \mathscr{E}_i\rangle\langle E_a, E_b\rangle - \langle (A_i - T_i^\sharp)E_a, E_b\rangle - \langle \mathfrak{T}_i E_a, E_b\rangle\big)$$

$$+ \langle S_a \mathscr{E}_i, E_b\rangle\big(\langle \tilde{H}_{\mathfrak{T}} - H, \mathscr{E}_i\rangle\langle E_a, E_b\rangle + \langle (A_i + T_i^\sharp)E_b, E_a\rangle - \langle \mathfrak{T}_i E_b, E_a\rangle\big)$$

$$+ \langle S_a \mathscr{E}_i, \mathscr{E}_j\rangle\big(\langle (\tilde{A}_a - \tilde{T}_a^\sharp)\mathscr{E}_i, \mathscr{E}_j\rangle - \langle (\tilde{A}_a + \tilde{T}_a^\sharp)\mathscr{E}_i, \mathscr{E}_j\rangle - \langle \mathfrak{T}_i \mathscr{E}_j + \mathfrak{T}_j^* \mathscr{E}_i, E_a\rangle\big)$$

$$+ \langle S_i \mathscr{E}_j, \mathscr{E}_k\rangle\big(\langle H_{\mathfrak{T}^*} - H, \mathscr{E}_k\rangle\langle \mathscr{E}_i, \mathscr{E}_j\rangle + \langle H_{\mathfrak{T}} + H, \mathscr{E}_j\rangle\langle \mathscr{E}_i, \mathscr{E}_k\rangle\big)$$

$$+ \langle S_i \mathscr{E}_j, E_a\rangle\big(\langle H_{\mathfrak{T}^*} + \tilde{H}, E_a\rangle\langle \mathscr{E}_i, \mathscr{E}_j\rangle - \langle (\tilde{A}_a + \tilde{T}_a^\sharp)\mathscr{E}_j, \mathscr{E}_i\rangle - \langle \mathfrak{T}_a \mathscr{E}_i, \mathscr{E}_j\rangle\big)$$

$$+ \langle S_i E_a, \mathscr{E}_j\rangle\big(\langle H_{\mathfrak{T}} - \tilde{H}, E_a\rangle\langle \mathscr{E}_i, \mathscr{E}_j\rangle + \langle (\tilde{A}_a + \tilde{T}_a^\sharp)\mathscr{E}_j, \mathscr{E}_i\rangle - \langle \mathfrak{T}_a \mathscr{E}_j, \mathscr{E}_i\rangle\big)$$

$$+ \langle S_i E_a, E_b\rangle\big(\langle (A_i - T_i^\sharp)E_a, E_b\rangle - \langle (A_i + T_i^\sharp)E_a, E_b\rangle - \langle \mathfrak{T}_a E_b + \mathfrak{T}_b^* E_a, \mathscr{E}_i\rangle\big) \big\}\,\mathrm{d}\,\mathrm{vol}_g.$$

Since no further assumptions are made about S or $\mathfrak{T}$, all the components $\langle S_\mu e_\lambda, e_\rho\rangle$ are independent and after replacement $\delta_{ab} = \langle E_a, E_b\rangle$ etc. the above formula gives rise to (4.91). $\qquad\square$

Remark 4.62. One may show that if $\mathfrak{T}$ is critical for (4.80) with $\mathscr{L} = 0$, i.e., the spin tensor vanishes in (4.91), then both distributions $\mathscr{D}$ and $\widetilde{\mathscr{D}}$ are totally umbilical. This follows from homogeneous system (4.91), i.e., $s = 0$. For example, summing the third and the fifth equations yields

$$2\langle A_U X, Y \rangle = -\langle \tilde{H}_{\mathfrak{T}^*} + \tilde{H}_{\mathfrak{T}}, U \rangle \langle X, Y \rangle.$$

Corollary 4.63. *For a space-time (M^{p+1}, g) endowed with $\widetilde{\mathscr{D}}$ spanned by a timelike unit vector field N, the system (4.91) reduces to*

$$
\begin{aligned}
&\langle \tilde{H}_{\mathfrak{T}^* + \mathfrak{T}}, N \rangle = -(\mathfrak{a}/2)\,\langle s(N,N), N \rangle, \\
&\langle H_{\mathfrak{T}^*} - H, W \rangle \langle U, V \rangle + \langle H_{\mathfrak{T}} + H, V \rangle \langle U, W \rangle = -(\mathfrak{a}/2)\,\langle s(U,V), W \rangle, \\
&\langle \tilde{H}_{\mathfrak{T}^*}, U \rangle - \langle \mathfrak{T}_U N, N \rangle = -(\mathfrak{a}/2)\,\langle s(N,N), U \rangle, \\
&(\langle H_{\mathfrak{T}^*}, N \rangle + \tilde{\tau}_1)\langle U, V \rangle - \langle (\tilde{A}_N - \tilde{T}_N^\sharp + \mathfrak{T}_N)U, V \rangle = -(\mathfrak{a}/2)\,\langle s(U,V), N \rangle, \\
&\langle \tilde{H}_{\mathfrak{T}}, U \rangle - \langle \mathfrak{T}_U N, N \rangle = -(\mathfrak{a}/2)\,\langle s(N,U), N \rangle, \\
&(\langle H_{\mathfrak{T}}, N \rangle - \tilde{\tau}_1)\langle U, V \rangle + \langle (\tilde{A}_N + \tilde{T}_N^\sharp - \mathfrak{T}_N)V, U \rangle = -(\mathfrak{a}/2)\,\langle s(N,U), V \rangle, \\
&\langle 2\tilde{T}_N(U, V) + \mathfrak{T}_U V + \mathfrak{T}_V^* U, N \rangle = (\mathfrak{a}/2)\,\langle s(N,U), V \rangle, \\
&\langle \mathfrak{T}_N N + \mathfrak{T}_N^* N, U \rangle = (\mathfrak{a}/2)\,\langle s(U,N), N \rangle,
\end{aligned}
$$

where $U, V, W \in \mathscr{D}$.

Together with arbitrary variations of connection, one may consider also variations belonging to the distinguished classes (for example, statistical and metric compatible) of connections. The following connection is metric compatible.

Definition 4.64. A linear connection $\bar{\nabla}$ on M is *semi-symmetric* if its torsion S satisfies $S(X,Y) = \omega(Y)X - \omega(X)Y$, where ω is a one-form on M. For (M,g) we have

$$\bar{\nabla}_X Y = \nabla_X Y + \langle U, Y \rangle X - \langle X, Y \rangle U, \tag{4.92}$$

where $U = \omega^\sharp$ is a dual vector field.

The Euler–Lagrange equation of (4.80) with varying g generalizes (4.83):

$$\overline{\mathrm{Ric}}_{\widetilde{\mathscr{D}}} - (1/2)\,\overline{\mathrm{Scal}}_{\widetilde{\mathscr{D}}} \cdot g + \Lambda\, g = \mathfrak{a}\Theta. \tag{4.93}$$

The tensor $\overline{\mathrm{Ric}}_{\widetilde{\mathscr{D}}}$ is more complicated than $\mathrm{Ric}_{\widetilde{\mathscr{D}}}$, and we omit its expression. However, varying $\mathfrak{T}$ among semi-symmetric connections, we express $\overline{\mathrm{Ric}}_{\widetilde{\mathscr{D}}}$ explicitly.

According to Lemma 2.70, set

$$Q = \mathrm{div}\left(H_{\mathfrak{T}-\mathfrak{T}^*}^{\perp} + \tilde{H}_{\mathfrak{T}-\mathfrak{T}^*}^{\top}\right) - 2\left(\overline{S}_{\mathrm{mix}} - S_{\mathrm{mix}}\right).$$

Lemma 4.65.

(a) *If $\bar{\nabla}$ satisfies (4.92), then Q reduces to*

$$Q = (n - p)\langle U, H - \tilde{H} \rangle + np\langle U, U \rangle - n\langle U^{\perp}, U^{\perp} \rangle - p\langle U^{\top}, U^{\top} \rangle. \tag{4.94}$$

(b) For any $g^\pitchfork$-variation of metric g and Q given by (4.94) we have

$$\partial_t Q(g_t)|_{t=0} = \langle B, -(n-p)\tilde{\delta}_{U^\perp} - (n-p)\langle \tilde{\alpha} - \tilde{\theta}, U^\perp \rangle + 2(p-n)\langle \theta, U^\top \rangle$$
$$- \frac{1}{2}(p-n)(\operatorname{div} U^\top)g^\perp + n(p-1)U^{\perp\flat} \otimes U^{\perp\flat} + 2p(n-1)U^{\top\flat} \odot U^{\perp\flat}\rangle. \quad (4.95)$$

Proof.

(a) From (4.92) we obtain

$$H_{\mathfrak{T}} = \sum_a \langle U, E_a \rangle E_a - \sum_a \langle E_a, E_a \rangle U = U^\top - nU.$$

Similarly, $\tilde{H}_{\mathfrak{T}} = U^\perp - pU$. We also have

$$\mathfrak{T}_a \mathcal{E}_i = \langle U, \mathcal{E}_i \rangle E_a, \quad \mathfrak{T}_i E_a = \langle U, E_a \rangle \mathcal{E}_i,$$

so we obtain $\langle \mathfrak{T}, \widehat{\mathfrak{T}} \rangle_{|V} = 0$. Next, we have

$$\langle H_{\mathfrak{T}} - \tilde{H}_{\mathfrak{T}}, H - \tilde{H} \rangle = (p-n-1)\langle U^\perp, H \rangle + (n-p-1)\langle U^\top, \tilde{H} \rangle$$

and $(\mathfrak{T}_i + \widehat{\mathfrak{T}}_i)E_a = \langle U, E_a \rangle \mathcal{E}_i + \langle U, \mathcal{E}_i \rangle E_a$. Also we have

$$\langle H_{\mathfrak{T}}, \tilde{H}_{\mathfrak{T}} \rangle = np\langle U, U \rangle - n\langle U^\perp, U^\perp \rangle - p\langle U^\top, U^\top \rangle.$$

Thus,

$$\langle \mathfrak{T} + \widehat{\mathfrak{T}}, \tilde{A} - \tilde{T}^\sharp + A - T^\sharp \rangle = \langle H + \tilde{H}, U \rangle.$$

From the above, (4.94) follows.

(b) By Lemma 4.32, and

$$g(\partial_t \tilde{H}, X) = \langle 2\langle \theta, X^\top \rangle, B \rangle - \frac{1}{2}X^\top(\operatorname{trace}_{\mathscr{D}} B),$$

see [RZ2, Eqn. (20)], we have:

$$\langle U^\perp, \partial_t U^\perp \rangle = \langle U^\perp, -B^\sharp(U^\perp) \rangle = 0,$$
$$\langle U^\top, \partial_t U^\top \rangle = \langle U^\top, B^\sharp(U^\perp) \rangle = \langle B, U^{\top\flat} \odot U^{\perp\flat} \rangle.$$

Therefore,

$$\langle \partial_t \tilde{H}, U \rangle = \operatorname{div}((\operatorname{trace}_{\mathscr{D}} B)U^\top) + \langle B, 2\langle \theta, U^\top \rangle - \frac{1}{2}(\operatorname{div} U^\top)g^\perp \rangle,$$
$$\langle \partial_t H, U \rangle = \operatorname{div}((B^\sharp(U^\perp))^\top) + \langle B, U^\perp \odot (\tilde{H}-H) - U^\top \odot H - \tilde{\delta}_{U^\perp} - \langle \tilde{\alpha} - \tilde{\theta}, U^\perp \rangle \rangle.$$

Omitting divergences of vector fields and using $B|_{\tilde{\mathscr{D}} \times \tilde{\mathscr{D}}} = 0$, we obtain

$$\partial_t Q(g_t)|_{t=0} = (n-p)B(U, H - \tilde{H}) + (n-p)\langle U, \partial_t H \rangle - (n-p)\langle \partial_t \tilde{H}, U \rangle$$
$$+ npB(U, U) - nB(U^\perp, U^\perp) - 2n\langle \partial_t U^\perp, U \rangle - pB(U^\top, U^\top) - 2p\langle \partial_t U^\top, U^\top \rangle,$$

that reduces to (4.95). $\square$

First, we will present Euler–Lagrange equations of (4.80) as a particular case of (4.91), considering variations of a semi-symmetric connection among connections also satisfying (4.92) for some vector field U.

Theorem 4.66. *A semi-symmetric connection* $\bar{\nabla}$ *on* $(M, g, \mathscr{D})$ *is critical for the action* (4.80) *with fixed g among all connections satisfying* (4.92) *if and only if*

$$2p(n-1)U^\top - (n-p)\tilde{H} = -(\mathfrak{a}/2)\,s^\top,$$
$$2n(p-1)U^\perp - (p-n)H = -(\mathfrak{a}/2)\,s^\perp, \tag{4.96}$$

where $s^\top = (s(\cdot,\cdot))^\top$ *and* $s^\perp = (s(\cdot,\cdot))^\perp$. *In particular, if* $n = p = 1$ *and* $s = 0$ *(no spin) then every semi-symmetric connection is critical among all such connections,*

Proof. Let U_t, $t \in (-\varepsilon, \varepsilon)$, be a family of compactly supported vector fields on M, and let $U = U_0$ and $\dot{U} = \partial_t U_t|_{t=0}$. Then for a fixed metric g, from (4.94) we get

$$\partial_t Q(U_t)|_{t=0} = (p-n)\langle \dot{U}, \tilde{H}\rangle + 2p(n-1)\langle U^\top, \dot{U}\rangle + (n-p)\langle \dot{U}, H\rangle + 2n(p-1)\langle U^\perp, \dot{U}\rangle.$$

Separating parts with $(\dot{U})^\top$ and $(\dot{U})^\perp$, we get

$$\partial_t Q(U_t)|_{t=0} = \langle \dot{U}, (p-n)\tilde{H} + 2p(n-1)U^\top\rangle + \langle \dot{U}, (n-p)H + 2n(p-1)U^\perp\rangle,$$

from which (4.96) follow.   □

Next, we will present Euler–Lagrange equations of (4.81) (with a fixed semi-symmetric connection) using volume-preserving $g^\pitchfork$-variations of metric.

Theorem 4.67. *A pair* $(g, \mathfrak{T})$, *where* $g \in \mathrm{Riem}(M, \tilde{\mathscr{D}}, \mathscr{D})$ *and* $\mathfrak{T}$ *corresponds to a semi-symmetric connection on M defined by* (4.92), *is critical for* (4.81) *with respect to volume-preserving* $g^\pitchfork$-variations of metric if and only if the following Euler–Lagrange equations are satisfied:

$$\mathrm{Ric}^\perp - \langle \tilde{h}, \tilde{H}\rangle + \widetilde{\mathscr{A}^\flat} - \widetilde{\mathscr{T}^\flat} + \Psi + \widetilde{\mathscr{K}^\flat} - \mathrm{Def}_\mathscr{D} H + H^\flat \otimes H^\flat - \frac{1}{2}\Upsilon_{h,h} - \frac{1}{2}\Upsilon_{T,T} \tag{4.97a}$$

$$-\frac{1}{2}\left(\mathrm{S}_{\mathrm{mix}} + \mathrm{div}(\tilde{H} - H)\right)g^\perp - \frac{p-n}{4}(\mathrm{div}\,U^\top)g^\perp + \frac{n(p-1)}{2}U^{\perp\flat} \otimes U^{\perp\flat} = \lambda\,g^\perp,$$

$$4\langle \theta, \tilde{H}\rangle + 2(\mathrm{div}(\alpha - \tilde{\theta}))_{|\mathrm{V}} + 2\langle \tilde{\theta} - \tilde{\alpha}, H\rangle + 2H^\flat \odot \tilde{H}^\flat - 2\tilde{\delta}_H$$

$$+4\Upsilon_{\tilde{\alpha}, \theta} + 2\Upsilon_{\alpha, \tilde{\alpha}} + 2\Upsilon_{\tilde{\theta}, \theta} + \frac{1}{2}(n-p)\tilde{\delta}_{U^\perp} + \frac{1}{2}(n-p)\langle \tilde{\alpha} - \tilde{\theta}, U^\perp\rangle$$

$$-(p-n)\langle \theta, U^\top\rangle - p(n-1)U^{\top\flat} \otimes U^{\perp\flat} = 0. \tag{4.97b}$$

Proof. By Proposition 2.70 and (4.95), we obtain

$$\partial_t \int_M (\bar{\mathrm{S}}_{\mathrm{mix}} - \mathrm{S}_{\mathrm{mix}})\,d\mathrm{vol}_g = \int_M \left\langle \frac{1}{4}(p-n)(\mathrm{div}\,U^\top)g^\perp - (p-n)\langle \theta, U^\top\rangle \right.$$

$$\left. -\frac{1}{2}n(p-1)U^{\perp\flat} \otimes U^{\perp\flat} - p(n-1)U^{\top\flat} \otimes U^{\perp\flat}, B\right\rangle d\mathrm{vol}_g.$$

Using (4.66a,b) gives rise to (4.97a,b). □

For semi-symmetric connections, according to Theorem 4.66 and (4.93), we present the mixed Ricci tensor explicitly as

$$
\overline{\mathrm{Ric}}_{\widetilde{\mathscr{D}}\,|\,\mathscr{D}\times\mathscr{D}} = \mathrm{Ric}_{\widetilde{\mathscr{D}}\,|\,\mathscr{D}\times\mathscr{D}}
$$
$$
+ \frac{1}{2}n(p-1)U^{\perp\flat}\otimes U^{\perp\flat} - \frac{1}{4}(p-n)(\operatorname{div}U^{\top})g^{\perp} + \frac{Z}{2-n-p}g^{\perp},
$$
$$
\overline{\mathrm{Ric}}_{\widetilde{\mathscr{D}}\,|\,V} = \mathrm{Ric}_{\widetilde{\mathscr{D}}\,|\,V}
$$
$$
- \frac{1}{2}(n-p)\big(\tilde{\delta}_{U^{\perp}} + \langle\tilde{\alpha} - \tilde{\theta}, U^{\perp}\rangle\big) + (p-n)\langle\theta, U^{\top}\rangle + p(n-1)U^{\top\flat}\otimes U^{\perp\flat},
$$
$$
\overline{\mathrm{Ric}}_{\widetilde{\mathscr{D}}\,|\,\widetilde{\mathscr{D}}\times\widetilde{\mathscr{D}}} = \mathrm{Ric}_{\widetilde{\mathscr{D}}\,|\,\widetilde{\mathscr{D}}\times\widetilde{\mathscr{D}}}
$$
$$
+ \frac{1}{2}p(n-1)U^{\top\flat}\otimes U^{\top\flat} - \frac{1}{4}(n-p)(\operatorname{div}U^{\perp})g^{\top} + \frac{Z}{2-n-p}g^{\top}, \qquad (4.98)
$$

where $\mathrm{Ric}_{\widetilde{\mathscr{D}}}$ and $\mathrm{Scal}_{\widetilde{\mathscr{D}}}$ are given in Section 4.4.1, $n+p>2$ and

$$
Z = \frac{n(p-1)}{2}\|U^{\perp}\|^{2} + \frac{p(n-1)}{2}\|U^{\top}\|^{2} - \frac{p(p-n)}{4}\operatorname{div}U^{\top} - \frac{n(n-p)}{4}\operatorname{div}U^{\perp}. \quad (4.99)
$$

From (4.98) and (4.99) with $n=1$ it follows that for a *space-time* (M^{p+1}, g) endowed with $\widetilde{\mathscr{D}}$ spanned by a timelike unit vector field N, the tensor $\overline{\mathrm{Ric}}_{\mathscr{D}}$ in (4.98) and its trace have the following particular form:

$$
\begin{cases}
\overline{\mathrm{Ric}}_{\widetilde{\mathscr{D}}\,|\,\mathscr{D}\times\mathscr{D}} = \mathrm{Ric}_{\widetilde{\mathscr{D}}\,|\,\mathscr{D}\times\mathscr{D}} + \frac{1}{2}(p-1)U^{\perp\flat}\otimes U^{\perp\flat} - \frac{1}{4}(p-1)(\operatorname{div}U^{\top})g^{\perp} + \frac{Z}{1-p}g^{\perp}, \\
\overline{\mathrm{Ric}}_{\widetilde{\mathscr{D}}\,|\,V} = \mathrm{Ric}_{\widetilde{\mathscr{D}}\,|\,V} - \frac{1}{2}(1-p)\big(\tilde{\delta}_{U^{\perp}} + \langle\tilde{\alpha} - \tilde{\theta}, U^{\perp}\rangle\big), \\
\overline{\mathrm{Ric}}_{\widetilde{\mathscr{D}}\,|\,\widetilde{\mathscr{D}}\times\widetilde{\mathscr{D}}} = \mathrm{Ric}_{\widetilde{\mathscr{D}}\,|\,\widetilde{\mathscr{D}}\times\widetilde{\mathscr{D}}} - \frac{1}{4}\varepsilon_N(1-p)(\operatorname{div}U^{\perp}) + \varepsilon_N\frac{Z}{1-p},
\end{cases}
$$

$$
\overline{\mathrm{S}}_{\widetilde{\mathscr{D}}} = \mathrm{S}_{\widetilde{\mathscr{D}}} + \frac{2\varepsilon_N}{1-p}Z,
$$

where $Z = \frac{1}{4}(p-1)(2\|U^{\perp}\|^{2} - p\operatorname{div}U^{\top} + \operatorname{div}U^{\perp})$.

Remark 4.68. In [R15], the total mixed scalar curvature of a Riemannian almost multi-product structure (see Definition 2.59) is considered, the Euler–Lagrange equations for the Einstein–Hilbert type action with respect to adapted (i.e., preserving the pairwise orthogonality of $k>2$ distributions) variations of metric are derived in a nice form of Einstein equation, see (4.83).

4.5 The Godbillon-Vey Type Invariant

The Godbillon-Vey cohomology class $\mathrm{gv}(\mathscr{F})$ of a transversely oriented codimension-one foliation $\mathscr{F}$ of a compact manifold M, see [GV], is the de Rham cohomology class of the 3-form $\eta \wedge d\eta$, where η is a 1-form satisfying $d\omega = \omega \wedge \eta$, ω being a 1-form defining the tangent bundle (distribution $T\mathscr{F} = \ker \omega$) of $\mathscr{F}$. The $\mathrm{gv}(\mathscr{F})$ was then extended for foliations of codimension $q \geq 1$, see [CC]. The complex Godbillon-Vey class (defined for transversely holomorphic foliations) is often referred as the Bott class, see [Asu]. The $\mathrm{gv}(\mathscr{F})$ is the simplest among characteristic classes of foliations, it measures some sort of "twisting" of the leaves, Figure 4.1 and plays a crucial role in the topology and dynamics of foliations [CC, GLW, HL, HL, RW] and [Hu, Problem 10], and in applications [WPAH]. For example, all codimension-one foliations of closed manifolds with $\mathrm{gv}(\mathscr{F}) \neq 0$ have positive entropy in the sense of [GLW] and contain resilient leaves [CC1], Figure 1.5.

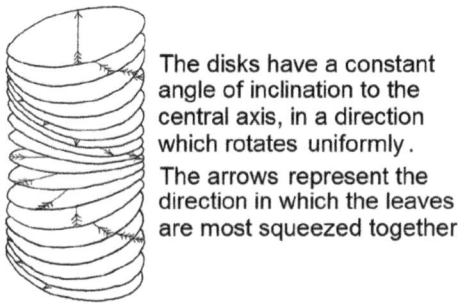

The disks have a constant angle of inclination to the central axis, in a direction which rotates uniformly.

The arrows represent the direction in which the leaves are most squeezed together.

Fig. 4.1 Twisting of the leaves (the helical wobble of a foliation)

Variations of Godbillon-Vey class under infinitesimal deformations of $\mathscr{F}$ have been studied in [Asu], where the formula for the variation was given in terms of a transversal projective structure.

If $\dim M = 3$, then the Godbillon-Vey class provides a number

$$\mathrm{gv}(\mathscr{F}) = \int_M \eta \wedge d\eta. \tag{4.100}$$

Integrability of the distribution $\mathscr{D} = T\mathscr{F}$ implies the existence of such η, while simple calculations show that $\mathrm{gv}(\mathscr{F})$ is well defined, that is it does not depend on choices of ω and η. If $f : \overline{M} \to M$ is a smooth map transverse to $\mathscr{F}$, then $\mathrm{gv}(f^*\mathscr{F}) = f^*\mathrm{gv}(\mathscr{F})$, thus concordant foliations have the same Godbillon-Vey classes.

Let a smooth three-dimensional manifold M be equipped with a vector field T transverse to a plane field $\mathscr{D}$—the kernel of a one form ω such that $\omega(T) = 1$. In this section, we construct a three-form analogous to that defining the Godbillon-Vey class of a foliation, show how does this form depend on ω and T. On a Riemannian manifold, this form is expressed in terms of the curvature and torsion of normal curves and the non-symmetric second fundamental form of $\mathscr{D}$. We derive Euler–Lagrange equations of the associated functional and characterize critical pairs

$(\mathscr{D}, T)$ and foliations for different types of variations, find sufficient conditions for critical pairs when variations are among foliations, find the index form of our variation problem, and provide examples with Roussarie and Reeb foliations and twisted products. Following ideas of Section 2.2, we consider singular foliations and forms, i.e., defined outside a "singularity set" under assumption of convergence of certain integrals.

4.5.1 Construction

There exists a one parameter family of foliations on S^3 with the Godbillon-Vey number taking all values in an interval [Th1] (while for the Reeb foliation this number is zero), hence $gv(\mathscr{F})$ is not a homotopy invariant, see also [Tam]. Variations of (4.100) under infinitesimal deformations of $\mathscr{F}$ have been studied in [Asu, Mas], and the variation was expressed in terms of a transversal projective structure.

Our variational approach differs from one mentioned above. We extend definition of $gv(\mathscr{F})$ by introducing an analogous *Godbillon–Vey invariant* $gv(\mathscr{D}, T)$ for a pair consisting of an arbitrary, *a priori* non-integrable, plane field $\mathscr{D}$ and a transverse to it vector field T on a Riemannian manifold (M^3, g), [RW8, RW9], and we examine its dependence on $\mathscr{D}$, T and g. The fundamental question is: *What are the best almost product structures on a manifold?* Such pairs (ω, T) (of the above question) are proposed to be among critical points of the action, see (4.103) below.

Let M^3 be equipped with a plane field $\mathscr{D}$ (two-dimensional subbundle of TM), and a vector field T transverse to $\mathscr{D}$. A priori, our $\mathscr{D}$ is not integrable. We assume that we have a 1-form ω such that $\mathscr{D} = \ker \omega$ and

$$\omega(T) = 1, \tag{4.101}$$

and then build a 3-form analogous to that defining $gv(\mathscr{F})$. In fact, a pair (ω, T) obeying (4.101) is the main geometric structure considered here.

Definition 4.69. A Riemannian metric g on M^3 is *compatible* with a pair $(\mathscr{D}, T)$ if T is the unit normal to $\mathscr{D}$, that is $T = \omega^\sharp$. Denote by $\mathrm{Riem}(M, \mathscr{D}, T)$ the space of all such metrics.

Given $g \in \mathrm{Riem}(M, \mathscr{D}, T)$, consider in the space $\Lambda^1(M)$ of 1-forms on M the subspace $\omega^\perp$ orthogonal to the line spanned by ω. Consider also, in the three-dimensional space $\Lambda^2(M)$ of 2-forms, the subspace $\omega \wedge \omega^\perp$ of all 2-forms $\omega \wedge \theta$, θ being a 1-form of $\omega^\perp$. Now, project $d\omega$ orthogonally onto the subspace $\omega \wedge \omega^\perp$ and obtain $\omega \wedge \eta$, η belonging to $\omega^\perp$, Figure 4.2.

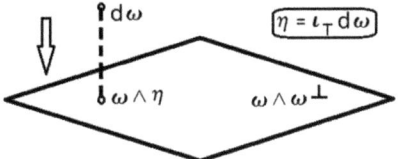

Fig. 4.2 Construction of η

Suppose that $g \in \text{Riem}(M, \mathcal{D}, T)$. The property $\eta \perp \omega$ means $\eta(T) = 0$. The property $d\omega - \omega \wedge \eta \perp \omega \wedge \omega^{\perp}$ means $(d\omega - \omega \wedge \eta)(T, \cdot) = 0$, that is the above η is unique and has the form

$$\eta = \iota_T \, d\omega. \tag{4.102}$$

The 3-form $\eta \wedge d\eta$ represents the *Godbillon-Vey type invariant* of a pair (ω, T):

$$\text{gv} : (\omega, T) \to \int_M \eta \wedge d\eta. \tag{4.103}$$

Note also that $d\eta = \mathscr{L}_T(d\omega)$. If $\mathcal{D}$ is integrable ($\mathcal{D} = T\mathscr{F}$ for a foliation $\mathscr{F}$), then

$$\text{gv}(\omega, \tilde{T}) = \text{gv}(\omega, T) = \text{gv}(\mathscr{F})$$

for any $\tilde{T}$ and T transverse to $\mathcal{D}$. If T is the Reeb vector field of a contact distribution $\mathcal{D} = \ker \omega$ on $M^3 \setminus \Sigma$, then $\omega(T) = 1$ and $\eta := \iota_T \, d\omega$ vanishes. Hence, $\eta \wedge d\eta = 0$.

Given a compatible Riemannian metric on our three-dimensional manifold M, the unit normal N, the binormal $B = T \times N$ and the *torsion* τ of T-curves are defined on an open subset U of M where the *curvature* k of T-curves is nonzero. Using Frenet formulae for T-curves,

$$\nabla_T T = kN, \quad \nabla_T N = -kT + \tau B, \quad \nabla_T B = -\tau N, \tag{4.104}$$

and the formula for the Levi-Civita ∇ connection of g, we obtain

$$k = \langle [N, T], T \rangle, \quad 2\tau = \langle [T, N], B \rangle - \langle [T, B], N \rangle.$$

By (4.104), we get equivalent definition of our 1-form:

$$\eta = \begin{cases} kN^{\flat} & \text{on } U, \\ 0 & \text{on } M \setminus U. \end{cases} \tag{4.105}$$

Proposition 4.70. *If* $\text{gv}(\omega, T) \neq 0$, *then there is no metric* $g \in \text{Riem}(M, \mathcal{D} = \ker \omega, T)$ *such that* T *is a geodesic field with respect to* g.

Proof. If T is geodesic (i.e., $k = 0$), then $\eta = 0$, see (4.105), hence $\text{gv}(\omega, T) = 0$. $\square$

Example 4.71 (Geodesible Vector Fields on 3-Manifolds). Recall the characterization of geodesic fields, see survey in [Gl]: Let T be a nonsingular vector field on a smooth manifold M, then there is a Riemannian metric making the orbits of T geodesics if and only if no nonzero foliation cycle for T can be arbitrarily well approximated by the boundary of a 2-chain tangent to N.

Due to Definition 3.13, a pair $(\mathscr{D} = \ker \omega, T)$, where ω is a 1-form on M^3 such that $\omega(T) = 1$, is (1) *geometrically taut* if T is geodesic and orthogonal to $\ker \omega$ for some Riemannian metric g on M; (2) *topologically taut* if the cone C^{fh} intersects trivially the smallest closed linear subspace P^{fh} containing B^{fh} and all Dirac currents z_v with $v \in \mathscr{D}_x$ and $x \in M$. The cone C^{fh} is generated by Dirac current z_T, and B^{fh} is the closed linear subspace generated by boundaries of Dirac currents z_v, where $v = w \wedge T$. By Theorem 3.17, a pair $(\mathscr{D}, T)$ on a closed manifold M^3 is geometrically taut if and only if it is topologically taut.

Let $\mathscr{D} = \ker \omega$ be a codimension-one distribution on M^3 defined by a 1-form ω, and T—a vector field transversal to $\mathscr{D}$. If $\mathrm{gv}(\omega, T) \neq 0$, then $(\mathscr{D}, \mathrm{span}\{T\})$ is not topologically taut. Of course, since geometric and topological tautenesses are equivalent, the statement follows from the Reinhart–Wood type formula of Section 4.5.6: if T is geodesic then the form $\eta \wedge d\eta$ representing the Godbillon-Vey class under consideration can be chosen to be identically zero.

4.5.2 Variations of (ω, T) and the Index Form

The Stokes' Theorem states that $\int_M d\beta = \int_{\partial M} \beta$, when β is a $(\dim M - 1)$-form on M. Thus, $\int_M d\beta = 0$, when M is closed; this is also true if M is open and β is supported in a relatively compact domain Ω. The Stokes' Theorem on a Riemannian manifold with the volume form $d\,\mathrm{vol}_g$ and $X = \beta^\sharp$ yields the Divergence Theorem (1.12).

Lemma 4.72. *If β is a $(\dim M - 1)$-form on $M \setminus \Sigma$ with Riemannian metric g obeying $\int_M \|\beta\|^2 d\,\mathrm{vol}_g < \infty$, then $\int_M d\beta = 0$.*

Proof. This follows from Lemma 2.23 applied to a vector field $X = \beta^\sharp$ satisfying $\beta = \iota_X d\,\mathrm{vol}_g$ and the equality $(\mathrm{div}\,X)\,d\,\mathrm{vol}_g = d\beta$. $\qquad\square$

For variable pairs (ω_t, T_t) denote by dot the t-derivative at $t = 0$ of any quantity on M. We will assume

$$\omega_t(T_t) = 1 \qquad\qquad (4.106)$$

in what follows; thus, variations satisfy

$$\dot{\omega}(T) + \omega(\dot{T}) = 0, \quad \omega(\ddot{T}) + 2\,\dot{\omega}(\dot{T}) + \ddot{\omega}(T) = 0.$$

Also, from (4.102) we get for $\eta_t = \iota_{T_t}\,d\omega_t$:

$$\dot{\eta} = \iota_T\,d\dot{\omega} + \iota_{\dot{T}}\,d\omega, \quad \ddot{\eta} = \iota_T\,d\ddot{\omega} + 2\iota_{\dot{T}}\,d\dot{\omega} + \iota_{\ddot{T}}\,d\omega.$$

Lemma 4.73. *Let (ω_t, T_t) $(|t| \leq \varepsilon)$ be a smooth family of 1-forms and vector fields on $M^3 \setminus \Sigma$ satisfying (4.106) and let $\mathscr{D}_t = \ker \omega_t$. Suppose that $g \in \mathrm{Riem}(M, \mathscr{D}, T)$ and*

$$\int_M \|\dot{\eta} \wedge \eta\|^2 \, d\,\mathrm{vol}_g < \infty. \tag{4.107}$$

Then

$$\dot{\mathrm{gv}}(\omega, T) = 2 \int_M \dot{\eta} \wedge d\eta. \tag{4.108}$$

Moreover, if the variation obeys

$$\int_M \|\ddot{\eta} \wedge \eta\|^2 \, d\,\mathrm{vol}_g < \infty, \tag{4.109}$$

then

$$\ddot{\mathrm{gv}}(\omega, T) = 2 \int_M (\ddot{\eta} \wedge d\eta + \dot{\eta} \wedge d\dot{\eta}). \tag{4.110}$$

Proof. We write $\eta_t = \eta + t\,\dot{\eta} + (t^2/2)\ddot{\eta} + O(t^3)$. By the above and equalities

$$d(\dot{\eta} \wedge \eta) = d\dot{\eta} \wedge \eta - \dot{\eta} \wedge d\eta, \quad d(\ddot{\eta} \wedge \eta) = d\ddot{\eta} \wedge \eta - \ddot{\eta} \wedge d\eta,$$

we have

$$\eta_t \wedge d\eta_t = \eta \wedge d\eta + t(2\,\dot{\eta} \wedge d\eta + d(\dot{\eta} \wedge \eta))$$
$$+ t^2(\ddot{\eta} \wedge d\eta + \dot{\eta} \wedge d\dot{\eta} + \frac{1}{2}d(\ddot{\eta} \wedge \eta)) + O(t^3). \tag{4.111}$$

From (4.111), using (4.107) and Lemma 4.72, we get (4.108). Next, (4.110) follows from the above and (4.109). □

Example 4.74. (i) Let T be the Reeb field of a contact distribution $\mathscr{D} = \ker \omega$ on $M \setminus \Sigma$. Then $\omega(T) = 1$ and $\eta := \iota_T \, d\omega$ vanishes, see definitions in Section 1.2.1, hence $\eta \wedge d\eta = 0$. (ii) Let T be a geodesic vector field on $M^3 \setminus \Sigma$ with a Riemannian metric g, and $\omega = T^\sharp$. Then $\eta = 0$, see (4.105). Thus $\eta \wedge d\eta = 0$.

In both cases, (i) and (ii), $d\eta = 0$ is valid, hence (ω, T) is critical for $\mathrm{gv}(\omega, T)$ in (4.103) for all variations (ω_t, T_t) obeying (4.107). In particular, two-dimensional transversely oriented Riemannian foliations of 3-manifolds are critical points for Godbillon-Vey integrals varying over all plane fields.

The next theorem presents Euler–Lagrange equations for (4.103) and integrable $\mathscr{D}$.

Theorem 4.75. *Let $\mathscr{D} = \ker \omega$ on $M^3 \setminus \Sigma$ be integrable and $g \in \mathrm{Riem}(M, \mathscr{D}, T)$. Then (ω, T) is critical for the functional (4.103) with respect to all variations obeying (4.106) and*

$$\int_M \|\dot{\eta} \wedge \eta - 2\omega \wedge \iota_T \, d\eta\|^2 \, d\,\mathrm{vol}_g < \infty, \tag{4.112}$$

if and only if

$$(\iota_T \, d)^3 \, \omega = 0. \tag{4.113}$$

Proof. Using (4.102), $d\omega = \omega \wedge \eta$ (integrability of $\mathscr{D}$) and $\dot{\eta} = \iota_{\dot{T}} d\omega + \iota_T d\dot{\omega}$, we obtain

$$\frac{1}{2}(\eta \wedge d\eta)^{\cdot} = (\iota_{\dot{T}} d\omega + \iota_T d\dot{\omega}) \wedge d\eta + d(\frac{1}{2}\dot{\eta} \wedge \eta)$$

$$= \iota_{\dot{T}}(\omega \wedge \eta) \wedge d\eta + \iota_T d\dot{\omega} \wedge d\eta + d(\frac{1}{2}\dot{\eta} \wedge \eta)$$

$$= \dot{\omega} \wedge (\eta \wedge \iota_T d\eta - d\iota_T d\eta) + d(\frac{1}{2}\dot{\eta} \wedge \eta - \dot{\omega} \wedge \iota_T d\eta). \qquad (4.114)$$

Here, we used also the identity $\dot{\omega}(T) + \omega(\dot{T}) = 0$ and the equalities

$$-\omega \wedge d\eta = d(d\omega) - d\omega \wedge \eta = -\omega \wedge \omega \wedge \eta = 0,$$

$$\iota_T(\eta \wedge d\eta) = -\eta \wedge \iota_T(d\eta), \quad \iota_{\dot{T}}(\omega \wedge \eta) = \omega(\dot{T})\eta - \eta(\dot{T})\omega.$$

From (4.114), using condition (4.112), Lemma 4.72 and Stokes' Theorem, we obtain

$$\frac{1}{2}\dot{\mathrm{gv}}(\omega, T) = \int_M \dot{\omega} \wedge (\eta \wedge (\iota_T d\eta) - d\iota_T d\eta) = \int_M \dot{\omega} \wedge \Omega, \qquad (4.115)$$

where

$$\Omega := \iota_T d\omega \wedge (\iota_T d)^2\omega - d(\iota_T d)^2\omega. \qquad (4.116)$$

For a critical pair (ω, T) for all variations $\dot{\omega}$, the above (4.115) yields point-wise equality

$$\Omega = 0. \qquad (4.117)$$

Note that (4.113) is equivalent to vanishing of the 1-form $\iota_T \Omega$. Next, we will show equivalence of (4.117) and (4.113). The implication (4.117) $\Rightarrow$ (4.113) is obvious.
    (4.113) $\Rightarrow$ (4.117). By our conditions, $d\omega = \omega \wedge \eta$ and $\omega \wedge d\eta = 0$, hence

$$d\eta = \omega \wedge \alpha \qquad (4.118)$$

for some 1-form α. We claim that $\Omega(X, Y) = 0$ for $X, Y \in \mathscr{D}$. Indeed, using $d\omega(X, Y) = 0$ we find

$$\Omega(X, Y) = (\eta \wedge \omega \wedge \alpha)(T, X, Y) + [d(\alpha(T)) \wedge \omega + \alpha(T) d\omega - d\alpha](X, Y)$$
$$= (\eta \wedge \alpha - d\alpha)(X, Y).$$

By (4.118), we get

$$0 = d^2\eta = d\omega \wedge \alpha - \omega \wedge d\alpha = \omega \wedge \eta \wedge \alpha - \omega \wedge d\alpha.$$

Therefore,

$$0 = (\omega \wedge \eta \wedge \alpha - \omega \wedge d\alpha)(T, X, Y) = (\eta \wedge \alpha)(X, Y) - d\alpha(X, Y) = \Omega(X, Y).$$

This proofs the claim. Thus, $\iota_T \Omega = 0$ yields $\Omega = 0$.   □

Remark 4.76.

(a) Conditions (4.107) and (4.112) (as well as similar conditions in what follows) are trivially satisfied for all variations compactly supported on $M \setminus \Sigma$, in particular, when $\Sigma = \emptyset$.

(b) When $\mathscr{D} = \ker \omega$ is tangent to a foliation, variations of the form (ω, T_t) do not change the functional, and only variations of type (ω_t, T) are essential.

(c) Since $\mathscr{L}_T = \iota_T d + d \iota_T$, we also have $(\mathscr{L}_T)^3 = (\iota_T d)^3 + (d \iota_T)^3$. Note that $(d \iota_T) \omega = 0$ when $\omega(T) = 1$. Thus, (4.113) are equivalent to

$$(\mathscr{L}_T)^3 \omega = 0.$$

In Lemma 4.73, we obtained a general formula (4.110) for the second variation of our Godbillon-Vey invariant. We will recalculate it at critical points of gv. The following bilinear form on $M \setminus \Sigma$ (which depends on T):

$$I_T(\alpha, \beta) = \int_M (\mathscr{L}_T)^2 d\alpha \wedge \beta,$$

is symmetric on the space of 1-forms α, β on $M \setminus \Sigma$ satisfying (for some metric g)

$$\int_M \|\gamma_1 - \gamma_2 + \gamma_3\|^2 d\mathrm{vol}_g < \infty,$$

where

$$\gamma_1 = \iota_T(\mathscr{L}_T d\alpha) \wedge \beta, \quad \gamma_2 = \iota_T d\alpha \wedge \iota_T d\beta, \quad \gamma_3 = \alpha \wedge \iota_T(\mathscr{L}_T d\beta).$$

This follows from Lemma 4.72 and the following calculation:

$$\begin{aligned}
\mathscr{L}_T(\mathscr{L}_T d\alpha) \wedge \beta &= d\gamma_1 - d\iota_T d\alpha \wedge \iota_T d\beta \\
&= d(\gamma_1 - \gamma_2) - d\alpha \wedge \iota_T(\mathscr{L}_T d\beta) \\
&= d(\gamma_1 - \gamma_2 + \gamma_3) + \mathscr{L}_T(\mathscr{L}_T d\beta) \wedge \alpha.
\end{aligned}$$

This I_T plays the role of the *index form* for our variational problem.

We have three independent cases for a pair (ω_t, T_t) obeying (4.106):

(i) $\dot{T}$ is parallel to T, thus $\dot{T} = \phi T$ and $\dot{\omega} = -\phi \omega$ for some $\phi : M \to \mathbb{R}$,

(ii) $\dot{T}$ is parallel to $\mathscr{D}$ and $\dot{\omega} = 0$, hence $\omega(\dot{T}) = 0$,

(iii) $\dot{T} = 0$ and $\dot{\omega}$ is a 1-form such that $\dot{\omega}(T) = 0$.

Proposition 4.77. *Let $g \in \mathrm{Riem}(M, \mathscr{D}, T)$ and $\mathscr{D} = \ker \omega$ be integrable on $M \setminus \Sigma$. For a critical pair (ω, T) for the functional (4.103) and all variations obeying (4.106) and*

$$\int_M \|\dot{\omega} \wedge (\mathscr{L}_T)^2 \dot{\omega} + \ddot{\omega} \wedge \mathscr{L}_T \eta\|^2 d\mathrm{vol}_g < \infty,$$

we have $\ddot{g}v = I_T(\dot{\omega}, \dot{\omega})$.

Proof. When $\mathscr{D}$ is integrable, variations (i)–(ii) do not change the functional, and only variations (iii) are essential. For a variation (ω_t, T), using $\ddot{\eta} = \iota_T d\ddot{\omega}$, $\ddot{T} = 0$ with $\ddot{\omega}(T) = 0$, and

$$(\ddot{\omega} \wedge d\iota_T d\eta)(T, N, B) = (\ddot{\omega} \wedge (\iota_T d)^2 \eta)(N, B) = 0,$$

we get $\ddot{\eta} \wedge d\eta = -d(\ddot{\omega} \wedge \iota_T d\eta)$. Thus, see the integrand in (4.110),

$$\dot{\eta} \wedge d\dot{\eta} = \iota_T d\dot{\omega} \wedge d(\iota_T d\dot{\omega}) = (d\iota_T)^2 d\dot{\omega} \wedge \dot{\omega} + d(\dot{\omega} \wedge (\iota_T d)^2 \dot{\omega}).$$

Next, we calculate

$$(\eta \wedge d\eta)^{\cdot\cdot} = (d\iota_T)^2 d\dot{\omega} \wedge \dot{\omega} + d(\dot{\omega} \wedge (\iota_T d)^2 \dot{\omega} + \ddot{\omega} \wedge \iota_T d\eta).$$

From the above and Lemma 4.72, the claim follows.   □

4.5.3 Integrability in Average

For "good" $\mathscr{F}$, if one thinks about the Godbillon-Vey number as "a function on the moduli space of foliations", then its critical points are foliations admitting a transversal projective structure, see [Asu, Mas]. It is rather difficult to find explicitly the derivative of the functional (4.103) for all variations among foliations. Therefore, we analyze here only sufficient conditions for being critical points (foliations) of (4.103) with respect to such variations.

Definition 4.78. The space $\Lambda^1_{\text{av}}(M^3)$ of 1-forms *integrable in average* is defined as the following extension of the space of 1-forms with integrable kernel distribution: $\omega \in \Lambda^1_{\text{av}}(M)$ if and only if

$$J_1(\omega) := \int_M \omega \wedge d\omega = 0. \tag{4.119}$$

Theorem 4.79. *Let* $\mathscr{D} = \ker \omega$ *on* $M^3 \setminus \Sigma$ *be integrable and* $g \in \text{Riem}(M, \mathscr{D}, T)$. *Then* (ω, T) *is critical for the functional* (4.103) *with respect to all variations obeying* (4.106), (4.119) *and inequalities*

$$\int_M \|\dot{\eta} \wedge \eta - 2\dot{\omega} \wedge \iota_T d\eta\|^2 d\text{vol}_g < \infty, \quad \int_M \|\omega \wedge \dot{\omega}\|^2 d\text{vol}_g < \infty, \tag{4.120}$$

if and only if the following holds for some $\lambda \in \mathbb{R}$:

$$(\mathscr{L}_T)^3 \omega = \lambda \mathscr{L}_T \omega. \tag{4.121}$$

Moreover, (4.121) yields (but is not equivalent to) the equality

$$\mathscr{L}_T(\eta \wedge d\eta) = 0, \tag{4.122}$$

telling that the 3-form representing gv(ω, T) *is invariant under the flow of* T.

Proof. The proof of Theorem 4.75 together with the formula $(\alpha, \beta)_{L^2} = \int_M \alpha \wedge \star\beta$ defining the inner product in the space of forms, shows that the form Ω of (4.116), or rather its Hodge star image $\star\Omega$, can be considered as the gradient of the functional gv at ω. Since $\mathscr{D}$ is tangent to a foliation, we may assume $\dot{T} = 0$. We have

$$(\omega \wedge d\omega)^{\cdot} = 2\dot{\omega} \wedge d\omega - d(\omega \wedge \dot{\omega}).$$

This shows that the 1-form $\star d\omega$ can be considered as the gradient (in the L^2-space $\Lambda^1(M^3)$) of J_1. By the Lagrange multipliers method, (and in view of (4.108), Lemma 4.72 with condition (4.120) and Stokes' Theorem), a point (ω, T) is critical for the functional gv with respect to variations obeying the conditions if and only if the gradient of gv coincides with the gradient of J_1 multiplied by a constant coefficient. We conclude that a foliation tangent to $\mathscr{D}$ is critical for (4.103) and all variations obeying (4.119) and (4.120) if and only if

$$\int_M \dot{\omega} \wedge (d\iota_T d\eta - \lambda d\omega) = 0$$

for some $\lambda \in \mathbb{R}$, i.e., $d\iota_T d\eta = \lambda d\omega$. Applying ι_T, we see that also the 1-forms $(\iota_T d)^2 \eta$ and η are parallel each to other, this is equivalent to (4.121).

Certainly, (4.121) yields

$$(\mathscr{L}_T)^2 \eta \wedge \eta = 0, \tag{4.123}$$

and (4.123) is equivalent to

$$\iota_T(\eta \wedge d(\iota_T d)\eta) = 0. \tag{4.124}$$

Then (using (4.118) and $2d\eta \wedge \iota_T d\eta = \iota_T(d\eta \wedge d\eta) = 0$) show that

$$\mathscr{L}_T(\eta \wedge d\eta) = \eta \wedge d(\iota_T d)\eta. \tag{4.125}$$

From the above, (4.122) follows:

$$0 \overset{(4.124)}{=} \iota_T(\eta \wedge d(\iota_T d)\eta)(X, Y)$$
$$= (\eta \wedge d(\iota_T d)\eta)(T, X, Y) \overset{(4.125)}{=} \mathscr{L}_T(\eta \wedge d\eta)(T, X, Y). \quad \square$$

Corollary 4.80. *Let $\mathscr{F}$ be a foliation of M^3. If (4.121) is valid for any (ω, T) such that $T\mathscr{F} = \ker \omega$ and $\omega(T) = 1$, then* gv$(\mathscr{F})$ *is infinitesimally rigid, i.e.,* $\dot{\text{gv}}(\mathscr{F}) = 0$ *for any infinitesimal deformation of $\mathscr{F}$.*

To write (4.113) and (4.122) in metric terms, in this section we define the *non-symmetric second fundamental form h* of $\mathscr{D}$, dual to the co-nullity tensor (1.26),

$$h_{X,Y} = \langle \nabla_X Y, T \rangle \qquad (X, Y \in \mathfrak{X}_{\mathscr{D}}),$$

and denote by $\sigma_1 = h_{N,N} + h_{B,B}$ its trace (the mean curvature of $\mathscr{D}$). Recall that a plane field $\mathscr{D}$ is *totally geodesic* if $\mathrm{Sym}(h) = 0$ (the symmetrization of h), *harmonic* if $\sigma_1 = 0$, and *totally umbilical* if $\mathrm{Sym}(h)$ coincides with $\frac{1}{2}\sigma_1 \cdot g_{|\mathscr{D}}$. In general non-self-adjoint (but self adjoint if $\mathscr{D}$ is integrable) *shape operator* $A : \mathscr{D} \to \mathscr{D}$ is given by $\langle AX, Y \rangle = h_{X,Y}$ for all $X, Y \in \mathfrak{X}_{\mathscr{D}}$.

Proposition 4.81. *Let* $\mathscr{D} = \ker \omega$ *on* $M^3 \setminus \Sigma$ *be integrable and* $g \in \mathrm{Riem}(M, \mathscr{D}, T)$. *Then* (ω, T) *is critical for (4.103) with respect to all variations obeying (4.106) and (4.112) if and only if the following is valid on* U:

$$T(T(k)) - 2T(k)h_{N,N} - k(T(h_{N,N}) - h_{N,N}^2 - h_{B,N}^2 + \tau^2) = 0,$$
$$\mathrm{div}(k^2(\tau - h_{B,N}) \cdot T) = 0. \qquad (4.126)$$

Proof. By (4.102), $(\iota_T d)^3 \omega = (\iota_T d)^2 \eta$ is valid. Using Frenet formulas (4.104), we obtain

$$(\iota_T d)^2 \eta(y) = (\iota_T d(\iota_T d\eta))(y) = T(d\eta(T, y)) - d\eta(T, [T, y])$$
$$= T(T(k \langle N, y \rangle)) - 2T(k \langle N, [T, y] \rangle) + k \langle N, [T, [T, y]] \rangle, \qquad (4.127)$$

where one may assume $y \in \mathscr{D}$. Here we used

$$d\eta(T, y) = T(k \langle N, y \rangle) - k \langle N, [T, y] \rangle,$$
$$d\eta(T, [T, y]) = T(k \langle N, [T, y] \rangle) - k \langle N, [T, [T, y]] \rangle.$$

By integrability of $\mathscr{D}$, $[N, B]^\perp = 0$. By the above (for $y = N$) we get $\langle N, [T, N] \rangle = h_{N,N}$ and

$$\langle N, [T, [T, N]] \rangle = T(h_{N,N}) + h_{N,N}^2 + h_{B,N}^2 - \tau^2;$$

thus (4.127) provides (4.126)$_1$. Similarly, for $y = B$, we have

$$\langle N, [T, B] \rangle = h_{B,N} - \tau,$$
$$\langle N, [T, [T, B]] \rangle = T(h_{B,N}) - T(\tau) + \sigma_1(h_{B,N} - \tau),$$

thus (4.127) provides

$$T(k^2(\tau - h_{B,N})) = \sigma_1 k^2(\tau - h_{B,N}),$$

which is equivalent to (4.126)$_2$. $\qquad \square$

Remark 4.82.

(i) By (4.140) in Section 4.5.6, $\eta \wedge d\eta = \alpha \, d\,\mathrm{vol}_g$, where $\alpha = -k^2(\tau - h_{B,N})$. By $(4.126)_2$, $T\alpha = \sigma_1 \alpha$. Using equality $\mathrm{div}(d\,\mathrm{vol}_g) = (\mathrm{div}\,T)\,d\,\mathrm{vol}_g$, we get

$$\mathscr{L}_T(\eta \wedge d\eta) = \mathscr{L}_T(\alpha \, d\,\mathrm{vol}_g) = T(\alpha)\,d\,\mathrm{vol}_g + \alpha(\mathrm{div}\,T)\,d\,\mathrm{vol}_g = \mathrm{div}(\alpha \cdot T)\,d\,\mathrm{vol}_g.$$

Thus, $(4.126)_2$ is equivalent to (4.122).

(ii) One may show that equality $\tau - h_{B,N} = 0$ means that the distribution of osculating planes $\{T,N\}$ is totally geodesic. Of course, from

$$\tau = \langle \nabla_T N, B \rangle, \quad h_{B,N} = \langle \nabla_N B, T \rangle = -\langle \nabla_N T, B \rangle$$

get the total geodesy condition $\langle \nabla_N T + \nabla_T N, B \rangle = 0$ for osculating planes.

(iii) Using Frenet formulas (4.104), we calculate

$$\nabla_T^\top (kA)(N) = \nabla_T^\top (kAN) - k\,\tau AB,$$
$$\nabla_T^\top \nabla_T^\top (kN) = (T(T(k)) - \tau^2 k)N + (T(k\tau) + \tau T(k))B,$$

and notice that

$$AN = h_{N,N} N + h_{N,B} B, \quad AB = h_{B,N} N + h_{B,B} B.$$

Therefore, the system (4.126) is equivalent to the equation

$$\nabla_T^\top \nabla_T^\top (kN) - \nabla_T^\top (kA)(N) + (kh_{N,N} - T(k))AN + k(h_{B,N} - 2\tau)AB = 0, \quad (4.128)$$

where $\nabla_T^\top X = (\nabla_T X)^\top$ for any vector field X. The N- and B- components of (4.128) are, respectively, $(4.126)_{1,2}$.

Corollary 4.83. *Let $\mathscr{D} = \ker \omega$ on $M^3 \setminus \Sigma$ be integrable and $g \in \mathrm{Riem}(M, \mathscr{D}, T)$. Then (ω, T) is critical for (4.103) with respect to all variations obeying (4.106) and conditions (4.119) and (4.120) if and only if the following is valid for some $\lambda \in \mathbb{R}$:*

$$\nabla_T^\top \nabla_T^\top (kN) - \nabla_T^\top (kA)(N) + (kh_{N,N} - T(k))AN + k(h_{B,N} - 2\tau)AB = \lambda kN.$$

4.5.4 Concordance and Homotopy

It is well known (e.g. [CC, vol. I, Section 3.6]) that the Godbillon-Vey class of foliations is invariant under the relation of concordance (in fact, cobordance).

The relation of concordance of foliations is stronger than concordance of distributions in the space of distributions: two non-concordant (among foliations) foliations seen as integrable plane fields $\mathscr{D}_i$, $i = 0, 1$, (or, corresponding to them pairs (ω_i, T_i)) may occur to be concordant in the space of all plane fields (in the sense of Definition 4.84 below). This situation can be compared with that of [ET, Theorem 2.4.1], where the authors prove that any confoliation (in particular, foliation) of any oriented 3-manifold M different from the product foliation of $S^2 \times S^1$ can be approximated by contact structures. Recall that two codimension-one foliations $\mathscr{F}_0$ and $\mathscr{F}_1$ of a manifold M are *concordant* when there exists a codimension-one foliation $\mathscr{F}$ of a 'cylinder' $M \times [0,1]$ which is transverse to the boundary $M \times \{0,1\}$ and induces $\mathscr{F}_i$ on $M \times \{i\}$, $i = 0,1$. If $\mathscr{F}$ is given by the equation $\omega = 0$ and $d\omega = \omega \wedge \eta$ on $M \times [0,1]$, then $\mathscr{F}_i$ is given by $\omega_i = 0$ and $d\omega_i = \omega_i \wedge \eta_i$, where $\omega_i = \phi_i^* \omega$, $\eta_i = \phi_i^* \eta$ and $\phi_i : M \to M \times [0,1]$ is given by $\phi_i(x) = (x,i)$, $i = 0,1$. Since the maps ϕ_0 and ϕ_1 are homotopic and $\eta_i \wedge d\eta_i = \phi_i^*(\eta \wedge d\eta)$, the cohomology classes of 3-forms $\eta_i \wedge d\eta_i$, $i = 0,1$, are equal.

Definition 4.84. We shall say that two pairs (ω_i, T_i), $i = 0,1$, consisting of 1-forms ω_i and vector fields T_i satisfying $\omega_i(T_i) = 1$ are *concordant* when there exists a pair (ω, T) consisting of a 1-form ω and a vector field T on $M \times [0,1]$ such that

$$\omega(T) = 1, \quad \omega_i = \phi_i^* \omega, \quad \phi_{i*}(T_i(x)) = T(\phi_i(x))$$

for all $x \in M$ and $i = 0,1$, where $\phi_i : M \to M \times [0,1]$ is given by $\phi_i(x) = (x,i)$.

If M ($\dim M = 3$) is closed and oriented, then it is parallelizable, so one can find triples (ω_j) and (T_j), $j = 1,2,3$, of 1-forms and vector fields satisfying $\omega_j(T_k) = \delta_{jk}$ for all j and $k \in \{1,2,3\}$. These fields and forms can be extended over $\tilde{M} = M \times [0,1]$ and completed by another vector field and another form, say $\partial/\partial t$ and dt, to get parallelizations of $T\tilde{M}$ and $T^*\tilde{M}$. Take on M any pair (ω, T) satisfying $\omega(T) = 1$ and write

$$\omega = \sum_i f_i \omega_i, \quad T = \sum_j h_j T_j.$$

Assume that ω and T are unit with respect to given parallelizations, that is that $\sum_i f_i^2 = \sum_j h_j^2 = 1$. The condition $\omega(T) = 1$ implies $f_i = h_i$ for all i's, that is such a pair is uniquely determined by a map $f = (f_1, f_2, f_3) : M \to S^2$. Since S^2 is contractible in S^3, f is homotopic to a constant map $f_0 : M \to S^3$. A homotopy between f and a constant map f_0, say, $f_0 = (0,0,0,1)$ everywhere on M, determines a pair $(\tilde{\omega}, \tilde{T})$ on $\tilde{M}$, which coincides with (ω, T) on, say, $M \times \{1\}$ and with $(\partial/\partial t, dt)$ on $M \times \{0\}$. Since any pair $(e^\phi \omega, e^{-\phi} T)$ is concordant to (ω, T) and the relation of concordance is transitive, we arrive at the following conclusion:

Any two pairs (ω, T), (ω', T') satisfying $\omega(T) = \omega'(T') = 1$ on a closed, oriented 3-manifold M are concordant in our sense.

Certainly, one can find such pairs with different Godbillon-Vey invariants. For example, on the unit tangent bundle $S\Sigma$ of a closed, oriented surface Σ of genus > 1, one has a foliation $\mathscr{F}$ (arising to a pair like that) with non-zero Godbillon-Vey class ([GV] or [CC, vol. I, Example 1.3.14]) and a contact structure (defined, for example, as the week-stable or week-unstable distribution of the geodesic flow on Σ equipped with a Riemannian metric of constant, negative curvature) arising to a pair (ω, ξ) which consists of a contact form ω and its Reeb field ξ and has zero as its Godbillon-Vey invariant (see Example 4.74). Finally, take into account the following, rather trivial, observation: for any f, the systems (ω, T) and $(e^f \omega, e^{-f} T)$ are homotopic (therefore, cobordant and concordant as well) but in general their gv classes are different. Therefore, unfortunately, *our Godbillon-Vey type invariant is not invariant under the concordance relation defined above.*

Remark 4.85. Example [Th1] of a family of smooth foliations $\{\mathscr{F}_t\}_{t>0}$ on the 3-sphere, for which $\mathrm{gv}(\mathscr{F}_t) = t$, is obtained from the weak stable foliation starting with a punctured surface and the leaves being weakly stable submanifolds of the geodesic flow. Therefore, if (T, N, B) is the Frenet frame of curves orthogonal to the leaves as in Section 4.5.6, then T corresponds to strongly unstable directions, while N and B can be determined from the Lie algebra description of $T^1(H^2)$. Due to [Pa, Section 1.3.3], one can define a contact 1-form α is whose characteristic (Reeb) flow T' coincides with the geodesic flow restricted on $T^1(S^3)$. Let $\mathscr{D}'$ be the distribution orthogonal (with respect to the Sasaki metric) to T'. Rotating T in the plane $\mathrm{span}(T', T)$, we obtain a deformation (homotopy) from Thurston's construction $(\mathscr{D}, T)$ to the contact structure $(\mathscr{D}', T')$. Consequently, $\mathrm{gv}(\mathscr{D}, T) \neq 0$ changes continuously to $\mathrm{gv}(\mathscr{D}', T') = 0$.

4.5.5 Critical Foliations

Here, we present several examples of critical foliations: Roussarie's foliation, Reeb foliations and twisted products.

The first example of a foliation with $\mathrm{gv}(\omega, T) \neq 0$ is due R. Roussarie. Suppose that $M = \Gamma \setminus G$, where $G = PSL(2, \mathbb{R})$ and Γ is a discrete, cocompact subgroup, see [CC, Vol. 1, Example 1.3.14]. The foliation of G by the left cosets of $H = \left\{ \begin{pmatrix} a & b \\ 0 & a^{-1} \end{pmatrix}, \ a > 0 \right\}$ projects to a foliation $\mathscr{F}$ of M. The Lie algebra of G can be identified with the vector space of 2×2 real matrices of trace 0, where the bracket operation is the commutator product $[A, B] = AB - BA$. One may take the basis of Lie algebra of G as

$$\left\{ X = \begin{pmatrix} 1 & 0 \\ 0 & -1 \end{pmatrix}, \ Y = \begin{pmatrix} 0 & 1 \\ 0 & 0 \end{pmatrix}, \ Z = \begin{pmatrix} 0 & 0 \\ 1 & 0 \end{pmatrix} \right\}.$$

The Lie brackets of the basic fields are given by $[X, Y] = 2Y$, $[X, Z] = -2Z$ and $[Y, Z] = X$. Then, $\{X, Y\}$ is a basis of the Lie subalgebra of H, it spans an integrable distribution $\mathscr{D}$.

Proposition 4.86. *The Roussarie's foliation is critical for* (4.103).

Proof. Set $\omega = Z^\flat$ and $T = Z$, then find

$$d\omega(X^\flat \wedge Y^\flat) = 0, \quad d\omega(X^\flat \wedge Z^\flat) = 2, \quad d\omega(Y^\flat \wedge Z^\flat) = 0.$$

It follows that $d\omega = 2X^\flat \wedge Z^\flat$ and

$$\eta = -2X^\flat, \quad d\eta = -2Y^\flat \wedge Z^\flat.$$

Hence $\eta \wedge d\eta = -4X^\flat \wedge Y^\flat \wedge Z^\flat$ is a nowhere-vanishing 3-form on G. Since M is closed, $\mathrm{gv}(\omega, T) \neq 0$. To verify (4.113), we use $dX^\flat = -Y^\flat \wedge Z^\flat$, to find

$$d\omega = \omega \wedge \eta = -2Z^\flat \wedge X^\flat, \quad \iota_T d\omega = -2\iota(Z^\flat \wedge X^\flat) = -2X^\flat,$$
$$d\iota_T d\omega = -2dX^\flat = 2Y^\flat \wedge Z^\flat, \quad (\iota_T d)^2 \omega = \iota_T(d\iota_T d\omega) = -2Y^\flat.$$

Note that $dY^\flat = 2X^\flat \wedge Y^\flat$ and

$$\iota_T dY^\flat(X) = 0, \quad \iota_T dY^\flat(Y) = 0, \quad \iota_T dY^\flat(Z) = 0,$$

hence $\iota_T dY^\flat = 0$. Finally,

$$d(\iota_T d)^2 \omega = -2dY^\flat, \quad (\iota_T d)^3 \omega = \iota_T(d(\iota_T d)^2 \omega) = -2\iota_T dY^\flat,$$

hence, (4.113) is valid.   □

The foliation of the Reeb component in the solid torus $D^2 \times S^1$, Figure 4.3 (see also Example 1.20), is given by the equation $\omega = 0$, where

$$\omega(r,t) = \cos \mu(r)\, dr + \sin \mu(r)\, dt, \quad \mu(r) = \arctan f'(r), \quad r \geq 0,$$

(r, θ) are polar coordinates in the disc D^2 and t is a parameter along S^1; the function $f(r)$ is convex, has vertical asymptote at $r = r_0 > 0$ and satisfies $f(0) = f'(0) = 0$. Gluing two foliated solid tori yields a Reeb foliation of a three-dimensional sphere.

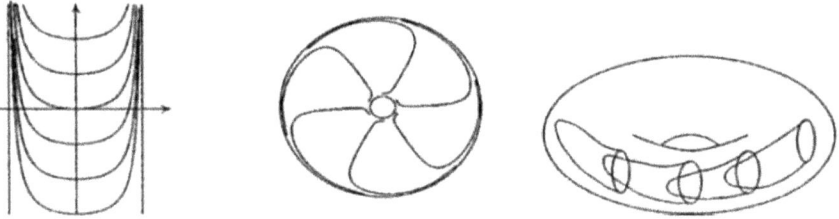

Fig. 4.3 Reeb foliation: strip, annulus and torus

Proposition 4.87. *The Reeb foliation of S^3 produced by a function $f = f(r)$ $(r \geq 0)$ is critical for* (4.103)

(i) in general if and only if f solves the following Cauchy's problem:

$$f''' = \frac{2((f')^2 - 1)}{(1 + (f')^2)f'}(f'')^2 + \frac{A_0(1 + (f')^2)^{5/2}}{(f')^3},$$
$$f(0) = 0, \quad f'(0) = A_1 \in \mathbb{R} \setminus 0, \quad f''(0) = A_2 \in \mathbb{R}. \tag{4.129}$$

(ii) *and all variations obeying* (4.119), *if and only if f solves the following Cauchy's problem with real parameter* λ :

$$f^{(4)} = \frac{(6(f')^2 - 7)}{f'(1 + (f')^2)} f''' f'' - \frac{2(3(f')^4 - 9(f')^2 + 2)}{(1 + (f')^2)^2 (f')^2} (f'')^3 + \lambda \frac{(1 + (f')^2)}{(f')^2} f'',$$

$$f(0) = 0, \quad f'(0) = A_1 \in \mathbb{R} \setminus 0, \quad f''(0) = A_2 \in \mathbb{R}, \quad f'''(0) = A_3 \in \mathbb{R}. \quad (4.130)$$

Proof. Set $T(r,t) = \cos \mu(r) \partial_t - \sin \mu(r) \partial_r$, then $\omega(T) \equiv 1$ in $M = D^2 \times S^1$. First we compute

$$d\omega = -\mu' \sin \mu (dr \wedge dt), \quad \iota_T(dr \wedge dt) = -\cos \mu \, dr - \sin \mu \, dt.$$

Then we observe that $gv(\omega, T) = 0$:

$$\eta = \iota_T d\omega = \mu' \sin \mu (\cos \mu \, dr + \sin \mu \, dt), \quad d\eta = (\mu' \sin^2 \mu)' dr \wedge dt,$$

therefore, $\eta \wedge d\eta = 0$. To verify (4.113), we then find

$$(\iota_T d)^2 \omega = -(\mu' \sin^2 \mu)' (\cos \mu \, dr + \sin \mu \, dt),$$

$$(\iota_T d)^3 \omega = ((\mu' \sin^2 \mu)' \sin \mu)' (\cos \mu \, dr + \sin \mu \, dt).$$

(i) Due to (4.113), a pair (ω, T) is critical for the action (4.103) if and only if $((\mu' \sin^2 \mu)' \sin \mu)' \equiv 0$ for $r \geq 0$, that is

$$(\mu' \sin^2 \mu)' \sin \mu = A_0 \quad (4.131)$$

for some $A_0 \in \mathbb{R}$. The ODE (4.131), using

$$\mu = \arctan f', \quad \mu' = f''/(1 + (f')^2),$$

can be rewritten in terms of Cauchy's problem (4.129), which has a unique solution. This way we get a family (which depends on $f'(0) = A_1 \neq 0$) of solutions of (4.129), see graphs (obtained by Maple program) on Figure 4.4 with the value r_0 depending on A_0. If $A_0 = 0$, then (4.131) reduces to

$$\mu' \sin^2 \mu = \tilde{A}_0 \quad (4.132)$$

for another constant $\tilde{A}_0 \in \mathbb{R}$. This ODE has the following integral:

$$2\mu - \sin(2\mu) = 4\tilde{A}_0 r + C.$$

Notice that $\mu \neq \text{const}$, hence $\tilde{A}_0 \neq 0$, because if $\mu = k = \text{const}$ then $f(r) = (\tan k)r$ has no asymptotes for $r > 0$ and does not produce critical foliation. For f, (4.132) provides the ODE,

$$f'' = \tilde{A}_0 \left(\frac{(1 + (f')^2)}{f'} \right)^2,$$

with similar to Figure 4.4 graphs of solutions with the value r_0 depending on $\tilde{A}_0$.

(ii) By the above, $(\iota_T d)^3 \omega$ is parallel to η, and (4.122) is valid if the ratio is constant,

$$\left(\left(\mu' \sin^2 \mu\right)' \sin \mu\right)' = \lambda \mu' \sin \mu \quad \text{for some } \lambda \in \mathbb{R}.$$

From this, with the little aid of Maple calculations, we yield (4.130). □

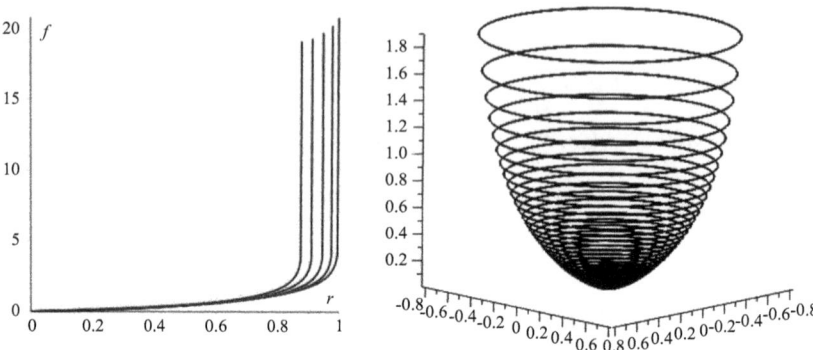

Fig. 4.4 Family of solutions $f(r)$ to (4.129) with $A_0 = 1, A_2 = 0$ and $A_1 = i/8$ $(i = 1 \ldots 5)$, producing singular Reeb foliations by rotation about f-axis

It is known [Ch] that

fibers of a twisted product $M = B \times_\phi F$ have parallel $\tilde{H}$ if and only if
$$\phi = \phi_1 \phi_2 \text{ with } \phi_1 \in C^2(B) \text{ and } \phi_2 \in C^2(F). \tag{4.133}$$

When $\dim B = 2$ and $\dim F = 1$, we have

$$h = 0, \quad A = 0, \quad \tilde{H} = kN,$$

and the system (4.126) on U reads as

$$T(T(k)) - \tau^2 k = 0, \qquad T(k^2 \tau) = 0, \tag{4.134}$$

with $\omega = T^\sharp$, $T = \partial_t$, t being the standard coordinate on $F \in \{S^1, \mathbb{R}\}$, and (4.128) reduces to

$$\nabla_T^\top \nabla_T^\top (kN) = 0. \tag{4.135}$$

Proposition 4.88. *Let (M, g) be a twisted product $B \times_\phi F$ with $\dim B = 2$ and $\dim F = 1$.*

(i) Suppose that $F = \mathbb{R}^1$. Then (ω, T) is critical for (4.103) if and only if

$$T(T(\|\nabla^\top \log \phi\|)) = c^2 / \|\nabla^\top \log \phi\|^3, \tag{4.136}$$

    where $c : M \to \mathbb{R}$ is constant along the T-curves.

(ii) *Suppose now that $F = S^1$. Then (ω, T) is critical for (4.103) if and only if ϕ obeys (4.136) with $c = 0$, that is ϕ is the product of functions $\phi_1 \in C^\infty(B^2)$ and $\phi_2 \in C^\infty(S^1)$.*

(iii) *Let $B \times_\phi F$ be a warped product. Then (ω, T) is critical for (4.103) and $\mathrm{gv}(\omega, T) = 0$.*

(iv) *Then (ω, T) is critical for (4.103) with respect to all variations obeying (4.119), in particular, if and only if*

$$\nabla_T^\top \nabla_T^\top (kN) = \lambda kN$$

is valid for some $\lambda \in \mathbb{R}$.

Proof. Notice that $k = \|\nabla^\top \log \phi\|$. Suppose that $k \neq 0$ (otherwise, (4.134) is satisfied trivially). By (4.134)$_2$, we get $T(\tau k^2) = 0$, hence $\tau = c/k^2$, where $T(c) = 0$, that is c is constant along the T-curves. Then by (4.134)$_1$ we get (4.136), which is integrable along T-curves, this completes the proof of (i). Since $T(T(k)) \geq 0$ when $k > 0$ and $T(T(k)) \leq 0$ when $k < 0$, the only periodic solutions of (4.134) are

$$T(k) = 0 = \tau, \tag{4.137}$$

that is the case of $F = S^1$. Notice that (4.137) is equivalent to the equation stronger than (4.135):

$$\nabla_T^\top (kN) = 0. \tag{4.138}$$

Applying (4.133) to (4.138) yields (ii), from which (iii) follows. Next, (iv) follows from the above. □

    A codimension-one foliation $\mathscr{F}$ (with the tangent distribution $\mathscr{D} = T\mathscr{F}$) is said to be *Riemannian* if T-curves are D^T-geodesics, that is if $D_T^T T = 0$.

    Example 4.74(ii) can be extended for Finsler metrics as follows.

Proposition 4.89. *Let $\mathscr{F}$ be a two-dimensional transversely oriented Riemannian foliation of M^3 with a Finsler metric F. Then a pair $(T\mathscr{F}, T)$, where T a unit F-normal to $\mathscr{F}$, is a critical point for Godbillon-Vey integral varying over all plane fields.*

Proof. Define a Riemannian metric $g := g_T$ on M, see Section 2.5.2, which is compatible with $(T\mathscr{F}, T)$ with the Levi-Civita connection ∇ and the Chern connection D^T, see (1.20). Observe that $k^T = F(D_T^T T)$ is the curvature of T-curves on (M, F). By (1.20), $D_T^T T = \nabla_T T$. Thus, T is a geodesic vector field for F if and only if it is geodesic for g. Hence, the claim follows from Example 4.74(ii) for metric g. □

4.5.6 Around the Reinhart–Wood Formula

We will show that the 2-form $d\eta$ attains the following values on $U = k^{-1}(\mathbb{R} \setminus \{0\})$:

$$d\eta(N, B) = -2\,\mathrm{div}(\mathscr{T}_{N,B}T), \quad d\eta(T, B) = k(\tau - h_{B,N}), \quad d\eta(T, N) = T(k) - kh_{N,N}, \tag{4.139}$$

where the *integrability tensor* $\mathscr{T}$ of $\mathscr{D}$ (vanishing when $\mathscr{D}$ is tangent to a foliation) is given by

$$2\mathscr{T}_{X,Y} = \langle [X,Y], T\rangle = h_{X,Y} - h_{Y,X}.$$

As far as $d\eta$ is concerned, see (4.139), one has

$$d\eta(N,B) = N(\eta(B)) - B(\eta(N)) - \eta([N,B])$$
$$= -B(\langle \nabla_T T, N\rangle) - \langle \nabla_T T, \nabla_N B - \nabla_B N\rangle = -B(k) - k\langle \nabla_N B, N\rangle.$$

Differentiating $\langle [N,B], T\rangle$ in the T-direction, after a lengthy calculation involving the use of symmetries of the curvature tensor R for the second order derivatives $\nabla_T \nabla_N B$ and $\nabla_T \nabla_B N$, yields

$$T(\langle [N,B], T\rangle) = B(k) + k\langle \nabla_N B, N\rangle + 2\sigma_1 \mathscr{T}_{N,B}.$$

Notice that $\operatorname{div} T = -\sigma_1$. From the above and

$$2(\nabla_T \mathscr{T})_{N,B} = T(\langle [N,B], T\rangle)$$

we deduce $(4.139)_1$:

$$B(k) + k\langle \nabla_N B, N\rangle = 2(\nabla_T \mathscr{T})_{N,B} - 2\sigma_1 \mathscr{T}_{N,B}.$$

Next,

$$d\eta(T,B) = T(\eta(B)) - B(\eta(T)) - \eta([T,B]) = k\langle [T,B], N\rangle,$$

from which $(4.139)_2$ follows. The proof of $(4.139)_3$ is also straightforward.

Using (4.139), Frenet formulas (4.104) and definition (4.105), we find

$$\eta \wedge d\eta(T,N,B) = \eta(N) d\eta(B,T) = -k^2(\tau - h_{B,N}).$$

Thus, applying the volume form $d\operatorname{vol}_g$ on (M,g), we get

$$\eta \wedge d\eta = -k^2(\tau - h_{B,N}) d\operatorname{vol}_g, \tag{4.140}$$

obtained for foliations in [RW] with opposite sign (due just to a convention).

When $\mathscr{D}$ is fixed and metric varies, only variations (i)–(ii) (for which T varies) are essential, and variations of type (iii) do not appear.

Lemma 4.90. *Suppose that $g \in \operatorname{Riem}(M, \mathscr{D}, T)$ and $\mathscr{D} = \ker \omega$.*
(i) If $T_t = T + \phi_t T$ for some $\phi_t \in C^1(M)$ ($|t| < \varepsilon$) and $\phi_0 \equiv 0$, then

$$(\eta_t \wedge d\eta_t)^{\boldsymbol{\cdot}} = 4T(\dot{\phi})\operatorname{div}(\mathscr{T}_{N,B} \cdot T) d\operatorname{vol}_g + d(\dot{\eta} \wedge \eta + 2\dot{\phi}\, d\eta)$$
$$= -4\dot{\phi}\operatorname{div}(\operatorname{div}(\mathscr{T}_{N,B} \cdot T) \cdot T) d\operatorname{vol}_g + 4\operatorname{div}(\dot{\phi}\operatorname{div}(\mathscr{T}_{N,B} \cdot T) \cdot T) d\operatorname{vol}_g$$
$$+ d((d\dot{\phi} - T(\dot{\phi})\,\omega) \wedge \eta + 2\dot{\phi}\, d\eta). \tag{4.141}$$

(ii) *If $T_t = T + X_t$, $X_t \in \mathfrak{X}_{\mathscr{D}}$ ($|t| < \varepsilon$) and $X_0 = 0$, then*

$$(\eta_t \wedge d\eta_t)^{\cdot} = 4 \langle k \operatorname{div}(\mathscr{T}_{N,B} \cdot T)N - (T(k) - k h_{N,N})(\mathscr{T}_{\cdot,B})^{\sharp}$$

$$+ k(\tau - h_{B,N})(\mathscr{T}_{\cdot,N})^{\sharp}, \dot{X} \rangle d\operatorname{vol}_g - d(\eta \wedge \iota_{\dot{X}} d\omega). \qquad (4.142)$$

Proof.

(i) We have $\dot{T} = \dot{\phi} T$ and $\dot{\omega} = -\dot{\phi} \omega$. The first equality of (4.141) is provided by equalities

$$(\eta_t \wedge d\eta_t)^{\cdot} = -2T(\dot{\phi}) \omega \wedge d\eta + d(\dot{\eta} \wedge \eta + 2\dot{\phi} d\eta),$$
$$(\omega \wedge d\eta)(T, N, B) = \omega(T) d\eta(N, B) = -2\operatorname{div}(\mathscr{T}_{N,B} \cdot T).$$

The second equality of (4.141) follows from the above, equality $\sigma_1 = -\operatorname{div} T$ and the following general formula applied to $Q = (\nabla_T \mathscr{T})_{N,B}$:

$$\operatorname{div}(\dot{\phi} Q T) = T(\dot{\phi})Q + \dot{\phi}\operatorname{div}(QT).$$

(ii) We have $\dot{T} = \dot{X}$ and $\dot{\omega} = 0$. Since

$$\eta_t \wedge d\eta_t = \eta \wedge d\eta - d(\eta_t \wedge \iota_{X_t} d\omega_t) + 2 d\eta_t \wedge (\iota_{X_t} d\omega) + (\iota_{X_t} d\omega) \wedge d(\iota_{X_t} d\omega),$$

we obtain

$$(\eta_t \wedge d\eta_t)^{\cdot} = 2 d\eta \wedge \iota_{\dot{X}} d\omega - d(\eta \wedge \iota_{\dot{X}} d\omega),$$

and (4.142) follows from the above and (4.139). $\qquad \qquad \square$

Let $g = g_0 \in \operatorname{Riem}(M, \mathscr{D}, T)$ and g_t ($|t| < \varepsilon$) be an arbitrary one-parameter family of metrics on $(M, \mathscr{D})$; hence, generically, $\dot{T} \neq 0$. The symmetric $(0,2)$-tensor $\dot{g}$ has six independent components $\dot{g}_{T,T}, \dot{g}_{T,N}, \dot{g}_{T,B}, \dot{g}_{N,N}, \dot{g}_{N,B}, \dot{g}_{B,B}$. A family g_t preserving a metric on $\mathscr{D}$ is called $g^{\pitchfork}$-*variation* for $(\mathscr{D}, \operatorname{Span}(T))$ (see definition in Section 4.3.1): its tensor $\dot{g}$ has three nonzero components $\dot{g}_{T,T}, \dot{g}_{T,N}$ and $\dot{g}_{T,B}$. Variations g_t, producing only nonzero components $\dot{g}_{N,N}, \dot{g}_{N,B}, \dot{g}_{B,B}$, preserve T and thus provide trivial Euler–Lagrange equations for the functional (which is constant when $\mathscr{D}$ is integrable), compare (4.140),

$$\operatorname{gv}_{\mathscr{D}} : g \to -\int_M k^2(\tau - h_{B,N}) d\operatorname{vol}_g. \qquad (4.143)$$

Theorem 4.91. *Let a vector field T be transverse to a plane field $\mathscr{D} = \ker \omega$ on $M^3 \setminus \Sigma$. Then $g \in \operatorname{Riem}(M, \mathscr{D}, T)$ is critical for (4.143) with respect to all variations g_t with $\dot{g}$ obeying*

$$\int_M \|4\dot{g}_{T,T}\,\mathrm{div}(\mathscr{T}_{N,B}T)\omega+(d(\dot{g}_{T,T})-T(\dot{g}_{T,T})\omega)\wedge\eta-\dot{g}_{T,T}\,d\eta\|^2\,d\mathrm{vol}_g<\infty,\qquad(4.144a)$$

$$\int_M \|k\dot{g}_{T,N}\,\omega\wedge\eta\|^2\,d\mathrm{vol}_g<\infty,\qquad\qquad\qquad\qquad\qquad\qquad(4.144b)$$

if and only the following system of equations hold on U:

$$\mathrm{div}(\mathrm{div}(\mathscr{T}_{N,B}\cdot T)\cdot T)=0,$$
$$\mathrm{div}(\mathscr{T}_{N,B}\cdot T)-(T(\log k)-h_{N,N})\mathscr{T}_{N,B}=0,$$
$$(\tau-h_{B,N})\,\mathscr{T}_{N,B}=0.\qquad\qquad(4.145)$$

For integrable $\mathscr{D}$, *equations* (4.145) *reduce to the expected trivial equalities.*

Proof. An arbitrary $g^\pitchfork$-variation for $(\mathscr{D},\mathrm{Span}(T)))$ of a Riemannian metric can be decomposed into two cases: 1) the metric varies along T only; 2) variations preserve the metric on $\mathscr{D}$ and T but disturb their orthogonality. Thus, nonzero components of corresponding $\dot{g}$ are divided into two sets: $\{\dot{g}_{T,T}\}$ and $\{\dot{g}_{T,N},\dot{g}_{T,B}\}$.

Case 1. Here, $T_t=e^{-\phi_t}T$ is the unit normal to $\mathscr{D}$ with respect to g_t for some smooth function ϕ_t with $\phi_0=0$. Differentiating $g_t(T_t,T_t)=1$ at $t=0$ we obtain $2\langle\dot{\phi}T,T\rangle+\dot{g}_{T,T}=0$. Hence,

$$\dot{\phi}=-\dot{g}_{T,T}/2.$$

By Lemma 4.90(i), we have

$$(\eta_t\wedge d\eta_t)^{\boldsymbol{\cdot}}=2\,\mathrm{div}(\mathrm{div}(\mathscr{T}_{N,B}\cdot T)\cdot T)\dot{g}_{T,T}\,d\mathrm{vol}_g-2\,\mathrm{div}\left(\dot{g}_{T,T}\,\mathrm{div}(\mathscr{T}_{N,B}\cdot T)\cdot T\right)d\mathrm{vol}_g$$
$$-\frac{1}{2}d\big((d(\dot{g}_{T,T})-T(\dot{g}_{T,T})\,\omega)\wedge\eta-\dot{g}_{T,T}\,d\eta\big).$$

By Stokes' theorem and (4.144a), the Euler–Lagrange equations are $(4.145)_1$.

Case 2. Now, $T_t=T+X_t$ is the unit normal to $\mathscr{D}$ with respect to g_t for some vector field $X_t\in\mathfrak{X}_\mathscr{D}$ with $X_0=0$. Differentiating $g_t(T+X_t,N)=0$ at $t=0$ we obtain $\langle\dot{X},N\rangle=-\dot{g}(T,N)=-\dot{g}_{T,N}$. Similarly, we get $\langle\dot{X},B\rangle=-\dot{g}_{T,B}$. Hence,

$$\dot{X}=-\dot{g}_{T,N}N-\dot{g}_{T,B}B.$$

By Lemma 4.90(ii) and using equalities

$$\mathscr{T}_{\dot{X},N}=\dot{g}_{T,B}\mathscr{T}_{N,B},\qquad\mathscr{T}_{\dot{X},B}=-\dot{g}_{T,N}\mathscr{T}_{N,B},\qquad i_{\dot{X}}\,d\omega=k\dot{g}_{T,N}\,\omega,$$

we find

$$
\begin{aligned}
(\eta_t \wedge d\eta_t)^{\cdot} &= 4\big\{ k\dot{g}_{X,N}\, \mathrm{div}(\mathscr{T}_{N,B}\cdot T) \\
&\quad - (T(k) - kh_{N,N})\mathscr{T}_{\dot{X},B} + k(\tau - h_{B,N})\mathscr{T}_{\dot{X},N}\big\} d\,\mathrm{vol}_g - d(\eta \wedge \iota_{\dot{X}}\, d\omega) \\
&= 4\big\{ ((T(k) - kh_{N,N})\mathscr{T}_{N,B} - k\,\mathrm{div}(\mathscr{T}_{N,B}\cdot T))\dot{g}_{T,N} \\
&\quad - k(\tau - h_{B,N})\,\mathscr{T}_{N,B}\,\dot{g}_{T,B}\big\} d\,\mathrm{vol}_g + d(k\dot{g}_{T,N}\,\omega \wedge \eta).
\end{aligned}
$$

Thus, Euler–Lagrange equations, provided by vanishing of $\dot{g}_{T,N}, \dot{g}_{T,B}$ components of

$$
\int_M \big\{ ((T(k) - kh_{N,N})\mathscr{T}_{N,B} - k\,\mathrm{div}(\mathscr{T}_{N,B}\cdot T))\dot{g}_{T,N} - k(\tau - h_{B,N})\,\mathscr{T}_{N,B}\,\dot{g}_{T,B}\big\} d\,\mathrm{vol}_g
$$

under conditions (4.144b), have the form of (4.145)$_{2,3}$.   □

Corollary 4.92. *Let $g \in \mathrm{Riem}(M, \mathscr{D}, T)$ and T be a geodesic vector field on $M \setminus \Sigma$. Then g is a critical point for the functional $\mathrm{gv}_{\mathscr{D}}$ for all variations obeying (4.144a).*

Proof. Since $k = 0$, then $U = \emptyset$ and (4.145) and (4.144b) become trivial under condition (4.144a).   □

4.5.7 The Bott Invariant

Let Y be a nonzero on $M^3 \setminus \Sigma$ vector field. The flow of Y is *transversely holomorphic* if there is a complex structure J on the 2-plane bundle $TM/\langle Y \rangle$ invariant under this flow. Assume that $TM/\langle Y \rangle$ is trivial and there exist pointwise linearly independent 1-forms ω_1, ω_2 on $M \setminus \Sigma$, whose common kernel $\ker \omega_1 \cap \ker \omega_2$ is spanned by Y, and such real 2-form $\omega_1 \wedge \omega_2$ defines the transverse orientation. Consider the complex-valued 1-form $\omega_c = \omega_1 + i\omega_2$. The following conditions are equivalent, see [GP]:

(C1) The complex-valued 1-form ω_c is *formally integrable*, i.e.,

$$
\omega_c \wedge d\omega_c = 0 \quad \Longleftrightarrow \quad
\begin{cases}
\omega_1 \wedge d\omega_1 = \omega_2 \wedge d\omega_2, \\
\omega_1 \wedge d\omega_2 = -\omega_2 \wedge d\omega_1.
\end{cases}
$$

(C2) The complex-valued 1-form ω_c defines a *transverse holomorphic structure* for the flow of Y, i.e., there is a function $h_c : M \to \mathbb{C}$ such that $\mathscr{L}_Y \omega_c = h_c\, \omega_c$. If (C1) is valid, then there is a complex-valued 1-form $\eta_c = \eta_1 + i\eta_2$ such that

$$
d\omega_c = \omega_c \wedge \eta_c. \tag{4.146}
$$

Definition 4.93. Then the complex number

$$
\mathrm{gv}(\omega_c, T_c) := \int_M \eta_c \wedge d\eta_c,
$$

is called the *Bott invariant* of the flow of Y for formally integrable ω_c.

Remark 4.94. One may show that $\mathrm{gv}(\omega_c, T_c)$ is independent of choices. Of course, the difference of $\eta_c \wedge d\eta_c$ for two choices of ω_c has the form $d\alpha$ for some 3-form α, hence the integral of this difference vanishes.

If (C2) is valid, then the flow of Y pulls back ω_c to a complex multiple of itself and preserves the complex structure on $TM/\langle Y \rangle$ defined by the dual basis to (ω_1, ω_2). Consider the vector field $T_c = T_1 + iT_2$ and assume $\omega_j(iT_k) = i\omega_j(T_k)$. Then

$$\iota_{T_c}\omega_c = 1 \quad \Longleftrightarrow \quad \begin{cases} \iota_{T_1}\omega_1 + \iota_{T_2}\omega_2 = 1, \\ \iota_{T_1}\omega_2 + \iota_{T_2}\omega_1 = 0. \end{cases} \tag{4.147}$$

Proposition 4.95. *Let ω_c be formally integrable and a complex-valued vector field $T_c = T_1 + iT_2$ obey (4.147). Then the 1-form η_c, see (4.146), can be chosen by*

$$\eta_c = \iota_{T_c}d\omega_c. \tag{4.148}$$

Proof. We have
$$0 = \iota_{T_c}(d\omega_c \wedge \omega_c) = \iota_{T_c}d\omega_c \wedge \omega_c + d\omega_c.$$
This and (4.148) yield (4.146). $\square$

The Bott invariant (of the flow above) is non-trivial and admits continuous variations. However, the Godbillon-Vey invariant is rigid under both actual and infinitesimal deformations in the category of transversely holomorphic foliations, see [Asu]. Some of the above results (e.g., Theorem 4.75) with ω, T, η replaced by complex-valued forms and vector fields ω_c, T_c, η_c remain valid.

4.5.8 Higher Dimensional Cases

The Godbillon-Vey class can be defined also for foliations of arbitrary codimension q: just, one has to replace the form $\eta \wedge d\eta$ by $\eta \wedge (d\eta)^q$, where $d\omega = \omega \wedge \eta$ and ω is a q-form defining the foliation, and it becomes the simplest one among so called *exotic* classes of foliations (see, for example, [CC, Vol. II, Part 2]).

Assume that $\dim M = 2n + 1 \geq 5$. Let ω, T and $\eta = \iota_T d\omega = kN^\flat$ be as above. Then the following *Godbillon-Vey type invariants* are well-defined:

$$\mathrm{gv}_s(\omega, T) = \int_M \eta \wedge (d\eta)^s \wedge (d\omega)^{n-s}, \qquad 0 \leq s \leq n,$$

see [FH, RW9]. If $\mathscr{D}$ is integrable then, then since $d\omega = \omega \wedge \eta$ and $d\eta = \omega \wedge \alpha$, we have $\mathrm{gv}_s(\omega, T) = 0$ for all $s \geq 1$. Let $\{N, Z_0 = B, Z_1, \dots, Z_{2n-2}\}$ be a local orthonormal basis of $\mathscr{D}$, and as before, h its second fundamental form, and k, τ the first and the second (among $2n$) curvatures of T-curves.

Let $\mathscr{T}^Z$ be the integrability tensor of the distribution $\mathscr{D}_Z$ orthogonal to $\{T,N\}$. Denote by S^i_{2n-2} $(0 \le i \le 2n-1)$ the set of all permutations $\boldsymbol{j} = \{j_1, j_2 \ldots, j_{2n-2}\}$ of $2n-2$ elements $\{0,1,\ldots,2n-2\} \setminus \{i\}$. Notice that $gv_0(\omega,T) = 0$ and

$$gv_s(\omega,T) = (-2)^{n-1} \int_M \{k^{s+1} \sum_{i=0}^{2n-2} [(h_{Z_i,N} - \tau\delta_{i0}) \times$$

$$\times \sum_{\boldsymbol{j} \in S^i_{2n-2}} \underbrace{\langle \mathscr{T}^Z_{j_1,j_2}, N\rangle \ldots \langle \mathscr{T}^Z_{j_{2s-1},j_{2s}}, N\rangle}_{s} \underbrace{\mathscr{T}_{j_{2s+1},j_{2s+2}} \cdots \mathscr{T}_{j_{2n-3},j_{2n-2}}}_{n-s}] \} d\,vol_g$$

for $s \ge 1$, see [RW9]. Now, let $\mathscr{D}_Z$ be integrable. Since

$$\eta(T) = 0, \quad \eta(N) = k, \quad d\omega(T,Z_i) = 0,$$

then $gv_0(\omega,T) = 0$. Next,

$$d\omega(Z_i,Z_j) = \langle T, [Z_i,Z_j]\rangle, \quad d\eta(Z_i,Z_j) = \langle N, [Z_i,Z_j]\rangle.$$

Thus, if $\mathscr{D}_Z$ is integrable then $gv_s(\omega,T) = 0$ for all $s \ge 1$.

Lemma 4.96. *Let* (ω_t,T_t) $(|t| \le \varepsilon)$ *be a smooth family of 1-forms and vector fields on* $M^{2n+1} \setminus \Sigma$ *satisfying* (4.106), *and let* $\mathscr{D}_t = \ker\omega_t$. *Suppose that* $g \in$ $\mathrm{Riem}(M,\mathscr{D},T)$ *and*

$$\int_M \|\dot{\eta} \wedge \eta \wedge (d\eta)^{s-1} \wedge (d\omega)^{n-s}\|^2 d\,vol_g < \infty.$$

Then

$$\dot{gv}_s(\omega,T) = \int_M ((s+1)\dot{\eta} \wedge d\omega + (n-s)d\dot{\omega} \wedge \eta) \wedge (d\eta)^s \wedge (d\omega)^{n-s-1}.$$

Proof. This is similar to the proof of Lemma 4.73. $\qquad\square$

Proposition 4.97. *A pair* (ω,T) *is critical for* gv_s $(s > 0)$ *in the following cases, see also Example 4.74:*

(i) $\mathscr{D} = \ker\omega$ *is a contact distribution on* M^{2n+1} *with the Reeb field* T,
(ii) T *is a geodesic vector field on* M^{2n+1} *such that* $\omega = T^\sharp$.

Proof.

(i) Let $\mathscr{D} = \ker\omega$ be a contact distribution on M^{2n+1} with the Reeb field T. Then $\omega(T) = 1$ and $\eta := \iota_T d\omega$ vanishes, see [Bl], hence $gv_s(\omega,T) = 0$.
(ii) Let M^{2n+1} be a Riemannian manifold with metric g and a geodesic vector field T such that $\omega = T^\sharp$. Then $\eta := \iota_T d\omega$ vanishes, hence $gv_s(\omega,T) = 0$.
In both cases, (i) and (ii), $d\eta = 0$ is valid, hence (ω,T) is critical for gv_s $(s>0)$ for all variations obeying (4.106).

Remark 4.98. The topic of Section 4.5 has been developed in [RW10, W7] for a $(2q+1)$-dimensional manifold M equipped with a $(q+1)$-dimensional, *a priori* non-integrable, distribution $\mathscr{D}$ and a q-vector field $\mathbf{T} = T_1 \wedge \ldots \wedge T_q$, where $\{T_i\}$ are linearly independent vector fields transverse to $\mathscr{D}$.

Chapter 5
Extrinsic Geometric Flows

By *geometric flow* we mean the evolution of a geometric structure on a manifold under a differential equation, usually associated with curvature. These correspond to dynamical systems in the infinite-dimensional space of all appropriate geometric structures on a given manifold. In the study of flows there are several important problems to consider, e.g. the limit sets and the stationary/fixed points. The heat flow, Ricci flow and the Mean Curvature flow are popular geometric flows in mathematics, see [AH, ES]. There exists just a few works considering flows of metrics on a foliated manifold, see [BHV, LMR, SWZ], which are called *transverse* Ricci, Sasaki-Ricci, etc., because the metric varies along normal to the leaves distribution, and $B(g)$ is elliptic along the same distribution. Results on geometric flows transverse to a Riemannian foliation [BHV] motivate the study of other evolution equations. In [R4, RW2, RWo1], the *extrinsic geometric flows of metrics on a foliation* are studied (i.e., the evolution depends on the second fundamental tensor of the foliation), the metric varies along normal to the leaves distribution, while the evolution operator is elliptic along the leaves.

5.1 Prescribing the Mean Curvature Vector

There is a definite interest in prescribing geometric quantities for objects and for foliations using geometrical flows [Au, CW, R6, RZe1, RZe3, SW] and Sections 3.3–3.4: *Given a foliation $\mathscr{F}$ and a geometric quantity Q (function or tensor) one may search for a metric g on M for which a given geometric invariant of $\mathscr{F}$, say, mean curvature, coincides with Q.* To find how to achieve the expected foliation property, we examine variations of metrics that preserve normal distribution (for a given foliation), define extrinsic geometric flows of metrics on foliations and supplement the above problem as follows: *Find a flow on a foliation $\mathscr{F}$ such that the solution metrics g_t $(t \geq 0)$ converge, as $t \to \infty$, to a metric, for which $\mathscr{F}$ enjoys a given extrinsic geometric property.* Here, we study extrinsic geometric flows on a codimension-one

© Springer Nature Switzerland AG 2021   223
V. Rovenski, P. Walczak, *Extrinsic Geometry of Foliations*, Progress
in Mathematics 339, https://doi.org/10.1007/978-3-030-70067-6_5

foliation, which yield the second order parabolic PDE's, for which the existence and uniqueness of a solution is shown, see [RW2, R4]. Such flow of metrics extends to the case of foliations of any codimension, when the flow rate is proportional to the divergence of the mean curvature of the leaves. The flow is reduced to the heat flow for one-forms, so it admits a unique solution converging to some metric, see [RWo1].

Let $\mathcal{M}$ be the space of Riemannian metrics of finite volume on M^{n+p} such that complementary distributions $\mathscr{D}^p$ and $\widetilde{\mathscr{D}}^n$ are orthogonal. Elements of $\mathcal{M}$ are called *adapted metrics*. Let $\mathcal{M}_1 \subset \mathcal{M}$ be the subspace of metrics of unit volume, and

$$\pi : \mathcal{M} \to \mathcal{M}_1, \quad \pi(g) = \hat{g} = \mathrm{Vol}(M,g)^{-2/n} g^{\top} \oplus g^{\perp},$$

the $\widetilde{\mathscr{D}}$-conformal projection. Assume that $\mathscr{D}$ is tangent to a foliation $\mathscr{F}$ with compact and orientable leaves, and the mean curvature vector field H. Here, we consider (normalized) flows of metrics g_t, $t \in [0, \varepsilon)$, of the form, respectively,

$$\partial_t g_t = -\frac{2}{n}(\mathrm{div}^{\perp} H_t) g_t^{\top}, \tag{5.1a}$$

$$\partial_t g_t = -\left(\frac{2}{n}\mathrm{div}^{\perp} H_t + \rho(t)\right) g_t^{\top}$$

$$\text{with } \rho(t) = -\frac{2}{n\,\mathrm{Vol}(M,g_t)} \int_M (\mathrm{div}^{\perp} H_t) \mathrm{d\,vol}_t. \tag{5.1b}$$

By Proposition 5.7 below, (5.1a,b) keep harmonicity and total geodesy of $\mathscr{F}$.

If $g_0 \in \mathcal{M}_1$, then metrics g_t ($t \geq 0$) of (5.1b) belong to $\mathcal{M}_1$, because, see (5.4),

$$\frac{d}{dt}\mathrm{Vol}(M,g_t) = \frac{n}{2}\rho(t)\mathrm{Vol}(M,g_t) - \frac{n}{2}\int_M \rho(t)\,\mathrm{d\,vol}_t = 0.$$

Using (1.33) with $X = H$, we get

$$\int_M (\mathrm{div}^{\perp} H)\,\mathrm{d\,vol} = \int_M \langle H, H \rangle \,\mathrm{d\,vol} \geq 0. \tag{5.2}$$

Substituting (5.2) in the definition (5.1b) of $\rho(t)$, we have

$$\rho(t) = -\frac{2}{n}\int_M |H_t|^2 \mathrm{d\,vol}_t / \mathrm{Vol}(M,g_t) \leq 0.$$

We will show the existence/uniqueness and convergence of a global solution g_t ($t \geq 0$). For this we will verify that

(i) the 1-form $H^{\flat}$ (dual to H) satisfies the heat equation along $\mathscr{D}$;
(ii) for appropriate g_0 the metrics g_t converge to a metric g_{∞} with harmonic $\widetilde{\mathscr{D}}$, and under some topological assumptions prescribe the mean curvature vector H.

5.1.1 $\widetilde{\mathscr{D}}$- and $\mathscr{D}$-Related Geometric Quantities

The notion of the $\widetilde{\mathscr{D}}$-truncated (r,k)-tensor field $B^\top$, where $r = 0, 1$, and $^\top$ is the $\widetilde{\mathscr{D}}$-component (and similarly for $\mathscr{D}$), will be helpful,

$$B^\top(X_1, \ldots, X_k) = B(X_1^\top, \ldots, X_k^\top) \quad (X_i \in TM).$$

Thus, $g^\top$ is the $\widetilde{\mathscr{D}}$-truncated metric tensor g on M. Let g_t $(0 \le t < \varepsilon)$ be a family of adapted metrics which are t-independent on $\mathscr{D}$, then the tensor $B_t = \partial_t g_t$ is $\widetilde{\mathscr{D}}$-truncated. If such metrics g_t are conformally equivalent on $\widetilde{\mathscr{D}}$, then

$$\partial_t g_t = s_t\, g_t^\top, \tag{5.3}$$

where $s_t \in C^k(M)$. Recall that $B^\sharp$ is the $(1,1)$-tensor on M, which is g-dual to a symmetric $(0,2)$-tensor B, i.e., $B(X,Y) = \langle B^\sharp(X), Y \rangle$ for all vectors X, Y. Recall that the volume form of g_t evolves by (4.15).

By (4.15) with $B = s\,g^\perp$, for (5.3) the volume form of g_t evolves by

$$\frac{d}{dt}\,\mathrm{vol}_t = (n/2)\,s_t\,\mathrm{vol}_t, \tag{5.4}$$

and the metrics $\hat{g}_t = \phi_t g_t^\top \oplus g_t^\perp$ with dilation $\phi_t = \mathrm{Vol}(M, g_t)^{-2/n}$ belong to $\mathscr{M}_1$.

We will use the following result on $\widetilde{\mathscr{D}}$-conformal metrics.

Proposition 5.1 (See Appendix A in [Br]). *Let* $\pi : M \to B$ *be a fiber bundle with compact fibers of a closed Riemannian manifold* (M, g_0), *and* $g_t \in \mathscr{M}$ *obey* (5.3). *Let the functions* $u_m(t) = \sup_M |(\nabla^{\perp,t})^m s_t|_{g(t)}$ *satisfy* $\int_0^\infty u_m(t)\,dt < \infty$ *for all* $0 \le m \le k$. *Then, as* $t \to \infty$, *the metrics* g_t *converge in* C^k *to a smooth limit metric* g_∞.

Proof. Our assumptions ensure that g_t converge in C^k to a symmetric $(0,2)$-tensor g_∞. The metrics are uniformly equivalent: $c^{-1}g_0^\top \le g_t^\top \le c g_0^\top$ for some $c > 0$ and all $t \ge 0$. Hence, g_∞ is positive definite. $\qquad\square$

Observe that our flow is equivalent to the heat flow of the 1-form dual to H, provided the initial 1-form is $\mathscr{D}$-closed, see also [MR].

Theorem 5.2 (The Heat Flow on 1-Forms). *Let* θ *be a 1-form on a closed oriented manifold* M *of class* C^k. *Then the heat flow (with the Hodge Laplacian* Δ_d*)*

$$\partial_t\,\theta_t = \Delta_d\,\theta_t, \qquad \theta_0 = \theta.$$

admits a unique solution θ_t *for all* $t \ge 0$. *As* $t \to \infty$, θ_t *converges exponentially in* C^k *towards a harmonic 1-form* θ_∞, *i.e.,* $\|\theta_t - \theta_\infty\| \le c \cdot e^{-\lambda t}$ *for some positive* c, λ *(*λ *is independent of* θ*). If the form* θ *is closed, then all* θ_t *are closed as well.*

Next, we develop variational formulae for $\widetilde{\mathscr{D}}$-conformal changes of metric. Let $\{E_a, \mathscr{E}_i\}_{1 \le a \le n,\, 1 \le i \le p}$ be a local g_0-orthonormal frame on TM adapted to $\widetilde{\mathscr{D}}$ and $\mathscr{D}$.

Lemma 5.3. *Let* $g_t \in \mathcal{M}$ *satisfy* (5.3) *and a local* $\widetilde{\mathcal{D}}$-*frame* $\{E_a\}$ *evolve according to*

$$\partial_t E_a(t) = -(s_t/2) E_a(t).$$

Then $\{E_a(t)\}$ *is a* g_t-*orthonormal frame of* $\widetilde{\mathcal{D}}$ *for all* t.

Proof. This is the particular case $B = s\, g^\top$ of Lemma 4.6.   □

Remark 5.4. We have the following:

(i) $\Pi_t := \partial_t \nabla^t$, see (4.8), is a symmetric $(1,2)$-tensor field on (M, g_t);
(ii) if the vector fields $X = X(t)$, $Y = Y(t)$ are t-dependent, then

$$\partial_t \nabla^t_X Y = \Pi_t(X,Y) + \nabla^t_X(\partial_t Y) + \nabla^t_{\partial_t X} Y. \tag{5.5}$$

Let $\nabla^\perp \phi$ be the $\mathcal{D}$-component of the gradient of a function $\phi \in C^1(M)$.

The second fundamental forms of $\widetilde{\mathcal{D}}$ with respect to metrics g and $\hat{g} = e^{2\phi} g^\top \oplus g^\perp$ are related by the following lemma.

Lemma 5.5. *Let* $(M, g = g^\top \oplus g^\perp)$ *be a Riemannian manifold with complementary orthogonal distributions* $\widetilde{\mathcal{D}}$ *and* $\mathcal{D}$. *Given* $\phi \in C^1(M)$, *define adapted metric* $\hat{g} = e^{2\phi} g^\top \oplus g^\perp$. *Then the second fundamental forms and the mean curvature vectors of* $\widetilde{\mathcal{D}}$ *with respect to* $\hat{g}$ *and* g *are related by*

$$\hat{h} = e^{2\phi}\left(h - (\nabla^\perp \phi) g^\top\right), \qquad \hat{H} = H - n\nabla^\perp \phi. \tag{5.6}$$

Proof. By (1.9), for any $X, Y \in \widetilde{\mathcal{D}}$ and $\xi \in \mathcal{D}$ we have

$$\langle \hat{\nabla}_X Y, \xi \rangle = e^{2\phi} \langle \nabla_X Y, \xi \rangle - e^{2\phi} \langle X,Y \rangle\, \xi(\phi) - \frac{1}{2}(e^{2\phi} - 1)\langle [X,Y], \xi \rangle.$$

From this and (1.27), formula $(5.6)_1$ follows. Since $H = \mathrm{trace}_g\, h$, we get $(5.6)_2$.   □

The following lemma completes (for the case of flows) Lemma 5.5.

Lemma 5.6. *Let* $g_t \in \mathcal{M}$ *obey* (5.3). *Then*

$$\partial_t h(X,Y) = s\, h(X,Y) - \frac{1}{2}\langle X,Y \rangle \nabla^\perp s + \frac{1}{2}(\nabla_X \partial_t Y + \nabla_{\partial_t X} Y + \nabla_Y \partial_t X + \nabla_{\partial_t Y} X)^\perp, \tag{5.7}$$

$$\partial_t H = -(n/2)\nabla^\perp s, \quad \partial_t(\mathrm{div}^\perp H) = -(n/2)\Delta^\perp s, \quad \partial_t H^\flat = (\partial_t H)^\flat, \tag{5.8}$$

$$\partial_t \tilde{h} = -s\tilde{h}, \qquad \partial_t \tilde{H} = -s\tilde{H}. \tag{5.9}$$

Proof. Let $B = \partial_t g_t$ be $\widetilde{\mathcal{D}}$-truncated. By (4.8), (1.27), equality $B(\cdot, X) = 0$ $(X \in \mathcal{D})$ and symmetry of B, we have

$$g_t(\partial_t h(X,Y), \xi) = g_t\big(\partial_t(\nabla^t_X Y) + \partial_t(\nabla^t_Y X),\, \xi\big)/2$$
$$= \big[(\nabla^t_X B)(Y,\xi) + (\nabla^t_Y B)(X,\xi) - (\nabla^t_\xi B)(X,Y)\big]/2 + Q$$

for all $\xi \in \mathcal{D}$ and t-dependent $X, Y \in \mathcal{D}$. Here, due to (5.5),

$$Q := g_t(\nabla^t_X \partial_t Y + \nabla^t_{\partial_t X} Y + \nabla^t_Y \partial_t X + \nabla^t_{\partial_t Y} X, \xi)/2.$$

Substituting $B = s g^\top$, we obtain the required (5.7):

$$g(\partial_t h(X,Y), \xi) = s g(h(X,Y), \xi) - \frac{1}{2} g^\top(X,Y) \xi(s) + Q.$$

By the above we have $(5.8)_1$:

$$\partial_t H = \sum_a \partial_t h(E_a(t), E_a(t)) - \sum_a s h(E_a(t), E_a(t)) = -(n/2) \nabla^\perp s.$$

To show $(5.8)_2$, let $B = \partial_t g_t$ be $\widetilde{\mathscr{D}}$-truncated (i.e., $B(\mathscr{D}, \cdot) = 0$). By (4.8) we get

$$\partial_t(\operatorname{div}^\perp H) = \sum_i \left[(\partial_t g)(\nabla_i H, \mathscr{E}_i) + g_t(\partial_t(\nabla_i H), \mathscr{E}_i) \right] = \operatorname{div}^\perp(\partial_t H).$$

Here, we used $B(\mathscr{E}_i, \cdot) = 0$ and

$$(\nabla_H B)(\mathscr{E}_i, \mathscr{E}_i) = H(B(\mathscr{E}_i, \mathscr{E}_i)) - 2B(\nabla_H \mathscr{E}_i, \mathscr{E}_i) = 0.$$

Now, assuming $B = s g^\top$, and using $(5.8)_1$, we obtain $(5.8)_2$:

$$\partial_t(\operatorname{div}^\perp H) = \operatorname{div}^\perp(\partial_t H) = -(n/2) \operatorname{div}^\perp(\nabla^\perp s) = -(n/2) \Delta^\perp s.$$

To show $(5.8)_3$, we use $(5.8)_1$ to calculate for any $\xi \in \mathscr{D}$:

$$\partial_t H^\flat(\xi) = \partial_t(g_t(H_t, \xi)) = -(n/2) g_t(\nabla^\perp s, \xi),$$

that proves $(5.8)_3$. We shall show for the more general setting $B = \partial_t g_t$ that

$$\partial_t \tilde{h} = -B^\sharp \circ \tilde{h}, \qquad \partial_t \tilde{H} = -B^\sharp(\tilde{H}). \tag{5.10}$$

Using (4.8), we compute for any $X \in \widetilde{\mathscr{D}}$ and $\mathscr{E}_i, \mathscr{E}_j \in \Gamma(\mathscr{D})$,

$$g_t(\partial_t \tilde{h}(\mathscr{E}_i, \mathscr{E}_j), X) = g_t(\partial_t(\nabla^t_i \mathscr{E}_j) + \partial_t(\nabla^t_j \mathscr{E}_i), X)/2 = -B(\tilde{h}(\mathscr{E}_i, \mathscr{E}_j), X).$$

From this $(5.10)_1$ follows. Next, for any $X \in \widetilde{\mathscr{D}}$, we have

$$g_t(\partial_t \tilde{H}, X) = \sum_i g_t(\partial_t(\nabla_i \mathscr{E}_i), X) = -\sum_i B(\nabla_i \mathscr{E}_i, X) = -B(\tilde{H}, X),$$

that confirms $(5.10)_2$. By (5.10) with a special choice $B = s g^\top$ we get (5.9). $\qquad \square$

Next, we show that $\widetilde{\mathscr{D}}$-conformal variations of metrics preserve some important geometric properties of $\widetilde{\mathscr{D}}$ and $\mathscr{D}$.

Proposition 5.7. *Let $g_t \in \mathscr{M}$ satisfy (5.3).*

(i) *If $\widetilde{\mathscr{D}}$ is totally umbilical for g_0, then $\widetilde{\mathscr{D}}$ is the same for any g_t.*
(ii) *If $\mathscr{D}$ is either umbilical or harmonic for g_0, then $\mathscr{D}$ is the same for any g_t.*

Proof.

(i)   Since $\widetilde{\mathscr{D}}$ is g_0-totally umbilical, we have $h = (1/n) H g^\top$ at $t = 0$. Applying
      to (5.7) the theorem on existence/uniqueness of a solution of ODE's, we con-
      clude that $h_t = (1/n) H'_t g_t^\top$ for all t and for some $H'_t \in \Gamma(\mathscr{D})$. Tracing this, we
      see that H'_t is the mean curvature vector of h_t, thus $\widetilde{\mathscr{D}}$ is totally umbilical for
      any g_t.
(ii) If $\mathscr{D}$ is g_0-totally umbilical, then $\tilde{h} = (1/p) \tilde{H} g^\perp$ at $t = 0$. Applying to $(5.9)_1$
      the theorem on existence and uniqueness of a solution of ODE's, we conclude
      that $\tilde{h}_t = (1/p) \tilde{H}'_t g_t^\perp$ for all t, where $\tilde{H}'_t \in \Gamma(\widetilde{\mathscr{D}})$. Tracing this, we show that $\tilde{H}'_t$
      is the mean curvature vector of $\tilde{h}_t$, hence $\mathscr{D}$ is totally umbilical for any g_t. The
      proof of remaining property, i.e., $\mathscr{D}$ is harmonic, is similar.   □

5.1.2 Existence and Uniqueness

An important step in the study of evolutionary PDE's is to show short-time existence
and uniqueness. If one has a solution u_0 to a given non-linear PDE, then it is possible
to linearise the equation by considering a smooth family $u = u(t)$ of solutions with
a variation $v = \partial_t u_{|t=0}$. Differentiating the PDE with respect to t, gives a linear PDE
for the function v.

Proposition 5.8. *Let $\mathscr{D}$ be integrable with all leaves compact and orientable. Then
(5.1a) has a unique leafwise smooth solution g_t defined on a positive t-interval $[0, \varepsilon)$.
Moreover, if the leaves define a fibration $\pi : (M, g) \to B$ then g_t is smooth on M.*

Proof. Let $g = g_0 + h$, where $h = s g_0^\top$ and $s \in C^2(M)$. By Lemma 5.6, the linearization of (5.1a) about g_0 is the linear PDE on the leaves/fibres:

$$\partial_t s = \Delta^\perp s - (2/n) (\operatorname{div}^\perp H)_{g_0} s$$

(with coefficients depending on g_0) and the function $s_{|t=0}$ is bounded. The result
follows from the theory of parabolic PDE's [Tay] and the "fibration" assumption. □

The next proposition shows that *unnormalized and normalized flow differ only
by rescaling along the distribution $\widetilde{\mathscr{D}}$*.

Proposition 5.9. *Let g_t be a solution of finite volume to (5.1a) on $(M, \widetilde{\mathscr{D}}, \mathscr{D})$. Then
the adapted metrics $\hat{g}_t = \phi_t\, g_t^\top \oplus g^\perp$, where $\phi_t = \operatorname{Vol}(M, g_t)^{-2/n}$, evolve according
to the normalized flow (5.1b).*

Proof. This is similar to the proof of Proposition 5.31. Since ϕ depends only on t,
by Lemma 5.5, $\hat{H}_t = H_t$ for metrics $\hat{g}_t$ and g_t. Hence, $\operatorname{div}^\perp \hat{H}_t = \operatorname{div}^\perp H_t$. From (5.4)
with $s_t = -(2/n)(\operatorname{div}^\perp H_t)$ we find the derivative of the volume function

$$\frac{d}{dt} \operatorname{Vol}(M, g_t) = -\int_M (\operatorname{div}^\perp H_t)\, d\operatorname{vol}_t .$$

Thus, $\phi_t = \mathrm{Vol}(M, g_t)^{-2/n}$ is a smooth function of variable t. By Lemma 5.5, we have $\hat{h}_t = \phi_t \cdot h_t$. Therefore,

$$\partial_t \hat{g}_t = \phi_t \partial_t g_t + \phi_t' g_t^\top = -\Big(\frac{2}{n}\,\mathrm{div}^\perp H_t - \frac{\phi_t'}{\phi_t}\Big)\hat{g}_t^\top.$$

Notice that $d\,\widehat{\mathrm{vol}}_t = \phi_t^{n/2}\,d\,\mathrm{vol}_t$. Using this and (5.4), we obtain

$$\frac{d}{dt}\widehat{\mathrm{vol}}_t = \frac{n}{2}\Big(\frac{\phi_t'}{\phi_t} + s_t\Big)\widehat{\mathrm{vol}}_t.$$

Let ρ_t be the average of s_t, see (5.1b). From the above we get

$$0 = 2\,\frac{d}{dt}\int_M d\,\widehat{\mathrm{vol}}_t = \int_M n\Big(\frac{\phi_t'}{\phi_t} + s_t\Big)d\,\widehat{\mathrm{vol}}_t = n\Big(\frac{\phi_t'}{\phi_t} + \rho_t\Big).$$

This shows that $\rho_t = -\phi_t'/\phi_t$. Hence, $\hat{g}$ evolves according to (5.1b). □

Proposition 5.10. *Let $\mathscr{D}$ be integrable. Then the mean curvature vector H of $\widetilde{\mathscr{D}}$ and its $\mathscr{D}$-divergence with respect to g_t of flow (5.1a) or (5.1b) satisfy PDE's*

$$\partial_t H = \nabla^\perp(\mathrm{div}^\perp H), \qquad \partial_t(\mathrm{div}^\perp H) = \Delta^\perp(\mathrm{div}^\perp H). \qquad (5.11)$$

Proof. By Lemma 5.6 with $s = -(2/n)\,\mathrm{div}^\perp H$ or $s = -(2/n)\,\mathrm{div}^\perp H - \rho(t)$, we obtain $(5.11)_1$. Similarly, from $(5.8)_2$ we deduce $(5.11)_2$. □

Denote by $\xi^\flat$ the 1-form on the leaves of $\mathscr{D}$ dual to the vector field ξ on $\mathscr{D}$. The operators d and δ defined for differential forms on the leaves of $\mathscr{D}$ will be denoted by $d^\perp$ and $\delta^\perp$. The 1-form $\xi^\flat$ is $\mathscr{D}$-harmonic (see Section 5.1.1) if and only if $\delta^\perp\xi^\flat = 0$ (i.e., $\mathrm{div}^\perp\xi = 0$) and $d^\perp\xi^\flat = 0$ (i.e., $\nabla^\perp\xi$ is symmetric). One may consider (5.1a) as a PDE on the vector bundle that is $TM|_L$ along a separate leaf L. We will give a method (by approximation) to provide metrics well adapted to foliations, our conditions are rather different (and seem to be stronger) of the known hypotheses to guarantee tautness in known non-constructive results.

In the following theorem, the distribution $\mathscr{D}$ is integrable and the proof is based on solving the heat equation for 1-forms on the leaves of $\mathscr{D}$.

Theorem 5.11. *Let $\mathscr{D}$ be integrable with compact orientable leaves $\{L\}$, and if $p > 1$ then suppose that $d^\perp H^\flat = 0$. Then the flow (5.1a) admits a unique smooth along any leaf solution g_t $(t \ge 0)$ that converges as $t \to \infty$ in C^1 (along the leaves) to the limit metric g_∞ for which $\widetilde{\mathscr{D}}$ is harmonic. Moreover, if $\{L\}$ are fibers of a fibration $\pi : M \to B$ then g_t $(t \ge 0)$ and g_∞ are smooth metrics on M.*

Proof. This is based on the heat flow for 1-forms, see Theorem 5.2.

The eigenvalue problem, $-\Delta^\perp u = \lambda u$, on any leaf L (of $\mathscr{D}$) has solution with a sequence of eigenvalues with repetition (each one as many times as the dimension of its finite-dimensional eigenspace) $0 = \lambda_0 < \lambda_1 \le \lambda_2 \le \cdots \uparrow \infty$. The eigenfunctions (ϕ_j) form an orthonormal system in $L^2(L)$, i.e., $\int_L \phi_i(x)\phi_j(x)\,d\,\mathrm{vol}_g = \delta_{ij}$. Then

$$G^\perp(t,x,y) = \sum_j e^{\lambda_j t} \phi_j(x)\,\phi_j(y)$$

is a fundamental solution of the heat equation on L. A solution satisfying $u(x,0) = u_0(x)$ is given by

$$u(x,t) = \int_{y \in L} G^\perp(t,x,y)u_0(y)\,\mathrm{d}\,\mathrm{vol}.$$

Moreover, if $u_0 \in L^2(L)$ then $u(x,t)$ converges as $t \to \infty$ uniformly on L to a $\mathscr{D}$-harmonic function (constant when L is closed). Let $H_t^\flat$ $(t \geq 0)$ be the dual 1-form on $\mathscr{D}$ to the mean curvature vector field H_t with respect to g_t. Using $\mathrm{div}^\perp H_t = -\delta^\perp H_t^\flat$ and $(5.8)_3$ with $s = -(2/n)\,\mathrm{div}^\perp H$, we show similarly to $(5.11)_1$ that

$$\partial_t H_t^\flat = -d^\perp \delta^\perp H_t^\flat, \tag{5.12}$$

where $H_0^\flat = H^\flat$ is known. We obtain the following inequality for L^2-product on $\{L\}$:

$$\frac{1}{2}\partial_t\left(\|H_t^\flat\|^2\right) = -\langle d^\perp \delta^\perp H_t^\flat,\, H_t^\flat\rangle = -\|\delta^\perp H_t^\flat\|^2 \leq 0.$$

By the above, we have uniqueness of a solution of the linear PDE (5.12).

By Theorem 5.2 in Section 5.1.1, the heat equation (considered along $\{L\}$)

$$\partial_t\,\omega = \Delta_d^\perp\,\omega, \qquad \text{where} \quad \omega_0 = H^\flat, \tag{5.13}$$

admits a unique solution ω_t $(t \geq 0)$. As $t \to \infty$, the 1-form ω_t converges exponentially to a $\mathscr{D}$-harmonic 1-form ω_∞, i.e., $\|\omega_t - \omega_\infty\| \leq c\,e^{-\lambda t}$ for some constants $c, \lambda > 0$. Since $\omega_0 = H^\flat$ is $\mathscr{D}$-closed, by Theorem 5.2, we have $d^\perp \omega_t = 0$ for all $t \geq 0$, hence ω_t satisfies (5.12). (Notice that for $p = 1$ the 1-form $H^\flat$ is always $\mathscr{D}$-closed). Comparing (5.12) with (5.13), we conclude that $H_t^\flat = \omega_t$ is a unique solution of (5.12). By the above, $\mathrm{div}^\perp H_\infty = 0$ (here the vector field $H_\infty \in \Gamma(\mathscr{D})$ is dual to $H_\infty^\flat = \omega_\infty$). Applying (5.2), we conclude that $\lim_{t\to\infty} \|H_t\|^2 = 0$, hence $H_\infty = 0$.

With known H_t, the PDE (5.1a) also has a unique smooth global solution

$$g_t^\top = g_0^\top \exp\left(-\frac{2}{n}\int_0^t (\mathrm{div}^\perp H_s)\,\mathrm{d}s\right).$$

By Lemma 5.6, we have

$$\tilde{h}_t = \tilde{h}_0 \exp\left(\frac{2}{n}\int_0^t (\mathrm{div}^\perp H_s)\,\mathrm{d}s\right)$$

for $t \geq 0$. Since the leaves $\{L\}$ are compact, and $\mathrm{div}^\perp H_t$ satisfies $(5.11)_2$ (on any leaf of $\mathscr{D}$), we get $|\mathrm{div}^\perp H_t| \leq e^{-\lambda_1 t}|\mathrm{div}^\perp H_0|$ and

$$\left|\int_0^t (\mathrm{div}^\perp H_x)\,\mathrm{d}x\right| \leq \int_0^t |\mathrm{div}^\perp H_x|\,\mathrm{d}x < \tilde{c}\int_0^t e^{-\lambda_1 x}\,\mathrm{d}x = \tilde{c}\,\frac{1 - e^{-\lambda_1 t}}{\lambda_1} < \frac{\tilde{c}}{\lambda_1}$$

for some $\tilde{c} > 0$ and all $t \geq 0$. Now, for some $c \geq 1$ and all $t \geq 0$, we have the bounds

$$c^{-1} g_0^\top \le g_t^\top \le c \, g_0^\top, \quad c^{-1} \tilde{h}_0 \le \tilde{h}_t \le c \tilde{h}_0.$$

Hence $g_t \to g_\infty$ and $\tilde{h}_t \to \tilde{h}_\infty$ as $t \to \infty$ in $\mathbf{C}^0$.

Consider the 1-form $\omega_1 = d^\perp \delta^\perp H^\flat$ on $\mathscr{D}$, and set $u_1(t) = \sup_M |\omega_1(t)|_{g(t)}$. Notice that $\omega_1 = (H_1)^\flat$, where $H_1 = -\nabla^\perp \operatorname{div}^\perp H$. We have $d^\perp \omega_1 = d^\perp (d^\perp \delta^\perp H^\flat) = 0$ and

$$\partial_t \omega_1 = \partial_t (d^\perp \delta^\perp H^\flat) = d^\perp \delta^\perp \partial_t H^\flat = -d^\perp \delta^\perp \omega_1.$$

Similarly to (5.12), we see that the solution ω_1 of above PDE is unique. By Theorem 5.2, ω_1 is also a solution of the heat equation $\partial_t \omega_1 = \Delta_d^\perp \omega_1$, and we have $\omega_1 \to \omega_1^\infty = d^\perp \delta^\perp H_\infty^\flat = 0$ as $t \to \infty$. By the above and Proposition 5.1, $g_t \to g_\infty$ and $\tilde{h}_t \to \tilde{h}_\infty$ as $t \to \infty$ in $\mathbf{C}^1$ along any leaf. $\qquad \square$

A foliation with totally umbilical normal distribution is locally conformally equivalent to a Riemannian foliation [Mon]. The next corollary relates both metrics by a homotopy.

Corollary 5.12. *Under assumptions of Theorem 5.11, let $\widetilde{\mathscr{D}}$ be g-totally umbilical. Then $\widetilde{\mathscr{D}}$ is g_∞-totally geodesic, where g_∞ is the limit as $t \to \infty$ of the solution metrics of (5.1a).*

Proof. By Theorem 5.11, the metrics g_t converge to a smooth metric g_∞ with $H_\infty = 0$. By Proposition 5.7, the flow preserves the total umbilicity of $\widetilde{\mathscr{D}}$; hence, $\widetilde{\mathscr{D}}$ is g_∞-totally umbilical. Certainly, a totally umbilical foliation with vanishing mean curvature is totally geodesic. $\qquad \square$

The following corollary (of Theorem 5.11) shows how to produce in some cases a family of metrics converging to the metric for which $\mathscr{F}$ is harmonic.

Corollary 5.13. *Let $\mathscr{F}$ be a foliation of codimension $p > 1$ on (M, g_0). Suppose that the normal distribution $\mathscr{D}$ is integrable and its leaves are compact and form a fibration $\pi : M \to B$, and the equality $d^\perp H_0^\flat = 0$ is valid. Then (5.1a) admits a unique solution g_t $(t \ge 0)$, converging in C^1, as $t \to \infty$, to the limit metric g_∞, for which $\mathscr{F}$ is harmonic.*

Proof. The leaves of orthogonal foliation (tangent to $\mathscr{D}$) are immersed compact cross-sections, hence $\mathscr{F}$ is taut. On the other hand, by Theorem 5.11, the solution metrics g_t converge as $t \to \infty$ to a smooth metric g_∞ with $H_\infty = 0$. $\qquad \square$

We will produce in some cases a one-parameter family of metrics on a foliation converging to the metric with prescribed mean curvature vector field of the leaves.

Theorem 5.14. *Let X be a smooth vector field tangent to integrable distribution $\mathscr{D}$ of dimension $p > 1$ with the property $d^\perp (H - X)^\flat = 0$. Then the PDE*

$$\partial_t g_t = -(2/n) \operatorname{div}^\perp (H_t - X) g_t^\top, \qquad g_0 = g \tag{5.14}$$

admits a unique, smooth along any leaf L of $\mathscr{D}$, solution g_t $(t \ge 0)$, converging (along L), as $t \to \infty$, in C^1 to the limit metric g_∞ with $\mathscr{D}$-harmonic 1-form $(H_\infty - X)^\flat$;

and if $H^1(L, \mathbb{R}) = 0$, then $H_\infty = X$. Moreover, if the leaves $\{L\}$ are fibers of a fibration $\pi : M \to B$ then g_t $(t \geq 0)$ and g_∞ are smooth metrics on M.

Proof. The vector field $\hat{H}_t = H_t - X$ satisfies PDE's of Proposition 5.10,

$$\partial_t \hat{H}_t = \nabla^\perp \operatorname{div}^\perp \hat{H}_t, \qquad \partial_t (\operatorname{div}^\perp \hat{H}_t) = \Delta^\perp (\operatorname{div}^\perp \hat{H}_t).$$

Since $\partial_t X = 0$, the 1-form $\hat{H}_t^\flat := H_t^\flat - X^\flat$ satisfies the PDE, see (5.12),

$$\partial_t \hat{H}^\flat = -d^\perp \delta^\perp \hat{H}^\flat \qquad (5.15)$$

with the initial value $\hat{H}_0^\flat = H^\flat - X^\flat$. As in the proof of Theorem 5.11, we have uniqueness of a solution of (5.15) for $t \geq 0$. By Theorem 5.2, the heat equation

$$\partial_t \omega = \Delta_d^\perp \omega \qquad (5.16)$$

(considered along $\{L\}$) admits a unique smooth solution ω_t $(t \geq 0)$ with the initial value $\omega_0 = H^\flat - X^\flat$. As $t \to \infty$, the 1-form ω_t converges exponentially to a $\mathscr{D}$-harmonic 1-form ω_∞, and $\|\omega_t - \omega_\infty\| \leq c e^{-\lambda t}$ for some constants $c, \lambda > 0$. Since ω_0 is $\mathscr{D}$-closed, again by Theorem 5.2, we have $d^\perp \omega_t = 0$ for all $t \geq 0$. Comparing (5.15) with (5.16), as in the proof of Theorem 5.11, we conclude that $\widetilde{H}^\flat = \omega_t$ is a unique solution of (5.15). By the above, $\operatorname{div}^\perp (H_\infty - X) = 0$ (here $H_\infty - X$ is dual to ω_∞). With known H_t, (5.14) also has a unique smooth global solution, and

$$g_t^\top = g_0^\top \exp \left(-\frac{2}{n} \int_0^t \operatorname{div}^\perp (H_s - X) \, ds \right).$$

By Lemma 5.6, we also have for all $t \geq 0$:

$$\tilde{h}_t = \tilde{h}_0 \exp \left(\frac{2}{n} \int_0^t \operatorname{div}^\perp (H_s - X) \, ds \right).$$

Using $|\operatorname{div}^\perp (H_t - X)| \leq e^{-\lambda_1 t} |\operatorname{div}^\perp (H_0 - X)|$, we get for some $\tilde{c} > 0$ and all $t \geq 0$:

$$\left| \int_0^t (\operatorname{div}^\perp (H_s - X)) \, ds \right| \leq \int_0^t |\operatorname{div}^\perp (H_s - X)| \, ds < \tilde{c} \int_0^t e^{-\lambda_1 t} \, ds = \tilde{c} \frac{1 - e^{-\lambda_1 t}}{\lambda_1} < \frac{\tilde{c}}{\lambda_1}.$$

As in the proof of Theorem 5.11, we show that the 1-form $\omega_1 = d^\perp \delta^\perp (H - X)^\flat$ converges exponentially to 0, thus g_t converges in C^1 as $t \to \infty$ to a metric g_∞ with $\mathscr{D}$-harmonic 1-form $(H_\infty - X)^\flat$. Certainly, if $H^1(L, \mathbb{R}) = 0$, then $H_\infty = X$. $\qquad \square$

Remark 5.15.

(a) By Bochner theorem, see [Jo], if the Ricci curvature of a leaf L (of $\mathscr{D}$) is nonnegative everywhere and positive at some point then $H^1(L, \mathbb{R}) = 0$, see assumptions of Theorem 5.14.

(b) The following example shows us that the condition $d^\perp (H - X)^\flat = 0$ (of Theorem 5.14) and the assumption $d^\perp H^\flat = 0$ (of Theorem 5.11 and Corollary 5.13) are needed. Let X be a divergence free (e.g., a Killing) vector field on the leaves

S^p of the product $M = M_1 \times S^p$ of a unit p-sphere and a Riemannian manifold (M_1, g_1). Let the distribution $\widetilde{\mathscr{D}}$ on M corresponds to TM_1. Then the product metric g has the mean curvature $H = 0$, and g is a fixed point of the dynamical system (5.14). Consequently, $H_t = 0$ for all $t \geq 0$ and $H_\infty = 0 \neq X$.

(c) Let the leaves $\{L\}$ be flat tori T^p. Any differential 1-form on T^p can be written as $\omega = \sum_i \omega_i \, dx^i$. The 1-form ω on T^p is harmonic if and only if the functions ω_i are harmonic, therefore constant. In this case, the vector field dual to ω is constant. Certainly, the space of constant vector fields on a flat torus T^p is isomorphic to $H^1(T^p, \mathbb{R}) \simeq \mathbb{R}^p$.

(d) To build a vector field X satisfying assumptions of Theorem 5.14, one can take any smooth function f on M and put $X = H - \nabla^\perp f$, $\nabla^\perp f$ being the $\mathscr{D}$ component of the gradient ∇f of f. Then, $(H - X)^\flat = (d^\perp)^2 f = 0$.

Theorem 5.14 applied to codimension-one foliations of S^1-fibrations yields

Corollary 5.16. *If $\mathscr{F}$ a codimension-one foliation of an S^1 fibration $\pi : M \to B$ is transverse to the fibers, then for any function $f : M \to \mathbb{R}$ there exists a Riemannian metric $\tilde{g}$ on M for which the mean curvature $\tilde{\tau}_1$ of the leaves of $\mathscr{F}$ equals $\tilde{\tau}_1 = f + \phi \circ \pi$, where ϕ is a smooth function on B.*

Proof. Choose a Riemannian metric g on M making the fibers orthogonal to the leaves. Take $X = fN$, where N is a g-normal of $\mathscr{F}$. Since fibers are one-dimensional, the condition $d^\perp (H - X) = 0$ is satisfied and one may put $\tilde{g} = g_\infty$, g_∞ being as in Theorem 5.14. Then, $(\tilde{\tau}_1 - f)N^\flat$ is fiber-wise harmonic, therefore $\tilde{\tau}_1 - f$ is constant along the fibers. In other words, $\tilde{\tau}_1 = f + \phi \circ \pi$ for some $\phi : B \to \mathbb{R}$. $\square$

Note that the above result can be derived also from Theorem 3.33: A foliation transverse to fibers is topologically taut, therefore any function F on M with $F(x) \cdot F(y) < 0$ for some $x, y \in M$ can become the mean curvature of the leaves (with respect to some Riemannian metric). It is obvious, that for any $f : M \to \mathbb{R}$ one can find $\phi : B \to \mathbb{R}$ such that $F = f + \phi \circ \pi$ satisfies the above sign condition.

5.1.3 The Codimension-One Case

Let (M, g) be a closed Riemannian manifold with a codimension-one distribution $\widetilde{\mathscr{D}}$ (i.e., $p = 1$). Let N be the unit vector field orthogonal to $\widetilde{\mathscr{D}}$, and h—the scalar second fundamental form of $\widetilde{\mathscr{D}}$ with respect to N. Certainly, $2h(X, Y) = \langle \nabla_X Y + \nabla_Y X, N \rangle$. Hence, $H = \tau_1 N$ where $\tau_1 = \langle N, H \rangle = \text{trace}\, A$ and $A = h^\sharp$ is the shape operator. By Lemma 5.6, we find the variations

$$\partial_t A = -(1/2) N(s) \, \text{id}^\top, \qquad \partial_t \tau_1 = -(n/2) N(s), \qquad (5.17)$$

where g_t obey (5.3). For a codimension-one case, (5.1a,b) read as:

$$\partial_t g_t = -\frac{2}{n} N(\tau_1) g_t^\top, \qquad (5.18a)$$

$$\partial_t g_t = -\left(\frac{2}{n} N(\tau_1) + \rho(t)\right) g_t^\top, \qquad (5.18b)$$

respectively, where $g_0 = g$ and

$$\rho(t) = -\frac{2}{n} \int_M N(\tau_1) \, d\,vol_t \,/\, Vol(M, g_t).$$

The (1.33) with $X = fN$, where f has a compact support, reduces to the formula,

$$\int_M N(f) \, d\,vol = \int_M \tau_1 f \, d\,vol. \qquad (5.19)$$

By the above with $f = \tau_1$, we have $\rho(t) = -(2/n) \int_M \tau_1^2 \, d\,vol_t \,/\, Vol(M, g_t) \le 0$.

Proposition 5.17. *Let M and all N-curves be closed. Then (5.18a) or (5.18b) has a unique smooth along N-curves global solution g_t $(t \ge 0)$. The metrics g_t approach in C^∞ as $t \to \infty$ to a smooth along N-curves metric g_∞ for which $\widetilde{\mathscr{D}}$ is harmonic. Moreover, if N-curves compose a circle bundle $\pi : M \to B$ then g_t $(t \ge 0)$ and g_∞ are smooth metrics on M.*

Proof. By (5.17) with $s = -(2/n)N(\tau_1)$ or $s = -(2/n)N(\tau_1) - \rho(t)$, the functions $\psi_m = \underbrace{N(N(\ldots N(\tau_1)\ldots))}_{m \text{ times}}$ (where $m \in \{0, 1\}$ and $\psi_0 = \tau_1$) obey the heat equation $\partial_t \psi_m = N(\partial_t \tau_1) = N(N(\psi_m))$ on N-curves, see also (5.11). This has a unique global solution ψ_m^t $(t \ge 0)$ (in particular, τ_1^t), and $\psi_m^t \to \psi_m^\infty$ (in particular, $\tau_1^t \to \tau_1^\infty$) as $t \to \infty$. Hence, $N(\tau_1^\infty) = $ const on any N-curve. In case of (5.18b), we have $\rho(t) \to 0$ as $t \to \infty$. Since N-curves are closed, we obtain $N(\tau_1^\infty) = 0$. With known τ_1^t, (5.18a) has a unique smooth solution

$$g_t^\top = g_0^\top \exp\left(-\frac{2}{n} \int_0^t N(\tau_1^s) \, ds\right), \quad t \ge 0.$$

We have the uniform bounds

$$c^{-1} g_0^\top \le g_t^\top \le c g_0^\top, \quad c^{-1} \tilde{h}_0 \le \tilde{h}_t \le c \tilde{h}_0$$

for some $c \ge 1$ and all $t \ge 0$. By Proposition 5.1 with $u_m(t) = \sup_M |\psi_m^t|$, we obtain convergence in C^∞ of g_t to the limit metric g_∞ and also $\tilde{h}_t \to \tilde{h}_\infty$ as $t \to \infty$. By the integral formula

$$\int_M [N(\tau_1^\infty) - (\tau_1^\infty)^2] \, d\,vol_\infty = 0,$$

see (5.19) with $f = \tau_1$, we find that $\tau_1^\infty = 0$ on M. $\qquad \square$

5.1.4 The Doubly Twisted Products

By Proposition 5.7, the flows (for $\widetilde{\mathscr{D}} = \ker \pi_{2*}$ and $\mathscr{D} = \ker \pi_{1*}$) preserve the doubly twisted product structure (see Section 1.3.2). From Theorem 5.11 we deduce

Proposition 5.18. Let $M_1 \times_{(f_1,f_2)} M_2$ be a doubly twisted product of closed Riemannian manifolds (M_i, g_i). Then (5.1a) has a unique smooth solution $g_t \in \mathscr{M}$ for all $t \geq 0$, consisting of doubly twisted product metrics on $M_1 \times_{(f_1(t),f_2)} M_2$. As $t \to \infty$, the metric g_t converges in C^1 to a doubly twisted product metric g_∞ corresponding to $M_1 \times_{(\bar{f}_1,f_2)} M_2$, where $\bar{f}_1(x) = \int_{y \in M_2} f_1(0,x,y) \, d\mathrm{vol}_g$. If $f_2 = 1$ (i.e., g is a twisted product), then g_∞ splits as the product $(M_1, \bar{f}_1^2 \cdot g_1) \times (M_2, g_2)$.

Proof. Define $\phi : M \to \mathbb{R}$ by $f_1 = e^\phi$. The mean curvature vector of $M_1 \times \{y\}$ is $H = -\nabla^\perp \phi$. The evolution $\partial_t g(t) = s(t) g_1(t) \oplus 0$ preserves the doubly twisted product structure. Denoting $g_t^\top = g_1(t)$, we obtain $g(t) = f_1(t)^2 g_1^\top \oplus f_2^2 g_2$, where f_2 and g_2 do not depend on t. Assuming $f_1(t) = e^{\phi(t)}$, we find $g_t^\top = e^{2\phi(t)} g_0^\top$. In this case, $\partial_t g(t) = 2 e^{2\phi(t)} \partial_t \phi(t) g_0^\top$. Hence, $\partial_t \phi(t) = s(t)/2$.

For $s = -(2/n) \mathrm{div}^\perp H$, we have $\mathrm{div}^\perp H = -\Delta^\perp \phi$, and (5.1a) reads as

$$\partial_t \phi = (1/n) \Delta^\perp \phi. \tag{5.20}$$

Replacing $t \to t/n$, we reduce (5.20) to the heat equation. By Proposition 5.9, there is a unique solution ϕ_t $(t \geq 0)$ obeying $\lim_{t \to \infty} \phi_t(x,y) = \phi_\infty(x)$. Certainly, $f_1^\infty = e^{\phi_\infty}$. $\square$

Example 5.19.

(a) Let $M = M_1 \times M_2$ be the product of smooth manifolds, and M_2 admits a metric g_2 of quasi-positive Ricci curvature. Then for any vector field X on M, satisfying the condition $d_2(H_0 - X)^\flat = 0$ (derivation along M_2), there is a doubly twisted product structure $M = M_1 \times_{(f_1,f_2)} M_2$ for which X is the mean curvature vector of a foliation $M_1 \times \{y\}$. To show this, for any metric g_1 on M_1, consider the metric $g = g_1 \oplus g_2$ on M, and apply Proposition 5.18.

(b) Let $S^1 \times_{(f,1)} S^1$ be a twisted product of circles. Then (5.1a) has a unique smooth solution $g_t \in \mathscr{M}$ for all $t \geq 0$, consisting of twisted product metrics on the torus T^2. As $t \to \infty$, the metric g_t converges in C^1 to a flat metric on the torus.

5.2 Flows of Metrics on Codimension-One Foliations

Let M^{n+1} be a connected smooth manifold with a codimension-one foliation $\mathscr{F}$, and a vector field N transverse to $\mathscr{F}$. Set $\widetilde{\mathscr{D}} = T\mathscr{F}$ and $\mathscr{D} = \mathrm{span}(N)$. In this section, a metric g on M will be called *adapted* if $N \perp \widetilde{\mathscr{D}}$ and $\|N\| = 1$. Denote by $\mathscr{M} = \mathscr{M}(\widetilde{\mathscr{D}}, N)$ the *space of adapted metrics of finite volume* for $(M, \widetilde{\mathscr{D}}, N)$. Important example of $\widetilde{\mathscr{D}}$-truncated $(1,2)$-tensor field is the scalar second fundamental form and the shape operator

$$h(X,Y) = \langle \nabla_X Y, N \rangle, \quad A(X) = -(\nabla_X N)^\top, \quad X,Y \in T\mathscr{F},$$

of $\mathscr{F}$, extended on TM by $h(N, \cdot) = 0$ and $A(N) = 0$. Let h_m $(m \in \mathbb{N})$ be the symmetric $(0,2)$-tensor dual to A^m, thus $h_m(X,Y) = \langle A^m(X), Y \rangle$.

A family g_t of adapted metrics on $(M, \mathscr{F}, N)$ satisfying

$$\partial_t g_t = B(h_t), \tag{5.21}$$

where $B(h)$ is a symmetric $\widetilde{\mathscr{D}}$-truncated $(0,2)$-tensor expressed in terms of h, is an *extrinsic geometric flow*. Observe that $[X,Y]^\perp$ does not depend on t. Let $\mathscr{C}(h) = \sum_{m=0}^{n-1} f_m h_m$ be a symmetric $\widetilde{\mathscr{D}}$-truncated $(0,2)$-tensor, see (2.14), where f_m are either (a) arbitrary functions on $M \times \mathbb{R}$, or (b) symmetric functions of the principal curvatures of $\mathscr{F}$. Flows of the form

$$\partial_t g_t = \mathscr{C}(h_t),$$

deal with the first order PDE's and yield quasi linear hyperbolic PDE's, see [RW2]. In the section, we study (5.21) of the form

$$\partial_t g_t = \nabla_N^t \mathscr{C}(h_t). \tag{5.22}$$

Since the symmetric tensor $\mathscr{C}(h)$ can be expressed in terms of the first partial derivatives of metric, thus $\nabla_N \mathscr{C}(h)$ is a second-order differential operator. Under certain conditions, (5.22) yields second order quasi linear parabolic PDE.

Flows (5.22) preserve several properties of codimension-one foliations: *umbilical* ($A = \lambda \, \mathrm{id}^\top$, Proposition 5.28), *totally geodesic* ($A = 0$) and *Riemannian* ($\nabla_N N = 0$). The flows serve as a tool for studying the following **question** on foliations: *Under what conditions on* (f_m) *and* $(M, \mathscr{F}, g_0)$ *solution metrics* g_t *converge to one with a given extrinsic geometric property of* $\mathscr{F}$, *e.g., being harmonic?*

The following condition will be used to show convergence of evolving conformal metrics (see, for example, [Br, Appendix A] for a general formulation).

Proposition 5.20. *Let a family* g_t *$(t \geq 0)$ of* $\widetilde{\mathscr{D}}$*-conformal metrics on a closed codimension-one foliated manifold M with a unit normal N satisfy* $\partial_t g_t = s_t \, g_t^\top$ *and*

$$\int_0^\infty s_{\max}(t) \, dt < \infty, \quad where \quad s_{\max}(t) := \sup_M |s_t|_{g(t)}.$$

Then, as $t \to \infty$, *the metrics* g_t *converge in* C^0 *to a smooth limit metric* g_∞.

Proof. Our assumptions ensure that the metrics g_t converge in C^0 to a symmetric $(0,2)$-tensor g_∞. Notice that the metrics are uniformly equivalent, that is, $c^{-1} g_0^\top \leq g_t^\top \leq c g_0^\top$ for some $c > 0$ and all $t \geq 0$, therefore they define the same topology on M. Hence, g_∞ is positive definite. $\qquad\square$

5.2.1 $g^\top$-Variations of Mean Curvatures

The section contains formulas for deformations of geometric quantities as the Riemannian metric varies along the leaves of a codimension-one foliation.

Let $\mathscr{M}_1 \subset \mathscr{M}$ be the *subspace of adapted metrics with unit volume*, and

$$\pi : \mathscr{M} \to \mathscr{M}_1, \qquad \pi(g) = \overline{g} = \mathrm{Vol}(M,g)^{-2/n} g^\top \oplus g^\perp$$

the $\widetilde{\mathscr{D}}$-conformal projection. Let $g_t \in \mathscr{M}$ (with $0 \leq t < \varepsilon$) be a family of metrics with $\mathrm{Vol}(M,g_0) = 1$. Set $B_t = \partial_t g_t$. The volume form of g_t evolves as (4.15).

Remark 5.21. From (5.4) it follows that if g_t are $\widetilde{\mathscr{D}}$-conformal metrics, i.e., $B_t = s_t\, g_t^\top$ with $s_t : M \to \mathbb{R}$. Then (5.4) is valid and metrics $\tilde{g}_t = \phi_t g_t^\top \oplus g_t^\perp$ with dilating factors $\phi_t = \mathrm{Vol}(M,g_t)^{-2/n}$ belong to $\mathscr{M}_1$.

Next, we find the variation of A, and apply it to symmetric functions τ_j and σ_j of eigenvalues of A (defined in Section 1.5.3).

Lemma 5.22. *Let $g_t \in \mathscr{M}$ and $B = \partial_t g_t$ be a $\widetilde{\mathscr{D}}$-truncated tensor. Then the shape operator A of $\mathscr{F}$ (with respect to N) and the symmetric functions τ_i and σ_i of A evolve by*

$$\partial_t A = \left([A, B^\sharp] - \nabla_N^t B^\sharp\right)/2, \tag{5.23}$$

$$\partial_t \tau_i = -(i/2)\,\mathrm{trace}(A^{i-1}\nabla_N^t B^\sharp), \quad \partial_t \sigma_i = -(1/2)\,\mathrm{trace}(T_{i-1}(A)\nabla_N^t B^\sharp), \quad i > 0. \tag{5.24}$$

Proof. Using (4.8) and $B(\cdot, N) = 0$ for $\mathscr{F}$-truncated tensors, we obtain

$$\begin{aligned}
\partial_t h(X,Y) &= \partial_t g_t(\nabla_X^t Y, N) = (\partial_t g_t)(\nabla_X^t Y, N) + g_t(\partial_t \nabla_X^t Y, N)\\
&= B(\nabla_X^t Y, N) + (1/2)\big((\nabla_X^t B)(Y,N) + (\nabla_Y^t B)(X,N) - (\nabla_N^t B)(X,Y)\big)\\
&= (1/2)\big(B(AX,Y) + B(AY,X) - (\nabla_N^t B)(X,Y)\big), \quad X,Y \in T\mathscr{F}.
\end{aligned}$$

As $B(AX,Y) = g_t(B^\sharp AX, Y)$ and $(\nabla_N^t B)(X,Y) = g_t((\nabla_N^t B^\sharp)X, Y)$, we have

$$\begin{aligned}
g_t((\partial_t A)X,Y) &= g_t(\partial_t(AX),Y) = \partial_t h(X,Y) - B(AX,Y)\\
&= \frac{1}{2}[g_t(B^\sharp AY, X) - g_t(B^\sharp AX, Y) - (\nabla_N^t B)(X,Y)]\\
&= \frac{1}{2}[g_t([A, B^\sharp]X, Y) - g_t((\nabla_N^t B^\sharp)X, Y)].
\end{aligned}$$

Formula (5.23) follows from the above and the freedom of choice of $X, Y \in T\mathscr{F}$.

Multiplying (5.23) from the left by A^{i-1}, we get

$$2A^{i-1}\partial_t A = A^{i-1}[A, B^\sharp] - A^{i-1}\nabla_N^t B^\sharp, \quad i > 0.$$

Notice that $\mathrm{trace}(A^{i-1} \cdot [A, B^\sharp]) = 0$. Then, using the identity, see (1.81),

$$i\,\mathrm{trace}(A^{i-1}\partial_t A) = \mathrm{trace}(\partial_t A^i) = \partial_t \tau_i,$$

we deduce $(5.24)_1$. Substituting $\partial_t A$ from (5.23) into the formula (1.82), we obtain

$$\partial_t \sigma_i = \frac{1}{2}\, \text{trace} \left(T_{i-1}(A)([A, B^\sharp] - \nabla_N^t B^\sharp) \right) = -\frac{1}{2}\, \text{trace}(T_{i-1}(A)\nabla_N^t B^\sharp),$$

that proves $(5.24)_2$. For $B = s\hat{g}$, we have, respectively, $\nabla_N^t B^\sharp = N(s)\,\widehat{\text{id}}$, and

$$\text{trace}(A^{i-1}\nabla_N^t B^\sharp) = \tau_{i-1}N(s), \quad \text{trace}(T_{i-1}(A)\nabla_N^t B^\sharp) = (n-i+1)\,\sigma_{i-1}N(s). \quad \square$$

Remark 5.23. Let $(M, g = g^\top \oplus g^\perp)$ be a Riemannian manifold with a codimension-one foliation $\mathscr{F}$ and a unit normal N. Given $\varphi \in C^1(M)$, define a metric $\hat{g} = e^{2\varphi}g^\top \oplus g^\perp$. Then, see (5.6) when $p = 1$, the second fundamental forms, shape operators and τ's of $\mathscr{F}$ with respect to $\tilde{g}$ and g are related by

$$\hat{h} = e^{2\phi}(h - N(\varphi)g^\top), \quad \hat{A} = A - N(\varphi)\,\text{id}^\top, \quad \hat{\tau}_1 = \tau_1 - nN(\varphi).$$

Lemma 5.24. *The vector field* $Z = \nabla_N^t N$ *is evolved by* $g_t \in \mathscr{M}$ *with* $\partial_t g_t = B$ *as*

$$(i)\ \partial_t Z = -B^\sharp(Z), \quad (ii)\ \partial_t Z = -sZ \quad \text{for} \quad B = sg^\top. \tag{5.25}$$

In particular, all variations $g_t \in \mathscr{M}$ *preserve Riemannian foliations.*

Proof. We use (4.8) to compute for any $X \in T\mathscr{F}$

$$g_t(\partial_t Z, X) = \frac{1}{2} \left(2(\nabla_N^t B)(X, N) - (\nabla_X^t B)(N, N) \right) = -g_t(B^\sharp(Z), X).$$

From this, all of (5.25) follow. If $Z = 0$ at $t = 0$, then by uniqueness of a solution to the linear ODE (5.25)(i) along N-curves, we have $Z = 0$ for all t. $\quad \square$

5.2.2 The Extrinsic Geometric Flow Depending on $\{f_m\}$

Here, we study two types of evolution of metrics on $(M, \mathscr{F})$, depending on functions f_m $(m < n)$, at least one of them is not identically zero:

$$(a)\ f_m \in C^2(M \times \mathbb{R}), \qquad (b)\ f_m = f_m(\vec{\tau}) \in C^2(\mathbb{R}^n).$$

Definition 5.25. Given functions f_m of type either (a) or (b), consider *extrinsic geometric flow* g_t, $t \in [0, \varepsilon)$ of adapted metrics on $(M, \mathscr{F}, N)$ satisfying along $\mathscr{F}$ the PDE (5.22). The *normalized* flow is defined along $\mathscr{F}$ by

$$\partial_t g_t = \nabla_N^t \mathscr{C}(h_t) - \frac{1}{n}\rho(t)\,g_t^\top \text{ with } \rho(t) = \int_M N(\text{trace}\,\mathscr{C}(h))\,d\text{vol}_t\,/\text{Vol}(M, g_t). \tag{5.26}$$

Lemma 5.26. *Let* g_t *be the solution to (5.22). Then*

$$\partial_t A = \sum_{m=0}^{n-1} \left(\frac{1}{2}f_m[A, \nabla_N^t(A^m)] - N(N(f_m))A^m - N(f_m)\nabla_N^t A^m - f_m\nabla_N^t\nabla_N^t A^m \right). \tag{5.27}$$

Proof. Substituting $B^{\sharp} = \nabla_N^t \mathscr{C}(A)$ into (5.23), we obtain (5.27). $\qquad\square$

Let g_t be a family of metrics of finite volume on $(M, \mathscr{F})$. Metrics $\hat{g}_t = \phi_t g_t^{\top} \oplus g_t^{\perp}$ with $\phi_t = \mathrm{Vol}(M, g_t)^{-2/n}$ have unit volume: $\int_M d\widehat{\mathrm{vol}}_t = 1$, see Remark 5.21.

We show that *unnormalized and normalized flows differ by rescaling along* $\mathscr{F}$.

Proposition 5.27. *Let $(M, \mathscr{F})$ be a foliation, and g_t a solution (of finite volume) to (5.22) with $\mathscr{C}(h) = \sum_{m=0}^{n-1} f_m(\vec{\tau}) h_m$. Then the metrics $\hat{g}_t = \phi_t g_t^{\top} \oplus g^{\perp}$, where $\phi_t = \mathrm{Vol}(M, g_t)^{-2/n}$, evolve according to the normalized flow (5.26).*

Proof. By Remark 5.23, $\hat{A} = A$, thus $\hat{\tau}_j = \tau_j$ and $\mathscr{C}(\hat{A}) = \mathscr{C}(A)$ (for $\hat{g}_t$ and g_t, respectively). Hence, trace $\mathscr{C}(\hat{A}) = \sum_{m=0}^{n-1} f_m(\vec{\tau}) \tau_m$. From (4.15) with $B = \nabla_N^t \mathscr{C}(h)$ we get the derivative of the volume function

$$\frac{d}{dt} \mathrm{Vol}(M, g_t) = \frac{d}{dt} \int_M d\mathrm{vol}_t = \frac{1}{2} \int_M N(\mathrm{trace}\, \mathscr{C}(A)) \, d\mathrm{vol}_t \,.$$

Thus $\phi_t = \mathrm{Vol}(M, g_t)^{-2/n}$ is a smooth function. By Remark 5.23, we have $\mathscr{C}(\hat{h}) = \phi_t \cdot \mathscr{C}(h)$. By the above and $\hat{g}_t^{\top} = \phi_t g_t^{\top}$, we have

$$\partial_t \hat{g}_t = \phi_t \partial_t g_t + \phi_t' g_t^{\top} = \mathscr{C}(\hat{h}) + \phi_t'/\phi_t \, \hat{g}_t^{\top} \,.$$

Using (4.15) and $d\widehat{\mathrm{vol}}_t = \phi_t^{n/2} d\mathrm{vol}_t$, we obtain

$$\frac{d}{dt} \widehat{\mathrm{vol}}_t = \frac{d}{dt}(\phi_t^{\frac{n}{2}} \mathrm{vol}_t) = \frac{1}{2}\Big(n \frac{\phi_t'}{\phi_t} + N(\mathrm{trace}\, \mathscr{C}(A))\Big) \widehat{\mathrm{vol}}_t \,.$$

Let ρ_t be the average of $N(\mathrm{trace}\, \mathscr{C}(A))$, see (5.26). From the above we get

$$0 = 2 \frac{d}{dt} \int_M d\widehat{\mathrm{vol}}_t = \int_M \Big(n \frac{\phi_t'}{\phi_t} + N(\mathrm{trace}\, \mathscr{C}(A))\Big) d\widehat{\mathrm{vol}}_t = n \frac{\phi_t'}{\phi_t} + \rho_t.$$

Thus, $\rho_t/n = -\phi_t'/\phi_t$. Hence, $\hat{g}$ evolves by (5.26) with g_t replaced by $\hat{g}_t$. $\qquad\square$

By Lemma 5.24, flows preserve Riemannian foliations. Obviously, totally geodesic foliations are the fixed points of (5.22). We shall show that flows preserve the umbilicity of $\mathscr{F}$. Define the function of one variable by

$$\psi(\lambda) = -\sum_{m=0}^{n-1} f_m(n\lambda, n\lambda^2, \ldots, n\lambda^n)\lambda^m.$$

Proposition 5.28. *Given g_0 on $(M, \mathscr{F})$, let g_t $(0 \leq t < \varepsilon)$ be a unique smooth solution to (5.22) with f_m of type (b). If $\mathscr{F}$ is totally umbilical for g_0 and $\psi'(\lambda) > 0$, then $\mathscr{F}$ is umbilical for any g_t.*

Proof. Since $\mathscr{F}$ is g_0-umbilical, we have $A_0 = \lambda_0 \, \mathrm{id}^{\top}$ for some function $\lambda_0 : M \to \mathbb{R}$. Hence, $\mathscr{C}(A_0) = -\psi(\lambda_0) \mathrm{id}^{\top}$. If $\hat{A}_t = \lambda_t \mathrm{id}$ correspond to g_t, where $\lambda_t : M \to \mathbb{R}$ are smooth functions, then $\mathscr{C}(\hat{A}_t) = -\psi(\lambda_t) \mathrm{id}^{\top}$. By Lemma 5.22 with $s = -\psi(\lambda_t)$ we will get

$$2 \partial_t \lambda_t = N(N(\psi(\lambda_t))) = N(\psi'(\lambda_t) N(\lambda_t)). \tag{5.28}$$

Since $\psi'(\lambda) > 0$, the PDE (5.28) is parabolic and admits a unique local solution. Now, (5.22) for $\hat{g}_t$ reads as

$$\partial_t \hat{g}_t = -N(\psi(\lambda_t)) \hat{g}_t^\top, \tag{5.29}$$

from which we find $\hat{g}_t^\top = \exp(-\int_{s=0}^t N(\psi(\lambda_s)) \, ds) \hat{g}_0^\top$ along $\mathscr{F}$. The shape operator of certain metrics $\hat{g}_t$ is conformal: $\hat{A}_t = \lambda_t$ id. Thus, $\hat{g}_t$ is the solution of (5.22). Our assumptions ensure that a solution is unique, hence $g_t = \hat{g}_t$ and $A_t = \hat{A}_t$ for all t. □

5.2.3 The Generalized Companion Matrix

Assume that a polynomial $P_n = k^n - p_1 k^{n-1} - \ldots - p_{n-1} k - p_n$ with real coefficients p_i has n real roots $k_1 \le k_2 \le \ldots \le k_n$. Then $p_i = (-1)^{i-1} \sigma_i$, where σ_i are elementary symmetric functions of the roots k_i, see Section 1.5.3. The following *generalized companion matrix* plays a key role in this section:

$$B_{n,1} = \begin{pmatrix} 0 & \frac{1}{2} & 0 & \cdots & 0 \\ 0 & 0 & \frac{2}{3} & \cdots & 0 \\ \cdots & \cdots & \cdots & \cdots & \cdots \\ 0 & 0 & 0 & \cdots & \frac{n-1}{n} \\ (-1)^{n-1} \frac{n}{1} \sigma_n & (-1)^{n-2} \frac{n}{2} \sigma_{n-1} & \cdots & -\frac{n}{n-1} \sigma_2 & \sigma_1 \end{pmatrix}. \tag{5.30}$$

Lemma 5.29. *The matrix (5.30) has the following properties:*

(a) The characteristic polynomial of $B_{n,1}$ is P_n.

(b) $v_j = (1, 2k_j, 3k_j^2, \ldots, nk_j^{n-1})$ is the eigenvector of $B_{n,1}$ for the eigenvalue k_j.

(c) $B_{n,1} V = VD$, where $V = \{\frac{n}{i} k_j^{i-1}\}_{1 \le i,j \le n}$ is the Vandermonde type matrix, and $D = \mathrm{diag}(k_1 \ldots k_n)$ is a diagonal matrix. (If k_i's are distinct, then $B_{n,1} = VDV^{-1}$).

Proof. (a) One may show by induction for n that $\det|\lambda\, \mathrm{id}_n - B_{n,1}| = P_n$, hence the eigenvalues $\lambda_1 \le \ldots \le \lambda_n$ of $B_{n,1}$ coincide with the roots of P_n. Expanding by cofactors down the first column, we obtain

$$\det|\lambda\, \mathrm{id}_n - B_{n,1}| = \lambda\, P_{n-1} - (-1)^n c_n p_n \prod_{i=1}^{n-1} (-c_i/c_{i+1}),$$

where $P_{n-1} = \lambda^{n-1} - p_1 \lambda^{n-2} - \ldots - p_{n-2} \lambda - p_{n-1}$ (by the induction assumption) is a certain polynomial of degree $n-1$. As $c_n \prod_{i=1}^{n-1} \frac{c_i}{c_{i+1}} = 1$ and $P_n + p_n = \lambda P_{n-1}$, the claim follows. (b) One may show that $(\lambda_j\, \mathrm{id}_n - B_{n,1}) v_j = 0$. (c) Hence, $B_{n,1} V = VD$, where $D = \mathrm{diag}(\lambda_1, \ldots, \lambda_n)$ is a diagonal matrix. If $\{\lambda_j\}$ are pairwise distinct, then $\det V \ne 0$ and $V^{-1} B_{n,1} V = D$. □

Remark 5.30. Given $m > 0$, consider the infinite system of linear PDE's

$$\partial_t \tau_i = -\frac{im}{2(i+m-1)} \partial_x \tau_{i+m-1}, \qquad i \in \mathbb{N}, \tag{5.31}$$

where $x \in \mathbb{R}$, $t \geq 0$ and $\tau_i = k_1^i + \ldots + k_n^i$ are the power sums of smooth functions $k_i(t,x)$. The n-truncated (5.31), i.e., τ_{n+j}'s, $j > 0$, are eliminated using Newton formulas (1.79), is, see also [RW2, p. 62],

$$\partial_t \vec{\tau} = -\frac{m}{2}(B_{n,1})^{m-1} \partial_x \vec{\tau},$$

where $\vec{\tau} = (\tau_1, \ldots, \tau_n)$. For $m = 1$ the system (5.31) has diagonal form, and for $m = 2$ the n-truncated system (5.31) reads as $\partial_t \tau_i = -B_{n,1} \partial_x \tau_i$.

Using the generalized companion matrix, we obtain the following result about truncation of infinite system of PDE's.

Proposition 5.31. *Given $m > 0$, consider the infinite system of linear PDE's*

$$\partial_t \tau_i = \frac{im}{2(i+m-1)} \partial_{xx}^2 \tau_{i+m-1}, \qquad i \in \mathbb{N}, \tag{5.32}$$

where τ_i are the power sums of smooth functions $k_i(t,x)$. Then the n-truncated system (5.32) has the form

$$\partial_t \vec{\tau} = \frac{m}{2}(B_{n,1})^{m-1} \partial_{xx}^2 \vec{\tau} + a_m(\vec{\tau}, \partial_x \vec{\tau}),$$

where a_m are smooth functions of $2n$ real variables.

Proof. By Remark 5.30, $\partial_x \tau_{n+i} = \sum_{j=1}^n \tilde{b}_{ij} \partial_x \tau_j$, where $\tilde{b}_{ij}$ are elements of the matrix $B_{n,m-1}$. Derivation of this yields $\partial_{xx}^2 \tau_{n+i} = \sum_{j=1}^n \tilde{b}_{ij} \partial_{xx}^2 \tau_j + \partial_x \tilde{b}_{ij} \partial_x \tau_j$. $\square$

Example 5.32.

(i) One may show that for $m = 1$ the system (5.32) has diagonal form: $\partial_t \tau_i = \frac{1}{2} \partial_{xx}^2 \tau_i$, and for $m = 2$ the n-truncated system (5.32) reads as $\partial_t \vec{\tau} = B_{n,1} \partial_{xx}^2 \vec{\tau}$. For $m = 1, 2$ the terms $\partial_{xx}^2 \tau_{n+m}$ are given by

$$\frac{1}{n+1} \partial_{xx}^2 \tau_{n+1} = \sum_{i=1}^n (-1)^{n-i} \frac{1}{i} \sigma_{n-i+1} \partial_{xx}^2 \tau_i + \tilde{a}_1, \tag{5.33}$$

$$\frac{1}{n+2} \partial_{xx}^2 \tau_{n+2} = \sum_{i=1}^n (-1)^{n-i} \frac{1}{i} (\sigma_1 \sigma_{n-i+1} - \sigma_{n-i+2}) \partial_{xx}^2 \tau_i + \tilde{a}_2. \tag{5.34}$$

The proof of (5.33)–(5.34) is similar to proof of (3.22)–(3.23) in [RW2]. By Proposition 5.31, the last row of the matrix $B_{n,1}$ or $\frac{3}{2}(B_{n,1})^2$ consists of the coefficients at $\partial_{xx}^2 \tau_i$'s on the rhs of (5.33), respectively, of (5.34), and so on.

(ii) For $f_j = -2\delta_{j1}$, (5.32) reduces to the heat equations $\partial_t \tau_i = \partial_{xx}^2 \tau_i$ ($i \in \mathbb{N}$), having explicit solution, see Section 5.5.1. Consider more complicated cases.

1. For $f_j = -\delta_{j2}$, (5.32) reduces to the system

$$\partial_t \tau_i = \frac{i}{i+1} \partial_{xx}^2 \tau_{i+1}, \qquad i \in \mathbb{N}, \tag{5.35}$$

whose n-truncated version is: $\partial_t \overrightarrow{\tau} = B_{n,1} \partial_{xx}^2 \overrightarrow{\tau} + a_2$. For $n = 2$, we have

$$\partial_t \tau_1 = (1/2)\partial_{xx}^2 \tau_2,$$
$$\partial_t \tau_2 = (2/3)\partial_{xx}^2 \tau_3 = (\tau_2 - \tau_1^2)\partial_{xx}^2 \tau_1 + \tau_1 \partial_{xx}^2 \tau_2 + 2[(\partial_x \tau_1)(\partial_x \tau_2) - \tau_1 (\partial_x \tau_1)^2],$$

and the matrix $B_{2,1} = \begin{pmatrix} 0 & \frac{1}{2} \\ -2\sigma_2 & \sigma_1 \end{pmatrix}$. If the roots $k_1 \neq k_2$, then the eigenvectors of $B_{2,1}$ are $v_j = (1, 2k_j)$, $j = 1, 2$. For $n = 3$, (5.35) reduces to the quasilinear system of three PDE's with the matrix

$$B_{3,1} = \begin{pmatrix} 0 & \frac{1}{2} & 0 \\ 0 & 0 & \frac{2}{3} \\ 3\sigma_3 & -\frac{3}{2}\sigma_2 & \sigma_1 \end{pmatrix},$$

whose eigenvalues are k_j, and the eigenvectors are $v_j = (1, 2k_j, 3k_j^2)$.

2. For $f_j = -\delta_{j3}$, (5.32) reduces to the system

$$\partial_t \tau_i = \frac{3i}{2(i+2)}\partial_{xx}^2 \tau_{i+2} \quad (i \in \mathbb{N}) \iff \partial_t \overrightarrow{\tau} = -\frac{3}{2}(B_{n,1})^2 \partial_{xx}^2(\overrightarrow{\tau}) + a_3.$$

For $n = 3$, the system has the following matrix:

$$\frac{3}{2}(B_{3,1})^2 = \begin{pmatrix} 0 & 0 & \frac{1}{2} \\ 3\sigma_3 & -\frac{3}{2}\sigma_2 & \sigma_1 \\ \frac{9}{2}\sigma_1\sigma_3 & \frac{9}{4}(\sigma_3 - \sigma_1\sigma_2) & \frac{3}{2}(\sigma_1^2 - \sigma_2) \end{pmatrix},$$

whose eigenvalues are $\frac{3}{2}k_j^2$. These examples can be continued as long as one desires.

5.2.4 Searching for Power Sums

Although (5.22) consists of (second order) non-linear PDE's, the corresponding power sums τ_i ($i > 0$) satisfy an infinite quasilinear system.

Proposition 5.33. *Power sums $\{\tau_i\}_{i \in \mathbb{N}}$ of (5.22) with f_m of type (a) satisfy the infinite system*

$$\partial_t \tau_i = -\frac{i}{2}\sum_{m=1}^{n-1} \frac{m f_m}{i+m-1} N(N(\tau_{i+m-1})) + a_i, \tag{5.36}$$

where $a_i = -\frac{i}{2}\sum_{m=1}^{n-1}[N(N(f_m))\tau_{i+m-1} - f_m \operatorname{trace}(\nabla_N^t(A^{i-1})\nabla_N^t(A^m))]$. The n-truncated system (5.36) is

$$\partial_t \overrightarrow{\tau} = -\sum_{m=1}^{n-1}(m/2) f_m (B_{n,1})^{m-1} N(N(\overrightarrow{\tau})) + a(\overrightarrow{\tau}, N(\overrightarrow{\tau})). \tag{5.37}$$

Proof. Since (5.22), we substitute $B = \nabla^t_N \mathscr{C}(h)$ into $(5.24)_1$ and obtain

$$\partial_t \tau_i = -(i/2)\,\text{trace}\left(A^{i-1}\nabla^t_N\nabla^t_N\left(\sum_{m=0}^{n-1} f_m A^m\right)\right), \quad i > 0.$$

The desired system (5.36) follows from the above and the identity

$$\frac{i+m-1}{m}\,\text{trace}(A^{i-1}\nabla^t_N A^m) = \text{trace}(\nabla^t_N A^{i+m-1}) = N(\tau_{i+m-1}). \tag{5.38}$$

By Proposition 5.31, the system (5.36) is equivalent to (5.37). $\qquad\square$

Remark 5.34. By (5.38), power sums $\{\tau_i\}_{i\in\mathbb{N}}$ of (5.22) with f_m of type (b) satisfy

$$\partial_t \tau_i = -\frac{i}{2}\left[\sum_{m=1}^{n-1}\frac{m f_m}{i+m-1}N(N(\tau_{i+m-1}))\right.$$
$$\left. + \sum_{m=0}^{n-1}\tau_{i+m-1}\sum_{s=1}^{n} f_{m,s}N(N(\tau_s))\right] + a_i, \tag{5.39}$$

where $N(f_m) = \sum_s f_{m,s} N(\tau_s)$ and

$$a_i = -\frac{i}{2}\sum_{m=1}^{n-1}\left[\frac{2m}{i+m-1}N(f_m)N(\tau_{i+m-1}) - f_m \,\text{trace}(\nabla^t_N(A^{i-1})\nabla_N(A^m))\right.$$
$$\left. + \tau_{i+m-1}\sum_{a,b=1}^{n} f_{m,ab}N(\tau_a)N(\tau_b)\right].$$

By Proposition 5.31, the system (5.39) equals to (5.40). The n-truncated (5.39) is

$$\partial_t \overrightarrow{\tau} = -(\widetilde{B}_n + \widetilde{A})N(N(\overrightarrow{\tau})) + a(\overrightarrow{\tau}, N(\overrightarrow{\tau})), \tag{5.40}$$

where a is a smooth vector-function, the matrices $\widetilde{B}_n$ (based on the generalized companion matrix) and $\tilde{A} = (\tilde{A}_{ij})$ are defined by

$$\widetilde{B}_n = \sum_{m=1}^{n-1}(m/2) f_m(\overrightarrow{\tau})(B_{n,1})^{m-1}, \quad \tilde{A}_{ij} = (i/2)\sum_{m=0}^{n-1}\tau_{i+m-1} f_{m,\tau_j}(\overrightarrow{\tau}). \tag{5.41}$$

Example 5.35. Among flows (5.22), the *Newton transformation flow*,

$$\partial_t g_t = \nabla^t_N T_k(h_t) \tag{5.42}$$

depends on h_m $(m \le k)$ and has $f_m = (-1)^m \sigma_{k-m}$ for some $0 < k \le n$. Since $\text{trace}\,T_k(A) = (n-k)\sigma_k$, the corresponding to (5.42) normalized flow is

$$\partial_t g_t = \nabla^t_N T_k(h_t) - \frac{1}{n}\rho(t)g_t^{\top}, \quad \text{where } \rho(t) = (n-k)\int_M N(\sigma_k)\,d\,\text{vol}_t / \text{Vol}(M, g_t).$$

For (5.42), denoting $T_j = T_j(A)$, by Lemma 5.22, we have

$$\partial_t \sigma_i = -\frac{1}{2} \left[N(\operatorname{trace}(T_{i-1} \cdot \nabla_N^t T_k)) - \operatorname{trace}((\nabla_N^t T_{i-1})(\nabla_N^t T_k)) \right]$$

$$= \frac{1}{2} \operatorname{trace}((\nabla_N^t T_{i-1})(\nabla_N^t T_k))$$

$$- \frac{1}{2} \sum_{j=0}^k (-1)^j \left[N(\sigma_{k-j} \operatorname{trace}(T_{i-1} \cdot \nabla_N^t (A^j))) + N(\sigma_{k-j}) \operatorname{trace}(T_{i-1} \cdot A^j)) \right].$$

One may express the above (by induction) through σ's and their N-derivative.

5.2.5 Existence and Uniqueness

Here, we show that under suitable assumptions, flow (5.22) yields the second order parabolic PDE's, for which the existence/uniqueness and in some cases convergence of a solution are shown. For brevity we shall omit the index t for time-dependent tensors A, a, b_j and functions τ_i, σ_i. Generally, flows with f_m of type (a) are solvable under additional conditions (e.g., for A and the auxiliary functions f_m).

The coordinate system described in the following lemma, see [CC, Section 5.1], is here called a *biregular foliated chart*.

Lemma 5.36. *Let M be a differentiable manifold with a codimension-one foliation $\mathscr{F}$ and a vector field N transverse to $\mathscr{F}$. Then for any $q \in M$ there exists a coordinate system $(x_0, x_1, \ldots x_n)$ on a neighborhood $U_q \subset M$ (centered at q) such that the leaves on U_q are given by $\{x_0 = c\}$ (hence, the coordinate vector fields $\partial_i = \partial_{x_i}$, $i \geq 1$, are tangent to leaves), and N is directed along $\partial_0 = \partial_{x_0}$ (one may assume $N = \partial_0$ at q).*

On a foliated Riemannian manifold $(M, \mathscr{F}, g)$ with a unit normal N to $\mathscr{F}$, in biregular foliated coordinates $(x_0, \ldots, x_n)$, one may present the metric g in the form

$$g = g_{00}\, dx_0^2 + \sum_{i,j>0} g_{ij}\, dx_i dx_j, \tag{5.43}$$

where one may assume $g_{00} = 1$ along the axis $(x_0, 0, \ldots, 0)$. Let g^{ij} be the entries of the matrix inverse to (g_{ij}), and $g_{ij,k}$ the derivative of g_{ij} in the ∂_k-direction. The second fundamental form is $h_{ij} = \langle \nabla_{\partial_i} \partial_j, N \rangle$.

Lemma 5.37. *Prove that for a pseudo-Riemannian metric g in orthogonal biregular foliated coordinates of a codimension-one foliation $\mathscr{F}$, one has*

$$N = \partial_0 / \sqrt{|g_{00}|} \quad \text{(the unit normal)},$$

$$h_{ij} = \Gamma_{ij}^0 \sqrt{g_{00}} = -\frac{1}{2} \varepsilon_N \, \delta_{ij}\, g_{ii,0} / \sqrt{|g_{00}|} \quad \text{(the second fundamental form)},$$

$$A_i^j = -\Gamma_{i0}^j / \sqrt{|g_{00}|} = -\frac{1}{2\sqrt{|g_{00}|}} \delta_i^j \frac{g_{ii,0}}{g_{ii}} \quad \text{(the Weingarten operator)},$$

$$\tau_1 = -\frac{1}{2\sqrt{|g_{00}|}} \sum_{i>0} \frac{g_{ii,0}}{g_{ii}}, \quad \tau_2 = \frac{1}{4|g_{00}|} \sum_{i>0} \left(\frac{g_{ii,0}}{g_{ii}} \right)^2, \quad \text{etc.}$$

Proof. This is similar to the proof of [RW2, Lemma 2.2] for Riemannian case. For convenience, observe that the formula for A follows from that for h and $A_i^j = \sum_s h_{is} g^{sj}$. Notice that

$$(A^m)_i^j = \sum_{\{s_l\}} A_{s_2}^j A_{s_3}^{s_2} \dots A_{s_m}^{s_{m-1}} A_i^{s_m}.$$

Formulae for h_m follow from the above and $(A^m)_i^s g_{sj} = g(A^m e_i, e_j) = (h_m)_{ij}$. Formulae for τ's follow directly from the above and the equality $\tau_i = \text{trace}(A^i)$. $\square$

Define the quantities $\mu_{m;ij}$, depending on the principal curvatures (k_i) of $\mathscr{F}$, by

$$\mu_{m;ij} = \sum_{\alpha=0}^{m-1} k_i^\alpha k_j^{m-1-\alpha}, \qquad 1 \le i \le j \le n, \quad 1 \le m < n.$$

Note that $\mu_{1;ij} = 1$ for all i, j. The *ellipticity condition* for (5.22) takes the form

$$\sum_{m=1}^{n-1} f_m \mu_{m;ij} < 0, \qquad 1 \le i \le j \le n. \tag{5.44}$$

For umbilical foliations ($A = \lambda \, \text{id}^\top$), (5.44) reads as $\sum_{m=1}^{n-1} m f_m \lambda^{m-1} < 0$.

The next theorem concerns the local existence and uniqueness of (5.22) with f_m of type (a) and is used in the proof of Theorem 5.40 about flows with f_m of type (b).

Theorem 5.38. *Let (M, g_0) be a closed Riemannian manifold with a codimension-one foliation $\mathscr{F}$, and let $f_m \in C^2(M \times \mathbb{R})$. If (5.44) is satisfied on M for $t = 0$, then there is a unique smooth solution g_t of (5.22) defined on a positive t-interval $[0, \varepsilon)$.*

Proof. Given $q \in M$, by Lemma 5.36, there exist biregular foliated coordinates $(x_0, x_1, \dots x_n)$ on $U_q \subset M$ (with center at q) and the metric has the form (5.43). Then the unit normal to $\mathscr{F}$ on U_q is $N = g_{00}^{-1/2} \partial_0$. Denote $F_{ij} := \mathscr{C}(h)(t, x, g_{ab}, \psi_{ab})_{ij}$ and $\psi_{ab} = \partial_0 g_{ab}$. The system (5.22) with f_m of type (a) has the following form along N-curve $\gamma : x \to \gamma(x)$ (on U_q):

$$\partial_t g_{ij} = g_{00}^{-1/2} \left(\partial_0 F_{ij} - \frac{1}{2} \sum_{s,k} (\psi_{is} g^{sk} F_{kj} + \psi_{is} g^{sk} F_{ik}) \right) \quad (i, j = 1, \dots, n),$$

where $g_{00} = 1$ along the axis $\gamma_0 = (x_0, 0, \dots, 0)$. For example, if $f_m = 0$ ($m \ge 2$) then $F_{ij} = f_0 g_{ij} + f_1 h_{ij}$ ($i, j > 0$), and (5.45) reads as:

$$\partial_t g_{ij} = -\frac{1}{2} g_{00}^{-1/2} f_1 \partial_{00}^2 g_{ij} + a_{ij}(t, x, g_{ab}, \psi_{ab}),$$

which is parabolic when $f_1 < 0$, see also (5.44). Using Theorem 5.108 completes the proof in this case. Now let $f_m \ne 0$ for some $m \ge 2$. Computing the derivative

$$\partial_0 F_{ij} = \sum_{a \le b} \frac{\partial F_{ij}}{\partial \psi_{ab}} \partial_{00}^2 g_{ab} + \sum_{a \le b} \frac{\partial F_{ij}}{\partial g_{ab}} \psi_{ab} + \frac{\partial F_{ij}}{\partial x_0},$$

we rewrite (5.45) as the quasi linear system

$$\partial_t g_{ij} = g_{00}^{-1/2} \sum_{a \leq b} \frac{\partial F_{ij}}{\partial \psi_{ab}} \partial_{00}^2 g_{ab} + c_{ij}(t, x, g_{ab}, \psi_{ab}). \tag{5.45}$$

In general, the square matrix $d_\psi F = \left\{ \frac{\partial F_{ij}}{\partial \psi_{ab}} \right\}$ $(i \leq j, \; a \leq b)$ of order $\frac{1}{2} n(n+1)$ is not symmetric. We claim that (5.45) is strong parabolic. If we change the local coordinate system on M, then the components F_{ij} $(i \leq j)$ and ψ_{ab} $(a \leq b)$ at q will be transformed by the same tensor low. Hence, $d_\psi F$ is a $(1,1)$-tensor on the vector bundle of symmetric $(0,2)$-tensors on $T\mathscr{F}$. So, $d_\psi F(q)$ can be seen as the linear endomorphism of the space of symmetric $(0,2)$-tensors on $T_q \mathscr{F}$.

The strong parabolicity is a pointwise property, so can be considered at any point $q \in M$ in a special biregular foliated chart around q, for example, such that $g_{ij} = \delta_{ij}$ and $A_{ij} = k_i \delta_{ij}$ at q for $t = 0$. (Recall that (k_i) are the principal curvatures of $\mathscr{F}$.) In this chart, the matrix $d_\psi F$ is diagonal, so has real eigenvalues at q, and

$$\frac{\partial F_{ij}}{\partial \psi_{ab}} = \sum_{m \geq 1} f_m(q, 0) \, \mu_{m;ij} \, \delta_{\{a,b\}}^{\{i,j\}}.$$

By (5.44), the system (5.45) is strong parabolic: the constant in the condition (5.143) is $c = \inf_q \inf_{i,j} \left\{ \sum_{m=1}^{n-1} f_m(q,0) \, \mu_{m;ij} \right\} > 0$. By Theorem 5.108, given $q \in M$ there exists a unique solution to (5.45) which is defined in U_q along the N-curve through q for some time interval $[0, \varepsilon_q)$ and satisfies the initial conditions $g_{ij}(0, x) = g_{0,ij}$. Moreover, ε_q depends continuously on $q \in M$. The claim follows from the above and compactness of M. $\qquad\square$

Example 5.39. For $\mathscr{C}(h) = h_2$, the matrix $d_\psi F(q)$ in an orthonormal frame at any point is

$$\frac{\partial (h_2)_{ij}}{\partial \psi_{ab}} = \frac{1}{4} \begin{bmatrix} 2\psi_{11} & 2\psi_{12} & 2\psi_{13} & 0 & 0 & 0 \\ \psi_{12} & \psi_{11} + \psi_{22} & \psi_{23} & \psi_{12} & \psi_{13} & 0 \\ \psi_{13} & \psi_{23} & \psi_{11} + \psi_{33} & 0 & \psi_{12} & \psi_{13} \\ 0 & 2\psi_{12} & 0 & 2\psi_{22} & 2\psi_{23} & 0 \\ 0 & \psi_{13} & \psi_{12} & \psi_{23} & \psi_{22} + \psi_{33} & \psi_{23} \\ 0 & 0 & 2\psi_{13} & 0 & 2\psi_{23} & 2\psi_{33} \end{bmatrix}.$$

At a point q for $t = 0$ (i.e., $\psi_{ab} = 0$, $a \neq b$ and $\psi_{aa} = k_a$) it is diagonal with the elements $\mu_{2;ab} = k_a + k_b$. For instance, let $A_0 = [\frac{1}{2}, 1, 1, \frac{1}{2}, 1, \frac{1}{2}]$ be the diagonal matrix of order six. Then the matrix $A_1 = A_0 d_\psi(h_2)$ is symmetric, and the system (5.45) for $\mathscr{C}(h) = h_2$ is "symmetrizable": $A_0 \partial_t g_{ij} = A_1 \partial_{00}^2 g_{ij} + \{\text{low-order terms}\}$.

For flows with general f_m of type (b), the solution procedure requires additional assumptions (e.g., for the auxiliary functions f_m and the matrix $\widetilde{B}_n$).

The next Theorem 5.40 about flows with f_m of type (b) is based on Theorem 5.38.

Theorem 5.40 (Short Time Existence). *Let (M, g_0) be a closed Riemannian manifold with a codimension-one foliation $\mathscr{F}$ and a unit normal N. Given $f_m \in C^2(\mathbb{R}^n)$, suppose that at $t = 0$ the condition (5.44) is satisfied and the matrix $\widetilde{B}_n + \widetilde{A}$, see (5.41), is negative definite on M. Then there is a unique smooth solution g_t of (5.22) defined on a positive t-interval $[0, \varepsilon)$.*

Proof. Let A_0 and $\vec{\tau}^0$ be the values of extended shape operator and power sums of the principal curvatures k_i of $\mathscr{F}$ determined on $(M, \mathscr{F})$ by a given metric g_0.

(a) *Uniqueness.* Let $g_t^{(1)}$ and $g_t^{(2)}$ be two solutions to (5.22) with the same initial metric g_0. Functions $\overrightarrow{\tau}^{t,1}$ and $\overrightarrow{\tau}^{t,2}$, corresponding to $g_t^{(1)}$ and $g_t^{(2)}$, satisfy (5.39) and have the same initial value $\overrightarrow{\tau}^0$. By Theorem 5.108, a solution of the parabolic system (5.39) is unique, and we have $\overrightarrow{\tau}^{t,1} = \overrightarrow{\tau}^{t,2} = \overrightarrow{\tau}^t$ on some positive time interval $[0, \varepsilon_1)$. Hence, both $g_t^{(1)}$ and $g_t^{(2)}$ satisfy (5.22) with f_m of type (a) with known coefficients $\tilde{f}_j(p,t) := f_j(\overrightarrow{\tau}^t(p))$. By Theorem 5.38, $g_t^{(1)} = g_t^{(2)}$ on some positive time interval $[0, \varepsilon_2)$.

(b) *Existence.* By Proposition 5.34 and Theorem 5.108, (5.39) admits a unique solution $\overrightarrow{\tau}^t$ on a positive time interval $[0, \varepsilon_1)$. By Theorem 5.38, (5.22) with f_m of type (a) with known functions $\tilde{f}_j(\cdot, t) := f_j(\overrightarrow{\tau}^t)$ has a unique solution g_t^* ($g_0^* = g_0$) for $0 \leq t < \varepsilon^*$. The shape operator A_t^* ($A_0^* = A_0$) of $(M, \mathscr{F}, g_t^*)$ satisfies (5.27), hence the power sums of its eigenvalues, $\overrightarrow{\tau}^{t,*}$ ($\overrightarrow{\tau}^{0,*} = \overrightarrow{\tau}^0$), satisfy (5.36) with f_m of type (a) with the same coefficient functions $\tilde{f}_j$. By Proposition 5.33 and Theorem 5.108, the solution of the problem is unique, hence $\overrightarrow{\tau}^t = \overrightarrow{\tau}^{t,*}$, i.e., $\overrightarrow{\tau}^t$ are power sums of eigenvalues of A_t^*. Finally, g_t^* is a solution to (5.22) such that $\overrightarrow{\tau}^t$ are power sums of the principal curvatures of $\mathscr{F}$ in this metric. □

Example 5.41. Assume that $f(\overrightarrow{\tau}) \geq c > 0$. Consider $\partial_t g_t = -f(\overrightarrow{\tau}) \nabla_N^t h_t$. The corresponding system for τ's (see Lemma 5.22 with $B = f(\overrightarrow{\tau}) \nabla_N h$) is

$$\partial_t \tau_i = (1/2) f(\overrightarrow{\tau}) N(N(\tau_i)) + a_i, \qquad i > 0,$$

where $a_i = \frac{1}{2} N(f(\overrightarrow{\tau})) N(\tau_i) - \frac{i}{2} f(\overrightarrow{\tau}) \operatorname{trace}(\nabla_N(A^{i-1}) \nabla_N A)$. The above parabolic system for τ's on N-curves has a unique solution for $0 \leq t \leq \varepsilon$.

A foliation $\mathscr{F}$ is *geometrically taut* if there exists metric on M for which $\mathscr{F}$ is harmonic: $\tau_1 = 0$, see Section 3.2. The known proofs of the existence of such metrics use the Hahn–Banach Theorem and are not constructive. In particular, $\mathscr{F}$ is taut if and only if there is a closed transverse curve intersecting each leaf, see [CC, PRSW].

The next theorem (based on Theorem 5.38) helps us to find in some cases a one-parameter family of metrics converging to a metric for which $\mathscr{F}$ is harmonic.

Theorem 5.42. *The following flow on a closed Riemannian manifold (M, g_0):*

$$\partial_t g_t = -2\mathscr{L}_N h_t \tag{5.46}$$

admits a unique smooth global solution g_t ($t \geq 0$). Moreover, if N-curves compose a fibration $S^1 \xhookrightarrow{i} M \xrightarrow{\pi} B$, then $\mathscr{F}$ is topologically taut, see Corollary 3.34, and g_t converge as $t \to \infty$ to a smooth Riemannian metric $\bar{g}$, and $\mathscr{F}$ is $\bar{g}$-harmonic.

Proof. In biregular foliated coordinates, (5.46) has the form of parabolic PDE's

$$\partial_t g_{ij} = \frac{1}{\sqrt{g_{00}}} \partial_0 \left(\frac{1}{\sqrt{g_{00}}} \partial_0 g_{ij} \right) = \frac{1}{g_{00}} \partial_{00} g_{ij} + \frac{1}{2} \partial_0 \left(\frac{1}{g_{00}} \right) \partial_0 g_{ij}.$$

One may assume $g_{00} = 1$ along the central N-curve $\gamma_0(s)$, hence $\partial_t g_{ij} = \partial_{00} g_{ij}$ along γ_0. This linear system has a unique global solution $g_{ij}(t)$ for all $t \geq 0$, which converges as $t \to \infty$ (along any N-curve) to $\overline{g} = (\overline{g}_{ij}) \geq 0$. Hence, there is a global solution g_t $(t \geq 0)$ on M. If N-curves compose a S^1-fibration, then from $\partial_t(\det g_{ij}) = \frac{1}{2}(\text{trace}_g B) \det g_{ij}$ for $B = -2\mathscr{L}_N h$, see (4.15), we conclude that all g_t are positive definite. Since $\text{trace}\, B = -2N(\tau_1)$, the integral $\int_0^\infty \text{trace}\, B_t\, dt = -2(\overline{\tau}_1 - \tau_1^0)$ is finite. Thus, the limit $\overline{g} = \lim_{t \to \infty} g_t$ exists and is a smooth Riemannian metric on M.

The certain system for τ's, see (5.37) with $\widetilde{B}_n = \frac{1}{2}\,\text{id}^\top$, is diagonal parabolic,

$$\partial_t \tau_i = N(N(\tau_i)) - i\,\text{trace}(\nabla_N(A^{i-1})\nabla_N A).$$

In particular, $\partial_t \tau_1 = N(N(\tau_1))$; from this easily follows that $N(\tau_1)$ satisfies the heat equation $\partial_t(N(\tau_1)) = N(N(N(\tau_1)))$. Hence, $N(\overline{\tau}_1) = \text{const}$ on N-curves. Since $\overline{\tau}_1$ is bounded on a closed manifold M, we get $N(\overline{\tau}_1) = 0$. Using $\text{div}(\overline{\tau}_1 N) = \overline{\tau}_1\,\text{div}\, N + N(\overline{\tau}_1)$, where $\text{div}\, N = -\overline{\tau}_1$, see (1.14), and the Divergence Theorem, we get $\int_M [N(\overline{\tau}_1) - \overline{\tau}_1^2]\, d\overline{\text{vol}} = 0$. By the above, $\overline{\tau}_1 = 0$ on M.   □

Remark 5.43. By Example 5.48 (in Section 5.2.6), the claim in Theorem 5.42 is false without condition "N-curves compose a fibration".

Next, we give applications to the problem of prescribing mean curvature function and examples with harmonic and umbilical foliations and with twisted products.

Proposition 5.44. *The flow (5.46) on a twisted product $M_1 \times_f S^1$ admits a unique smooth solution $g_t \in \mathscr{M}$ for all $t \geq 0$, consisting of twisted product metrics on $M_1 \times_{f_t} S^1$. As $t \to \infty$, the metric g_t converges to the metric g_∞ of the product $(M_1, f_\infty^2 g_1) \times (S^1, g_2)$, where $f_\infty(x) = \int_{y \in S^1} f(0, x, y)\, d\,\text{vol}_g$.*

Proof. The foliation $\mathscr{F} = M_1 \times \{y\}$ is totally umbilical with the *mean curvature* $\tau_1 = -nN(\log f)$ (see Remark 5.23). Define $\phi : M \to \mathbb{R}$ by the equality $f = e^\phi$ and get $\tau_1 = -N(\phi)$. The evolution $\partial_t g(t) = s(t)\, g^\top$ preserves twisted product structure, where $g_t^\top$ corresponds to $g_1(t)$. Thus $g^\top(t) = f(t)^2 g^\top$. Assuming $f(t) = e^{\phi(t)}$, we find $g_t^\top = e^{2\phi(t)} g_0^\top$. In this case, $\partial_t g(t) = 2e^{2\phi(t)}\partial_t \phi_t\, g_0^\top$. Hence, $\partial_t \phi_t = \frac{1}{2}\, s(t)$. For $s = -(2/n)N(\tau_1)$, equation (5.46) reads as

$$\partial_t \phi = (1/n)N(N(\phi)). \tag{5.47}$$

Replacing $t \to t/n$, we reduce (5.47) to the heat equation. By Proposition 5.27, there is a unique solution ϕ_t $(t \geq 0)$, satisfying $\lim_{t \to \infty} \phi_t(x, y) = \phi_\infty(x)$. Then, $f_\infty = e^{\phi_\infty}$.   □

Example 5.45. Given $f : T^2 \to \mathbb{R}_+$, let $S^1 \times_f S^1$ be a twisted product of two circles (S^1, g_i). Then (5.46) has a unique smooth solution $g_t \in \mathscr{M}$ for all $t \geq 0$, consisting of twisted product metrics on the torus T^2. As $t \to \infty$, the metric g_t converges in C^0 to a flat metric g_∞ on T^2.

5.2.6 Extrinsic Geometric Flow on a Foliated Surface

In this section, (M^2, g_0) is a surface equipped with a transversely orientable foliation $\mathscr{F}$ by curves of geodesic curvature λ. Given $\psi \in C^3(\mathbb{R})$, let a 1-parameter family of metrics g_t ($0 \le t < \varepsilon$) satisfies the PDE, see (5.29),

$$\partial_t g_t = -N(\psi(\lambda_t)) g_t^\top. \tag{5.48}$$

Applying (4.15) with $B = -N(\psi(\lambda_t)) g_t^\top$, we have $\partial_t (d\operatorname{vol}_t) = -\frac{1}{2}N(\psi(\lambda_t)) d\operatorname{vol}_t$. Hence, the volume of a closed surface (M^2, g_t) satisfies

$$\frac{d}{dt}\operatorname{Vol}(M, g_t) = -\frac{1}{2}\int_M N(\psi(\lambda_t)) d\operatorname{vol}_t = -\frac{n}{2}\int_M \lambda_t \psi(\lambda_t) d\operatorname{vol}_t. \tag{5.49}$$

Proposition 5.46. *If $0 < C_0(\lambda) \le \psi'(\lambda) \le C_1(\lambda) < \infty$, then (5.48) on a surface $(M^2, \mathscr{F})$ has a unique global smooth solution g_t ($t \ge 0$).*

Proof. The g_t-geodesic curvature λ_t of $\mathscr{F}$ (with respect to N—the g_t-unit normal to $\mathscr{F}$) satisfies the PDE

$$\partial_t \lambda = \frac{1}{2}N(\psi'(\lambda)N(\lambda)), \tag{5.50}$$

see (5.28). By Proposition 5.110, (5.50) (with λ_0 determined by g_0) admits a unique smooth solution λ_t for all $t \ge 0$. In this case, (5.48) has a unique smooth solution g_t on M^2 for all $t \ge 0$, moreover, $g_t^\top = g_0^\top \exp(-\int_0^t N(\psi(\lambda_t)) dt)$. $\square$

Example 5.47. Assume that $\partial_t g = -2N(\lambda_t) g_t^\top$ and conditions of Proposition 5.46 are satisfied. One may show that (5.50) is the heat equation $\partial_t \lambda_t = N(N(\lambda_t))$ on N-curves, and $g_t^\top = g_0^\top \exp(-\int_0^t N(\lambda_\xi) d\xi)$ is its solution. One may show that for a closed surface M^2 the function $\operatorname{Vol}(M, g_t)$ is decreasing. By (5.49),

$$\frac{d}{dt}\operatorname{Vol}(M, g_t) = \int_0^t \lambda_\xi^2 d\xi \le 0.$$

Example 5.48. Assume that a function $f \in C^2(-1,1)$ has vertical asymptotes $x = \pm 1$ and exactly one strong minimum at $x = 0$. For example, $f = \frac{1}{10}\left(e^{x^2/(1-x^2)} - 1\right)$. Consider a *Reeb foliation* $\mathscr{F}$ in the strip $\Pi = [-1,1] \times \mathbb{R}$ (equipped with the flat metric g_0) whose leaves are, see e.g. [Tam],

$$L_\pm = \{x = \pm 1\}, \quad L_s(x) = \{(x, f(x)+s), |x| < 1\}, \quad \text{where} \quad s \in \mathbb{R}.$$

Identifying the points $(-1,y) \simeq (1,y)$, we get a foliated cylinder $S^1 \times \mathbb{R}$. Then, identifying $(1,y) \simeq (1,y+n)$ for $n \in \mathbb{Z}$, we get a foliated flat torus $\Pi' = S^1 \times S^1$.

Denote $\alpha(x)$ the angle between the leaves L_s and the x-axis at the intersection points. That is, f and α are related by

$$f'(x) = \tan \alpha(x) \quad \Leftrightarrow \quad \cos \alpha = [1 + (f')^2]^{-1/2}, \quad \sin \alpha = f'[1 + (f')^2]^{-1/2}.$$

The geodesic curvature of L_s is

$$\lambda_0(x) = f''(x)[1 + (f'(x))^2]^{-3/2} = \alpha'(x) \cdot |\cos\alpha(x)|, \quad |x| < 1.$$

The tangent and unit normal to $\mathscr{F}$ vectors (on the strip) are $X = [\cos\alpha(x), \sin\alpha(x)]$, and $N = [-\sin\alpha(x), \cos\alpha(x)]$. The unit normal N at the origin is directed along y-axis. We use X and N to represent the standard frame in Π,

$$e_1 = \cos\alpha(x)X - \sin\alpha(x)N, \quad e_2 = \sin\alpha(x)X + \cos\alpha(x)N.$$

Taking $\alpha(x) = \frac{\pi}{2}x$, we obtain the Reeb foliation with $\lambda_0(x) = \frac{\pi}{2}\cos(\frac{\pi}{2}x) > 0$.
    Let the metrics g_t for $t \in [0, \varepsilon)$ obey (5.48) on Π'. Thus $g_t^\top = g_0^\top e^{-U}$, where

$$U = \int_0^t N(\psi(\lambda_\xi(x)))\,d\xi = -\sin\alpha(x)\int_0^t \partial_x\psi(\lambda_\xi(x))\,d\xi.$$

So, $g_t(X,X) = e^{-U}$, $g_t(X,N) = 0$, and $g_t(N,N) = 1$. The scalar products $g_{11}(t) = g_t(e_1,e_1)$, $g_{12}(t) = g_t(e_1,e_2)$ and $g_{22}(t) = g_t(e_2,e_2)$ are

$$g_{11}(t) = \sin^2\alpha + \cos^2\alpha\,e^{-U}, \quad g_{12}(t) = \sin\alpha\cos\alpha(e^{-U} - 1),$$
$$g_{22}(t) = \cos^2\alpha + \sin^2\alpha\,e^{-U}.$$

In particular, $|g_t| = g_{11}(t)g_{22}(t) - g_{12}(t)^2 = e^{-U}$. The N-curves $\gamma(s) = (x(s), y(s))$ satisfy the system of ODE's $\frac{d}{ds}x = -\sin\alpha(x)$, $\frac{d}{ds}y = \cos\alpha(x)$. From the first of ODE's we deduce the implicit formula $s = -\int_x^{\phi_s(x)} \frac{d\xi}{\sin\alpha(\xi)}$ for local diffeomorphisms ϕ_s ($s \geq 0$) of the interval $(-1, 1)$. (Indeed, moving along N-curve by the length s from a point (x,y) to a point (x',y'), we find that $x' = \phi_s(x)$.)
    Assume that $\psi(\lambda) = 2\lambda$. By Proposition 5.46, $\partial_t g_t = -2N(\lambda_t)g_t^\top$ has on Π' a unique global solution g_t ($t \geq 0$), and (by Example 5.47) the function $\mathrm{vol}(\Pi', g_t)$ is decreasing. Hence $g_t^\top = g_0^\top e^{-U}$, where $U = -\sin\alpha(x)\int_0^t \partial_x\lambda_\xi(x)\,d\xi$, and λ evolves along N-curves with arc-length parameter $s \in \mathbb{R}$ by the heat equation $\partial_t\lambda = \partial_{ss}^2\lambda$. In terms of x, we have $\partial_t\lambda = \sin\alpha(x)^2\partial_{xx}^2\lambda$. This heat conduction PDE in a material with variable "thermal conductivity" $\sin\alpha(x)^2$ on a circle is parabolic everywhere except one (singular) point $x = 0$. Using Example 5.111 (i), we write solution along N-curves as

$$\lambda_t(\phi_0(s)) = \int_{\mathbb{R}} \lambda_0(\phi_0(\xi))G(t,s,\xi)\,d\xi.$$

Since $\lambda_0(\phi_0(s)) \to 0$ quickly as $|s| \to \infty$, we have uniform convergence $\lambda_t(x) \to 0$ as $t \to \infty$ for any $x \in (-1, 1)$. Along the leaves $L_{\pm\infty}$ the metric does not change, hence $\lambda_t(\pm 1) = 0$ for all $t \geq 0$. It is known that a curve on a surface with $\lambda = 0$ is a geodesic. From the above we conclude that g_t do not converge as $t \to \infty$ to a metric on a torus Π', see Remark 5.43 (otherwise $\mathscr{F}$ becomes a foliation by geodesics, that contradicts to the fact that Reeb foliation is not geodesible). Since g_t do not depend on the y-coordinate, the Gaussian curvature is given by (see [RW2])

$$K_t = -\frac{1}{2\sqrt{|g_t|}}\,\partial_x\Big(\frac{\partial_x g_{22}}{\sqrt{|g_t|}}\Big) = -\frac{1}{2}\,e^U\big(\partial_{xx}^2 g_{22} + \frac{1}{2}\,(\partial_x U)(\partial_x g_{22})\big).$$

Substituting $g_{22}(t) = \cos^2\alpha + \sin^2\alpha \cdot e^{-U}$ into the above formula yields

$$2e^{-U}K_t = -\big(2\cos(2\alpha)\,(\alpha')^2 + \sin(2\alpha)\,\alpha''\big)(e^{-U}-1) + \sin^2\alpha\,e^{-U}\partial_{xx}^2 U$$
$$+ \frac{3}{2}\sin(2\alpha)\,\alpha'e^{-U}\partial_x U - \frac{1}{2}\sin(2\alpha)\,\alpha'\partial_x U - \frac{1}{2}\sin^2\alpha\,e^{-U}(\partial_x U)^2. \tag{5.51}$$

Because $\alpha(0) = 0$ and $U(0) = 0$, by (5.51) one has $K_t(0) \equiv 0$. For $\alpha = \frac{\pi}{2}x$ and small x we have

$$e^{-U}K_t = \frac{3}{8}\,\pi^3 V(0)\,x + o(x), \quad \text{where} \quad V(x) = \int_0^t \partial_x\lambda_\xi(x)\,d\xi.$$

Thus $K_t(x)$ (for $t > 0$) changes its sign when we cross the line $x = 0$.

5.3 The Partial Ricci Flow

This chapter describes an approach of *extrinsic geometric flows* to Toponogov's problem and around it. The Partial Ricci flow, introduced for (co)dimension one foliation as well as for any codimension foliations, [R8, RS], is proposed as the main tool to prescribe the mixed curvature of a foliation. The flow preserves total umbilicity, total geodesy and harmonicity of foliations and is proposed as the tool to prescribe the partial Ricci and mixed curvature of a foliation (Conjecture 1.57 in Section 1.4.4). We prove local existence/uniqueness theorem, leafwise parabolic evolution equations for the curvature tensor, study the case of (co)dimension-one foliations and show that for the warped product initial metric the solution to the normalized flow converges, as $t \to \infty$, to the metric with $\mathrm{Ric}^\perp = \Phi g^\perp$ and leafwise constant Φ. (Co)dimension-one foliations are also discussed, and it is shown that for the initial metric of warped product type the solution for the normalized flow converges to a metric with leaf wise constant S_{mix}.

5.3.1 Preliminaries

One may try to attack Conjecture 1.57 by deforming the metric in directions orthogonal to leaves. The candidate for such a deformation is the partial Ricci flow.

Definition 5.49. The *Partial Ricci Flow* (PRF) is a family of metrics g_t, $t \in [0, \varepsilon)$, satisfying the PDE

$$\partial_t g = -2\,\mathrm{Ric}^\perp(g). \tag{5.52a}$$

The *normalized PRF* is defined by

$$\partial_t g = -2\operatorname{Ric}^{\perp}(g) + 2\,\Phi\,g^{\perp}, \tag{5.52b}$$

where $\Phi : M \to \mathbb{R}$ is a leafwise constant. For a foliated surface, (5.52b) reads as

$$\partial_t g = -2(K - \Phi)\,g^{\perp}.$$

Observe that $\operatorname{Ric}^{\perp}(X,Y) = 0$ if either X or Y is tangent to $\mathscr{F}$. Thus, the PRF preserves $\mathscr{D}$—the orthogonal distribution to $\mathscr{F}$, does not change the geometry of the leaves, and keeps them to be totally umbilical, totally geodesic or harmonic (minimal) submanifolds, see Proposition 5.65.

The fixed points of (5.52b) are metrics with $\operatorname{Ric}^{\perp} = \Phi g^{\perp}$, simple examples are Hopf fibrations of odd-dimensional spheres, see also Section 1.4.1.

In the case of a general foliation, the topology of the leaf through a point can change crucially with the point, this gives many difficulties in studying such PDE's. Thus, we assume (e.g., in Theorem 5.8) the following condition:

> a closed manifold M is fibered instead of being just foliated. $\qquad$ (5.53)

Observe that if (5.53) holds than all the leaves (fibers) are compact.

Theorem 5.50. *Let $\mathscr{F}$ be a smooth foliation of a closed Riemannian manifold (M, g_0). Then (5.52b) under assumptions (5.53) has a unique smooth solution g_t defined on a positive time interval $[0, t_0)$.*

Proof. We use variations of the form $g(t) = g_0 + tB$ with a $\mathscr{D}$-truncated symmetric $(0,2)$-tensor B. By Lemma 5.59, the linearization of $-2\operatorname{Ric}^{\perp}(g)$ is the second-order leafwise elliptic differential operator

$$D(-2\operatorname{Ric}^{\perp}(g))_{ik} = \Delta_{\mathscr{F}} B_{ik} + \tilde{B}_{ik},$$

where $\tilde{B}_{ik}$ consists of the first- and zero-order terms. Thus, $\Delta_{\mathscr{F}} B_{ik}$ yields the principal symbol of order two, and other terms are of order < 2. The claim follows from the theory of parabolic PDE's on vector bundles [AH, Section 5.1] and (5.53). $\qquad\square$

One may try to study the PRF along with the same line as the classical Ricci flow is applied in the proof of the smooth $1/4$-pinching sphere theorem, see e.g. [AH].

In [R8] it was conjectured the following: *Let $\mathscr{F}$ be a p-dimensional totally geodesic foliation of a closed Riemannian manifold (M^{n+p}, g). Assume all mixed curvatures to be sufficiently close to a positive constant. Then the PRF evolves the metric g to a limit metric whose mixed curvature is a positive function of a point.*

The conjecture seems to be an analogue of the result by C. Böhm and B. Wilking.

Theorem 5.51 (See e.g. [AH]). *On a compact manifold the Ricci flow evolves a Riemannian metric with 2-positive curvature operator R (i.e., the sum of the first two eigenvalues of R is positive) to a limit metric with constant sectional curvature.*

Observe the following difference in statements of the conjecture and Theorem 5.51: the sectional curvature of the limit metric is constant in Theorem 5.51, while the mixed curvature can depend on a point in the conjecture. The difference is caused by the absence of Schur's lemma for the mixed curvature. Nevertheless, the statement of the above conjecture implies inequality (1.67).

Theorem 5.98 and Corollaries 5.102 and 5.104 (Section 5.4.4) confirm the conjecture for a special case of warped product metrics when the leaves are space forms.

5.3.2 Time-Dependent Adapted Metrics

In this section, we describe the behavior of the basic tensors of the extrinsic geometry of foliations under $\mathscr{D}$-variations of a metric. Let $B(g)$ be a $\mathscr{D}$-truncated symmetric $(0,2)$-tensor on a foliated Riemannian manifold (M,g). Consider a family of adapted metrics g_t on M (with $0 \le t < \varepsilon$) satisfying PDE

$$\partial_t g = B(g). \tag{5.54}$$

Let $B^\sharp : TM \to TM$ be the $(1,1)$-tensor dual to B, that is $\langle B^\sharp(X), Y \rangle = B(X,Y)$.

Lemma 5.52. *Let the local $\mathscr{D}$-frame $\{\mathscr{E}_i\}$ evolve by (5.54) according to*

$$\partial_t \mathscr{E}_i = -(1/2) B_t^\sharp(\mathscr{E}_i).$$

Then $\{\mathscr{E}_i(t)\}$ is a g_t-orthonormal frame of $\mathscr{D}$ for all t.

Proof. In view of $B_t^\sharp(\mathscr{E}_i) \in \mathscr{D}$, the claim follows from Remark 4.50.   □

Lemma 5.53. *For (5.54) with a $\mathscr{D}$-truncated tensor B and vectors $X, Y \in \mathscr{D}, \xi \in T\mathscr{F}$ we have*

$$2\partial_t \tilde{A}_\xi = -\nabla_\xi B^\sharp + [\tilde{A}_\xi - \tilde{T}_\xi^\sharp, B^\sharp], \qquad \partial_t \tilde{T}_\xi^\sharp = -B^\sharp \tilde{T}_\xi^\sharp, \tag{5.55a}$$

$$2\partial_t \tilde{C}_\xi = -\nabla_\xi B^\sharp + [\tilde{C}_\xi, B^\sharp] - 2\tilde{T}_\xi^\sharp B^\sharp, \tag{5.55b}$$

$$2\partial_t \tilde{H} = -(\nabla \operatorname{trace} B^\sharp)^\top, \tag{5.55c}$$

$$\partial_t h = -B^\sharp \circ h, \qquad \partial_t H = -B^\sharp(H). \tag{5.55d}$$

Proof. Note that $\partial_t T = 0$. For all $X, Y \in \mathscr{D}$, using (4.8) and (1.27), we have

$$2\langle \partial_t(\nabla_X^t Y), \xi \rangle = (\nabla_X^t B)(Y, \xi) + (\nabla_Y^t B)(X, \xi) - (\nabla_\xi^t B)(X,Y)$$
$$= -(\nabla_\xi^t B)(X,Y) - B(Y, \nabla_X^t \xi) - B(X, \nabla_Y^t \xi).$$

From this and symmetry of $\partial_t \nabla^t$, we have

$$2\langle \partial_t \tilde{h}(X,Y), \xi \rangle = -(\nabla_\xi B)(X,Y) + B(Y, C_\xi(X)) + B(X, C_\xi(Y)). \tag{5.56}$$

Using (1.28), we then find

$$\langle \partial_t \tilde{A}_\xi(X), Y \rangle = \partial_t \langle \tilde{h}(X, Y), \xi \rangle - (\partial_t g)(\tilde{A}_\xi(X), Y),$$
$$\langle \partial_t \tilde{T}_\xi^\sharp(X), Y \rangle = -(\partial_t g)(\tilde{T}_\xi^\sharp(X), Y).$$

The above, (1.30) and (5.56) yield (5.55a). Using $\partial_t C_\xi = \partial_t \tilde{A}_\xi + \partial_t \tilde{T}_\xi^\sharp$ and (5.55a), we obtain (5.55b). Next, using Lemma 5.52, we deduce (5.55c):

$$2 \langle \partial_t \tilde{H}, \xi \rangle = 2 \sum_i \langle \partial_t (\tilde{h}(\mathscr{E}_i, \mathscr{E}_i)), \xi \rangle = 2 \sum_i \langle \partial_t \tilde{h}(\mathscr{E}_i, \mathscr{E}_i) + 2\tilde{h}(\partial_t \mathscr{E}_i, \mathscr{E}_i), \xi \rangle$$
$$= \sum_i [-(\nabla_\xi B)(\mathscr{E}_i, \mathscr{E}_i) + 2B(C_\xi(\mathscr{E}_i), \mathscr{E}_i) - \langle \tilde{h}(B^\sharp(\mathscr{E}_i), \mathscr{E}_i), \xi \rangle] = -\xi(\operatorname{trace} B^\sharp).$$

Finally, from (4.8) we have

$$2 \langle \partial_t h(\xi_1, \xi_2), X \rangle = \langle \partial_t(\nabla_{\xi_1}^t \xi_2) + \partial_t(\nabla_{\xi_2}^t \xi_1), X \rangle$$
$$= -B(\nabla_{\xi_1}^t \xi_2, X) + B(\nabla_{\xi_2}^t \xi_1, X) = -2B(h(\xi_1, \xi_2), X).$$

Hence $2 \langle \partial_t h(\xi_1, \xi_2), X \rangle = -2 \langle B^\sharp \circ h(\xi_1, \xi_2), X \rangle$ for $X \in TM$, i.e., the first equation in (5.55d). Since $\partial_t E_a = 0$, we have the second equation in (5.55d): $\partial_t H = \sum_a \partial_t h(E_a, E_a) = -B^\sharp(H)$.   □

Corollary 5.54. *For* (5.54), *the tensors* $\partial_t \tilde{A}_\xi$, $\partial_t \tilde{T}_\xi^\sharp$, *where* $\xi \in T\mathscr{F}$, *and* $\partial_t \operatorname{Ric}^\perp$ *may be not self-adjoint:*

$$(\partial_t \tilde{A}_\xi)^* - \partial_t \tilde{A}_\xi = [B^\sharp, \tilde{A}_\xi], \quad (\partial_t \tilde{T}_\xi^\sharp)^* + \partial_t \tilde{T}_\xi^\sharp = [\tilde{T}_\xi^\sharp, B^\sharp], \qquad (5.57a)$$
$$(\partial_t \operatorname{Ric}^\perp)^* - \partial_t \operatorname{Ric}^\perp = [B^\sharp, \operatorname{Ric}^\perp]. \qquad (5.57b)$$

Proof. From (5.55a) two formulae (5.57a) follow. Notice that

$$\partial_t \operatorname{Ric}^\perp(X, Y) = \partial_t \langle \operatorname{Ric}^\perp(X), Y \rangle = B(\operatorname{Ric}^\perp(X), Y) + \langle (\partial_t \operatorname{Ric}^\perp)(X), Y \rangle \qquad (5.58)$$

for $X, Y \in \mathscr{D}$. From this and symmetry of $\partial_t \operatorname{Ric}^\perp$, the equality (5.57b) follows.   □

The metrics g_t on M in (5.54) can be interpreted as a *natural bundle metric* on the *spatial tangent bundle* E, that is, the pull-back of TM under the projection $M \times (0, \varepsilon) \to M$, $(q, t) \to q$. The fiber of E over a point (q, t) is given by $E_{(q,t)} = T_q M$ and is endowed with the metric g_t. A connection ∇ on a vector bundle E over M is a map $\nabla : \mathscr{X}(M) \times \Gamma(E) \to \Gamma(E)$, written as $(X, \sigma) \to \nabla_X \sigma$, such that, see [AH],

1. ∇ is $C^\infty(M)$-linear in X: $\nabla_{f_1 X_1 + f_2 X_2} \sigma = f_1 \nabla_{X_1} \sigma + f_2 \nabla_{X_2} \sigma$,
2. ∇ is $\mathbb{R}$-linear in σ: $\nabla_X(\lambda_1 \sigma_1 + \lambda_2 \sigma_2) = \lambda_1 \nabla_X \sigma_1 + \lambda_2 \nabla_X \sigma_2$,
3. ∇ satisfies the product rule: $\nabla_X(f\sigma) = X(f)\sigma + f \nabla_X \sigma$.

A connection ∇ on a vector bundle E is said to be *compatible with a metric* g on E if for any $\xi, \eta \in \Gamma(E)$ and $X \in \mathscr{X}(M)$, we have $X\langle \xi, \eta \rangle = \langle \nabla_X \xi, \eta \rangle + \langle \xi, \nabla_X \eta \rangle$. Compatibility by itself is not enough to determine a unique connection. There is a natural connection $\widehat{\nabla}$ on E, which extends the Levi-Civita connection on TM. We need to specify only the covariant time derivative $\widehat{\nabla}_{\partial_t}$. Given any section X of the vector bundle E, we define $\widehat{\nabla}_{\partial_t}$ by

$$\widehat{\nabla}_{\partial_t} X = \partial_t X + (1/2) B^\sharp(X) \text{ for } X \in \mathscr{D}, \qquad \widehat{\nabla}_{\partial_t} \xi = 0 \text{ for } \xi \in T\mathscr{F}. \quad (5.59)$$

Lemma 5.55. *The connection $\widehat{\nabla}$ on E is compatible with the natural bundle metric:*

$$\widehat{\nabla}_{\partial_t} g = 0. \tag{5.60}$$

Proof. One may assume that $X, Y \in \mathscr{D}$ are constant in time. In this case, we have $\widehat{\nabla}_{\partial_t} X = \frac{1}{2} B^\sharp(X)$ and $\widehat{\nabla}_{\partial_t} Y = \frac{1}{2} B^\sharp(Y)$. Since $\partial_t g = B$, this and (5.54) imply (5.60):

$$(\widehat{\nabla}_{\partial_t} g)(X,Y) = \partial_t \langle X,Y \rangle - \langle \widehat{\nabla}_{\partial_t} X, Y \rangle - \langle X, \widehat{\nabla}_{\partial_t} Y \rangle = (\partial_t g)(X,Y) - B(X,Y) = 0. \quad \square$$

This connection is not symmetric: in general, $\widehat{\nabla}_{\partial_t} X \neq 0$, while $\widehat{\nabla}_X \partial_t = 0$ always for $X \in \mathscr{D}$. Clearly, the *torsion tensor* $\mathrm{Tor}(X,Y) := \widehat{\nabla}_X Y - \widehat{\nabla}_Y X - [X,Y]$ (of $\widehat{\nabla}$) vanishes if both arguments are spatial; so, the only nonzero components are

$$\mathrm{Tor}(\partial_t, X) = \widehat{\nabla}_{\partial_t} X - \widehat{\nabla}_X \partial_t = (1/2) B^\sharp(X) \qquad (X \in \mathscr{D}).$$

However, each submanifold $M \times \{t\}$ is totally geodesic; so, computing derivatives of spatial tangent vector fields gives the same result as computing for sections of $T(M \times [0, \varepsilon))$. In particular, the corresponding shape operators satisfy $\tilde{A}_\xi = A_\xi$.

Using connection (5.59), we also have

$$\langle (\widehat{\nabla}_{\partial_t} \tilde{h})(X,Y), \xi \rangle = \langle \partial_t \tilde{h}(X,Y) - \tilde{h}(\widehat{\nabla}_{\partial_t} X, Y) - \tilde{h}(X, \widehat{\nabla}_{\partial_t} Y), \xi \rangle$$
$$= -(1/2)(\nabla_\xi^t B)(X,Y),$$
$$(\widehat{\nabla}_{\partial_t} \tilde{A}_\xi)(X) = (\partial_t \tilde{A}_\xi)(X) - \tilde{A}_\xi(\widehat{\nabla}_{\partial_t} X) = -(1/2)(\nabla_\xi^t B^\sharp)(X) - (1/2)[\tilde{T}_\xi^\sharp, B^\sharp].$$

Since $T\mathscr{F}$ is integrable then, compare with (5.55a), $\widehat{\nabla}_{\partial_t} A_\xi = -(1/2) \widehat{\nabla}_\xi B^\sharp$.

5.3.3 The Leafwise Laplacian of the Curvature Tensor

Here, we find an expression for the $\mathscr{F}$-Laplacian of the curvature tensor. In analogy with [AH, Sect. 4.2.1], we define the (quadratic in the curvature R) tensor $\mathscr{R} \in \Lambda_0^4(M)$ as

$$\mathscr{R}(X,Y,Z,V) = \sum_a \langle R(X, \cdot, Y, E_a), R(Z, \cdot, V, E_a) \rangle \quad \text{for all} \quad X,Y,Z,V \in TM,$$

where $\{E_a\}$ is a local orthonormal frame on $T\mathscr{F}$. Although generally we have $\mathscr{R}(X,Y,Z,V) \neq \mathscr{R}(Y,X,V,Z)$, the tensor $\mathscr{R}$ has some symmetries of the curvature tensor, as

$$\mathscr{R}(X,Y,Z,V) = \mathscr{R}(Z,V,X,Y). \tag{5.61}$$

The leafwise rough Laplacian is defined by $\Delta_{\mathscr{F}} = \text{trace}_{\mathscr{F}}(\nabla^2) = \sum_a \nabla^2_{a,a}$.

Proposition 5.56 (See [RS] for $p = 1$). *On a Riemannian manifold (M,g) endowed with a smooth foliation $\mathscr{F}$, the $\mathscr{F}$-Laplacian of the curvature tensor R satisfies*

$$\Delta_{\mathscr{F}} R(X,Y,Z,V) = \sum_a [\nabla^2_{X,Z} R(Y,E_a,V,E_a) - \nabla^2_{Y,Z} R(X,E_a,V,E_a) \quad (5.62)$$
$$+ \nabla^2_{Y,V} R(X,E_a,Z,E_a) - \nabla^2_{X,V} R(Y,E_a,Z,E_a)] - (\mathscr{R}(X,Y,Z,V)$$
$$- \mathscr{R}(X,Y,V,Z) - \mathscr{R}(Y,X,Z,V) + \mathscr{R}(Y,X,V,Z) - 2\mathscr{R}(Z,Y,V,X)$$
$$+ 2\mathscr{R}(Z,X,V,Y)) + \langle R(\cdot,Y,Z,V), \sum_a R(X,E_a,\cdot,E_a)\rangle$$
$$- \langle R(\cdot,X,Z,V), \sum_a R(Y,E_a,\cdot,E_a)\rangle.$$

Proof. Using $\nabla_a R(X,Y,Z,V) + \nabla_X R(Y,E_a,Z,V) + \nabla_Y R(E_a,X,Z,V) = 0$ (the second Bianchi identity)—together with the linearity over $\mathbb{R}$ of ∇ on the space of tensor fields [AH]—we find that

$$\Delta_{\mathscr{F}} R(X,Y,Z,V) = \sum_a \nabla^2_{a,a} R(X,Y,Z,V)$$
$$= \sum_a \left(\nabla^2_{a,X} R(E_a,Y,Z,V) - \nabla^2_{a,Y} R(E_a,X,Z,V) \right). \tag{5.63}$$

It suffices to express first two terms on rhs of (5.63) using lower order terms. To compute the first term on the rhs of (5.63), we transpose ∇_a and ∇_X,

$$\nabla^2_{a,X} R(E_a,Y,Z,V) = \nabla^2_{X,a} R(E_a,Y,Z,V) + (R(X,E_a)R)(E_a,Y,Z,V). \tag{5.64}$$

Using the second Bianchi identity

$$\nabla_a R(Z,V,E_a,Y) + \nabla_Z R(V,E_a,E_a,Y) + \nabla_V R(E_a,Z,E_a,Y) = 0,$$

we transform the first term on the rhs of (5.64),

$$\nabla^2_{X,a} R(E_a,Y,Z,V) = \nabla^2_{X,Z} R(Y,E_a,V,E_a) - \nabla^2_{X,V} R(E_a,Y,E_a,Z). \tag{5.65}$$

Next, we transform the second term on the rhs of (5.64), using the identity

$$(R(X,Y)R)(Z,U,V,W) = -R(R(X,Y)Z,U,V,W) - R(Z,R(X,Y)U,V,W)$$
$$- R(Z,U,R(X,Y)V,W) - R(Z,U,V,R(X,Y)W)$$

and noting that $R(X,Y)f = 0$ for any function, in our case, $f = R(Z,U,V,W)$ and $R(X,E_a)(R(E_a,Y,Z,V)) = 0$,

$$(R(X,E_a)R)(E_a,Y,Z,V) = -R(R(X,E_a)(E_a,Y,Z,V) - \ldots - R(E_a,Y,Z,R(X,E_a)V)$$
$$= \langle R(X,E_a,\cdot,E_a), R(\cdot,Y,Z,V)\rangle + \langle R(E_a,X,Y,\cdot) R(E_a,\cdot,Z,V)\rangle$$
$$+ \langle R(E_a,X,Z,\cdot), R(E_a,Y,\cdot,V)\rangle + \langle R(E_a,X,V,\cdot), R(E_a,Y,Z,\cdot)\rangle. \tag{5.66}$$

The first term on the rhs of (5.66) yields

$$\langle \sum_a R(X, E_a, \cdot, E_a), R(\cdot, Y, Z, V) \rangle = -\langle R(Y, \cdot, Z, V), \sum_a R(X, E_a, \cdot, E_a) \rangle.$$

We transform the second term on the rhs of (5.66), using the first Bianchi identity,

$$\sum_a \langle R(E_a, X, Y, \cdot) R(E_a, \cdot, Z, V) \rangle = -\sum_a \langle R(\cdot, Y, X, E_a) R(Z, V, \cdot, E_a) \rangle$$
$$= \sum_a \left[\langle R(\cdot, Y, X, E_a) R(\cdot, Z, V, E_a) \rangle + \langle R(\cdot, Y, X, E_a) R(V, \cdot, Z, E_a) \rangle \right]$$
$$= \mathscr{R}(Y, X, Z, V) - \mathscr{R}(Y, X, V, Z).$$

Similarly, the third and the fourth terms in the rhs of (5.66) are transformed as

$$\langle R(E_a, X, Z, \cdot), R(E_a, Y, \cdot, V) \rangle + \langle R(E_a, X, V, \cdot), R(E_a, Y, Z, \cdot) \rangle$$
$$= -\mathscr{R}(Z, X, V, Y) + \mathscr{R}(V, X, Z, Y).$$

Hence, (5.66) takes the following form:

$$\sum_a (R(X, E_a)R)(E_a, Y, Z, V) = \mathscr{R}(Y, X, Z, V) - \mathscr{R}(Y, X, V, Z) - \mathscr{R}(Z, X, V, Y)$$
$$+ \mathscr{R}(V, X, Z, Y) - \langle R(Y, \cdot, Z, V), \sum_a R(X, E_a, \cdot, E_a) \rangle. \tag{5.67}$$

Substituting expressions of (5.65) and (5.67) into (5.64), we have

$$\sum_a \nabla^2_{a,X} R(E_a, Y, Z, V) = \sum_a \left[\nabla^2_{X,Z} R(Y, E_a, V, E_a) - \nabla^2_{X,V} R(Y, E_a, Z, E_a) \right]$$
$$- (\mathscr{R}(Y, X, V, Z) - \mathscr{R}(Y, X, Z, V) + \mathscr{R}(Z, X, V, Y) - \mathscr{R}(V, X, Z, Y))$$
$$- \langle R(Y, \cdot, Z, V), \sum_a R(X, E_a, \cdot, E_a) \rangle.$$

In the same way, we also have

$$\sum_a \nabla^2_{a,Y} R(E_a, X, Z, V) = \sum_a \left[\nabla^2_{Y,Z} R(X, E_a, V, E_a) - \nabla^2_{Y,V} R(X, E_a, Z, E_a) \right]$$
$$- (\mathscr{R}(X, Y, V, Z) - \mathscr{R}(X, Y, Z, V) + \mathscr{R}(Z, Y, V, X) - \mathscr{R}(V, Y, Z, X))$$
$$- \langle R(X, \cdot, Z, V), \sum_a R(Y, E_a, \cdot, E_a) \rangle.$$

By the above, (5.63) reduces to

$$\sum_a \nabla^2_{a,a} R(X, Y, Z, V) = \sum_a \left[\nabla^2_{X,Z} R(Y, E_a, V, E_a) - \nabla^2_{X,V} R(Y, E_a, Z, E_a) \right.$$
$$\left. - \nabla^2_{Y,Z} R(X, E_a, V, E_a) + \nabla^2_{Y,V} R(X, E_a, Z, E_a) \right]$$
$$+ \langle R(\cdot, Y, Z, V), \sum_a R(X, E_a, \cdot, E_a) \rangle - \langle R(\cdot, X, Z, V), \sum_a R(Y, E_a, \cdot, E_a) \rangle$$
$$- (\mathscr{R}(Y, X, V, Z) - \mathscr{R}(Y, X, Z, V) + \mathscr{R}(Z, X, V, Y) - \mathscr{R}(V, X, Z, Y) - \mathscr{R}(X, Y, V, Z)$$
$$+ \mathscr{R}(X, Y, Z, V) - \mathscr{R}(Z, Y, V, X) + \mathscr{R}(V, Y, Z, X)).$$

Using the symmetry (5.61) of $\mathscr{R}$, from the above, we obtain (5.62). □

Remark 5.57. The normal distribution $\mathscr{D}$ to a totally geodesic foliation $\widetilde{\mathscr{F}}$ has the property $R(E_a, X)E_a \in \mathscr{D}$ for $X \in \mathscr{D}$. (This holds, for example, when $\widetilde{\mathscr{D}} = T\widetilde{\mathscr{F}}$ is *curvature-invariant*, i.e., $R(X,Y)(\widetilde{\mathscr{D}}) \subset \widetilde{\mathscr{D}}$ for any $X,Y \in \widetilde{\mathscr{D}}$). In this case, for any vectors $X \in \mathscr{D}, Y \in TM$ we have (using the duality relation)

$$\sum_a R(X,E_a,Y,E_a) = \langle \mathrm{Ric}^\perp(X), Y \rangle = \mathrm{Ric}^\perp(X,Y), \qquad (5.68)$$

and (5.62) reads as

$$\Delta_{\mathscr{F}} R(X,Y,Z,V) = \sum_a \left(\nabla^2_{X,Z} R(Y,E_a,V,E_a) - \nabla^2_{Y,Z} R(X,E_a,V,E_a) \right.$$
$$+ \nabla^2_{Y,V} R(X,E_a,Z,E_a) - \nabla^2_{X,V} R(Y,E_a,Z,E_a) \right)$$
$$- \left(\mathscr{R}(X,Y,Z,V) - \mathscr{R}(X,Y,V,Z) - \mathscr{R}(Y,X,Z,V) + \mathscr{R}(Y,X,V,Z) - 2\mathscr{R}(Z,Y,V,X) \right.$$
$$\left. + 2\mathscr{R}(Z,X,V,Y) \right) + \langle R(\cdot,Y,Z,V), \mathrm{Ric}^\perp(X,\cdot) \rangle - \langle R(\cdot,X,Z,V), \mathrm{Ric}^\perp(Y,\cdot) \rangle.$$

5.3.4 Toward the Linearization of the Partial Ricci Flow

To linearize the differential operator $g \to -2\mathrm{Ric}^\perp(g)$, see (5.52a), on the space $\mathscr{M}$, we need the following.

Proposition 5.58 (See e.g. [AH]). *Let g_t be a family of metrics on a manifold M such that $\partial_t g = B$. Then*

$$2\partial_t R(X,Y,Z,V) = \nabla^2_{X,V} B(Y,Z) + \nabla^2_{Y,Z} B(X,V) - \nabla^2_{X,Z} B(Y,V) - \nabla^2_{Y,V} B(X,Z)$$
$$+ B(R(X,Y)Z, V) - B(R(X,Y)V, Z). \qquad (5.69)$$

Recall that the first and second derivatives of a $(0,2)$-tensor B can be expressed as

$$\nabla_Z B(Y,V) = Z(B(Y,V)) - B(\nabla_Z Y, V) - B(Y, \nabla_Z V),$$
$$\nabla^2_{X,Z} B(Y,V) = \nabla_X(\nabla_Z B)(Y,V) - \nabla_{\nabla_X Z} B(Y,V)$$
$$= \nabla_X(\nabla_Z B(Y,V)) - \nabla_Z B(\nabla_X Y, V) - \nabla_Z B(Y, \nabla_X V) - \nabla_{\nabla_X Z} B(Y,V). \quad (5.70)$$

The tensors C, A and H are given in Section 1.3.1.

Lemma 5.59. *Let (M,g) be a Riemannian manifold with a smooth foliation $\mathscr{F}$. Then the tensor r evolves by (5.54) (with a $\mathscr{D}$-truncated symmetric $(0,2)$-tensor $B(g)$) according to*

$$2\partial_t \mathrm{Ric}^\perp(X,Z) = -\Delta_{\mathscr{F}} B(X,Z)$$
$$+ \sum_a \left[\nabla_a B(C_a(X),Z) + \nabla_a B(C_a(Z),X) + B(C_a^2(X),Z) + B(C_a^2(Z),X) \right.$$
$$\left. - 2B(C_a(X),C_a(Z)) \right] + B(\mathrm{Ric}^\perp(Z),X) + B(\mathrm{Ric}^\perp(X),Z) - \nabla_Z B(H,X)$$
$$- \nabla_X B(H,Z) - B(\nabla_X H,Z) - B(\nabla_Z H,X) + \mathrm{trace}(A_{B^\sharp(X)} A_Z). \qquad (5.71)$$

If, in addition, $\mathscr{F}$ is totally geodesic, then (5.71) has zero 3rd line, and yields

$$2\partial_t \mathrm{Ric}^{\perp} = -\Delta_{\mathscr{F}} B^{\sharp} + \sum_a \left[(\nabla_a B^{\sharp})C_a + C_a^* \nabla_a B^{\sharp} + B^{\sharp} C_a^2 + (C_a^*)^2 B^{\sharp} - 2 C_a^* B^{\sharp} C_a \right]$$
$$+ \mathrm{Ric}^{\perp} B^{\sharp} - B^{\sharp} \mathrm{Ric}^{\perp},$$
$$2\partial_t S_{\mathrm{mix}}(g) = -\Delta_{\mathscr{F}}(\mathrm{trace}\, B^{\sharp}) + 2\sum_a [\mathrm{trace}((\nabla_a B^{\sharp})A_a) + 2\,\mathrm{trace}(B^{\sharp}(T_a^{\sharp})^2)]. \quad (5.72)$$

Proof. Since the tensor $\mathrm{Ric}^{\perp}$ is $\mathscr{D}$-truncated, one may assume $X, Z \in \mathscr{D}$, and then calculate the time derivative $\partial_t \mathrm{Ric}^{\perp}(X,Z) = \sum_a \partial_t R(X, E_a, Z, E_a)$. By Proposition 5.58 with $Y = V = E_i$, we then have

$$2\partial_t \mathrm{Ric}^{\perp}(X,Z) = B(\mathrm{Ric}^{\perp}(X), Z)$$
$$+ \sum_a \left[\nabla^2_{X,a} B(E_a, Z) + \nabla^2_{a,Z} B(X, E_i) - \nabla^2_{X,Z} B(E_a, E_a) - \nabla^2_{a,a} B(X,Z) \right]. \quad (5.73)$$

By definition (1.26), we have $(\nabla_X E_a)^{\perp} = -C_{E_a}(X)$, and we can take a local vector field X with the property $C_{E_a}(X) = -\nabla_a X$ at a fixed point $x \in M$. Define the bilinear form $F_{\xi} : \mathscr{D} \times \mathscr{D} \to \mathbb{R}$ for $\xi \in T\mathscr{F}$ ($F_{\xi} = 0$ for a totally geodesic foliation) by

$$F_{\xi}(Z, X) = \langle ((\nabla_Z A)_X - A_X A_Z)(\xi), \xi \rangle.$$

One may calculate, $\sum_a F_{E_a}(Z, X) = \langle \nabla_Z H, X \rangle - \mathrm{trace}_{\mathscr{F}}(A_X A_Z)$. Note that

$$\sum_a F_{E_a}(Z, B^{\sharp}(X)) = B(\nabla_Z H, X) - \mathrm{trace}_{\mathscr{F}}(A_{B^{\sharp}(X)} A_Z).$$

By the above and (5.70), for a $\mathscr{D}$-truncated symmetric $(0,2)$-tensor B we have $\nabla_Z B(X, E_a) = B(X, C_a(Z))$ and $\nabla_Z B(E_a, E_a) = 0$; hence,

$$\nabla^2_{X,a} B(E_a, Z) = \nabla_X(\nabla_a B(E_a, Z)) - \nabla_a B(\nabla_X E_a, Z)$$
$$- \nabla_a B(E_a, \nabla_X Z) - \nabla_{\nabla_X E_a} B(E_a, Z)$$
$$= \nabla_a B(C_a(X), Z) + B(C_a^2(X), Z) - \nabla_X B(\nabla_a E_a, Z) - B(\nabla_X(\nabla_a E_a), Z),$$
$$\nabla^2_{a,Z} B(X, E_a) = \nabla_a(\nabla_Z B(X, E_a)) - \nabla_Z B(\nabla_a X, E_a)$$
$$- \nabla_Z B(X, \nabla_a E_a) - \nabla_{\nabla_a Z} B(X, E_a)$$
$$= \nabla_a B(C_a(Z), X) - \nabla_Z B(\nabla_a E_a, X) + B(\nabla_a(C_a(Z)), X) + B(C_a^2(Z), X)$$
$$\overset{(1.40)}{=} \nabla_a B(C_a(Z), X) + B(C_a^2(Z), X) + B(R(E_a)(Z), X)$$
$$- F_{E_a}(Z, B^{\sharp}(X)) - \nabla_Z B(\nabla_a E_a, X),$$
$$\nabla^2_{X,Z} B(E_a, E_a) = \nabla_X(\nabla_Z B(E_a, E_a)) - 2\nabla_Z B(\nabla_X E_a, E_a) - \nabla_{\nabla_X Z} B(E_a, E_a)$$
$$= 2 B(C_a(X), C_a(Z)).$$

By the above and $H = \sum_a \nabla_a E_a$, (5.73) reduces to (5.71). $\qquad\square$

5.3.5 Evolution of the Curvature Tensor

In this section, we derive evolution equations for the Riemann curvature tensor, the partial Ricci curvature, and S_{mix} along the PRF equations. Define the difference tensor

$$Q(X,Z;Y,V) = \sum_a \nabla^2_{X,Z} R(Y,E_a,V,E_a) - \nabla^2_{X,Z} \mathrm{Ric}^{\perp}(Y,V).$$

Lemma 5.60. *For $Y,V \in \mathscr{D}$ and $X,Z \in TM$ we have*

$$\begin{aligned}
Q(X,Z;Y,V) = \sum_a &\big[\nabla_X R(Y,C_a(Z),V,E_a) + \nabla_X R(Y,E_a,V,C_a(Z)) \\
&+ \nabla_Z R(Y,C_a(X),V,E_a) + \nabla_Z R(Y,E_a,V,C_a(X)) + R(Y,\nabla_X C_a(Z),V,E_a) \\
&+ R(Y,E_a,V,\nabla_X C_a(Z)) - R(Y,C_a(Z),V,C_a(X)) - R(Y,C_a(X),V,C_a(Z)) \big].
\end{aligned}$$

Proof. We calculate

$$\begin{aligned}
\sum_a \nabla^2_{X,Z} R(Y,E_a,V,E_a) &= \sum_a \big[\nabla_X(\nabla_Z R)(Y,E_a,V,E_a) - \nabla_{\nabla_X Z} R(Y,E_a,V,E_a) \big] \\
&= \nabla^2_{X,Z} \mathrm{Ric}^{\perp}(Y,V) + \sum_a \big[\nabla_X R(Y,C_a(Z),V,E_a) + \nabla_X R(Y,E_a,V,C_a(Z)) \\
&+ \nabla_Z R(Y,C_a(X),V,E_a) + \nabla_Z R(Y,E_a,V,C_a(X)) + R(Y,\nabla_X C_a(Z),V,E_a) \\
&- R(Y,C_a(Z),V,C_a(X)) - R(Y,C_a(X),V,C_a(Z)) + R(Y,E_a,V,\nabla_X C_a(Z)) \big]
\end{aligned}$$

using

$$\begin{aligned}
\sum_a \nabla_Z R(Y,E_a,V,E_a) &= \nabla_Z \mathrm{Ric}^{\perp}(Y,V) - \sum_a \big[R(Y,\nabla_Z E_a,V,E_a) + R(Y,E_a,V,\nabla_Z E_a) \big], \\
\nabla^2_{X,Z} \mathrm{Ric}^{\perp}(Y,V) &= \nabla_X(\nabla_Z \mathrm{Ric}^{\perp}(Y,V)) - \nabla_Z \mathrm{Ric}^{\perp}(Y,\nabla_X V) - \nabla_Z \mathrm{Ric}^{\perp}(\nabla_X Y,V) \\
&- \nabla_{\nabla_X Z} \mathrm{Ric}^{\perp}(Y,V).
\end{aligned}$$

The above yields the claim.   □

By Lemma 5.60, the tensor

$$\tilde{Q}(X,Y,Z,V) := Q(X,Z;Y,V) - Q(Y,Z;X,V) + Q(Y,V;X,Z) - Q(X,V;Y,Z) \tag{5.74}$$

does not contain second-order derivatives when at least two vectors of $\{X,Y,Z,V\}$ belong to $\mathscr{D}$. Using Gauss and Codazzi equations for submanifolds, one may study the remaining case and show that $\tilde{Q}$ does not contain the second-order derivatives when at most one vector belongs to $\mathscr{D}$. By Lemma 5.60, we find (when at least three vectors of $\{X,Y,Z,V\}$ belong to $\mathscr{D}$)

$$\tilde{Q}(X,Y,Z,V) = \sum_a [\nabla_{C_a(Z)} R(X,Y,E_a,V) + \nabla_{C_a(Y)} R(X,E_a,Z,V) + \nabla_{C_a(X)} R(E_a,Y,Z,V)$$
$$+ \nabla_{C_a(V)} R(X,Y,Z,E_a) + \nabla_a R(X,C_a(Y),Z,V) + \nabla_a R(C_a(X),Y,Z,V)$$
$$+ \nabla_a R(X,Y,Z,C_a(V)) + \nabla_a R(X,Y,C_a(Z),V) + R(Y,\nabla_X C_a(Z),V,E_a)$$
$$+ R(Y,E_a,V,\nabla_X C_a(Z)) - R(Y,C_a(Z),V,C_a(X)) - R(Y,C_a(X),V,C_a(Z))$$
$$- R(Y,\nabla_X C_a(V),Z,E_a) - R(Y,E_a,Z,\nabla_X C_a(V)) + R(Y,C_a(V),Z,C_a(X))$$
$$+ R(Y,C_a(X),Z,C_a(V)) + R(X,\nabla_Y C_a(V),Z,E_a) + R(X,E_a,Z,\nabla_Y C_a(V))$$
$$- R(X,C_a(V),Z,C_a(Y)) - R(X,C_a(Y),Z,C_a(V)) - R(X,\nabla_Y C_a(Z),V,E_a)$$
$$- R(X,E_a,V,\nabla_Y C_a(Z)) + R(X,C_a(Z),V,C_a(Y)) + R(X,C_a(Y),V,C_a(Z))].$$

Theorem 5.61. *Let $\mathscr{F}$ be a smooth foliation of (M,g), and at least two vectors of $\{X,Y,Z,V\}$ belong to $\mathscr{D}$. Then the curvature tensor evolves by (5.52a) according to a leafwise heat equation*

$$\partial_t R(X,Y,Z,V) = \Delta_{\mathscr{F}} R(X,Y,Z,V) + \mathscr{R}(X,Y,Z,V) - \mathscr{R}(X,Y,V,Z)$$
$$- \mathscr{R}(Y,X,Z,V) + \mathscr{R}(Y,X,V,Z) - 2\mathscr{R}(Z,Y,V,X) + 2\mathscr{R}(Z,X,V,Y)$$
$$- \mathrm{Ric}^\perp(R(X,Y)V,Z) - \mathrm{Ric}^\perp(R(X,Y)Z,V) - \langle R(\cdot,Y,Z,V), \sum_a R(X,E_a,\cdot,E_a)\rangle$$
$$+ \langle R(\cdot,X,Z,V), \sum_a R(Y,E_a,\cdot,E_a)\rangle - \tilde{Q}. \tag{5.75}$$

Proof. Applying (5.69) with $B = -2\,\mathrm{Ric}^\perp$, we have

$$\partial_t R(X,Y,Z,V) = \nabla^2_{X,Z} \mathrm{Ric}^\perp(Y,V) - \nabla^2_{Y,Z} \mathrm{Ric}^\perp(X,V) + \nabla^2_{Y,V} \mathrm{Ric}^\perp(X,Z)$$
$$- \nabla^2_{X,V} \mathrm{Ric}^\perp(Y,Z) - \mathrm{Ric}^\perp(R(X,Y)V,\,Z) - \mathrm{Ric}^\perp(R(X,Y)Z,\,V). \tag{5.76}$$

Comparing (5.76) with (5.62), and using (5.74), completes the proof.   □

Example 5.62. Let $\mathscr{F}$ be a one-dimensional foliation spanned by a unit vector field N. Then

$$\mathscr{R}(X,Y,V,Z) = \langle R(X,\,\cdot\,,Y,N),\, R(V,\,\cdot\,,Z,N)\rangle \quad \text{for all} \quad X,Y,V,Z \in TM.$$

Formula (5.62) takes the form, see also [RS],

$$\nabla^2_{N,N} R(X,Y,Z,V) = [\nabla^2_{X,Z} R(Y,N,V,N) - \nabla^2_{Y,Z} R(X,N,V,N)$$
$$+ \nabla^2_{Y,V} R(X,N,Z,N) - \nabla^2_{X,V} R(Y,N,Z,N)] - (\mathscr{R}(X,Y,Z,V)$$
$$- \mathscr{R}(X,Y,V,Z) - \mathscr{R}(Y,X,Z,V) + \mathscr{R}(Y,X,V,Z) - 2\mathscr{R}(Z,Y,V,X)$$
$$+ 2\mathscr{R}(Z,X,V,Y)) + \langle R(\cdot,Y,Z,V),\, \mathrm{Ric}^\perp(X,\cdot)\rangle - \langle R(\cdot,X,Z,V), \mathrm{Ric}^\perp(Y,\cdot)\rangle.$$

Note that $Q(X,Z;Y,V) = \nabla^2_{X,Z} R(Y,N,V,N) - \nabla^2_{X,Z} \mathrm{Ric}^\perp(Y,V)$. Hence, the curvature tensor evolves by (5.52a) according to a heat-type equation along N-curves

$$\partial_t R(X,Y,Z,V) = \nabla^2_{N,N} R(X,Y,Z,V) + (\mathscr{R}(X,Y,Z,V) - \mathscr{R}(X,Y,V,Z)$$
$$- \mathscr{R}(Y,X,Z,V) + \mathscr{R}(Y,X,V,Z) - 2\mathscr{R}(Z,Y,V,X) + 2\mathscr{R}(Z,X,V,Y))$$
$$- \langle R(\cdot,Y,Z,V), \operatorname{Ric}^\perp(X,\cdot)\rangle + \langle R(\cdot,X,Z,V), \operatorname{Ric}^\perp(Y,\cdot)\rangle - \tilde{Q}.$$

Theorem 5.63. *Let $\mathscr{F}$ be a smooth foliation of (M,g). Then the tensor $\operatorname{Ric}^\perp$ evolves by (5.52a) according to*

$$\partial_t \operatorname{Ric}^\perp(X,Z) = \Delta_{\mathscr{F}} \operatorname{Ric}^\perp(X,Z) - 2\operatorname{Ric}^\perp(X,\operatorname{Ric}^\perp(Z))$$
$$- \sum_a \big[\nabla_a \operatorname{Ric}^\perp(C_a(X),Z) + \nabla_a \operatorname{Ric}^\perp(X,C_a(Z)) + \operatorname{Ric}^\perp(X,C_a^2(Z))$$
$$+ \operatorname{Ric}^\perp(C_a^2(X),Z) + 2\operatorname{Ric}^\perp(C_a(X),C_a(Z))\big] - \operatorname{trace}_{\mathscr{F}}(A_{\operatorname{Ric}^\perp(X)}A_Z)$$
$$+ \nabla_X \operatorname{Ric}^\perp(H,Z) + \nabla_Z \operatorname{Ric}^\perp(H,X) + \operatorname{Ric}^\perp(\nabla_X H,Z) + \operatorname{Ric}^\perp(\nabla_Z H,X), \quad (5.77)$$

where $X,Z \in \mathscr{D}$. For a totally umbilical foliation $\mathscr{F}$, we have

$$\operatorname{trace}_{\mathscr{F}}(A_{\operatorname{Ric}^\perp(X)}A_Z) = (1/p)\,\langle H,Z\rangle \operatorname{Ric}^\perp(H,X)$$

in (5.77), and for a totally geodesic $\mathscr{F}$ we obtain

$$\partial_t \operatorname{Ric}^\perp(X,Z) = \Delta_{\mathscr{F}} \operatorname{Ric}^\perp(X,Z) - 2\operatorname{Ric}^\perp(X,\operatorname{Ric}^\perp(Z))$$
$$- \sum_a \big[\nabla_a \operatorname{Ric}^\perp(C_a(X),Z) + \nabla_a \operatorname{Ric}^\perp(X,C_a(Z))$$
$$+ \operatorname{Ric}^\perp(X,C_a^2(Z)) + \operatorname{Ric}^\perp(C_a^2(X),Z) - 2\operatorname{Ric}^\perp(C_a(X),C_a(Z))\big], \quad (5.78a)$$
$$\partial_t S_{\operatorname{mix}} = \Delta_{\mathscr{F}} S_{\operatorname{mix}} - 2\sum_a \big[\operatorname{trace}(A_a \nabla_a R_{E_a}) + 2\operatorname{trace}((T_a^\sharp)^2 R_{E_a})\big]. \quad (5.78b)$$

Proof. Substituting $B = -2\operatorname{Ric}^\perp$ into (5.71), we obtain (5.77), which for $h = 0$ yields (5.78a). By (5.78a) and (5.58) we get

$$\partial_t \operatorname{Ric}^\perp = \Delta_{\mathscr{F}} \operatorname{Ric}^\perp$$
$$- \sum_a \big[(\nabla_a R_{E_a})C_a + (C_a)^*(\nabla_a R_{E_a}) + R_{E_a}C_a^2 + (C_a^2)^* R_{E_a} - 2(C_a)^* R_{E_a}C_a\big]. \quad (5.79)$$

Tracing (5.79) and using

$$\partial_t(\operatorname{trace}\operatorname{Ric}^\perp) = \operatorname{trace}(\partial_t\operatorname{Ric}^\perp), \quad \operatorname{trace}(T^\sharp \nabla_a R_{E_a}) = 0,$$

yield

$$\partial_t S_{\operatorname{mix}}(g) = \Delta_{\mathscr{F}} S_{\operatorname{mix}}(g) - 2\sum_a \big[\operatorname{trace}(A_a \nabla_a R_{E_a}) + 2\operatorname{trace}(C_a T_a^\sharp R_{E_a})\big].$$

By this, the skew-symmetry of $T_a^\sharp$ and $\operatorname{trace}(S_1 S_2) = \operatorname{trace}(S_2 S_1)$ we get (5.78b). $\square$

We apply Uhlenbeck's trick (see [AH]) to remove a group of terms in (5.75) with a 'change of variables'.

Corollary 5.64. *Let $\mathscr{F}$ be a totally geodesic foliation of a Riemannian manifold (M,g). Then the curvature tensor evolves by (5.52a) according to*

$$\widehat{\nabla}_{\partial_t} R(X,Y,Z,V) = \Delta_{\mathscr{F}} R(X,Y,Z,V) + \mathscr{R}(X,Y,Z,V) - \mathscr{R}(X,Y,V,Z) - \mathscr{R}(Y,X,Z,V)$$
$$+ \mathscr{R}(Y,X,V,Z) - 2\mathscr{R}(Z,Y,V,X) + 2\mathscr{R}(Z,X,V,Y) - \tilde{Q}, \quad (5.80)$$

where $X,Y,Z,V \in \mathscr{D}$, $\widehat{\nabla}$ is a natural connection on the spatial tangent bundle E and $\tilde{Q}$ is given in (5.74).

Proof. Using definition $\widehat{\nabla}_{\partial_t} X = \partial_t X - \mathrm{Ric}^{\perp}(X) = -\mathrm{Ric}^{\perp}(X)$, we obtain

$$\widehat{\nabla}_{\partial_t} R(X,Y,Z,V) = \partial_t R(X,Y,Z,V) - R(-\widehat{\nabla}_{\partial_t} X, Y, Z, V) - \ldots - R(X,Y,Z,-\widehat{\nabla}_{\partial_t} V)$$
$$= \partial_t R(X,Y,Z,V) + \langle R(\cdot, Y, Z, V),\ \mathrm{Ric}^{\perp}(X, \cdot) \rangle + \langle R(X, \cdot, Z, V),\ \mathrm{Ric}^{\perp}(Y, \cdot) \rangle$$
$$+ \langle R(X, Y, \cdot, V),\ \mathrm{Ric}^{\perp}(Z, \cdot) \rangle + \langle R(X, Y, Z, \cdot),\ \mathrm{Ric}^{\perp}(V, \cdot) \rangle.$$

This, (5.75) and (5.68) provide (5.80). $\qquad\qquad\qquad\qquad\qquad\qquad\qquad\qquad\square$

5.3.6 Evolution of the Extrinsic Geometry

In this section, we deduce the system of evolution equations (that are parabolic along the leaves) for the extrinsic geometry, e.g., the co-nullity operator, shape operator, second fundamental form and so on.

Proposition 5.65. *The normalized PRF (5.52b), preserves the metric of $\widetilde{\mathscr{D}} = T\mathscr{F}$ and the orthogonality of vectors to $\mathscr{F}$. If $\mathscr{F}$ is either totally umbilical, totally geodesic or harmonic foliation for $t = 0$, then it has the same property for all $t > 0$.*

Proof. Since $\mathrm{Ric}^{\perp}(\xi, \cdot) = 0$ for any $\xi \in \widetilde{\mathscr{D}}$, (5.52b) preserves the metric on $\widetilde{\mathscr{D}}$ and the scalar product $\langle \xi, X \rangle$ for any $X \in \mathscr{D}$. Hence the normalized PRF preserves the distribution $\mathscr{D}$ orthogonal to our p-dimensional foliation $\mathscr{F}$. By (5.55d) with $B = -2\mathrm{Ric}^{\perp} + 2\Phi g^{\perp}$, we have

$$\partial_t h = 2\mathrm{Ric}^{\perp} \circ h - 2\Phi h, \qquad \partial_t H = 2\mathrm{Ric}^{\perp}(H) - 2\Phi H. \qquad (5.81)$$

Hence (5.52b) preserves total geodesy and harmonicity of foliations. By (5.81) we have

$$\partial_t (h - (H/p)\, g^{\top}) = 2(\mathrm{Ric}^{\perp} - \Phi\, \mathrm{id}^{\top}) \circ (h - (H/p)\, g^{\top}).$$

By the local theorem of existence and uniqueness of a solution to ODE, if $h = (H/p)\, g^{\top}$ ($\mathscr{F}$ is totally umbilical) for $t = 0$ then $h = (H/p)\, g^{\top}$ for all $t > 0$. $\quad\square$

Example 5.66. Let $M = B \times_{\varphi} \bar{M}$ be a warped product. Then $\mathrm{Ric}^{\perp} = -(\Delta_{\mathscr{F}}\, \varphi/\varphi)\, g^{\perp}$. One may apply the existence/uniqueness Theorem 5.8 to conclude that (5.52b) preserves total umbilicity of foliations with integrable orthogonal distribution. In particular, the flow (5.52b) preserves warped product metrics.

By Corollary 5.54 with $B = -2\mathrm{Ric}^{\perp} + 2\Phi g^{\perp}$ and any $\xi \in \widetilde{\mathscr{D}}$, we have the following symmetries of the PRF:

$$(\partial_t \tilde{A}_\xi)^* - \partial_t \tilde{A}_\xi = 2[\tilde{A}_\xi, \mathrm{Ric}^\perp], \quad (\partial_t \tilde{T}^\sharp_\xi)^* + \partial_t \tilde{T}^\sharp_\xi = -2[\tilde{T}^\sharp_\xi, \mathrm{Ric}^\perp],$$
$$(\partial_t \mathrm{Ric}^\perp)^* = \partial_t \mathrm{Ric}^\perp.$$

Proposition 5.67. *Let $\mathscr{F}$ be a harmonic foliation. Then the tensor $\tilde{h}$ and the mean curvature vector $\tilde{H}$ of $\mathscr{D}$ evolve by (5.52b) according to*

$$\partial_t \tilde{h}(X,Y) = \nabla^{\mathscr{F}} \mathrm{div}_{\mathscr{F}}\, \tilde{h}(X,Y) - \mathrm{div}_{\mathscr{F}}\left(\tilde{h}(X,C_\circ(Y)) + \tilde{h}(C_\circ(X),Y)\right) + \Phi\tilde{h}(X,Y)$$
$$+ \langle[C_\circ^*(\mathscr{A}+\mathscr{T}) + (\mathscr{A}+\mathscr{T})C_\circ - \nabla^{\mathscr{F}}(\mathscr{A}+\mathscr{T})](X),Y\rangle$$
$$+ \mathrm{trace}(A_{C_\circ(Y)}A_X + A_{C_\circ(X)}A_Y) - \nabla^{\mathscr{F}}\mathrm{trace}(A_Y A_X), \qquad (5.82a)$$
$$\partial_t \tilde{H} = \nabla^{\mathscr{F}}(\mathrm{div}_{\mathscr{F}}\,\tilde{H}) + \nabla^{\mathscr{F}}(\|\tilde{T}\|^2 - \|\tilde{h}\|^2 - \|h\|^2). \qquad (5.82b)$$

The tensors $\tilde{A}_\xi$, $\tilde{T}^\sharp_\xi$ and $\tilde{C}_\xi = \tilde{A}_\xi + \tilde{T}^\sharp_\xi$ evolve by (5.52b) according to

$$\partial_t \tilde{A}_\xi = \nabla_\xi \mathrm{Ric}^\perp + [\tilde{T}^\sharp_\xi - \tilde{A}_\xi,\ \mathrm{Ric}^\perp], \quad \partial_t \tilde{T}^\sharp_\xi = 2(\mathrm{Ric}^\perp - \Phi\,\mathrm{id}^\perp)\tilde{T}^\sharp_\xi,$$
$$\partial_t \tilde{C}_\xi = \nabla_\xi \mathrm{Ric}^\perp + [\mathrm{Ric}^\perp, \tilde{C}_\xi] + 2\tilde{T}^\sharp_\xi \mathrm{Ric}^\perp - 2\Phi\tilde{T}^\sharp_\xi. \qquad (5.83)$$

Proof. From (5.56) (with $B = -2\,\mathrm{Ric}^\perp + 2\Phi g^\perp$) we get (5.82a). From (5.55c) we get

$$\partial_t \tilde{H} = \nabla^{\mathscr{F}} S_{\mathrm{mix}}(g_t). \qquad (5.84)$$

Substituting S_{mix} from (1.34) into (5.84) and using (1.44) (or tracing (5.82a)) yield (5.82b). Indeed, from (5.55a,b) with $B = -2\,\mathrm{Ric}^\perp + 2\Phi g^\perp$ we get (5.83). $\quad\square$

5.3.7 Examples with (Co)Dimension One Foliations

Here, we show that PRF preserves several classes of foliations and under certain conditions prove existence/uniqueness of global leaf wise smooth solution metrics and convergence of a solution.

5.3.7.1 Totally Geodesic Foliations of Codimension One

Let $\mathscr{F}$ be a codimension-one totally geodesic foliation of a Riemannian manifold (M^{p+1}, g) with a unit normal vector field N. Then M is locally (globally, if M is simply connected and the leaves are complete) isometric to a product manifold $F^p \times \mathbb{R}$, with a *twisted* product metric $dx^2 + \varphi^2 dy^2$, where dx^2 is a fixed metric on F^p and $\varphi \in C^\infty(F^p \times \mathbb{R})$, see [PR, Theorem 1]. The curvature vector of N-curves is $\tilde{H} = -\nabla^{\mathscr{F}} \log \varphi$. Note that $T = 0$, $H = 0$, and $S_{\mathrm{mix}} = \mathrm{Ric}_{N,N}$, and (1.34) reads

$$\mathrm{Ric}_{N,N} = \mathrm{div}_{\mathscr{F}}\,\tilde{H} - \|\tilde{H}\|^2 = \mathrm{div}\,\tilde{H}. \qquad (5.85)$$

Since $\mathrm{Ric}^{\perp} = \mathrm{Ric}_{N,N}\,g^{\perp}$, (5.52b) with a leafwise constant $\Phi : M \to \mathbb{R}$ reduces to

$$\partial_t g = -2\,(\mathrm{Ric}_{N,N} - \Phi)\,g^{\perp}. \tag{5.86}$$

The spectrum of $\Delta_{\mathscr{F}}$ on a compact leaf F is an infinite sequence of isolated real eigenvalues $0 = \lambda_0 < \lambda_1 \le \lambda_2 \le \ldots$ counting their multiplicities, and $\lim_{j \to \infty} \lambda_j = \infty$. One may fix in $L^2(F)$ an orthonormal basis of corresponding eigenfunctions $\{e_j\}$, i.e., $-\Delta_{\mathscr{F}}(e_j) = \lambda_j e_j$, and $e_0 = \mathrm{const} > 0$.

Theorem 5.68. *Let $\mathscr{F}$ be a totally geodesic foliation of codimension-one with simply connected leaves and a unit normal N on a Riemannian manifold (M, g_0), and let assumptions (5.53) be satisfied. Then (5.86) has a unique global smooth solution g_t $(t \ge 0)$. If $\Phi = 0$, then, the metrics g_t converge, as $t \to \infty$, with the exponential rate λ_1 to the limit smooth metric g_{∞} and*

$$\mathrm{Ric}_{N,N}^{\infty} = 0, \qquad \tilde{H}_{\infty} = 0.$$

Proof. From (5.82b) with $\tilde{T} = 0$, $h = 0$ and $\|\tilde{h}\| = \|\tilde{H}\|$ we obtain the following Burgers-type equation for $\tilde{H}$

$$\partial_t \tilde{H} + \nabla^{\mathscr{F}}(\|\tilde{H}\|^2) = \nabla^{\mathscr{F}}(\mathrm{div}_{\mathscr{F}}\,\tilde{H}). \tag{5.87}$$

For any leaf (fiber) F, by Reeb Stability Theorem, see e.g. [CC], there is a simply connected neighborhood $U_F \simeq F \times \mathbb{R}$ such that $\tilde{H}_0 = -\nabla^{\mathscr{F}} \log u_0$ for a smooth function $u_0 > 0$ on U_F. One may take $\tilde{H} = -\nabla^{\mathscr{F}} \log u$, where the function $u(t,x) > 0$ obeys the heat equation

$$\partial_t u = \Delta_{\mathscr{F}}\, u, \qquad u(0, \cdot) = u_0. \tag{5.88}$$

The Cauchy's problem (5.88) has a unique global (smooth for $t > 0$) solution and $\lim_{t \to \infty} u = \bar{u} > 0$ is leafwise constant. Denote by $(\cdot, \cdot)_0$ and $\|\cdot\|_0$ the inner product and the norm in $L^2(F)$. Using Fourier series representation

$$u = (u_0, e_0)_0\, e_0 + e^{-\lambda_1 t} \sum_{j>1} e^{(\lambda_1 - \lambda_j)t}(u_0, e_j)_0\, e_j,$$

we find $\nabla^{\mathscr{F}} u = e^{-\lambda_1 t} \sum_{j>1} e^{(\lambda_1 - \lambda_j)t}(u_0, e_j)_0 \nabla^{\mathscr{F}} e_j$. Since the series above converge absolutely and uniformly with exponential rate, and $(u_0, e_0)_0 > 0$, we have $\lim_{t \to \infty} u = (u_0, e_0)_0\, e_0 > 0$ is leafwise constant and $\lim_{t \to \infty} \nabla^{\mathscr{F}} u = 0$, see [RZe1, Proposition 4]. Hence (5.87) admits a unique smooth solution H and

$$\lim_{t \to \infty} \tilde{H} = \lim_{t \to \infty} \frac{\nabla^{\mathscr{F}} u}{u} = \lim_{t \to \infty} \frac{e^{-\lambda_1 t} \sum_{j>1} e^{(\lambda_1 - \lambda_j)t}(u_0, e_j)_0 \nabla^{\mathscr{F}} e_j}{(u_0, e_0)_0\, e_0 + e^{-\lambda_1 t} \sum_{j>1} e^{(\lambda_1 - \lambda_j)t}(u_0, e_j)_0\, e_j} = 0.$$

Note that $\mathrm{div}_{\mathscr{F}}\,\tilde{H} = \mathrm{div}_{\mathscr{F}}(u^{-1}\nabla^{\mathscr{F}} u) = u^{-1}\Delta_{\mathscr{F}}\, u - |\nabla^{\mathscr{F}} u|^2/u^2 \to 0$ as $t \to \infty$. By (5.85), this corresponds to smooth functions $\mathrm{Ric}_t(N,N)$, and

$$\lim_{t\to\infty} \mathrm{Ric}_{N,N}(t) = \lim_{t\to\infty}(\mathrm{div}_{\mathscr{F}}\,\tilde{H} - \|\tilde{H}\|^2) = 0,$$

where convergence is exponential. Then we recover the metrics g_t $(t \geq 0)$ from (5.86). If $\Phi = 0$, then g_t converge exponentially to a smooth limit metric $\bar{g}$. □

5.3.7.2 Foliations by Geodesics

Let $\mathscr{F}$ be a one-dimensional geodesic foliation of M^{n+1} spanned by a unit vector field N. Then $\mathrm{Ric}^\perp = R_N$—the Jacobi operator. For short, denote by $C := \tilde{C}_N$ the co-nullity operator, $T^\sharp := \tilde{T}^\sharp_N$ the integrability tensor, and $A := \tilde{A}_N$ the shape operator of normal distribution $\mathscr{D}$, see also Example 1.38. By Theorem 5.63 and Lemma 5.53 with $\dim \mathscr{F} = 1$ and $B = -2\,\mathrm{Ric}^\perp$, we have the following (see details in [RS]).

Proposition 5.69. *The curvature evolves by* (5.52a) *according to*

$$\partial_t \mathrm{Ric}^\perp(X,Y) = \nabla^2_{N,N}\mathrm{Ric}^\perp(X,Y) - \nabla_N\mathrm{Ric}^\perp(C(X),Y) - \nabla_N\mathrm{Ric}^\perp(X,C(Y))$$
$$- \mathrm{Ric}^\perp(C^2(X),Y) - \mathrm{Ric}^\perp(X,C^2(Y)) + 2\,\mathrm{Ric}^\perp(C(X),C(Y)) - 2\,\mathrm{Ric}^\perp(X,R_N(Y)),$$
$$\partial_t R_N = \nabla^2_{N,N}R_N - (\nabla_N R_N)C - C^*\nabla_N R_N - R_N C^2 - (C^*)^2 R_N + 2C^*R_N C,$$
$$\partial_t \mathrm{Ric}_{N,N} = N(N(\mathrm{Ric}_{N,N})) - 2\,\mathrm{trace}(A\,\nabla_N R_N) - 4\,\mathrm{trace}((T^\sharp)^2 R_N). \qquad (5.89)$$

For the normalized PRF (5.52b), *we also have*

$$\partial_t C = \nabla_N(\nabla_N C) - (C + C^*)\nabla_N C - (C - C^*)C^2 - 2\,\Phi T^\sharp,$$
$$\partial_t T^\sharp = 2(\nabla_N A)T^\sharp - 2A^2 T^\sharp - 2(T^\sharp)^3 - 2\,\Phi T^\sharp,$$
$$\partial_t A = \nabla_N(\nabla_N A) - 2A\nabla_N A + [A^2,\,T^\sharp] - 2(T^\sharp)^2 A - 2T^\sharp A T^\sharp,$$
$$\partial_t H = \nabla_N(\nabla_N H) - \nabla_N\,\mathrm{trace}(A^2) - 4\,\mathrm{trace}((T^\sharp)^2 A). \qquad (5.90)$$

Remark 5.70. Let the normal distribution to N be integrable. Using $(5.90)_3$ and definition (5.59), one may show that the operator A evolves by (5.52a) according to

$$\widehat{\nabla}_{\partial_t} A = \nabla_N(\nabla_N A) - \nabla_N(A^2).$$

Due to (5.59), we obtain $\widehat{\nabla}_{\partial_t} = \partial_t - R_N$. If $\mathscr{D}$ is integrable, then (5.83) reads as

$$\partial_t A = \nabla_N(\nabla_N A) - 2A\nabla_N A. \qquad (5.91)$$

One may assume that $X \in \mathscr{D}$ is constant in time. In this case, replacing R_N due to $(1.42)_1$, we have

$$(\widehat{\nabla}_{\partial_t} A)(X) = \widehat{\nabla}_{\partial_t}(A(X)) - A(\widehat{\nabla}_{\partial_t} X) = \partial_t(A(X)) - R_N A(X) - A(-R_N(X))$$
$$= (\partial_t A - (\nabla_N A)A + A\nabla_N A)(X)$$

for any $X \in \mathscr{D}$. Applying (5.91), we obtain the required equation.

Example 5.71. Let $\mathscr{F}$ be a one-dimensional Riemannian foliation of M^{n+1} by geodesics. One may show, using the existence and uniqueness Theorem 5.8, that the flow (5.52b) preserves these properties (that is similar to Remark 5.66).

Next theorem deals with Riemannian foliations by geodesics such that $T \neq 0$, certain examples are provided by Hopf fibrations of odd-dimensional spheres.

Recall [Bl] that a *K-contact structure* is a contact metric structure for which the characteristic field is Killing.

Theorem 5.72. *Let $\mathscr{F}$ be a Riemannian foliation by geodesics with nowhere integrable orthogonal distribution. Suppose that $\mathrm{Ric}^\perp \leq \Phi g^\perp$ and $\mathrm{Ric}^\perp_{|\mathscr{D}} > 0$ at $t = 0$. Then (5.52b) admits a unique solution g_t ($t \in \mathbb{R}$) such that $\lim\limits_{t \to \infty} R_N(t) = 0$, $\lim\limits_{t \to -\infty} R_N(t) = \Phi\,\mathrm{id}^\perp$ and there exists metric $\hat{g} = \lim\limits_{t \to -\infty} g_t$. Moreover, $(M, \hat{g})$ is a K-contact manifold for $\Phi = 1$.*

Proof. By Proposition 5.71, we have $A = 0$ for $t \geq 0$; hence, $C = T^\sharp$. By (1.43), $\nabla_N T^\sharp = 0$ and $R_N = -(T^\sharp)^2 \geq 0$. This yields

$$\nabla_N R_N = 0, \quad \nabla_N \mathrm{Ric}^\perp = 0, \quad N(\mathrm{Ric}_{N,N}) = 0$$

for $t \geq 0$; hence, (5.89)–(5.90) reduce to ODE's in the variable t. By (5.72) with $p = 1$ and $B^\sharp = -2R_N + 2\,\Phi\,\mathrm{id}^\perp$, we obtain

$$\partial_t R_N = -R_N(T^\sharp)^2 - (T^\sharp)^2 R_N - 2T^\sharp R_N T^\sharp - 4\Phi R_N = 4R_N(R_N - \Phi\,\mathrm{id}^\perp),$$
$$\partial_t \mathrm{Ric}_{N,N} = -4\,\mathrm{trace}((T^\sharp)^2 R_N) - 4\Phi \mathrm{Ric}_{N,N} = 4\,\mathrm{trace}(R_N^2) - 4\Phi \mathrm{Ric}_{N,N}$$
$$\geq \frac{4}{n}(\mathrm{Ric}_{N,N})^2 - 4\Phi \mathrm{Ric}_{N,N}.$$

Thus, (5.52b) preserves condition $\mathrm{Ric}_{N,N} > 0$ for $t > 0$. In our case $\mathrm{Ric}^\perp_{|\mathscr{D}} > 0$, the dimension n (of $\mathscr{D}$) should be even. Otherwise, the skew-symmetric operator $T^\sharp$ has zero eigenvalues; thus, $R_N = -(T^\sharp)^2$ also is singular—a contradiction. Let $\mu_i(t) > 0$ be the eigenvalue and $e_i(t)$ the eigenvector of $R_N(t)$ under the flow (5.52b). Then

$$\partial_t e_i = (\mu_i - \Phi)e_i, \quad \partial_t(\log \mu_i) = 4(\mu_i - \Phi),$$

where $g_t(e_i(t), e_j(t)) = \delta_{ij}$. Thus, $\mu_i(t) = \frac{\mu_i(0)\Phi}{\mu_i(0) + \exp(4\Phi t)(\Phi - \mu_i(0))}$ with $\mu_i(0) > 0$ and the PRF preserves directions of the eigenvectors of R_N. Since $e_i(t) = z_i(t)e_i(0)$ with $z_i(0) = 1$, then $\partial_t \log z_i = \mu_i - 1$. The above provide $z_i = (\mu_i(t)/\mu_i(0))^{1/4}$. Hence

$$g_t(e_i(0), e_j(0)) = z_i^{-1} z_j^{-1}(t)\, g_0(e_i(0), e_j(0)) = \delta_{ij}(\mu_i(t)\mu_j(t)/(\mu_i(0)\mu_j(0)))^{-1/4}.$$

As $t \to -\infty$, g_t converges to a metric given by $\hat{g}(e_i(0), e_j(0)) = \delta_{ij}(\mu_i(0)\mu_j(0))^{\frac{1}{4}}$. By [Bl, Proposition 7.4], M is K-contact for metric $\hat{g}$ with $\hat{R}_N = \mathrm{id}^\perp$. $\qquad\square$

5.3.8 Around an Almost Contact Structure

A foliation $\mathscr{F}$ on a Riemannian manifold (M, g) is called a *tangentially Lie foliation* or $\mathfrak{g}$-foliation, if there is a complete Lie parallelism $\{\xi_1, \ldots, \xi_p\}$ on its leaves, that preserves the horizontal subbundle $\mathscr{H} = \mathscr{F}^{\perp}$. Here, we introduce new metric structures on $\mathfrak{g}$-foliations [RWo2], that form wider and less rigid classes than classical structures in Section 1.4.1, i.e., their identity operator is replaced by a nonsingular $(1,1)$-tensor. We discuss briefly the properties of novel structures that demonstrate their similarity with the corresponding classical structures. Then we prove, using the partial Ricci flow of metrics, which in the case of a $\mathfrak{g}$-foliation reduces to ODE's, that our structures with positive partial Ricci curvature retract to the classical structures (with positive constant partial Ricci curvature).

5.3.8.1 Weak Almost Contact Structure

Here, we study structures, which generalize contact metric structures, and show convergence of the PRF for these structures with positive partial Ricci curvature.

Definition 5.73. A *weak almost contact structure* is an odd-dimensional manifold M endowed with a $(1,1)$-tensor field ϕ, a characteristic vector field ξ, a dual 1-form η, and a nonsingular $(1,1)$-tensor field Q satisfying

$$\phi^2 = -Q + \eta \otimes \xi, \quad \eta(\xi) = 1,$$
$$Q\xi = \xi. \tag{5.92}$$

If for a weak almost contact structure there exists a metric g such that

$$g(\phi X, \phi Y) = g(X, QY) - \eta(X)\eta(Y), \tag{5.93}$$

then we get a *weak almost contact metric structure*. A weak almost contact manifold is *normal* if $N_\phi + 2d\eta \otimes \xi = 0$, where N_ϕ is the Nijenhuis torsion of ϕ. A *weak Sasakian structure* is a normal weak almost contact metric structure, whose fundamental 2-form F defined by $F(X, Y) = g(X, \phi Y)$ is $d\eta$. This condition ensures that $T_\xi^\sharp = \phi|_{\mathscr{D}}$.

The following proposition generalizes [Bl, Theorem 4.1].

Proposition 5.74. *Let* (ϕ, ξ, η, Q) *be a weak almost contact structure on* M^{2n+1}. *Then* ϕ *has rank* $2n$ *and the following equalities hold:*

$$\phi\xi = 0, \quad \eta \circ \phi = 0, \quad [Q, \phi] = 0.$$

Proof. By (5.92), $\phi^2\xi = 0$, hence, either $\phi\xi = 0$ or $\phi\xi$ is a nontrivial vector of $\ker \phi$. Applying (5.92) to $\phi\xi$, we get $Q(\phi\xi) = \eta(\phi\xi)\xi$. If $\phi\xi = \mu\xi$ for some nonzero function $\mu : M \to \mathbb{R}$, then $0 = \phi^2\xi = \mu \cdot \phi\xi = \mu^2\xi \neq 0$—a contradiction.

Assuming $\phi\xi = \mu\xi + X$ for some $\mu : M \to \mathbb{R}$ and nonzero $X \in \ker\eta$, again by (5.92) we get $QX = 0$—a contradiction. Thus, $\phi\xi = 0$.

Next, since $\phi\xi = 0$ everywhere, $\operatorname{rank}\phi < 2n + 1$. If a vector field $\overline{\xi}$ satisfies $\phi\overline{\xi} = 0$, then (5.92) gives $Q\overline{\xi} = \eta(\overline{\xi})\xi$. One may write $\overline{\xi} = \mu\xi + X$ for some $\mu : M \to \mathbb{R}$ and $X \in \ker\eta$. This yields $QX = 0$, hence, $\overline{\xi}$ is collinear with ξ, and so $\operatorname{rank}\phi = 2n$.

To show $\eta \circ \phi = 0$, observe that from (5.92) we get $[Q, \phi] = 0$. Since $\phi\xi = 0$, we also have, applying (5.92),

$$\eta(\phi X) = \phi^3 X + Q(\phi X) = Q\phi X - \phi QX = [Q, \phi](X) = 0$$

for any X, that proves the claim. $\qquad\square$

Proposition 5.75. *For a weak almost contact metric structure, the tensor ϕ is skew-symmetric and the tensor Q is self-adjoint,*

$$g(\phi X, Y) = -g(X, \phi Y), \quad g(QX, Y) = g(X, QY). \tag{5.94}$$

Proof. Setting $Y = \xi$ in (5.93) and using $Q\xi = \xi$, we get $\eta(X) = g(X, \xi)$. By (5.93), the tensor field Q is self-adjoint, For any $Y \in \mathscr{D}$ there is $\tilde{Y} \in \mathscr{D}$ such that $\phi Y = \tilde{Y}$. Thus, $(5.94)_1$ follows from (5.93) and $(5.94)_2$ for $X \in \mathscr{D}$ and $\tilde{Y}$. $\qquad\square$

Definition 5.76. We say that an endomorphism $F : \mathscr{D} \to \mathscr{D}$ has a *skew-symmetric representation* if there exists an isomorphism $j : \mathscr{D} \to \mathscr{D}^*$ such that the (0,2)-tensor $\tilde{F} : (X, Y) \to j(Y)(F(X))$ is skew-symmetric; or equivalently, for any $x \in M$ there exist a neighborhood $U_x \subset M$ and a frame $\{e_i\}$ on U_x, for which F has a skew-symmetric matrix.

Next, we generalize the fact that any almost contact structure admits a compatible metric.

Proposition 5.77. *If ϕ in (5.92) has a skew-symmetric representation, then the weak almost contact structure admits a compatible metric.*

Proof. Let (5.92) holds and ϕ has a skew-symmetric matrix in a local frame $\{e_i\}$ on a domain $U \subset M$. There exists metric g_U on U such that $\{e_i\}$ is orthonormal. Thus, (5.94) holds for g_U, in particular, $g_U(\phi(X), X) = 0$ for all $X \in TU$. The last property is preserved when summing a finite number of metrics. Hence, using a partition of unity, we get a metric g on M with the same property $g(\phi(X), X) = 0$ for all $X \in \mathfrak{X}_M$, i.e., ϕ is skew-symmetric for g. Thus, (5.93) is valid. $\qquad\square$

Example 5.78. Let (M, g, ϕ, ξ, η) be an almost contact metric manifold. For any $(1,1)$-tensor ϕ' on M commuting with ϕ, define $\tilde{\phi} = \phi + \phi'$ and $Q = \operatorname{id}_{TM} - (\phi\phi' + \phi'\phi) - (\phi')^2$ on $\mathscr{D}$. Then $(M, g, \tilde{\phi}, \xi, \eta, Q)$ is a *weak almost contact manifold* when $|\phi'|$ is sufficiently small.

Next, we extend Definition 5.73 for a finite set of weak almost contact structures.

Definition 5.79. A set of p weak almost contact structures with the same tensor Q on an $(n + p + np)$-dimensional manifold M, satisfying the following conditions:

$$\phi_i \circ \phi_j = -\delta_{ij} Q + \eta^j \otimes \xi_i + \sum_k \varepsilon_{ijk} \phi_k, \quad i, j, k \in \{1, \ldots, p\},$$
$$Q \xi_i = \xi_i, \quad i \in \{1, \ldots, p\}, \tag{5.95}$$

where ε_{ijk} is the completely antisymmetric symbol, is called a *weak almost p-contact structure*. We obtain a *weak almost p-contact metric structure*, if there exists a metric g *compatible* with each of our weak almost contact structures,

$$g(\phi_i X, \phi_i Y) = g(X, QY) - \eta^i(X) \eta^i(Y). \tag{5.96}$$

For $p = 3$, it generalizes an almost 3-contact (metric) structure. We get a *weak p-Sasakian structure* if each of the structures $(\phi_i, \xi_i, \eta^i, Q)$ is weak Sasakian.

Theorem 5.80. *Let $(\phi_i, \xi_i, \eta^i, Q)$ be weak p-Sasakian structures on (M, g_0), such that $\widetilde{\mathscr{D}} = \mathrm{span}(\xi_1, \ldots, \xi_p)$ determines a tangentially Lie foliation. If $\mathrm{Ric}^\perp(g_0) > 0$ on the normal distribution $\mathscr{D}$, then (5.52b) with $\Phi = p$ has a unique solution g_t ($t \in \mathbb{R}$); moreover, there exists metric $\hat{g} = \lim_{t \to -\infty} g_t$ with $r(\hat{g}) = p \hat{g}^\perp$, that corresponds to an almost p-contact metric structure.*

Proof. Observe that $(\phi_i(t), \xi_i, \eta^i, Q(t))$ is a weak p-Sasakian structure on (M, g_t). In our case, we have $\mathrm{Ric}^\perp = -\sum_i \phi_i^2 = p Q$ on $\mathscr{D}$, thus

$$\sum_i T_{\xi_i}^\sharp \mathrm{Ric}^\perp T_{\xi_i}^\sharp = -p Q^2 = -(\mathrm{Ric}^\perp)^2 / p.$$

By the above, we obtain the ODE

$$\partial_t \mathrm{Ric}^\perp = 2 \mathrm{Ric}^\perp \left((1 + 1/p) \mathrm{Ric}^\perp - 2 \Phi \, \mathrm{id}^\perp \right).$$

Each eigenvalue μ_j of $\mathrm{Ric}^\perp$ satisfies $\dot{\mu}_j = 2 \frac{p+1}{p} \mu_j (\mu_j - p)$, which has solution $\mu_j(t) = \frac{\mu_j(0) \, p}{\mu_j(0) + \exp(2(p+1)t)(p - \mu_j(0))}$ with $\mu_j(0) > 0$ (functions on M) and $\lim_{t \to -\infty} \mu_j(t) = \Phi$. Thus, the metric $\hat{g}$ determines an almost p-contact metric structure. $\square$

Proposition 5.81. *Let (ϕ, ξ, η, Q) be a weak Sasakian structure on (M, g_0) such that ξ is a unit Killing vector. If the sectional curvature $K_0(\xi, X) > 0$ for $X \perp \xi$ and metric g_0, then (5.52b) with $\Phi = 1$ has a unique global solution g_t ($t \in \mathbb{R}$); moreover, there exists metric $\hat{g} = \lim_{t \to -\infty} g_t$ that gives a Sasakian structure.*

Proof. By Theorem 5.80 with $p = 1$, the metric $\hat{g}$ determines a Sasakian structure with $R_\xi = \mathrm{id}$. $\square$

5.3.8.2 Weak Almost f-Structure

In this subsection we study a structure, which generalizes an almost f-structure and show the convergence of the partial Ricci flow for this structure.

Definition 5.82. A *weak f-structure* on a manifold M^{2n+p} is defined by a $(1,1)$-tensor field f of rank $2n$ and a nonsingular $(1,1)$-tensor field Q satisfying

$$f^3 + fQ = 0,$$
$$Q\xi = \xi, \quad \xi \in \ker f. \tag{5.97}$$

A *weak globally framed f-structure* occurs when there exist vector fields ξ_i, $i \in \{1,\dots,p\}$, with their dual 1-forms η^i, satisfying

$$f^2 = -Q + \sum_i \eta^i \otimes \xi_i, \quad \eta^i(\xi_j) = \delta^i_j. \tag{5.98}$$

A pseudo-Riemannian metric g is *compatible* if

$$g(fX, fY) = g(X, QY) - \sum_i \varepsilon_i \eta^i(X) \eta^i(Y) \tag{5.99}$$

with $\varepsilon_i = g(\xi_i, \xi_i) = \pm 1$ (and we get a *metric weak f-structure*).

Notice that TM splits into complementary subbundles $\mathscr{D} = f(TM)$ and $\widetilde{\mathscr{D}} = \ker f = \text{span}\{\xi_1,\dots,\xi_p\}$. From (5.99) we get $g(X,\xi_i) = \varepsilon_i \eta^i(X)$. Similarly to Example 5.78, we can build an example of a weak almost f-manifold.

Proposition 5.83. *Let (f,ξ_i,η^i,Q) be a weak f-structure on M^{2n+p}, see (5.98). Then $f\xi_i = 0$ and $\eta^i \circ f = 0$ for any $i \in \{1,\dots,p\}$.*

Proof. By (5.98), $f^2\xi_i = 0$. Applying (5.97) to ξ_i, we get $f\xi_i = 0$. To show $\eta^i \circ f = 0$, observe that from (5.98) and $Q\xi_i = \xi_i$ we get $[Q, f] = 0$. Since $f\xi_i = 0$, we also have, applying (5.98),

$$\eta^i(fX) = f(f^2 X) + Q(fX) = Q(fX) - f(QX) = [Q, f](X) = 0$$

for any X, that proves the claim. $\qquad\qquad\qquad\qquad\qquad\qquad\square$

Remark 5.84. Similarly to Proposition 5.75, one may prove the following.

(a) For a metric weak almost f-manifold, the Reeb vector fields $\{\xi_i\}$ are orthonormal with respect to g, the tensor Q is self-adjoint and the tensor f is skew-symmetric:

$$g(QX, Y) = g(X, QY), \quad g(fX, Y) = -g(X, fY).$$

(b) If the tensor field f in (5.98) has a skew-symmetric representation then the weak f-structure admits a compatible metric.

Set $F(X,Y) = g(X, f(Y))$ for $X, Y \in \mathfrak{X}_M$. A metric weak globally framed f-manifold will be called a *weak almost $\mathscr{S}$-manifold* if $d\eta^i = F$ for any i. A normal (i.e., $N_f + 2\sum_i d\eta^i \otimes \xi_i = 0$) weak almost $\mathscr{S}$-manifold is called a *weak $\mathscr{S}$-manifold*.

Theorem 5.85. *Let (f,ξ_i,η^i,Q) be a metric weak $\mathscr{S}$-structure on (M,g) such that $\widetilde{\mathscr{D}} = \text{span}(\xi_1,\dots,\xi_p)$ determines a tangentially Lie foliation. Consider the partial Ricci flow (5.52b) of metrics g_t on M, and redefine on normal distribution $\mathscr{D}$,*

$$Q(t) = (1/p)\mathrm{Ric}_t^{\perp}, \quad f(t) = T_{\xi_i}^{\sharp}(t).$$

Then $(f(t), \xi_i, \eta^i, Q(t))$ is a weak almost $\mathscr{S}$-structure on (M, g_t). If $\mathrm{Ric}^{\perp}(g) > 0$ on $\mathscr{D}$, then (5.52b) with $\Phi > 0$ has a unique global solution g_t ($t \in \mathbb{R}$); and there exists metric $\hat{g} = \lim_{t \to -\infty} g_t$ for which

$$\lim_{t \to -\infty} \mathrm{Ric}^{\perp}(g_t) = \Phi \hat{g}^{\perp}, \quad \lim_{t \to -\infty} Q(t)|_{\mathscr{D}} = (\Phi/p)\,\mathrm{id}^{\perp}.$$

In particular, for $\Phi = p$ the limit yields a metric almost $\mathscr{S}$-structure.

The proof is similar to the proof of Theorem 5.80.

Remark 5.86. One can define a *weak para-f-structure* on a manifold M^{2n+p} by a $(1,1)$-tensor field f of rank $2n$ and a nonsingular $(1,1)$-tensor field Q satisfying

$$f^3 - fQ = 0,$$
$$Q\xi = \xi, \quad \xi \in \ker f.$$

If there exist vector fields ξ_i, $1 \le i \le p$, and dual 1-forms $\{\eta^i\}$, satisfying

$$f^2 = Q - \sum_i \eta^i \otimes \xi_i, \quad \eta^i(\xi_j) = \delta_j^i,$$

then M is a *weak almost para-f-manifold with complemented frames*. We get a *metric weak para-f-manifold* if there is a *compatible* pseudo-Riemannian metric g,

$$g(fX, fY) = -g(X, QY) + \sum_i \varepsilon_i\, \eta^i(X)\eta^i(Y).$$

Similar statements (as for a weak f-structure) are valid for a weak para-f-structure.

5.4 Prescribing the Mixed Scalar Curvature

Geometrical problems of prescribing curvature of a Riemannian manifold (M, g) using a conformal change of metric g have been popular for a long time, e.g., the problem of prescribing Gauss curvature on a closed surface seems to be important, see [KW]. The study of constancy of the scalar curvature was began by Yamabe in 1960 and completed by several mathematicians in 1986, see [Au]. This geometrical problem is expressed in terms of the existence and multiplicity of solutions of a given semilinear elliptic PDE in (M, g).

Here, we explore the problem of prescribing the constant S_{mix} of a foliated Riemannian and more general Riemann-Cartan manifolds by conformal change of the structure in tangent and normal to the leaves directions. Under certain geometrical assumptions and in two special cases: along a compact leaf and for a closed fibered manifold, the problem reduces to solution of a nonlinear leafwise elliptic equation for the conformal factor. We are looking for such its solutions that are stable stationary for the associated parabolic equation. Stable stationary solutions of the associated parabolic equation are expressed using spectral parameters of the Schrödinger operator. Our main tool consists in the use of majorizing and minorizing nonlinear heat equations with constant coefficients and application of comparison theorems for solutions of Cauchy's problem for parabolic equations.

5.4.1 Leafwise Constant Mixed Scalar Curvature

Let M^{n+p} be a connected closed manifold equipped with a pseudo-Riemannian metric g and a n-dimensional foliation $\mathscr{F}$. Let a metric g be non-degenerate on complementary distributions $\widetilde{\mathscr{D}}$ (tangent to $\mathscr{F}$) and $\mathscr{D}$ (orthogonal to $\mathscr{F}$) on M. The following question was posed in [Hu, Problem 16]:

Q_1 *Given a foliated Riemannian manifold (M,g), do there exist a smooth function $u > 0$ and a leafwise constant Φ such that a $\mathscr{D}$-conformal metric $g' = g^\top \oplus u^2 g^\perp$ has the mixed scalar curvature equal to Φ?*

Here, we consider the following two approaches [R12, RZe1, RZe2, RZe3]:

- evolving the metric by the $\mathscr{D}$-conformal flow g_t,
- exploring the factor u of $\mathscr{D}$-conformal change of metric.

Both approaches reduce the problem to studying nonlinear leafwise elliptic or parabolic PDE's. By the same reason as in Section 5.3.1, we examine two formulations of the question:

(i) *to prescribe constant S_{mix} on a given compact leaf F;*
(ii) *to prescribe leafwise constant S_{mix} on a closed M, satisfying (5.53).*

Recall that in the last case, the leaves (fibers) are compact. The results for Riemannian case, in shortened form, are as follows.

Theorem A [RZe2] *Let $\mathscr{F}$ be a foliation of a closed Riemannian manifold (M,g) obeying (5.53) and any of conditions:*

(a) *$\mathscr{F}$ is harmonic and nowhere totally geodesic,*
(b) *$\mathscr{F}$ is totally geodesic with integrable $\mathscr{D}$.*

Then there exists on M a $\mathscr{D}$-conformal metric g' with leafwise constant S'_{mix}.

One may extend the question **Q_1** to metric-affine manifolds, in particular, Riemann–Cartan manifolds (i.e., with a linear connection $\overline{\nabla}$ such that $\overline{\nabla}g = 0$ instead of ∇) by a conformal change of the structure in $\widetilde{\mathscr{D}}$ and $\mathscr{D}$ directions:

Q_2 *Given a foliated Riemann–Cartan space $(M,g,\overline{\nabla})$ do there exist a smooth function $u > 0$ and a leafwise constant Φ such that $(\widetilde{\mathscr{D}},\mathscr{D})$-conformal Riemann-Cartan structure,*

$$g' = g^\top \oplus u^2 g^\perp, \quad \mathfrak{T}' = u^2 \mathfrak{T}^\top \oplus \mathfrak{T}^\perp, \tag{5.100}$$

has the mixed scalar curvature equal to Φ?

Here, we denote $\mathfrak{T}^\top(X,Y) = (\mathfrak{T}(X,Y))^\top$ and $\mathfrak{T}^\perp(X,Y) = (\mathfrak{T}(X,Y))^\perp$ for $X,Y \in \mathfrak{X}_M$. We apply the algebraic analogue of S_{mix} for a $(1,2)$-tensor $\mathscr{K}$ on a foliation. (The sectional curvature of a symmetric $\mathscr{K}$ has been defined in [Op]).

Definition 5.87. The *mixed scalar curvature of a* $(1,2)$-*tensor* $\mathscr{K}$ on a foliation is an averaged sectional curvature of tensor $\mathscr{K}$ over all mixed planes:

$$S_{\mathscr{K}} := \frac{1}{2}\sum_{a,i}\varepsilon_a\varepsilon_i\big(\langle[\mathscr{K}_i,\mathscr{K}_a]E_a,\mathscr{E}_i\rangle + \langle[\mathscr{K}_a,\mathscr{K}_i]\mathscr{E}_i,E_a\rangle\big). \qquad (5.101)$$

The definition (5.101) in independent on the choice of the local frame.

For $\mathscr{K} = \mathfrak{T}$ in Riemann–Cartan case we have $S_{\mathfrak{T}} := \sum_{a,i}\varepsilon_a\varepsilon_i\langle[\mathfrak{T}_a,\mathfrak{T}_i]\mathscr{E}_i,E_a\rangle$. Since $\mathfrak{T}^\top$ obeys $\langle[\mathfrak{T}_i^\top,\mathfrak{T}_a^\top]E_a,\mathscr{E}_i\rangle = 0$, then

$$S_{\mathfrak{T}^\top} := \sum_{a,i}\varepsilon_a\varepsilon_i\langle[\mathfrak{T}_a^\top,\mathfrak{T}_i^\top]\mathscr{E}_i,E_a\rangle. \qquad (5.102)$$

The leaves of a foliated manifold $(M,g,\overline{\nabla})$ are submanifolds with induced metric $g^\top$ and metric connection $\overline{\nabla}_X^\top Y := (\overline{\nabla}_X Y)^\top$ for $X,Y \in \mathfrak{X}^\perp$. Since, see (1.16),

$$g^\top(\mathfrak{T}_X^\top Y,Z) + g^\top(\mathfrak{T}_X^\top Z,Y) = \langle\mathfrak{T}_X Y,Z\rangle + \langle\mathfrak{T}_X Z,Y\rangle = 0 \quad (X,Y,Z \in \mathfrak{X}^\top),$$

the leaves with metric $g^\top$ and connection $\overline{\nabla}^\top$ are themselves Riemann-Cartan spaces. For traces of any $(1,2)$-tensor $\mathfrak{A}$ we use notations

$$\text{trace}^\perp\mathfrak{A} := \sum_i\varepsilon_i\mathfrak{A}_i\mathscr{E}_i, \quad \text{trace}^\top\mathfrak{A} := \sum_a\varepsilon_a\mathfrak{A}_aE_a.$$

In shortened form, the results for Riemann–Cartan case are as follows.

Theorem B [RZe3] *Let* $(M,g,\nabla+\mathfrak{T})$ *be a foliated closed Riemann-Cartan space with the space-like leaves. (i) If*

$$\tilde{H} = 0 = H \quad \text{and} \quad (\text{trace}^\top\mathfrak{T})^\top = 0, \qquad (5.103)$$

then there exist smooth solutions of $\mathbf{Q}_2$*: with* $u > 0$ *and constant* Φ *on a leaf* F.
(ii) If the following conditions hold: $S_{\mathfrak{T}^\top} \leq 0$, (5.53) *and*

$$\tilde{H} = 0 = H, \quad \text{trace}^\top\mathfrak{T} = 0 \quad \text{and} \quad (\text{trace}^\perp\mathfrak{T})^\perp = 0, \qquad (5.104)$$

then there exist smooth solutions of $\mathbf{Q}_2$*: with* $u > 0$ *and some leafwise constant* Φ. *Under some conditions, for any such* Φ *the solution* u *is unique in certain domain.*

Under assumptions (5.103), the factor u in (5.100) obeys a leafwise elliptic PDE

$$\mathscr{H}(u) = \Psi_1 u^{-1} - \Psi_2 u^{-3} + \Psi_3 u^3 \qquad (5.105)$$

with known functions Ψ_i, see (5.112), and the Schrödinger operator

$$\mathscr{H} = -\Delta^\top - (\beta^\top + \Phi)\,\text{id}^\top. \qquad (5.106)$$

Observe that for $\overline{\nabla} = \nabla$ (Riemannian case) we have $\Psi_3 = 0$. The key role in our study play spectral parameters of $\mathscr{H}$ (5.106). The spectrum $\lambda_0 \le \lambda_1 \le \ldots$ of the Schrödinger operator on compact leaves is discrete, the least eigenvalue λ_0 of $\mathscr{H}$ is simple, its eigenfunction e_0 (called the *ground state*) can be chosen positive, and

$$- \max{}_F (\beta^\top + \Phi) \le \lambda_0 \le - \min{}_F (\beta^\top + \Phi). \tag{5.107}$$

One may add to Φ a real constant to provide $\beta^\top + \Phi < 0$ without change of e_0; then $\mathscr{H}$ becomes invertible in $L^2(F)$ with bounded $\mathscr{H}^{-1}$. In case of (5.53), the leafwise constant λ_0 and the eigenfunction e_0 on M are smooth. We are looking for such solutions of (5.105) that are stable stationary solutions (attractors) of the associated Cauchy's problem:

$$\partial_t u + \mathscr{H}(u) = \Psi_1\, u^{-1} - \Psi_2\, u^{-3} + \Psi_3\, u^3, \qquad u|_{t=0} = u_0 > 0. \tag{5.108}$$

5.4.2 $\mathscr{D}$-Conformal Change of Metric

Here, we discuss how the extrinsic geometry is transformed under $\mathscr{D}$-*conformal* change of a metric. We derive the transformation of S_{mix} under $\mathscr{D}$-conformal change of Riemann–Cartan structure. This yields, under assumptions (5.103) and in light of the question $\mathbf{Q_2}$, the elliptic PDE (5.105) on a leaf. The shape operator $\tilde{A}_U$ of $\mathscr{D}$ and the skew-symmetric operator $\tilde{T}_U^\sharp$ are given by, see (1.28),

$$\langle \tilde{A}_U(X), Y \rangle = \langle \tilde{h}(X,Y), U \rangle, \qquad \langle \tilde{T}_U^\sharp(X), Y \rangle = \langle \tilde{T}(X,Y), U \rangle.$$

Given a foliation $\mathscr{F}$ on (M,g) and $\phi \in C^1(M)$, put $g' = g^\top + e^{2\phi} g^\perp$. As a special case of Lemma 4.9, we have

$$\begin{aligned}
&h' = e^{-2\phi} h, \quad H' = e^{-2\phi} H, \\
&\tilde{h}' = e^{2\phi}\big(\tilde{h} - (\nabla^\top \phi)\, g^\perp\big), \quad \tilde{H}' = \tilde{H} - p\nabla^\top \phi, \\
&\tilde{A}'_U = \tilde{A}_U - U(\phi)\, \mathrm{id}^\perp, \quad \tilde{T}_U'^\sharp = e^{-2\phi}\, \tilde{T}_U^\sharp \quad (U \in \tilde{\mathscr{D}}).
\end{aligned}$$

Hence, $\mathscr{D}$-conformal variations of metric preserve total umbilicity, harmonicity, and total geodesy of $\mathscr{F}$, and preserve total umbilicity of the normal distribution $\mathscr{D}$.

Lemma 5.88. *Let $\mathscr{F}$ be a foliation of a pseudo-Riemannian manifold (M,g). Then, after transformation $g' = g^\top \oplus u^2 g^\perp$, the mixed scalar curvature along any harmonic leaf F becomes*

$$S'_{\mathrm{mix}} = S_{\mathrm{mix}} - p u^{-1} \Delta^\top u + 2 u^{-1} \langle \tilde{H}, \nabla u \rangle + (u^{-4}-1)\|\tilde{T}\|^2_g - (u^{-2}-1)\|h\|^2_g. \tag{5.109}$$

The proof of Lemma 5.88 is based on Remark 5.4.2 and the calculations

$$\|h'\|^2_{g'} = e^{-2\phi}\|h\|^2_g, \quad \|\tilde{T}'\|^2_{g'} = e^{-4\phi}\|\tilde{T}\|^2_g,$$
$$\|\tilde{h}'\|^2_{g'} = \|\tilde{h}\|^2_g + p\|\nabla^\top \phi\|^2_g - 2\tilde{H}(\phi),$$
$$\|\tilde{H}'\|^2_{g'} = \|\tilde{H}\|^2_g + p^2\|\nabla^\top \phi\|^2_g - 2p\tilde{H}(\phi), \quad \mathrm{div}'^\top \tilde{H}' = \mathrm{div}^\top \tilde{H} - p\Delta^\top \phi.$$

Now, let $\mathscr{F}$ be a foliation of a Riemann–Cartan space $(M, g, \overline{\nabla} = \nabla + \mathfrak{T})$ with the space-like leaves (i.e., $g^\top > 0$). By (1.16),

$$g'(\mathfrak{T}'_X Y, Z) + g'(\mathfrak{T}'_X Z, Y) = u^2\left[\langle \mathfrak{T}_X Y, Z\rangle + \langle \mathfrak{T}_X Z, Y\rangle\right] = 0.$$

Hence, g' is $(\nabla' + \mathfrak{T}')$-parallel, where ∇' is the Levi-Civita connection of g'. Put

$$b_\mathfrak{T} = -\sum_{i,a} \varepsilon_i\,\varepsilon_a \langle \tilde{T}(\mathfrak{T}_i E_a + \mathfrak{T}_a \mathscr{E}_i, \mathscr{E}_i), E_a\rangle.$$

If either $\mathscr{D}$ is integrable or $\overline{\nabla}$ and ∇ are projectively equivalent (see Section 1.2.1), then $b_\mathfrak{T} = 0$. Using Remark 5.4.2 and Lemma 5.88, we obtain the following.

Proposition 5.89. *After transformation* (5.100), *the mixed scalar curvature* $\overline{S}'_{\mathrm{mix}}$ *of the Riemann–Cartan space* $(M, g', \overline{\nabla}' = \nabla' + \mathfrak{T}')$ *on any* ∇-*harmonic leaf* F *becomes*

$$\overline{S}'_{\mathrm{mix}} = \overline{S}_{\mathrm{mix}} + p\langle \mathrm{trace}^\top \mathfrak{T}\rangle^\perp, \nabla u\rangle u^{-1} + \langle(\mathrm{trace}^\perp \mathfrak{T})^\perp, \nabla u\rangle u^{-1}$$
$$+ pu\langle(\mathrm{trace}^\top \mathfrak{T})^\top, \nabla u\rangle - (u^2 - 1)\langle \mathrm{trace}^\top \mathfrak{T}, \tilde{H}\rangle - pu^{-1}\Delta^\top u + 2u^{-1}\langle \tilde{H}, \nabla u\rangle$$
$$+ (u^{-4} - 1)\|\tilde{T}\|^2_g - (u^{-2} - 1)(\|h\|^2_g - b_\mathfrak{T}) - (u^2 - 1)S_{\mathfrak{T}^\top}. \tag{5.110}$$

Observe that (5.110) is the second order PDE for the function $u > 0$,

$$-\Delta^\top u + \frac{2}{p}\langle \tilde{H}, \nabla u\rangle - (\beta^\top + \Phi)u = \Psi_1(x)u^{-1} - \Psi_2(x)u^{-3} + \Psi_3 u^3 - \langle(\mathrm{trace}^\top \mathfrak{T})^\perp, \nabla u\rangle$$

$$-\frac{1}{p}\langle(\mathrm{trace}^\perp \mathfrak{T})^\perp, \nabla u\rangle - u^2\langle(\mathrm{trace}^\top \mathfrak{T})^\top, \nabla u\rangle + \frac{u^3 - u}{p}\langle \mathrm{trace}^\top \mathfrak{T}, \tilde{H}\rangle, \tag{5.111}$$

where $p\Phi = \overline{S}'_{\mathrm{mix}}$ is the mixed scalar curvature after transformation (5.100) and

$$\beta^\top = \Psi_2 - \Psi_1 - \overline{S}_{\mathrm{mix}}/p - S_{\mathfrak{T}^\top}/p,$$
$$\Psi_1 = (\|h\|^2_g - b_\mathfrak{T})/p, \quad \Psi_2 = \|\tilde{T}\|^2_g/p \geq 0, \quad \Psi_3 = S_{\mathfrak{T}^\top}/p. \tag{5.112}$$

Example 5.90. Let $\widetilde{\mathscr{D}}$ be spanned by a unit vector field N: $g(N, N) = \varepsilon_N \in \{-1, 1\}$. Then $S_{\mathrm{mix}} = \varepsilon_N \mathrm{Ric}_{N,N}$ and $\overline{S}_{\mathrm{mix}} = \varepsilon_N \overline{\mathrm{Ric}}_{N,N}$. For the Riemann–Cartan case we get

$$\overline{\mathrm{Ric}}_{N,N} = \mathrm{Ric}_{N,N}$$
$$+ \sum_i \varepsilon_i\left[\langle(\nabla_N \mathfrak{T})_i \mathscr{E}_i, N\rangle + \langle(\nabla_{\mathscr{E}_i} \mathfrak{T})_N N, \mathscr{E}_i\rangle + \langle \mathfrak{T}_i N, \mathfrak{T}_N \mathscr{E}_i\rangle - \langle \mathfrak{T}_N N, \mathfrak{T}_i \mathscr{E}_i\rangle\right].$$

Put $\tilde{\tau}_i = \mathrm{trace}\,(\tilde{A}^i)$ $(i \geq 0)$, where the shape operator $\tilde{A}: \mathscr{D} \to \mathscr{D}$ obeys $\langle \tilde{A}(X), Y\rangle = \langle \tilde{h}(X, Y), N\rangle$. Let $H = 0$ on a compact leaf $F \in \mathscr{F}$ (a closed geodesic). The Ricci curvature of ∇ on F in the N-direction is transformed by (5.100) as

$$\mathrm{Ric}'_{N,N} = \mathrm{Ric}_{N,N} - p\,u^{-1}N(N(u)) + 2\,u^{-1}\tilde{\tau}_1 N(u) + (u^{-4} - 1)\|\tilde{T}\|^2_g,$$

see (5.109). Note that the vector field $\mathfrak{T}_X N$ belongs to $\mathfrak{X}^\perp$ for any $X \in \mathfrak{X}_M$. Hence,

$$S_{\mathfrak{T}^\top} \equiv 0, \quad b_{\mathfrak{T}} = -\varepsilon_N \sum_i \varepsilon_i \langle \tilde{T}(\mathfrak{T}_i N + \mathfrak{T}_N \mathscr{E}_i, \mathscr{E}_i), N \rangle,$$

where $\{\mathscr{E}_i\}_{i \le n}$ is a local orthonormal frame on $\mathscr{D}$. Extending u from the leaf F onto M with the property $(\nabla u)^\perp = 0$ along F, we reduce (5.111) to

$$-N(N(u)) + (2/p)\,\tilde{\tau}_1 N(u) - (\beta^\top + \Phi)\,u = \Psi_1 u^{-1} - \Psi_2 u^{-3},$$

see (5.112), where

$$\beta^\top = \Psi_2 - \Psi_1 - \frac{1}{p}\overline{\mathrm{Ric}_{N,N}}, \quad \Psi_1 = -\frac{1}{p}b_{\mathfrak{T}}, \quad \Psi_2 = \frac{1}{p}\|\tilde{T}\|^2_g \ge 0, \quad \Phi = \frac{1}{p}\overline{\mathrm{Ric}'_{N,N}}.$$

5.4.3 $\mathscr{D}$-Conformal Flows of Metrics

This section is aimed to prescribing S_{mix} by a $\mathscr{D}$-conformal flow of metrics. Evolution equations provide an important tool to study physical phenomena. The prototype for non-linear advection-diffusion processes is the *Burgers equation*

$$v_{,t} + (v^2)_{,x} = a\,v_{,xx}$$

for a scalar function $v(t,x)$ (a constant $a > 0$ is the kinematic viscosity). It serves as the model equation for solitary waves, and is used for describing wave processes in gas and fluid dynamics. The Cole-Hopf type transformation reduces the non-linear Burgers equation to the heat equation.

Example 5.91. The metric on a rotation surface in $\mathbb{R}^3$ belongs to the class of warped products, see Section 1.3.2. Metrics on a rotation surfaces M_t^2 in $\mathbb{R}^3$,

$$r(x,\theta,t) = [\varphi(x,t)\cos\theta, \ \varphi(x,t)\sin\theta, \ h(x,t)] \quad (0 \le x \le l, \ -\pi \le \theta \le \pi)$$

belong to a special class of warped products. The equation

$$\partial_t g = -2K\hat{g}. \tag{5.113}$$

on a surface of revolution (with Gaussian curvature K) provides geometrical interpretation of the classical relation between Burgers and heat equations. We are looking for a one-parameter family of surfaces of revolution, which are foliated by profile curves, and the induced metric g_t obeys (5.113). The profile of M_0^2 is XZ-plane curve $\gamma_0 = [\varphi(\cdot,0),0,h(\cdot,0)]$ (the fiber), and θ-curves are circles in $\mathbb{R}^3$. Let x be the natural parameter of $\gamma_t = \mathrm{Ric}^\perp(\cdot,t)$, i.e., $(\varphi_{,x})^2 + (h_{,x})^2 = 1$. Thus, $N = r_{,x}$ is the unit normal to θ-curves on M_t^2. The geodesic curvature of θ-curves obeys Burgers equation $k_{,t} + (k^2)_{,x} = k_{,xx}$, while the radius of θ-curves (as Euclidean circles) satisfies the heat equation $\varphi_{,t} = \varphi_{,xx}$; both functions are related by the *Cole-Hopf transformation* $k = -(\log \varphi)_{,x}$. When k and K are known, the metrics may be recovered as $g_t^\perp = g_0^\perp \exp(-2\int_0^t (K(s,t) - \Phi)\,ds)$.

A $\mathscr{D}$-*conformal* flow $\partial_t g = s(g)\,g^\perp$ on a foliation depends on a function $s(g)$ on the space of metrics. The flow preserves total umbilicity, total geodesy and harmonicity of the leaves. Using the inequality $p\|\tilde{h}\|^2 \geq \|\tilde{H}\|^2$ (with the equality when $\mathscr{D}$ is totally umbilical), we define the following *measure of non-umbilicity* of $\mathscr{D}$:

$$\beta^\top := p^{-2}\big(p\|\tilde{h}\|^2 - \|\tilde{H}\|^2\big) \geq 0.$$

For $p = 1$, we have $\beta^\top = p^{-2}\sum_{i<j}(k_i - k_j)^2$, where k_i are the principal curvatures of $\mathscr{D}$, see Section 2.1.2.2. The normalized flow on a harmonic foliation has the form

$$\partial_t g = -2\big(S_{\mathrm{mix}}(g) - \Phi\big)g^\perp, \tag{5.114}$$

with leafwise constant $\Phi : M \to \mathbb{R}$. Note that (5.114) with $\Phi = 0$ reduces to (5.113), that looks like the normalized Ricci flow on surfaces, but uses the truncated metric $g^\perp$ instead of g. The $\mathscr{D}$-conformal flow (5.114) is 'Yamabe-type' analogue to the partial Ricci flow. This yields the leafwise *forced Burgers equation* for $\tilde{H}$,

$$\partial_t \tilde{H} + \nabla^\top \|\tilde{H}\|^2 = p\nabla^\top(\mathrm{div}^\top \tilde{H}) + X \tag{5.115}$$

with $X = p\nabla^\top(\|\tilde{T}\|^2 - \|h\|^2 - n\beta^\top)$. If the vector $\tilde{H}$ is leafwise conservative,

$$\tilde{H}_0 = -p(\nabla \log u_0)^\top, \tag{5.116}$$

for a potential function $u_0 > 0$ (compare Example 1.42), then (5.115) yields the non-linear heat equation (5.108)$_1$ with $\Psi_3 = 0$.

Example 5.92. For the Hopf fibration $\pi : S^{2m+1} \to \mathbb{C}P^m$ of a sphere we have $S_{\mathrm{mix}} = 2m$. Thus, the canonical metric on S^{2m+1} is a fixed point of (5.114) with $\Phi = 2m$.

Under certain assumptions about spectral parameters of $\mathscr{H}$, (5.114) has a unique global solution, whose S_{mix} converges exponentially fast to a leafwise constant.

Based on variational formulae (for $\mathscr{D}$-conformal metrics $g_t = s_t g_t^\perp$),

$$\partial_t h = -s h, \quad \partial_t H = -s H, \quad \partial_t \tilde{H} = -(p/2)\nabla^\top s,$$

see also Remark 5.4.2, we get the following.

Proposition 5.93 (Conservation Laws). *Let g_t ($t \geq 0$) be $\mathscr{D}$-conformal metrics on a foliated manifold such that $\tilde{H}_0$ is leafwise conservative for u_0. Then the functions $\beta^\top$, $\|h\|^2/\|\tilde{T}\|$ and the vector field $\tilde{H} - (p/2)(\nabla \log \|\tilde{T}\|)^\top$ are t-independent.*

Let e_0 be the ground state of $\mathscr{H}$, see (5.106). Define

$$d_{u_0,e_0} := \min_F(u_0/e_0)/\max_F(u_0/e_0) \in (0,1].$$

Theorem 5.94. *Let $\mathscr{F}$ be a harmonic foliation of a closed Riemannian manifold (M,g_0) with assumptions (5.53) and (5.116). If Φ obeys*

$$\Phi \geq p\lambda_0 + d_{u_0,e_0}^{-4}\max_M\|\tilde{T}\|_{g_0}^2, \tag{5.117}$$

then (5.114) has a unique smooth solution g_t $(t \geq 0)$, and for any α in the interval $(0,\min\{\lambda_1-\lambda_0, 2(\frac{\Phi}{p}-\lambda_0)\})$ we have the leafwise convergence in C^∞, as $t \to \infty$, with the exponential rate $n\alpha$:

$$S_{\mathrm{mix}}(g_t) \to p\lambda_0 - \Phi \leq 0, \quad \tilde{H}_t \to -p\nabla^\top \log e_0, \quad h(g_t) \to 0.$$

For $\tilde{T} = 0$, condition (5.117) becomes $\Phi \geq p\lambda_0$, and we have the following.

Corollary 5.95. *Let $\mathscr{F}$ be a harmonic foliation with integrable normal distribution on a closed Riemannian manifold (M,g_0) with assumptions (5.53) and (5.116). If $\Phi \geq p\lambda_0$, then the statement of Theorem 5.94 is valid.*

Theorem 5.96. *Let $\mathscr{F}$ be a harmonic foliation of a closed Riemannian manifold (M,g_0) with assumptions (5.53) and (5.116). Suppose that*

$$d_{u_0/e_0}^2 > \sqrt{2}\max_M\|\tilde{T}\|_{g_0}/\min_M\|h\|_{g_0}.$$

Then the interval

$$I_0 = \Big(\max\big\{0,\, 3d_{u_0,e_0}^{-4}\max_M\|\tilde{T}\|_{g_0}^2 - \min_M\|h\|_{g_0}^2\big\},\ \frac{1}{4}d_{u_0,e_0}^4\min_M\|h\|_{g_0}^4/\max_M\|\tilde{T}\|_{g_0}^2\Big)$$

is nonempty, and for any Φ satisfying the condition $p\lambda_0 - \Phi \in I_0$, (5.114) admits a unique smooth solution g_t $(t \geq 0)$, and it converges in C^∞ exponentially fast to a limit metric $g_* = \lim_{t\to\infty} g_t$; moreover, we have the exponential convergence $S_{\mathrm{mix}}(g_t) \to \Phi$, as $t \to \infty$, in C^∞ on the leaves.

If $\tilde{T} = 0$ and $h \neq 0$, then $I_0 = (0,+\infty)$, and we have the following.

Corollary 5.97. *Let $\mathscr{F}$ be a harmonic foliation of a closed Riemannian manifold (M,g_0) with assumptions (5.53). Suppose that the normal distribution is integrable, $h \neq 0$, and $\tilde{H}_0$ obeys (5.116). If $\Phi \leq p\lambda_0$, then the claim of Theorem 5.96 holds.*

5.4.4 Prescribing S_{mix} on Warped Products

Here, we will illustrate our approach to $\mathbf{Q}_2$ in the case of generalized products (see Section 1.3.2). Namely, we examine, when for the warped product initial metric on $B \times_\varphi \bar{M}$ the global solution of (5.114) converges to the limit metric with leafwise constant partial Ricci curvature. Let's now see what happens when B has a boundary (e.g., B is a ball in $\mathbb{R}^p$) and $\varphi > 0$ inside B. Suppose that $\mu(t,x) := \varphi(t,x)_{|\partial B}$ is twice continuously differentiable in t, and there exist limits

$$\lim_{t \to \infty} \mu(t,x) = \tilde{\mu}(x), \quad \lim_{t \to \infty} \partial_t \mu(t,x) = 0, \quad \lim_{t \to \infty} \partial_t^2 \mu(t,\cdot) = 0 \qquad (5.118)$$

uniformly for $x \in \partial B$. Define the function

$$v(t) := \max\{\|\mu(t,\cdot)\|_{C^0(\partial B)}, \|\partial_t \mu(t,\cdot)\|_{C^0(\partial B)}\}.$$

By the maximum principle, see e.g. [Au, Section 3.73], the problem

$$\Delta_{\mathscr{F}} u = 0, \quad u_{|\partial B} = 0$$

has only zero solution; hence, $\lambda = 0$ is not the eigenvalue of $\Delta_{\mathscr{F}}$. Let $0 < \lambda_1 \le \ldots \le \lambda_i \ldots$ be eigenvalues and $\{e_i\}_{1 \le i < \infty}$ the unit L^2-norm eigenfunctions of the eigenvalue problem $-\Delta_{\mathscr{F}} e_i = \lambda_i e_i$ in B and $e_i = 0$ on ∂B. Note that λ_1 has multiplicity 1 and we may assume $e_1 > 0$ in the interior of B.

Theorem 5.98. *Let the metrics $g_t = dx^2 + \varphi_t^2(x)\bar{g}$ on $B^p \times \overline{M}^n$ solve (5.114). If any of conditions (i)–(iii) are satisfied:*
(i) $\Phi < 0$,   (ii) $0 \le \Phi < \lambda_1$, $p < 4$,
(iii) $\Phi = \lambda_1$, $p < 4$ and

$$\tilde{\mu} \equiv 0, \quad \int_0^\infty v(\tau) d\tau < \infty,$$
$$v(t) := \max\{\|\mu(t,\cdot) - \tilde{\mu}\|_{C^0(\partial B)}, \|\partial_t \mu(t,\cdot)\|_{C^0(\partial B)}\}, \qquad (5.119)$$

then g_t exist for all $t \ge 0$, as $t \to \infty$, g_t converge in the C^0-norm uniformly on $B \times \overline{M}^n$ to the limit metric $g_\infty = dx^2 + \varphi_\infty^2(x)\bar{g}$ with $\mathrm{Ric}^\perp(g_\infty) = \Phi\bar{g}$.

Moreover, in cases (i) and (ii), (5.114) has a global single point attractor, while in case (iii) the limit metric g_∞ depends on initial and boundary conditions.

Proof. By Lemma 1.40, $\partial_t(\varphi_t^2) = 2(p\Delta_{\mathscr{F}} \varphi_t/\varphi_t) + 2\Phi\varphi_t^2$. This yields the leafwise parabolic Cauchy's problem with Dirichlet boundary conditions,

$$\partial_t \varphi = p\Delta_{\mathscr{F}} \varphi + \Phi\varphi, \quad \varphi(0,\cdot) = \varphi_0, \quad \varphi(t,\cdot)_{|\partial B} = \mu(t,\cdot). \qquad (5.120)$$

Linear problem (5.120) has a unique solution $\varphi : [0,\infty) \times B \to \mathbb{R}$. By Lemma 5.99 in what follows, φ converges, as $t \to \infty$, to a stationary state, i.e., to a solution $\tilde{\varphi} : B \to \mathbb{R}$ of the problem

$$-\Delta_{\mathscr{F}} \tilde{\varphi} = (\Phi/p)\tilde{\varphi}, \quad \tilde{\varphi}_{|\partial B} = \tilde{\mu}, \qquad (5.121)$$

that completes the proof. $\square$

Lemma 5.99. *Let φ solve (5.120) on B^p with $\Phi \le \lambda_1$. If $\Phi > \lambda_1$, then φ_t diverges as $t \to \infty$. Otherwise, $\varphi(t,\cdot)$ converges in the C^0-norm uniformly on B to the limit* (*i*) $\tilde{\varphi}$, see (5.121), when $\Phi < 0$ and (5.118)$_{1,2}$ hold,

(*ii*) $\tilde{\varphi}$, when $0 \leq \Phi < \lambda_1$, $p < 4$, and (5.118) hold,

(*iii*) $\varphi_\infty = (v_1^0 + \int_0^\infty f_1(\tau)d\tau)e_1$ when $\Phi = \lambda_1$, $p < 4$, (5.118) and (5.119) on ∂B hold.

For $p = 1$, cases (ii) and (iii) are valid under assumption (5.118)$_{1,2}$ only.

Proof. The linear problem (5.120) has a unique C^2-regular solution $\varphi : [0,\infty) \times B \to \mathbb{R}$. We shall study convergence of φ as $t \to \infty$ to a stationary state, i.e., to a solution $\tilde{\varphi} : B \to \mathbb{R}$ of the problem (5.121). Let $U : [0,\infty) \times B \to \mathbb{R}$ solve the Dirichlet problem on B,

$$\Delta_{\mathscr{F}} U = 0, \qquad U_{|\partial B} = \delta(t, \cdot),$$

where $\delta(t, \cdot) = \mu(t, \cdot) - \tilde{\mu}$ and t plays role of a parameter. Since $U(t, \cdot)$ is harmonic on B, by the maximum principle, see e.g. [Au, Section 3.73], we have $\|U(t, \cdot)\|_{C^0} = \|\delta(t, \cdot)\|_{C^0(\partial B)}$ for any $t > 0$. It is easy to check that the function

$$v(t,x) = \varphi(t,x) - \tilde{\varphi}(x) - U(t,x), \quad (t,x) \in [0,\infty) \times B, \qquad (5.122)$$

solves the Cauchy's problem

$$\partial_t v = \Delta_{\mathscr{F}} v + \Phi v + f, \quad v(0, \cdot) = v_0, \quad v(t, \cdot)_{|\partial B} = 0, \qquad (5.123a)$$

where

$$v_0 := \varphi_0 - \tilde{\varphi} - U(0, \cdot), \qquad f := \Phi U - \partial_t U. \qquad (5.123b)$$

Since $\mu(t, \cdot)$ is twice differentiable in t, the functions $\partial_t U(t, \cdot)$ and $\partial_t^2 U(t, \cdot)$ are harmonic on B with boundary values $\partial_t U_{|\partial B} = \partial_t \delta(t, \cdot)$ and $\partial_t^2 U_{|\partial B} = \partial_t^2 \delta(t, \cdot)$. We have $\|\partial_t U\|_{C^0} = \|\partial_t \delta(t, \cdot)\|_{C^0(\partial B)}$ and $\|\partial_t^2 U\|_{C^0} = \|\partial_t^2 \delta(t, \cdot)\|_{C^0(\partial B)}$ for $t > 0$. Hence

$$\|f(t, \cdot)\|_{C^0} \leq (|\Phi| + 1)v(t), \qquad \|\partial_t f(t, \cdot)\|_{C^0} \leq (|\Phi| + 1)\tilde{v}(t). \qquad (5.124)$$

Consider Fourier series representations

$$v(t,x) = \sum_{j=1}^\infty v_j(t)e_j(x), \quad f(t,x) = \sum_{j=1}^\infty f_j(t)e_j(x), \quad v_0(x) = \sum_{j=1}^\infty v_j^0 e_j(x), \qquad (5.125)$$

where $\int_B e_i(s)e_j(s)\,ds = \delta_{ij}$ and

$$v_j = \int_B v(\cdot, s)e_j(s)\,ds, \quad f_j = \int_B f(\cdot, s)e_j(s)\,ds, \quad v_j^0 = \int_B v_0(s)e_j(s)\,ds.$$

We will get $\varphi_\infty = \tilde{\varphi}$ for cases (*i*) and (*ii*), and $\varphi_\infty := (v_1^0 + \int_0^\infty f_1(\tau)\,d\tau)e_1$ for (*iii*).

(*i*) By (5.122), we obtain

$$|\varphi(t,x) - \tilde{\varphi}(x)| = |v(t,x) + U(t,x)| \leq |v(t,x)| + |\delta(t,x)|. \qquad (5.126)$$

Then, from (5.123a) and (5.124)$_1$ we get the estimate

$$\partial_t v - \Delta_{\mathscr{G}} v + \Phi v - (|\Phi| + 1)v(t) \leq \partial_t v - \Delta_{\mathscr{G}} v + \Phi v + (|\Phi| + 1)v(t).$$

By the maximum principle for parabolic equations with Dirichlet's boundary conditions, see e.g. [Au, Section 4.46], we have $\|v(t, \cdot)\|_{C^0} \leq \bar{v}(t)$, where $\bar{v}(t)$ is a solution of the Cauchy's problem for the ODE, $d\bar{v}/dt = \Phi \bar{v} + (|\Phi| + 1)v(t)$, $\bar{v}(0) = \|v_0\|_{C^0}$. Using (5.126) and Lemma 5.100 below with $a = \Phi < 0$ and $s(t) = (|\Phi| + 1)v(t)$, we prove the case (i).

(ii) Substitution (5.125) into (5.123a) and comparison of coefficients yield

$$v'_j = (\Phi - \lambda_j)v_j + f_j(t), \qquad v_j(0) = v^0_j \tag{5.127}$$

—the Cauchy's problem for $v_j(t)$, whose solution is

$$v_j(t) = v^0_j e^{(\Phi - \lambda_j)t} + \int_0^t e^{(\Phi - \lambda_j)(t-\tau)} f_j(\tau) \, d\tau. \tag{5.128}$$

Denote by $w(t, \cdot) = (-\Delta_{\mathscr{G}} - \Phi \, \mathrm{id})v(t, \cdot)$ and $w_j(t) = (w(t, \cdot), e_j)_0$. By Elliptic Regularity Theorem (see Section 5.5.2) the operator $(-\Delta_{\mathscr{G}} - \Phi \, \mathrm{id})^{-1}$ maps L_2 into H^2 and, since $p < 4$, by Sobolev Embedding Theorem, see e.g. [Au], the space H^2 is embedded continuously into C^0. Therefore, the operator $(-\Delta_{\mathscr{G}} - \Phi \, \mathrm{id})^{-1}$ acts continuously from L_2 into C^0, hence

$$\begin{aligned}
\|v(t, \cdot)\|_{C^0} &\leq \|(-\Delta_{\mathscr{G}} - \Phi \, \mathrm{id})^{-1} w(t, \cdot)\|_{C^0} \\
&\leq \|(\Delta_{\mathscr{G}} + \Phi \, \mathrm{id})^{-1}\|_{\mathscr{B}(L_2, C^0)} \cdot \|w(t, \cdot)\|_0. \tag{5.129}
\end{aligned}$$

We denote by $\mathscr{B}(L_2, C^0)$ the Banach space of all bounded linear operators $A : L_2 \to C^0$ with the norm $\|A\|_{\mathscr{B}(L^2, C^0)} = \sup_{v \in L^2 \setminus 0} \|A(v)\|_C / \|v\|_0$. Furthermore, we get

$$\begin{aligned}
v_j(t) = (v(t, \cdot), e_j)_0 &= (\lambda_j - \Phi)^{-1}(v(t, \cdot), (\lambda_j - \Phi)e_j)_0 \\
&= (\lambda_j - \Phi)^{-1}(v(t, \cdot), (-\Delta_{\mathscr{G}} - \Phi \, \mathrm{id})e_j)_0 \\
&= (\lambda_j - \Phi)^{-1}((-\Delta_{\mathscr{G}} - \Phi \, \mathrm{id})v(t, \cdot), e_j)_0 = (\lambda_j - \Phi)^{-1} w_j(t).
\end{aligned}$$

In particular, $v^0_j = (\lambda_j - \Phi)^{-1} w_j(0)$. By (5.128), using integration by parts, we get

$$w_j(t) = w_j(0) e^{(\Phi - \lambda_j)t} + f_j(t) - e^{(\Phi - \lambda_j)t} f_j(0) - \int_0^t e^{(\Phi - \lambda_j)(t-\tau)} f'_j(\tau) \, d\tau.$$

Applying Schwarz's inequality to the integral, we get the following:

$$\left(\int_0^t e^{(\Phi - \lambda_j)(t-\tau)} f'_j(\tau) \, d\tau \right)^2 \leq \int_0^t e^{(\Phi - \lambda_j)(t-\tau)} \, d\tau \times \int_0^t e^{(\Phi - \lambda_j)(t-\tau)} (f'_j(\tau))^2 \, d\tau$$

$$= \frac{1 - e^{(\Phi - \lambda_j)t}}{\lambda_j - \Phi} \int_0^t e^{(\Phi - \lambda_j)(t-\tau)} (f'_j(\tau))^2 \, d\tau \leq \frac{1}{\lambda_j - \Phi} \int_0^t e^{(\Phi - \lambda_j)(t-\tau)} (f'_j(\tau))^2 \, d\tau.$$

Taking into account these circumstances and using Parseval's equality, we have

$$\|w(t,\cdot)\|_0 \leq \|w(0,\cdot)\|_0 e^{(\Phi-\lambda_1)t} + \|f(t,\cdot)\|_0 + \|f(0,\cdot)\|_0 e^{(\Phi-\lambda_1)t}$$
$$+ \frac{1}{\sqrt{\lambda_1-\Phi}} \left(\int_0^t e^{(\Phi-\lambda_j)(t-\tau)} \|\partial_t f(\tau,\cdot)\|_0^2 d\tau \right)^{1/2}. \tag{5.130}$$

Then using estimate $(5.124)_2$ and the arguments from the proof of Lemma 5.100 below, we get

$$\|w(t,\cdot)\|_0 \leq \|w(0,\cdot)\|_0 e^{(\Phi-\lambda_1)t} + \sqrt{\mathrm{vol}\,B}\,(|\Phi|+1)\big(v(t)+v(0)\,e^{(\Phi-\lambda_1)t}\big)$$
$$+ \frac{\sqrt{\mathrm{vol}\,B}}{\lambda_1-\Phi}\big(e^{(1-\theta)(\Phi-\lambda_1)t} \sup_{\tau\in[0,\theta t]} \tilde{v}^2(\tau) + \sup_{\tau\in[\theta t,t]} \tilde{v}^2(\tau)\big)^{1/2}, \quad \theta \in (0,1). \tag{5.131}$$

Using the above, (5.126), (5.129) and equality $\mathrm{Sc}_{\mathrm{mix}}(g_\infty) = -(n/\tilde{\varphi})\Delta_{\mathscr{F}}\,\tilde{\varphi} = n\,\Phi$, we prove the case (ii).

(iii) Assume that $\Phi = \lambda_1$. Since $\tilde{\mu} = 0$, we can choose $\tilde{\varphi} = 0$ as a solution of (5.121). Hence $v(t,x) = \varphi(t,x) - U(t,x)$, see (5.122), and $v_0(x) = \varphi_0(x) - U(0,x)$, see (5.123b). By (5.122), we have

$$|\varphi(t,x) - \varphi_\infty(x)| = |v(t,x) - \varphi_\infty(x)| + \|\delta(t,\cdot)\|_{C^0(B)}, \tag{5.132}$$

where $\varphi_\infty := (v_1^0 + \int_0^\infty f_1(\tau)\,d\tau)e_1$. For $j = 1$, we obtain from (5.127):

$$v_1' = f_1(t), \quad v_1(0) = v_1^0 \quad \Rightarrow \quad v_1(t) = v_1^0 + \int_0^\infty f_1(\tau)\,d\tau - \int_t^\infty f_1(\tau)\,d\tau,$$

where the improper integrals converge in view of condition (5.119) and definition (5.123b). Hence

$$\left| v_1(t) - v_1^0 - \int_0^\infty f_1(\tau)\,d\tau \right| = \left| \int_t^\infty f_1(\tau)\,d\tau \right| \leq (\mathrm{vol}\,B)^{1/2} \int_t^\infty v(\tau)\,d\tau, \tag{5.133}$$

which converges, as $t \to \infty$, to 0. Define

$$\tilde{w}(t,\cdot) = (-\Delta_{\mathscr{F}} - \gamma\,\mathrm{id})(v(t,\cdot) - \varphi_\infty)$$

with fixed $\gamma < \lambda_1$. Since $p < 4$, we get as in the proof of (ii):

$$\|v(t,\cdot) - \varphi_\infty\|_{C^0} \leq \|(-\Delta_{\mathscr{F}} - \Phi\,\mathrm{id})^{-1}\|_{\mathscr{B}(L^2,C^0)} \cdot \|\tilde{w}(t,\cdot)\|_0,$$

where $\tilde{w}_j(t) = (\tilde{w}(t,\cdot), e_j)$. Clearly, $\tilde{w}_j(t)$ for $j > 1$ coincides with $w_j(t)$ defined in the proof of claim (ii). Then as in this proof of (ii), we obtain for $j > 1$:

$$\tilde{w}_j(t) = \tilde{w}_j(0)\,e^{(\lambda_1-\lambda_j)t} + f_j(t) - e^{(\lambda_1-\lambda_j)t} f_j(0) - \int_0^t e^{(\lambda_1-\lambda_j)(t-\tau)} f_j'(\tau)\,d\tau.$$

Using the above and estimates (5.132), (5.133), we complete the proof of case (*iii*) similarly as the proof of case (*ii*) (see (5.130), (5.131) and all further arguments). In this case, $\text{Sc}_{\text{mix}}(g_\infty) = -(n/e_1)\Delta_{\mathscr{F}}\, e_1 = n\lambda_1$. □

Lemma 5.100. *Let $y(t)$ solve the Cauchy's problem (for the ODE)*

$$y' = \alpha(t)y + s(t), \quad y(0) = y_0,$$

where $\alpha, v \in C[0, \infty)$, $\alpha(t) \le a < 0$ and $s(t)$ is bounded. Then

$$|y(t)| \le |y_0|e^{at} + |a|^{-1}e^{(1-\theta)at}\sup_{\tau\in[0,\theta t]}|s(\tau)| + |a|^{-1}\sup_{\tau\in[\theta t,t]}|s(\tau)| \quad (5.134)$$

for any $\theta \in (0,1)$. In particular, if $\lim_{t\to\infty}s(t) = 0$ then $\lim_{t\to\infty}y(t) = 0$.

Proof. As is known, $y(t) = y_0 e^{\int_0^t \alpha(\xi)\,d\xi} + \int_0^t e^{\int_\tau^t \alpha(\xi)\,d\xi}s(\tau)\,d\tau$. Hence,

$$|y(t)| = |y_0|e^{at} + \int_0^{\theta t}e^{a(t-\tau)}|s(\tau)|\,d\tau + \int_{\theta t}^t e^{a(t-\tau)}|s(\tau)|\,d\tau$$

$$\le |y_0|e^{at} + \sup_{\tau\in[0,\theta t]}|s(\tau)|\int_0^{\theta t}e^{a(t-\tau)}\,d\tau + \sup_{\tau\in[\theta t,t]}|s(\tau)|\int_{\theta t}^t e^{a(t-\tau)}\,d\tau.$$

The above and equalities

$$\int_0^{\theta t}e^{a(t-\tau)}\,d\tau = (e^{at} - e^{(1-\theta)at})/a, \quad \int_{\theta t}^t e^{a(t-\tau)}\,d\tau = (e^{(1-\theta)at} - 1)/a$$

yield (5.134). □

Example 5.101. Let $M_t^2 \subset \mathbb{R}^3$ be a one-parameter family of rotation surfaces of Example 5.91, such that $(\partial_x\varphi)^2 + (\partial_x\psi)^2 = 1$. The profile curves $\theta = \text{const}$ are geodesics tangent to the vector field $N = r_{,x}$. The θ-curves are circles in $\mathbb{R}^3$ of geodesic curvature $k = -(\log\varphi)_{,x}$. The metric $g_t = dx^2 + \varphi^2(t,x)\,d\theta^2$ is rotational symmetric of Gaussian curvature $K = -\varphi_{,xx}/\varphi$. Let g_t satisfy (5.113). Then φ solves the Cauchy's problem, compare with (5.120),

$$\partial_t\varphi = \varphi_{,xx} + \Phi\varphi, \quad \varphi(0,x) = \varphi_0(x), \quad \varphi|_{x=0} = \mu_0 \ge 0, \quad \varphi|_{x=l} = \mu_1 \ge 0, \quad (5.135)$$

where $\varphi(x) > 0$ for $x \in (0,l)$, $\mu_0, \mu_1 \in C^1([0,\infty))$ and there exists $\lim_{t\to\infty}\mu_j(t) = \tilde{\mu}_j \in [0,\infty)$. The solution of stationary problem with $B = [0,l]$ and $\lambda_1 = (\pi/l)^2$ is

$$\tilde{\varphi}(x) = \begin{cases} \dfrac{\tilde{\mu}_1\sin(\sqrt{\Phi}x) + \tilde{\mu}_0\sin(\sqrt{\Phi}(l-x))}{\sin(\sqrt{\Phi}l)} & \text{if } 0 < \Phi < \lambda_1, \\[2mm] \tilde{\mu}_0 + (\tilde{\mu}_1 - \tilde{\mu}_0)(x/l) & \text{if } \Phi = 0, \\[2mm] \dfrac{\tilde{\mu}_1\sinh(\sqrt{-\Phi}x) + \tilde{\mu}_0\sinh(\sqrt{-\Phi}(l-x))}{\sinh(\sqrt{-\Phi}l)} & \text{if } \Phi < 0. \end{cases}$$

For the resonance case, $\Phi = \lambda_1 = (\pi/l)^2$, the stationary problem is solvable if and only if $\tilde{\mu}_0 = \tilde{\mu}_1 = 0$, and in this case the solutions are $\tilde{\varphi}(x) = C\sin(\pi x/l)$, where $C > 0$ is constant. By Theorem 5.98, if $\Phi > (\pi/l)^2$ then g_t diverge as $t \to \infty$, otherwise

g_t converge to the limit metric $g_\infty = \mathrm{d}x^2 + \varphi_\infty^2(x)\mathrm{d}\theta^2$ with $K(g_\infty) = \Phi$. Certainly, if $\Phi = (\pi/l)^2$ and, in addition, see Theorem 5.98(iii),

$$\int_0^\infty (|\mu_j(\tau)| + |\mu_j'(\tau)|)\,\mathrm{d}\tau < \infty \quad (j = 0, 1), \tag{5.136}$$

then $\varphi_\infty = (v_1^0 + \int_0^\infty f_1(\tau)\,\mathrm{d}\tau)\sin(\pi x/l)$, and if $\Phi < (\pi/l)^2$ then $\varphi_\infty = \tilde\varphi$. If a solution $\varphi\ (t \geq 0)$ of (5.135) is known and $|\varphi_{,x}| \leq 1$, then $\psi = \psi(t,0) + \int_0^x \sqrt{1-(\varphi_{,x})^2}\,\mathrm{d}x$.

Remark that rotation surfaces in $\mathbb{R}^3$ of constant Gaussian curvature are locally classified. Assume for simplicity that $\mu_j(t) \equiv \tilde\mu_j$. Then $\delta(t) \equiv 0$ and $U \equiv 0$. Hence, $f \equiv 0$, $v_0 = \varphi_0 - \tilde\varphi_0$ and (5.136) is satisfied. For $\Phi = (\pi/l)^2$ we get

$$\varphi_\infty = C\sin(\pi x/l), \quad \psi_\infty = (l/\pi)\,\mathrm{EllipticE}\,(\cos(\pi x/l), C\pi/l),$$

where $C = v_1^0 = \int_0^l v_0(s)\,\mathrm{d}s$ and $\mathrm{EllipticE}(z,k) = \int_0^z \sqrt{(1-k^2s^2)/(1-s^2)}\,\mathrm{d}s$ is the incomplete elliptic integral. To provide a numerical example for $\Phi = (\pi/l)^2$, let $l = \pi$, $C = 1$ and $\mu_j = 0$. In this case, the limit profile curve is a semi-circle $[\sin x, \cos x]$, $(0 \leq x \leq \pi)$, and the limit rotation surface is a unit sphere.

Example 5.102 (See Theorem 6 in [RS]). Let the metrics $g_t = \mathrm{d}x^2 + \varphi_t^2(x)\overline{g}$ on $M = [0,l] \times \overline{M}$ solve (5.52b) and (5.118) with $B = [0,l]$ hold. Suppose that $\Phi < (\pi/l)^2$ (see conditions (i)–(ii) of Theorem 5.98). Then g_t converge uniformly for $x \in [0,l]$ to the limit metric g_∞, whose mixed sectional curvature is constant (equal to Φ).

The following result will be used in the proof of Corollary 5.104.

Theorem 5.103 ([Re]). *Let (B,g) be a compact p-dimensional Riemannian manifold with totally geodesic boundary. Assume that there exists a function $\varphi \neq \mathrm{const}$ satisfying $\mathrm{Hess}_\varphi = -k^2\varphi\,\mathrm{id}$ on B, $\varphi = 0$ on ∂B, for a positive number k. Then (B,g) is isometric to the upper hemisphere of radius $1/k$ in $\mathbb{R}^{p+1}$.*

Corollary 5.104. *Let conditions of Theorem 5.98(iii) hold for (5.52b) with $\Phi = \lambda_1$, $p < 4$, and let $(B,\mathrm{d}x^2)$ be a hemisphere of radius $\sqrt{p/\lambda_1}$ in $\mathbb{R}^{p+1}$. Then, the mixed sectional curvature of the metric g_∞ is constant:*

$$K(N,X) = \Phi/p \quad \text{for any unit vectors } N \in \widetilde{\mathscr{D}},\ X \in \mathscr{D}. \tag{5.137}$$

Proof. By Theorem 5.98 (iii), φ_∞ is proportional to e_1. By Theorem 5.103, the Hessian of first eigenfunction e_1 (of Laplacian on a hemisphere) obeys $\mathrm{Hess}_{e_1}^{\mathscr{F}} = -ke_1\,\mathrm{id}_{\mathscr{F}}$. Hence, $\mathrm{Hess}_\varphi^{\mathscr{F}}$ is also proportional to $\mathrm{id}_{\mathscr{F}}$. Thus, (5.137) is valid. $\qquad\square$

5.4.5 Prescribing $\mathrm{S}_{\mathrm{mix}}$ by a $\mathscr{D}$-Conformal Change of Metric

Here, we use $\mathscr{D}$-conformal change of metrics and are supported by results about stable stationary solutions to the non-linear heat equation (5.108) associated with the elliptic PDE on a leaf F—a closed Riemannian manifold, and compactness in $C(F)$ of the set of all such solutions. Our approach to question $\mathbf{Q}_2$ using a $\mathscr{D}$-conformal change of metrics is based on using spectral parameters of the Schrödinger operator on compact leaves, and exploring stable solutions of the elliptic PDE (5.105) (that are stable stationary solutions of its parabolic counterpart), one of them (for $S_{\mathfrak{X}^{\top}} \equiv 0$) corresponds to the pseudo-Riemannian case. Assume the 'regularity' properties:

- *either $\mathscr{D}$ is nowhere integrable or $\tilde{T} \equiv 0$ ($\mathscr{D}$ is integrable)*,
- *either $S_{\mathfrak{X}^{\top}} \neq 0$ or $S_{\mathfrak{X}^{\top}} \equiv 0$*, see definition in (5.102).

As promised in Section 5.4.1, we find solutions for two formulations of the problem of prescribing leafwise constant $\overline{S}_{mix}$. Let $S'_{mix} = p\Phi$ be the mixed scalar curvature after transformation (5.100). For $f \in C(F)$ define

$$\delta(f) := (\min_F |f|)/(\max_F |f|) \in (0, 1].$$

For a compact leaf F, introduce the quantities

$$K_1 = \frac{\psi_3^+}{4\psi_2^-} \max\{18\psi_1^+ \psi_2^+, \, 4(\psi_1^+)^3 + 27(\psi_2^+)^2 \psi_3^+\},$$

$$K_2 = \frac{1}{8\psi_2^+(3\psi_3^- - \psi_3^+)} \max\{36\psi_1^+ \psi_2^+ \psi_3^- (\psi_3^- + \psi_3^+),$$
$$27\psi_3^+(\psi_2^+)^2(\psi_3^-)^2 + 3(\psi_3^-)^2 + (\psi_1^+)^3(\psi_3^+ + 3\psi_3^-)^2\}, \qquad (5.138)$$

where

$$\psi_k^+ = \max_F |\Psi_k|, \quad \psi_k^- = \min_F |\Psi_k| \quad (k = 1, 2),$$
$$\psi_3^+ = \max_F |S_{\mathfrak{X}^{\top}}|/p, \quad \psi_3^- = \min_F |S_{\mathfrak{X}^{\top}}|/p.$$

The following assumptions are helpful:

$$27(\psi_2^+)^2 \psi_3^+(\psi_1^-)^{-3} < \delta^8(e_0) \quad \text{when } S_{\mathfrak{X}^{\top}} > 0, \qquad (5.139)$$
$$\delta(|S_{\mathfrak{X}^{\top}}/p|)\delta^2(e_0) > 1/3 \quad \text{when } S_{\mathfrak{X}^{\top}} < 0. \qquad (5.140)$$

The equality $S_{\mathfrak{X}^{\top}} \equiv 0$ appears in the case of pseudo-Riemannian manifolds, see Theorem 5.107. Define polynomials $P_{\phi+}(z)$ and $P_{\phi-}(z)$ with constant coefficients,

$$P_{\phi\pm}(z) = \begin{cases} \psi_3^{\pm} z^3 - \lambda_0 z^2 + \psi_1^{\pm} z - \psi_2^{\mp} & \text{if } S_{\mathfrak{X}^{\top}} > 0, \, \|h\|_g^2 > b_{\mathfrak{X}} \text{ and (5.139)}, \\ -\psi_3^{\mp} z^3 - \lambda_0 z^2 - \psi_1^{\mp} z - \psi_2^{\mp} & \text{if } S_{\mathfrak{X}^{\top}} < 0 \text{ and } \|h\|_g^2 < b_{\mathfrak{X}}, \\ -\lambda_0 z^2 + \psi_1^{\mp} z - \psi_2^{\pm} & \text{if } S_{\mathfrak{X}^{\top}} \equiv 0 \text{ and } \|h\|_g^2 > b_{\mathfrak{X}}. \end{cases}$$

For $S_{\mathfrak{X}^{\top}} > 0$, positive roots of $f_{\pm}(y) := P_{\phi\pm}(y^2)$ and of $f'_{\pm}(y)$ are ordered as $y_5^- < y_2^- < y_4^- < y_1^-$ and $y_3^+ < y_5^+ < y_2^+ < y_4^+ < y_1^+$, see Figure 5.1; for $S_{\mathfrak{X}^{\top}} \leq 0$, these roots are ordered as $y_1^- < y_3^- < y_2^-$ and $y_1^+ < y_3^+ < y_2^+$, see Figure 5.2.

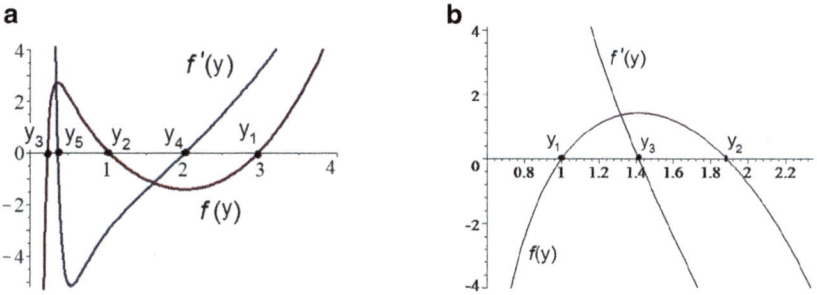

Fig. 5.1 Graphs of functions $f_\pm(y) = P_{\phi\pm}(y^2)/y^3$ and f'. (**a**) $S_{\mathfrak{T}^\top} > 0$, $\Psi_1 > 0$ and $\Psi_2 > 0$; (**b**) $S_{\mathfrak{T}^\top} < 0$, $\Psi_1 < 0$ and $\Psi_2 > 0$

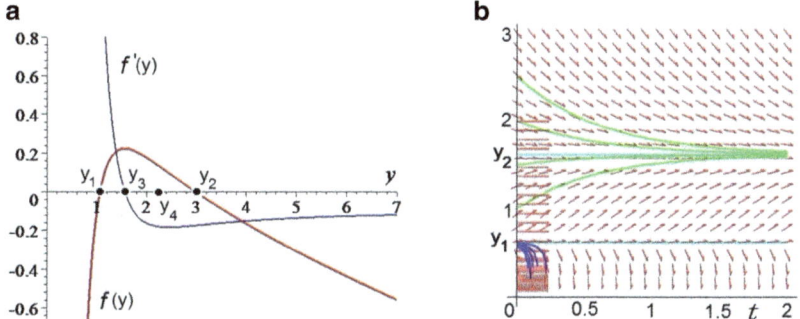

Fig. 5.2 $f_\pm(y) = P_{\phi\pm}(y^2)/y^3$, with $\beta < 0$ and $4|\beta|\Psi_2 < \Psi_1^2$. (**a**) Graphs of $f_\pm, f'_\pm$ for $S_{\mathfrak{T}^\top} \equiv 0$, $\Psi_1 > 0$, $\Psi_2 > 0$. (**b**) y_1—unstable, y_2—stable

Theorem 5.105. *Let $(M, g, \overline{\nabla})$ be a foliated Riemann–Cartan manifold with conditions $g^\top > 0$, $\|\tilde{T}\|_g^2 > 0$ and (5.103) on a compact leaf F. Given $\Phi \in C^\infty(M)$ such that*

$$\Phi_{|F} \in \begin{cases} (-\infty, -\beta^\top) & \text{if } S_{\mathfrak{T}^\top} > 0, \|h\|_g^2 > b_{\mathfrak{T}} \text{ and } (5.139), \\ (-\beta^\top + 1 + \delta^{-4}(e_0)\sqrt{K_1}, \infty) & \text{if } S_{\mathfrak{T}^\top} < 0 \text{ and } \|h\|_g^2 < b_{\mathfrak{T}}, \\ (-\beta^\top - \delta^4(e_0)(\psi_1^-)^2/(4\psi_2^+), -\beta^\top) & \text{if } S_{\mathfrak{T}^\top} \equiv 0 \text{ and } \|h\|_g^2 > b_{\mathfrak{T}}, \end{cases}$$

there exists a positive function $u_* \in C^\infty(M)$ obeying (5.105) such that M with a Riemann–Cartan structure $(g' = g^\top \oplus u_*^2 g^\perp, \mathfrak{T}' = u_*^2 \mathfrak{T}^\top \oplus \mathfrak{T}^\perp)$ has $\overline{S}'_{\text{mix}} = p\Phi$ on F; moreover, $y_2^- \le u_*/e_0 \le y_2^+$, and the set $\{u_{*|F}\}$ of such functions is compact in $C(F)$.

Proof. The required conformal factor u should satisfy on F the nonlinear elliptic PDE (5.105) with the Schrödinger operator (5.106) and $\beta^\top, \Psi_i$ ($i = 1, 2, 3$) given in (5.112). We are looking for solutions of (5.105), which are stationary solutions of the associated Cauchy's problem (5.108). Denote by $f(x, y)$ ($x \in F$) the rhs of (5.108). Then $f_\pm(y) = P_{\phi\pm}(y^2)/y^3$ are the majorizing and minorizing functions for f. By assumptions, the order of roots of $f(x, y)$ with any $x \in F$ is the same as for $P_{\phi\pm}$. Consider three cases according to sign of $S_{\mathfrak{T}^\top}$.

Case 1. Assume that $S_{\mathfrak{T}^\top} > 0$, $\|h\|_g^2 > b_{\mathfrak{T}}$ and (5.139) hold. Then $\lambda_0 > 0$ (the least eigenvalue of $\mathcal{H}$ on F); hence, each of bicubic polynomials $P_{\phi+}(y^2)$ and $P_{\phi-}(y^2)$ has three positive roots: $y_3^- < y_2^- < y_1^-$ and $y_3^+ < y_2^+ < y_1^+$, which can be expressed by Cardano or trigonometric formulas. Since (5.107) and (5.139) yield $(\psi_1^-)^3 > 27(\psi_2^+)^2\psi_3^+$, there exists $u_* \in C^\infty(M)$ obeying (5.105), that is (u_*, Φ) solves $\mathbf{Q}_2$ along F, and $y_2^- \le u_*/e_0 \le y_2^+$ is valid.

Case 2. Assume that $S_{\mathfrak{T}^\top} < 0$ and $\|h\|_g^2 < b_{\mathfrak{T}}$. Then, each of polynomials $P_{\phi+}(y^2)$ and $P_{\phi-}(y^2)$ has two positive roots: $y_1^- < y_2^-$ and $y_1^+ < y_2^+$, Figure 5.1b. By (5.138) there exists $u_* \in C^\infty(M)$ obeying (5.105).

Case 3. Assume that $S_{\mathfrak{T}^\top} \equiv 0$ and $\|h\|_g^2 > b_{\mathfrak{T}}$. The problem amounts to finding a positive solution of the elliptic PDE (5.105) with $\Psi_3 = 0$. For $\Psi_1 > 0$ and $\Psi_2 \neq 0$ each of polynomials $P_{\phi+}(y^2)$, $P_{\phi-}(y^2)$ has two positive roots $y_1^- < y_2^-$ and $y_1^+ < y_2^+$, Figure 5.2. By (5.107), there is a function $u_* \in C^\infty(M)$ obeying (5.105). $\square$

Remark 5.106. Under stronger geometric conditions, the solution $u_{*|F}$, obtained in Theorem 5.105 (Case 2), is unique in the set $\{\tilde{u} \in C(F) : y_3^- < \tilde{u}/e_0 < y_1^+\}$. Moreover, if $\Phi > -\beta^\top + 1 + \delta^{-4}(e_0)\sqrt{K_2}$ then the solution $u_{*|F}$ is unique in $\mathscr{U}_1 = \{\tilde{u} \in C(F) : \tilde{u}/e_0 > y_1^-\}$. In Theorem 5.105 (Case 3), the solution $u_{*|F}$ is unique in $\mathscr{U}_1$.

In the next theorem, we consider two cases: $S_{\mathfrak{T}^\top} < 0$ and $S_{\mathfrak{T}^\top} \equiv 0$. We omit the case $S_{\mathfrak{T}^\top} > 0$, having technical explicit conditions for uniqueness of a solution.

Theorem 5.107. *Let* $(M, g, \overline{\nabla})$ *be a foliated closed Riemann-Cartan manifold with conditions* $g^\top > 0$, $\|\tilde{T}\|_g^2 > 0$, (5.53) *and* (5.104). *Given a smooth leafwise constant function*

$$\Phi \in \begin{cases} (1 - \beta^\top + \delta^{-4}(e_0)\sqrt{K_2}, \infty) & \text{if } S_{\mathfrak{T}^\top} < 0, \ \|h\|_g^2 < b_{\mathfrak{T}} \text{ and } (5.140), \\ \left(-\beta^\top - \delta^4(e_0)(\psi_1^-)^2/(4\psi_2^+), -\beta^\top\right) & \text{if } S_{\mathfrak{T}^\top} \equiv 0 \text{ and } \|h\|_g^2 > b_{\mathfrak{T}}, \end{cases}$$

there exists a leafwise smooth $u_* \in C(M)$ *obeying* (5.105) *and unique in* $\mathscr{U}_1 = \{\tilde{u} \in C(M) : \tilde{u}/e_0 > y_1^-\}$, *such that* M *with a Riemann–Cartan structure* $(g' = g^\top \oplus u_*^2 g^\perp, \mathfrak{T}' = u_*^2 \mathfrak{T}^\top \oplus \mathfrak{T}^\perp)$ *has* $\overline{S}'_{\mathrm{mix}} = p\Phi$; *moreover,* $y_2^- \le u_*/e_0 \le y_2^+$.

Proof. Consider two cases for (5.105) and (5.108).

Case 1. Let $S_{\mathfrak{T}^\top} < 0$, $\|h\|_g^2 < b_{\mathfrak{T}}$ and (5.140) hold. As in Case 2 of Theorem 5.105, we apply Theorem 5.116.

Case 3. Assume that $S_{\mathfrak{T}^\top} \equiv 0$ and $\|h\|_g^2 > b_{\mathfrak{T}}$. Then we apply Theorem 5.118. $\square$

5.5 The Nonlinear Heat Equation

5.5.1 Parabolic PDE's

Recall [Tay] some facts about parabolic quasi linear PDE's. Let $A = (a_{ij}(t,x,u))$ be an $n \times n$-matrix, $a = (a_i(t,x,u,\partial_x u))$—an n-vector, and $t, x \in \mathbb{R}$. Consider the *quasilinear* system of PDE's, n equations in n unknown functions $u = (u_1, \ldots, u_n)$,

$$\partial_t u = A(t, x, u) \, \partial^2_{xx} u + a(t, x, u, \partial_x u). \tag{5.141}$$

PDE's (5.141) are *homogeneous* if $a \equiv 0$. When A depends on t and x only, the system is *semilinear*. The *initial value problem* for (5.141) with given smooth data A, a and u_0,

$$u(0, x) = u_0(x), \tag{5.142}$$

consists in finding smooth function $u(t, x)$ satisfying (5.141)–(5.142).

The *parabolicity condition* for (5.141) says that there is $c = \text{const} > 0$ such that

$$\langle A v, v \rangle \geq c \langle v, v \rangle, \quad \forall v \in \mathbb{R}^n. \tag{5.143}$$

Theorem 5.108 (See Section 15.8 of [Tay]). *Suppose that A of class C^∞ satisfies (5.143) and $a = (a_i(t, x, u))$. Then the system (5.141)–(5.142) with $u_0 \in H^s(\mathbb{R})$ (the Sobolev space) for some $s > 1$, admits a unique solution $u \in C([0, T), H^s(\mathbb{R})) \cap C^\infty((0, T) \times \mathbb{R})$, which persists as long as $\|u\|_{C^r}$ is bounded, given $r > 0$.*

Next, we will briefly discuss the homogeneous *heat equation* on a closed Riemannian manifold (M, g),

$$\partial_t u = \Delta u. \tag{5.144}$$

The *eigenvalue problem* $-\Delta u = \lambda u$ on (M, g) has solution with a sequence of eigenvalues with repetition (each one as many times as the dimension of its finite-dimensional eigenspace) $0 = \lambda_0 < \lambda_1 \leq \lambda_2 \leq \cdots \uparrow \infty$. The corresponding eigenfunctions (ϕ_j) form an orthonormal system in $L^2(M)$: $\int_M \phi_i(x) \phi_j(x) \, d\text{vol}_g = \delta_{ij}$. The smallest eigenvalue is $\lambda_0 = 0$; its eigenfunction (normalized to have unit L^2-norm) is the constant $\phi_0 = \text{Vol}(M, g)^{-1/2}$. The fundamental solution of (5.144),

$$G(t, x, y) = \sum_j e^{-\lambda_j t} \phi_j(x) \, \phi_j(y),$$

is called the *heat kernel*. A solution of (5.144), satisfying $u(x, 0) = u_0(x)$ is given by

$$u(x, t) = \int_{y \in M} G(t, x, y) \, u_0(y) \, d\text{vol}_g. \tag{5.145}$$

Moreover, if $u_0 \in L^2(M)$, the solution converges uniformly, as $t \to \infty$, to a constant function (harmonic function when M is open). Since $\lim_{t \to \infty} G(t, x, y) = 1/\text{Vol}(M, g)$, from (5.145) it follows that the "equilibrium temperature" is the average of u_0:

$$\lim_{t \to \infty} u(x, t) = \int_M u_0(x) \, d\text{vol}_g \, / \text{Vol}(M, g).$$

Example 5.109.

(a) Over (semi-)infinite spatial interval $\mathbb{R}$ (and similarly over $\mathbb{R}^p$ or a non-compact flat manifold), all solutions of (5.144) are obtained from the fundamental solution for the *Dirac delta function* $u_0(x) = \delta_x(\xi)$, which is the function $G(t, x, y) =$

$\frac{1}{(4\pi t)^{1/2}} e^{-(x-y)^2/(4t)}$. For any function $u_0 \in L^2(\mathbb{R})$, a unique solution of (5.144), $u(t,x) = \int_{\mathbb{R}} u_0(y) G(t,x,y) \, dy$, converges uniformly to a linear function, as $t \to \infty$. In particular, if u_0 is bounded then the linear function is constant.

(b) The eigenvalues of $-\Delta$ on a standard flat n-torus $\mathbb{R}^n/\mathbb{Z}^n$ are $\lambda_{l_1 \cdots l_n} = l_1^2 + \ldots + l_n^2$ with eigenfunctions $\phi_{l_1 \cdots l_n} = \frac{1}{(2\pi)^{n/2}} e^{-i(l_1 x_1 + \ldots + l_n x_n)}$. Here $l_1, \cdots, l_n$ take all possible positive and negative integer values. For the unit circle S^1, we have $\lambda_l = l^2$ and $\phi_l(x) = \frac{1}{\sqrt{2\pi}} e^{-ilx}$, where $l \in \mathbb{Z} \setminus \{0\}$. The Cauchy's problem on this circle

$$\partial_t u = \partial_{xx}^2 u, \qquad u(0,\cdot) = u_0 \in H^2(S^1) \tag{5.146}$$

has a unique solution in the class of functions $C([0,\infty), H^2(S^1)) \cap C^1((0,\infty), L^2(S^1))$. By Sobolev Embedding Theorem, see e.g. [Au], $H^2(S^1) \subset C^1(S^1)$. A unique solution of (5.146) has the property $u(t,\cdot) \in C^\infty(S^1)$ for all $t > 0$. Moreover, $u(t,\cdot) \to \bar{u}_0 = \frac{1}{2\pi} \int_{S^1} u_0(x) \, dx$ as $t \to \infty$ and $\|u(t,\cdot) - \bar{u}_0\| \le e^{-t} \|u_0 - \bar{u}_0\|$.

Applying the above and the maximum principle, we obtain the existence and uniqueness result for a scalar parabolic equation ($n = 1$ and $A_{11} = k(u) > 0$) on S^1.

Proposition 5.110. Let a "thermal diffusivity" function $k(u) \in C^\infty(\mathbb{R})$ satisfy $0 < c_1 \le k(u) \le c_2 < \infty$ for some real $c_1 \le c_2$. Then the quasi linear heat equation on a circle S^1

$$\partial_t u = \partial_x(k(u) \partial_x u), \qquad u(0,x) = u_0(x) \in C^\infty(S^1) \tag{5.147}$$

admits a unique solution $u \in C^\infty([0,\infty) \times S^1)$. Moreover, there exists $\lim_{t \to \infty} u(t,x) = u_\infty \in \mathbb{R}$, and for some real $\alpha, K > 0$ the following inequalities are satisfied:

$$\|u(t,\cdot) - u_\infty\|_{S^1} \le K e^{-\alpha t} \|u_0 - u_\infty\|_{S^1}. \tag{5.148}$$

Proof. By Proposition 9.11 in [Tay, Chapter 15], there is a unique solution $u \in C^\infty([0,\infty) \times S^1)$. By the maximum principle, $\|u(t,\cdot)\|_{S^1} \le \|u_0\|_{S^1}$ for all $t > 0$. Define the function $v = \varphi(u)$ of variables (t,x), where $\varphi'(u) = k(u)$. The monotone function φ (of one variable) satisfies inequalities

$$c_1 \le \varphi' \le c_2. \tag{5.149}$$

The PDE $(5.147)_1$ reads as $\partial_t u = \partial_{xx}^2 v$, hence (in view of derivation $\partial_t v = \varphi'(u) \partial_t u$) is equivalent to

$$\partial_t v = k(u) \partial_{xx}^2 v, \qquad v(0,x) = \varphi(u_0(x)).$$

Let us compare it with the linear problem on S^1,

$$\partial_t \tilde{v} = \tilde{k}(t,x) \partial_{xx}^2 \tilde{v}, \qquad \tilde{v}(0,x) = \varphi(u_0(x)), \tag{5.150}$$

where $\tilde{k}(t,x) = k(u(t,x))$ is given. Indeed, $c_1 \le \tilde{k}(t,x) \le c_2$ for all $t > 0$ and $x \in S^1$. From the existence and uniqueness of a solution to (5.150) we conclude that $\tilde{v} = v$. Denote $u_\infty \in \mathbb{R}$ the average of u_0 over S^1, and set $v_\infty = \varphi(u_\infty)$. The function $w = v(t,x) - v_\infty$ solves the linear problem on S^1

$$\partial_t \widetilde{w} = \widetilde{k}(t,x)\,\partial_{xx}^2 \widetilde{w}, \qquad \widetilde{w}(0,x) = \varphi(u_0(x)) - v_\infty, \qquad (5.151)$$

By the theory of linear PDE's [Tay], (5.151) possesses a fundamental solution $\widetilde{G}(t,x,y)$, which can be built by the classical parametrix method, and satisfies the inequalities $0 \leq \widetilde{G} \leq KG$ for some real $K > 0$. Here G is the fundamental solution of the heat equation $\partial_t u = \alpha \partial_{xx}^2 u$ for some constant $\alpha > 0$. Recall that $G(t,x,y) = \sum_{j \geq 0} e^{-\alpha j^2 t} \phi_j(x)\,\phi_j(y)$, where ϕ_j denotes the eigenfunction (of operator $-\alpha \partial_{xx}^2$ on S^1 with eigenvalue $\lambda_j = \alpha j^2$) satisfying $\int_{S^1} \phi_j^2(x)\,dx = 1$. Indeed, $\widetilde{w}(t,x) = \int_{S^1} \widetilde{G}(t,x,y)\,(\varphi(u_0(y)) - v_\infty)\,dy$. In particular,

$$\|\widetilde{v}(t,\cdot) - v_\infty\|_{S^1} \leq e^{-\alpha t}\|\widetilde{v}(0,\cdot) - v_\infty\|_{S^1}.$$

Thus, for all $t > 0$ and $x \in S^1$ we have a priori estimate

$$\|\varphi(u(t,\cdot)) - \varphi(u_\infty)\|_{S^1} \leq K e^{-\alpha t}\|\varphi(u_0) - \varphi(u_\infty)\|_{S^1}.$$

From this, using inequalities

$$c_1\|u(t,\cdot) - u_\infty\|_{S^1} \leq \|\varphi(u(t,\cdot)) - \varphi(u_\infty)\|_{S^1},$$
$$e^{-\alpha t}\|\varphi(u_0) - \varphi(u_\infty)\|_{S^1} \leq c_2\,e^{-\alpha t}\|u_0 - u_\infty\|_{S^1},$$

see (5.149), we deduce (5.148). $\qquad\qquad\qquad\qquad\qquad\qquad\qquad\qquad\qquad\square$

Example 5.111.

(i) Consider the *nonhomogeneous heat equation* over infinite (or semi-infinite) spatial interval $\mathbb{R}$,

$$\partial_t u = \partial_{xx}^2 u + a(t,x), \quad u(0,x) = u_0(x) \in L^2(\mathbb{R}). \qquad (5.152)$$

Here we assume that $u_0(x)$ is a bounded function. All the solutions of (5.152),

$$u(t,x) = \int_{\mathbb{R}} u_0(\xi) G(t,x,y)\,dy + \int_0^t \int_{\mathbb{R}} a(y,s) G(t-s,x,y)\,dy\,ds,$$

include the *heat kernel G*—the fundamental solution of homogeneous (5.152) for $u_0(x) = \delta_x(\xi)$—the Dirac delta function, see also Example 5.109(a).

(ii) For a unit circle S^1, the eigenvalues of $-\Delta$ are $\lambda_j = j^2$ with eigenfunctions $\phi_j(x) = \frac{1}{\sqrt{2\pi}} e^{-ijx}$, where $j \in \mathbb{Z} \setminus \{0\}$. The heat equation on S^1,

$$\partial_t u = \partial_{xx}^2 u, \quad u(0,\cdot) = u_0 \in H^2(S^1), \qquad (5.153)$$

has a unique solution for functions $u \in C([0,\infty), H^2(S^1)) \cap C^1((0,\infty), L^2(S^1))$. Again, by Sobolev Embedding Theorem, $H^2(S^1) \subset C^1(S^1)$. The solution of (5.153) satisfies $u(t,\cdot) \in C^\infty(S^1)$ for all $t > 0$. Moreover, $u(t,\cdot) \to \bar{u}_0$ as $t \to \infty$, where

$$\|u(t,\cdot) - \bar{u}_0\| \leq e^{-t}\|u_0 - \bar{u}_0\|, \quad \bar{u}_0 = \frac{1}{2\pi} \int_{S^1} u_0(x)\,dx.$$

Consider the Jacobi *theta function*

$$\theta(z,\tau) = \sum_{n=1}^{\infty} e^{\pi i n^2 \tau + 2\pi i n z} = 1 + 2\sum_{n=1}^{\infty} (e^{\pi i \tau})^{n^2} \cos(2\pi n z),$$

where $i^2 = -1$, z is a complex number and τ is confined to the upper half-plane. Taking $z = x \in \mathbb{R}$ and $\tau = 4\pi t\, i$ with real $t > 0$, we write

$$\theta(x, 4\pi t\, i) = 1 + 2\sum_{n=1}^{\infty} e^{-4\pi^2 n^2 t} \cos(2\pi n x),$$

which satisfies the heat equation $(5.153)_1$. Since

$$\lim_{t \to 0} \theta(x, 4\pi t\, i) = \sum_{n \in \mathbb{Z}} \delta(x - n),$$

the solution of (5.153) can be specified by convolving the periodic boundary condition at $t = 0$ with $\theta(x, 4\pi t\, i)$.

(iii) Denote $U(t,x) = \dfrac{\sin x}{\sqrt{\cos^2 x + e^{2t}}}$ $(t \geq 0)$ and $k(u) = \dfrac{1}{1+u^2}$, see [DV]. We have a family $u(t,x) = U(\omega^2(t + \beta_1), \omega(x + \beta_2))$ with three real parameters β_1, β_2 and ω of explicit solutions to the PDE $(5.147)_1$, $\partial_t u = \partial_x(k(u)\,\partial_x u)$ on S^1. Set $\omega = 1$ and $\beta_1 = \beta_2 = 0$, hence $u(t,x) = U(t,x)$ and $u_0(x) = \dfrac{\sin x}{\sqrt{\cos^2 x + 1}}$. Moreover, $\lim_{t \to \infty} u(t,x) = u_\infty = 0$ for all $x \in S^1$. Since $0 \leq u^2 \leq 1$, we conclude that $\frac{1}{2} \leq k(u) \leq 1$. Finally, we have $\|u(t, \cdot)\|_{S^1} \leq e^{-t}$ and $\|u_0\|_{S^1} = 1$, that is consistent with (5.148).

5.5.2 Stabilization of Solutions of the Nonlinear Heat Equation

In this section we study stabilization of solutions of the nonlinear heat equation, see (5.108) below, (to stationary ones) for the three cases: $\Psi_3 < 0$, $\Psi_3 > 0$ and $\Psi_3 = 0$. This is done using majorizing and minorizing nonlinear heat equations with constant coefficients and comparison theorems for solutions of Cauchy's problem for parabolic equations. Results of the section were applied in Section 5.4.5.

Let (F, g) be a closed p-dimensional Riemannian manifold, $H^l(F)$ the Hilbert space of differentiable by Sobolev real functions on F with the inner product $(\cdot, \cdot)_l$ and the norm $\|\cdot\|_l$; for example, $H^0(F) = L^2(F)$. Denote by $\|\cdot\|_{C^k}$ the norm in the Banach space $C^k(F)$ $(k \geq 1)$, and $\|\cdot\|_C$ for $k = 0$. Consider the nonlinear elliptic equation

$$\mathcal{H}(u) = \Psi_1 u^{-1} - \Psi_2 u^{-3} + \Psi_3 u^3, \tag{5.154}$$

see (5.105), where Ψ_i and β are any smooth functions on F, and $\Psi_2 \geq 0$.

The lhs of (5.154) is the Schrödinger operator $\mathcal{H} := -\Delta - \beta\,\mathrm{id}$ with domain $H^2(F)$. One can add a real constant to β such that $\mathcal{H}$ becomes invertible in L^2 (for example, $\lambda_0 > 0$) and $\mathcal{H}^{-1}$ is bounded in $L^2(F)$.

The **Elliptic regularity Theorem**, see e.g. [Au], tells:

If $0 \notin \sigma(\mathscr{H})$, then $\mathscr{H}^{-1} : H^k(F) \to H^{k+2}(F)$ for any integer $k \geq 0$.

For $k = 0$, we have $\mathscr{H}^{-1} : L^2(F) \to H^2(F)$, and the embedding of $H^2(F)$ into $L^2(F)$ is continuous and compact; hence, the operator $\mathscr{H}^{-1} : L^2(F) \to L^2(F)$ is compact. Thus, the spectrum of $\mathscr{H}$ is discrete, the least eigenvalue λ_0 of $\mathscr{H}$ is simple, its eigenfunction $e_0(x)$ (called the *ground state*) can be chosen positive. Since $(\beta(x)u, u)_0 \geq \beta^-(u, u)_0$, where $\beta^- = \min_F \beta$, we have

$$(\mathscr{H}u, u)_0 \leq \int_F (|\nabla u(x)|^2 - \beta^-|u(x)|^2)\, dx = (-\Delta u - \beta^- u, u)_0$$

for any $u \in \mathrm{Dom}(\mathscr{H})$. Since β^- is the maximal eigenvalue of the linear operator $\Delta + \beta^-\,\mathrm{id}$, by the variational principle for eigenvalues, we obtain $\lambda_0 \leq -\beta^-$, see (5.107). Similarly, $\lambda_0 \geq -\max_F \beta$. Solutions of (5.154) are stationary solutions of the Cauchy's problem for the heat equation (5.108). We look for attractor of (5.108) that is 'stable' solutions of (5.154). Let $\mathscr{C}_t = F \times [0, t)$, $(0 < t \leq \infty)$, be a cylinder with the base F. By [Au, Theorem 4.51], the problem (5.108) has a unique smooth solution in $\mathscr{C}_{t_0}$ for some $t_0 > 0$. Substituting $u = e_0 w$ into (5.108) and using

$$\Delta(e_0 w) = e_0 \Delta w + w \Delta e_0 + \langle 2\nabla e_0, \nabla w \rangle, \quad \Delta e_0 + \beta e_0 = -\lambda_0 e_0,$$

yields the Cauchy's problem for $w(x, t)$,

$$\partial_t w = \Delta w + \langle 2\nabla \log e_0, \nabla w \rangle + f(w, \cdot), \quad w(\cdot, 0) = u_0/e_0 > 0 \qquad (5.155)$$

where

$$f(w, \cdot) = -\lambda_0 w + (\Psi_1 e_0^{-2})\, w^{-1} - (\Psi_2 e_0^{-4})\, w^{-3} + (\Psi_3 e_0^2)\, w^3.$$

By (5.155) we get

$$\phi_-(w) \leq \partial_t w - \Delta w - \langle 2\nabla \log e_0, \nabla w \rangle \leq \phi_+(w), \qquad (5.156)$$

where the functions ϕ_- and ϕ_+ are defined for each case separately.

Define the parallelepiped $\mathscr{P} = \prod_{k=1}^3 [\Psi_k^-, \Psi_k^+] \subset \mathbb{R}_+^3$, where

$$\Psi_k^+ = \max_F (|\Psi_k| e_0^{-2k}), \quad \Psi_k^- = \min_F (|\Psi_k| e_0^{-2k}) \quad (k = 1, 2),$$
$$\Psi_3^+ = \max_F (|\Psi_3| e_0^2), \qquad \Psi_3^- = \min_F (|\Psi_3| e_0^2).$$

Then $\mathscr{P}_0 = \{(\Psi_1(x), \Psi_2(x), \Psi_3(x)) : x \in F\}$ is a closed subset of $\mathscr{P}$.

An important role in the theory of parabolic equations plays the following.

Proposition 5.112 (Maximum Principle, See [AH]). *Let X_t and g_t be smooth families of vector fields and metrics on a closed manifold F, and $f \in C^\infty(\mathbb{R} \times [0, T])$. Suppose that $u : F \times [0, T] \to \mathbb{R}$ is a C^∞ supersolution to the inequality*

$$\partial_t u \geq \Delta_t u - X_t(u) + f(u, t),$$

and $y : [0, T] \to \mathbb{R}$ *solves the Cauchy's problem for ODE's:*

$$y' = f(y(t), t), \quad y(0) = c.$$

If $u(\cdot, 0) \geq c$, *then* $u(\cdot, t) \geq y(t)$ *for* $t \in [0, T)$.

5.5.2.1 Comparison ODE

Among several tools of analysis on closed manifolds, used in our approach for pre-scribing S_{mix}, are majorizing and minorizing nonlinear heat equations with constant coefficients and comparison theorems for solutions of parabolic equations.

If Ψ_i $(i = 1, 2, 3)$ are real constants, then (5.108) belongs to *reaction-diffusion equations*, which are well understood and whose solutions can be written explicitly. Namely, leafwise constant solutions of (5.108) obey the Cauchy's problem:

$$y' = f(y) = P(y^2)/y^3, \qquad y(0) = y_0 > 0, \tag{5.157}$$

with $P(z) = \Psi_3 z^3 + \beta z^2 + \Psi_1 z - \Psi_2$. For $\Psi_3 \neq 0$ the polynomial $P(z)$ has three different real roots if and only if the discriminant

$$D_P := -\mathrm{Res}(P, P')/\Psi_3 \quad \text{(a cubic polynomial in } \beta)$$

is positive, where $\mathrm{Res}(P, P')$ is the resultant of two polynomials. Consequently, $P(z)$ has one real root if and only if $D_P < 0$. In a sense, $\Psi_3 = 0$ is the bifurcation point for our problem (5.157). We are looking for stable stationary solutions of (5.157), those are roots of P. If P has a real root $\bar{y} > 0$ such that $f'(\bar{y}) < 0$, then $y = \bar{y}$ is attractor for the semigroup associated to (5.157). The *attractor basin* is determined by other two positive roots which surround $\bar{y}$, Figures 5.1 and 5.2.

(a) Assume that $\Psi_3 > 0$. Then $P(z)$ obeys $P(0) = -\Psi_2 < 0$, $P(\infty) = \infty$ and $P(-\infty) = -\infty$. If $D_P > 0$, then all three real roots $z_3 < z_2 < z_1$ of $P(z)$ are positive. If

$$(\beta^\top)^2 - 3\Psi_1 \Psi_3 > 0, \quad \beta^\top < 0, \quad \Psi_1 > 0,$$

then both real roots $z_4 > z_5$ of $P'(z)$ are positive. Thus, conditions

$$\Psi_1 > 0, \quad \Psi_2 > 0, \quad \Psi_3 > 0, \quad \beta < 0, \quad D_P > 0$$

guarantee existence of a stable stationary solution $y_2 = z_2^2 > 0$ of (5.157), Figure 5.1a. The basin of a single-point attractor $y = y_2$ for the semigroup (5.157) is the invariant set of continuous functions $y(t)$, whose values belong to (y_3, y_1).

(b) Assume that $\Psi_3 < 0$. The polynomial $P(z)$ obeys $P(0) = -\Psi_2 < 0$, $P(\infty) = -\infty$. Its maximal real root z_2 is an attractor for (5.157). Considering $D_P > 0$ and the fact that the maximal root z_0 of P' is positive, we get $z_2 > 0$ (and $z_1 > 0$ is the minimal positive root of P). If

$$\beta > 0, \quad \Psi_1 < 0, \quad \beta^2 - 3\Psi_1\Psi_3 > 0$$

(the discriminant of P' is positive), then both roots of $P'(z) = 3\Psi_3 z^2 + 2\beta z + \Psi_1$ are real and the maximal root z_0 is positive. The condition $D_P > 0$ implies $\beta^2 - 3\Psi_1\Psi_3 > 0$. Thus,

$$\Psi_1 < 0, \quad \Psi_2 > 0, \quad \Psi_3 < 0, \quad \beta > 0, \quad D_P > 0$$

guarantee existence of a stable stationary solution $y_2 = z_2^2 > 0$ of (5.157), see Figure 5.1b. Note that $f(y)$ is concave for $y > 0$, and $f'(y)$ is monotone decreasing (with $f'(0+) = \infty$ and $f'(\infty) = -\infty$) and has one positive root. The basin of a single-point attractor $y = y_2$ for the semigroup of (5.157) is the (invariant) set of continuous functions $y(t) > y_1$.

(c) Assume that $\Psi_3 = 0$, then $P(z) = \beta z^2 + \Psi_1 z - \Psi_2$. A positive root $\tilde{z}$ of $P(z)$ corresponds to a stationary solution $\tilde{y} = \sqrt{\tilde{z}}$ of (5.157); moreover, if $P'(\tilde{z}) < 0$ then $\tilde{y}$ is a single-point attractor.

(c_1) Assume that $\beta < 0$, then $P(0) = -\Psi_2 < 0$ and $P(\infty) = -\infty$. Thus, $P(z)$ has real roots if and only if $P(z_0) > 0$, where $z_0 = -\Psi_1/\beta$ is a root of $P'(z) = 0$. The inequality $P(z_0) > 0$ is valid when $-(\Psi_1)^2/(4\Psi_2) < \beta < 0$. Maximal root y_2 of $f(y) = 0$ is asymptotically stable; $f'(y)$ has a unique positive root y_3, and $f'(y)$ takes minimum at y_4, Figure 5.2. If $-4\beta\Psi_2 = \Psi_1^2$, then (5.157) has one positive stationary solution, and has no stationary solutions if $-4\beta\Psi_2 > \Psi_1^2$.

(c_2) Assume that $\beta > 0$, then $P(0) = -\Psi_2 < 0$ and $P(\infty) = \infty$. Thus, $P(z)$ has one positive root z_2, which corresponds to unstable stationary solution of (5.157), because $P'(z_2) > 0$. For $\beta = 0$, a unique positive stationary solution of (5.157) is unstable.

(c_3) Assume that $\Psi_2 = 0$, then $f(y) = \beta y + \Psi_1 y^{-1}$. If $\beta \geq 0$, then (5.157) has no positive stationary solutions. If $\beta < 0$ and $\Psi_1 > 0$, then $f(y) = 0$ has one positive root $y_2 = (\Psi_1/|\beta|)^{1/2}$. The solution y_1 is stable (attractor) because $f'(y_2) < 0$.

Example 5.113. If F is a circle S^1 of length l, then (5.108) with $\Psi_3 = 0$ is the Cauchy's problem

$$u_{,t} = u_{,xx} + f(u), \quad u(x,0) = u_0(x) > 0 \quad (x \in S^1, \, t \geq 0), \tag{5.158}$$

with $f(u) = \beta u + \Psi_1 u^{-1} - \Psi_2 u^{-3}$. The stationary equation with $u(x)$ for (5.158) yields the problem

$$u'' + f(u) = 0, \quad u(0) = u(l), \quad u'(0) = u'(l), \quad l > 0. \tag{5.159}$$

We rewrite (5.159) as the dynamical system

$$u' = v, \quad v' = -f(u) \quad (u > 0). \tag{5.160}$$

Periodic solutions of (5.159) correspond to solutions of (5.160) with the same period. The system (5.160) is Hamiltonian, since $\partial_u v = \partial_v f(u)$, its Hamiltonian $H(u,v)$

(the first integral) solves the system

$$\partial_u H(u,v) = f(u), \quad \partial_v H(u,v) = v.$$

Thus, $H(u,v) = \frac{1}{2}(v^2 + \beta u^2) + \Psi_1 \ln u + \frac{1}{2}\Psi_2 u^{-2}$. The trajectories of (5.160) belong to level lines of $H(u,v)$. Consider the cases.

Case (a). Assume that $\beta < 0$. Then (5.160) has two fixed points: $(y_i, 0)$ ($i = 1, 2$) with $y_1 > y_2$. To clear up the type of fixed points, we linearize (5.160) at $(y_i, 0)$, $\vec{\eta}' = A_i \vec{\eta}$, where $A_i = \begin{pmatrix} 0 & 1 \\ -f'(y_i) & 0 \end{pmatrix}$. Since $f'(y_1) < 0$ and $f'(y_2) > 0$, the point $(y_1, 0)$ is a "saddle" and $(y_2, 0)$ is a "center". The separatrix is $H(u,v) = H(y_1, 0)$, see Figure 5.3a,

$$v^2 = |\beta|(u^2 - y_1^2) - 2\Psi_1 \ln(u/y_1) - \Psi_2(u^{-2} - y_1^{-2}).$$

The separatrix divides the half-plane $u > 0$ into three simply connected areas. Then $(y_2, 0)$ is a unique minimum point of H in $\Omega = \{(u,v): H(u,v) < H(y_1,0), 0 < u < y_1\}$. The phase portrait of (5.160) in Ω consists of the cycles surrounding the fixed point $(y_2, 0)$, all correspond to non-constant solutions of (5.159) with various l. Other two areas do not contain cycles, since they have no fixed points.

Case (b). Assume that $\beta \geq 0$. Then (5.160) has one fixed point $(y_1, 0)$ and $f'(y_1) > 0$. Hence, $(y_1, 0)$ is a "center". Since $(y_1, 0)$ is a unique minimum of $H(u,v)$ in the half-plane $u > 0$, the phase portrait of (5.160) consists of cycles surrounding $(y_1, 0)$, all correspond to non-constant solutions of (5.159) with various l, Figure 5.3b.

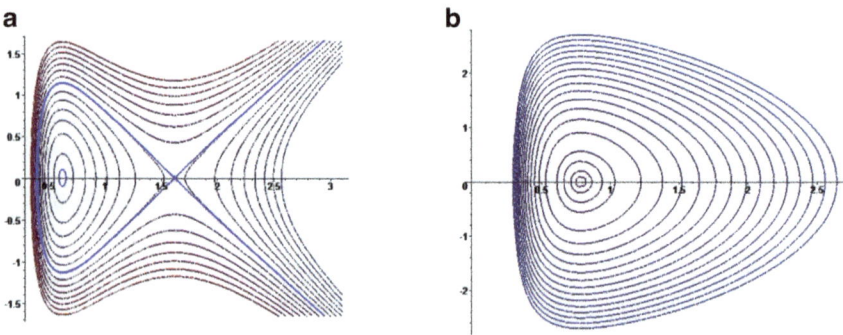

Fig. 5.3 Example 5.113: (a) $\beta < 0$, (b) $\beta > 0$

For $\Psi_2 = 0$ and $\Psi_1 > 0$, the Hamiltonian of (5.160) is $H(u,v) = \frac{1}{2}(v^2 + \beta u^2) + \Psi_1 \ln u$. Solving $H(u,v) = C$ with respect to v and substituting to $(5.160)_1$, we get

$$u' = \sqrt{-\beta u^2 - 2\Psi_1 \ln u + 2C}.$$

If $\beta \geq 0$, then (5.160) has no cycles (since it has no fixed points); hence, (5.159) has no solutions. If $\beta < 0$, then the separatrix $H(u,v) = H(u_*,0)$ is $v^2 = |\beta|(u^2 - u_*^2) - 2\Psi_1 \ln(u/u_*)$, (5.160) has a unique fixed point $(u_*,0)$ which is a "saddle". The separatrix divides the half-plane $u > 0$ into four simply connected areas with these lines. Each of these areas has no fixed points of (5.160), hence the system has no cycles. We conclude that u_* is a unique solution of (5.159).

Case (c). Consider (5.159) for $\Psi_1 = 0$, $\Psi_2 > 0$ and $l = 2\pi$. Set $p = u'$ and represent $p = p(u)$ as a function of u. Then $u'' = dp/du$, hence $(p^2)' = -2\beta u + 2\Psi_2 u^{-3}$ and $(u')^2 = C_1 - \beta u^2 - \Psi_2 u^{-2}$. After separation of variables and integration, we obtain

$$u = \begin{cases} \left(\frac{C_1}{2\beta} + \frac{1}{2\beta}\sqrt{C_1^2 - 4\beta\Psi_2}\sin(2\sqrt{\beta}(x+C_2))\right)^{1/2}, & (C_1^2 \geq 4\beta\Psi_2) \text{ for } \beta > 0, \\ \left(-\frac{C_1}{2|\beta|} + \frac{1}{2|\beta|}\sqrt{C_1^2 + 4|\beta|\Psi_2}\cosh(2\sqrt{|\beta|}(x+C_2))\right)^{1/2} & \text{for } \beta < 0, \\ \sqrt{\Psi_2/C_1 + C_1(x+C_2)^2} & \text{for } \beta = 0. \end{cases}$$

Thus, for $\beta \leq 0$, (5.159) has no positive solutions, while for $\beta > 0$ the solution is 2π-periodic and positive only if

- $\beta \neq k^2/4$ ($k \in \mathbb{N}$) and $C_1 = 2(\beta\Psi_2)^{1/2}$; a solution $u_* = (\Psi_2/\beta)^{1/4}$ is unique, or
  %noindent
- $\beta = k^2/4$ ($k \in \mathbb{N}$); all solutions form a two-dimensional manifold

$$u_0(C_1,C_2) = \frac{1}{k}\left(2C_1 + 2(C_1^2 - k^2\Psi_2)^{1/2}\sin(k(x+C_2))\right)^{1/2}.$$

Below we study the case $\Psi_3 \leq 0$ in (5.108) and omit more technical proof for $\Psi_3 > 0$.

5.5.2.2 Case of $\Psi_3 < 0$

Assume that $\Psi_3 < 0$, $\Psi_1 < 0$ and $\Psi_2 > 0$ in (5.108) and $\lambda_0 < 0$. For the function

$$\phi(y,\theta) = -\lambda_0 y - \theta_1 y^{-1} - \theta_2 y^{-3} - \theta_3 y^3 = P_\phi(y^2)/y^3, \quad y > 0,$$

where $P_\phi(z) = -\theta_3 z^3 - \lambda_0 z^2 - \theta_1 z - \theta_2$ and $\theta = (\theta_1, \theta_2, \theta_3) \in \mathscr{P}$, we have

$$\phi_-(y) \leq \phi(y,\theta) \leq \phi_+(y),$$

where $\phi_\pm(y) = P_{\phi_\pm}(y^2)/y^3$ and $P_{\phi_\pm}(z) = -\Psi_3^\mp z^3 - \lambda_0 z^2 - \Psi_1^\mp z - \Psi_2^\mp$. We calculate

$$\partial_y \phi = -\lambda_0 + \theta_1 y^{-2} + 3\theta_2 y^{-4} - 3\theta_3 y^2, \quad \partial_{yy}^2 \phi = -2\theta_1 y^{-3} - 12\theta_2 y^{-5} - 6\theta_3 y.$$

Since $\partial_{yy}^2 \phi < 0$ for $y > 0$ and $\phi(0+, \theta) = \phi(\infty, \theta) = -\infty$, the function ϕ is concave by y and "$\cap$"-shaped, and $\partial_y \phi$ is decreasing from ∞ to $-\infty$ for $y \in (0, \infty)$. Note that $\phi_-(y)$ and $\phi_+(y)$ are also concave. The discriminant of $P_\phi(z)$ is the following cubic polynomial in $-\lambda_0$:

$$D(P_\phi) = 4\,\theta_2(-\lambda_0)^3 + \theta_1^2(-\lambda_0)^2 - 18\,\theta_1\theta_2\theta_3\,(-\lambda_0) - (4\,\theta_1^3\theta_3 + 27\,\theta_2^2\theta_3^2)$$
$$\geq \overline{D} := 4\,\Psi_2^-(-\lambda_0)^3 + (\Psi_1^-)^2(-\lambda_0)^2 - 18\,\Psi_1^+\Psi_2^+\Psi_3^+(-\lambda_0) - 4(\Psi_1^+)^3\Psi_3^+$$
$$-27(\Psi_2^+\Psi_3^+)^2.$$

Maclaurin Method Suppose that the first m leading coefficients of the real polynomial $P_n(t) = a_0 t^n + a_1 t^{n-1} + \ldots + a_{n-1}t + a_n$ are nonnegative, that is $a_0 > 0$, $a_1 \geq 0, \ldots, a_{m-1} \geq 0$, and the next coefficient is negative, $a_m < 0$. Then $1 + (B/a_0)^{1/m}$ is an upper bound for the positive roots of this polynomial, where B is the largest of the absolute values of negative coefficients of $P_n(t)$. Note that $P_n(t) > 0$ for all $t \in [0,1]$ (so, $a_n > 0$) if

$$a_n > \sum_{0 \leq i < n} |a_i|, \quad \text{for all} \quad a_i < 0. \tag{5.161}$$

By Maclaurin method, the following condition is sufficient for $\overline{D} > 0$:

$$-\lambda_0 > \overline{K} := 1 + \left(\max\{18\Psi_1^+\Psi_2^+, 4(\Psi_1^+)^3 + 27(\Psi_2^+)^2\Psi_3^+\}\Psi_3^+/(4\Psi_2^-) \right)^{1/2}. \tag{5.162}$$

By the above, if (5.162) holds then $\phi(y, \theta)$ for any $\theta \in \mathscr{P}$ has two positive roots $y_2(\theta) > y_1(\theta)$, and $\partial_y \phi$ has a unique positive root $y_3(\theta) \in (y_1(\theta), y_2(\theta))$. Note that $\partial_y \phi|_{y=y_2(\theta)} < 0$ and $\partial_y \phi|_{y=y_1(\theta)} > 0$.

Let $y_1^+ < y_2^+$ be positive roots of $\phi_+(y)$, $y_1^- < y_2^-$ the positive roots of $\phi_-(y)$, and y_3^-, y_3^+ positive roots of decreasing functions

$$(\partial_y \phi)_\pm(y) = -\lambda_0 + \Psi_1^\pm y^{-2} + 3\Psi_2^\pm y^{-4} - 3\Psi_3^\mp y^2.$$

For all $\theta \in \mathscr{P}$ and $y > 0$ we have

$$(\partial_y \phi)_-(y) \leq \partial_y \phi(y, \theta) \leq (\partial_y \phi)_+(y).$$

Proposition 5.114. *If (5.162) is valid, then for any* $\theta \in \mathscr{P}$,

$$y_1^+ \leq y_1(\theta) \leq y_1^-, \quad y_2^- \leq y_2(\theta) \leq y_2^+, \quad y_3^- \leq y_3(\theta) \leq y_3^+.$$

If, in addition,

$$3\Psi_3^- > \Psi_3^+, \tag{5.163}$$
$$-\lambda_0 > 1 + \sqrt{K}, \quad \text{where}$$
$$K = \frac{1}{8\,\Psi_2^+(3\,\Psi_3^- - \Psi_3^+)}\,(\max\{36\,\Psi_1^+\Psi_2^+\Psi_3^-(\Psi_3^- + \Psi_3^+),$$
$$27\,\Psi_3^+(\Psi_2^+)^2((\Psi_3^+)^2 + 3(\Psi_3^-)^2) + (\Psi_1^+)^3(\Psi_3^+ + 3\Psi_3^-)^2\}), \tag{5.164}$$

hold, then there exist $K > \overline{K}$ *such that for all* $\lambda_0 < -K$ *we have*

$$y_1^+ < y_1^- < y_3^+ < y_2^- < y_2^+. \tag{5.165}$$

Proof. For implicit derivatives

$$\partial_{\theta_k} y_l = -(\partial_{\theta_k}\phi/\partial_y\phi)|_{y=y_l(\theta)}, \quad \partial_{\theta_k} y_3 = -(\partial^2_{\theta_k y}\phi/\partial^2_{yy}\phi)|_{y=y_3(\theta)},$$

where $l = 1,2$, $k = 1,2,3$, we calculate

$$\partial_{\theta_1}\phi = -y^{-1}, \quad \partial_{\theta_2}\phi = -y^{-3}, \quad \partial_{\theta_3}\phi = -y^3, \quad \partial_y\phi|_{y=y_2(\theta)} < 0,$$
$$\partial_y\phi|_{y=y_1(\theta)} > 0, \quad \partial^2_{\theta_1 y}\phi = y^{-2}, \quad \partial^2_{\theta_2 y}\phi = 3y^{-4}, \quad \partial^2_{\theta_3 y}\phi = -3y^2.$$

Note that $\partial^2_{yy}\phi < 0$ $(y > 0)$. Thus, the following inequalities hold:

$$\partial_{\theta_k} y_1(\theta) > 0, \quad \partial_{\theta_k} y_2(\theta) < 0, \quad (k = 1,2,3), \quad \partial_{\theta_k} y_3(\theta) > 0 \quad (k = 1,2),$$
$$\partial_{\theta_3} y_3(\theta) < 0.$$

The first claim follows from the above. For proving the second claim, is sufficient find $K > \overline{K}$ such that for all $\lambda_0 < -K$ we have $y_1^- < y_3^+ < y_2^-$. Consider the functions

$$\partial_y(\phi_-)(y) = P_{\partial_y(\phi_-)}(y^2)/y^4, \quad \text{where } P_{\partial_y(\phi_-)}(z) = -3\Psi_3^+ z^3 - \lambda_0 z^2 + \Psi_1^+ z + 3\Psi_2^+,$$
$$(\partial_y\phi)_+(y) = P_{(\partial_y\phi)_+}(y^2)/y^4, \quad \text{where } P_{(\partial_y\phi)_+}(z) = -3\Psi_3^- z^3 - \lambda_0 z^2 + \Psi_1^+ z + 3\Psi_2^+,$$

for $y > 0$, where $\partial_y(\phi_-)$ and $(\partial_y\phi)_+$ are decreasing. Notice that $\phi_-(y) > 0$ for $y \in (y_1^-, y_2^-)$, and $\phi_-(y) < 0$ for $y \in (0, \infty) \setminus [y_1^-, y_2^-]$, and we have $\phi_-(0+) = -\infty$ and $\phi_-(\infty) = -\infty$; moreover, $\phi_-(y)$ increases in $(0, y_3^-)$ and decreases in (y_3^-, ∞). The function $(\partial_y\phi)_+(y)$ decreases on $(0, \infty)$ from $+\infty$ to $-\infty$; moreover, $(\partial_y\phi)_+(y) > 0$ in $(0, y_3^+)$ and $(\partial_y\phi)_+(y) < 0$ in (y_3^+, ∞). Using that the positive root of $\partial_y(\phi_-)$ lied in (y_1^-, y_2^-), we will show that the resultant of two cubic polynomials

$$R_3(t) = -\text{Res}(P_{\phi_-}, (1-t)P_{\partial_y(\phi_-)} + t P_{(\partial_y\phi)_+})/\Psi_2^+$$

does not vanish (hence, they have no common roots) for any $t \in [0,1]$. Indeed,

$$R_3(0) = 8\Psi_3^+ D(P_{\phi_-}) \geq 8\Psi_3^+ \overline{D}(-\lambda_0) > 0,$$

where

$$D(P_{\phi_-}) = -4\Psi_2^+ \lambda_0^3 + (\Psi_1^+)^2\lambda_0^2 + 18\Psi_1^+ \Psi_2^+ \Psi_3^+ \lambda_0 - 4(\Psi_1^+)^3\Psi_3^+ - 27(\Psi_2^+\Psi_3^+)^2.$$

Assuming on the contrary that either $y_3^+ \geq y_2^-$ or $y_3^+ \leq y_1^-$, we get $R_3(1) \leq 0$; thus, a contradiction: $R_3(t_0) = 0$ for some $t_0 \in (0,1]$. Our $R_3(t)$ is a cubic polynomial with coefficients

$$a_0 = 27\delta_3^3(\Psi_2^+)^2, \quad a_1 = -18\delta_3^2(4\Psi_1^+\Psi_2^+(-\lambda_0) + (\Psi_1^+)^3 + 9(\Psi_2^+)^2\Psi_3^+),$$
$$a_2 = -12\delta_3 D(P_{\phi_-}), \quad a_3 = 8\Psi_3^+ D(P_{\phi_-}).$$

Hence, the condition (5.161) reads as $a_3 > |a_1| + |a_2|$ (since $a_0 > 0$),

$$2\,(3\,\Psi_3^- - \Psi_3^+)D(P_{\phi_-}) > 9\,\delta_3^2\left(4\Psi_1^+\Psi_2^+(-\lambda_0) + (\Psi_1^+)^3 + 9(\Psi_2^+)^2\Psi_3^+\right).$$

By (5.163), this is valid if $\delta_3 \geq 0$ is small or $P(-\lambda_0) = \sum_{i=0}^{3} b_{3-i}(-\lambda_0)^i > 0$, where

$$b_0 = 8\,\Psi_2^+(3\,\Psi_3^- - \Psi_3^+), \quad b_1 = 2\,(\Psi_1^+)^2(3\,\Psi_3^- - \Psi_3^+),$$
$$b_2 = -36\,\Psi_1^+\Psi_2^+\Psi_3^-(\Psi_3^- + \Psi_3^+) < 0,$$
$$b_3 = -27\,\Psi_3^+(\Psi_2^+)^2((\Psi_3^+)^2 + 3(\Psi_3^-)^2) - (\Psi_1^+)^3(\Psi_3^+ + 3\Psi_3^-)^2 < 0.$$

By Maclaurin method, the inequality $-\lambda_0 > K := 1 + (\max\{-b_2, -b_3\}/b_0)^{\frac{1}{2}}$ yields $P(-\lambda_0) > 0$ (if $\delta_3 \geq 0$ is small enough, then one may take $K = \overline{K}$). $\qquad\square$

Define closed in $C(F)$ nonempty sets

$$\mathscr{U}^{\,\varepsilon,\eta} = \{\tilde{u} \in C(F) : y_2^- - \varepsilon \leq \tilde{u}/e_0 \leq y_2^+ + \eta\}, \quad \varepsilon \in (0, y_2^- - y_1^-), \ \eta \in (0, \infty).$$

We have $\mathscr{U}^{\,\varepsilon,\eta} \subset \mathscr{U}^{\,\varepsilon,\infty} \subset \mathscr{U}_1$, where the set $\mathscr{U}_1 = \{\tilde{u} \in C(F) : \tilde{u}/e_0 > y_1^-\}$ is open.

Proposition 5.115 ([RZe3]). *Assume that (5.162) is satisfied. Then*

(i) *for any $u_0 \in \mathscr{U}^{\,\varepsilon,\eta}$, (5.108) admits a unique global solution, and $\mathscr{U}^{\,\varepsilon,\eta}$ are invariant sets for the associated semigroup $\mathscr{S}_t : u_0 \to u(\cdot, t)$ $(t \geq 0)$ in $\mathscr{C}_\infty$;*

(ii) *for any $\sigma \in (0, \varepsilon)$ there exists $t_1 > 0$ such that $\mathscr{S}_t(\mathscr{U}^{\,\varepsilon,\infty}) \subseteq \mathscr{U}^{\,\sigma,\infty}$ for $t \geq t_1$.*

By (5.165), we have $y_2^- - y_3^+ > 0$. Define the quantity for $\sigma \in (0, y_2^- - y_3^+)$:

$$\mu^+(\sigma) := -\sup\nolimits_{y \geq y_2^- - \sigma}(\partial_y\phi)_+(y) = -(\partial_y\phi)_+(y_2^- - \sigma) > 0.$$

Proposition 5.115 supports the following.

Theorem 5.116.

(i) *If (5.162) holds, then (5.154) has a solution $u_* \in \mathscr{U}_1 \cap C^\infty(F)$; moreover, the set U_* of all such solutions is compact in $C(F)$ and $U_* \subset \{\tilde{u} \in C(F) : y_2^- \leq \tilde{u}/e_0 \leq y_2^+\}$.*

(ii) *If, in addition, $\Psi_3^+ < 3\Psi_3^-$, then there exists $K > \overline{K}$ such that if $\lambda_0 < -K$ then the above solution is unique in $\mathscr{U}_1$, and $u_* = \lim_{t\to\infty} u(\cdot, t)$ where u solves (5.108) with $u_0 \in \mathscr{U}_1$; moreover, for any $\sigma \in (0, y_2^- - y_3^+)$, the set $\mathscr{U}^{\,\sigma,\infty}$ is attracted by the corresponding semigroup exponentially fast to the point u_* in C-norm:*

$$\|u(\cdot, t) - u_*\|_{C(F)} \leq \delta^{-1}(e_0)\,e^{-\mu^+(\sigma)t}\|u_0 - u_*\|_{C(F)} \quad (t > 0, \ u_0 \in \mathscr{U}^{\,\sigma,\infty}).$$

(iii) *Let β, Ψ_i be smooth functions on $F \times \mathbb{R}^n$ with a smooth metric $\langle \cdot, \cdot \rangle$. If (5.162)–(5.164) hold for any $F \times \{q\}$ $(q \in \mathbb{R}^n)$, then the solution u_* is smooth on $F \times \mathbb{R}^n$.*

5.5.2.3 Case of $\Psi_3 = 0$

Assume that $\Psi_3 = 0$, $\Psi_1 > 0$ and $\Psi_2 > 0$ in (5.108). Then (5.154) becomes

$$\mathscr{H}(u) = \Psi_1(x)\, u^{-1} - \Psi_2(x)\, u^{-3}, \tag{5.166}$$

where $\mathscr{H}(u) := -\Delta u - \beta\, u$. Certainly, Cauchy's problem (5.108) reads

$$\partial_t u + \mathscr{H}(u) = \Psi_1\, u^{-1} - \Psi_2\, u^{-3}, \qquad u(x,0) = u_0(x) > 0. \tag{5.167}$$

Then functions ϕ_- and ϕ_+ in (5.156) become

$$\phi_\pm(y) = P_{\phi_\pm}(y^2)/y^3, \quad \text{where} \quad P_{\phi_\pm}(z) = -\lambda_0 z^2 + \Psi_1^\mp z - \Psi_2^\pm,$$

and $f(w, \cdot) = -\lambda_0 w + (\Psi_1 e_0^{-2})\, w^{-1} - (\Psi_2 e_0^{-4})\, w^{-3}$. Also $\partial_w f(w, x) \le \partial_w \phi_-(w)$. Let

$$0 < \lambda_0 < (\Psi_1^-)^2/(4\Psi_2^+). \tag{5.168}$$

Each of functions $\phi_-(y)$ and $\phi_+(y)$ has two positive roots; moreover, $y_1^- < y_2^-$ and $y_1^+ < y_2^+$. Since $\phi_-(y) < \phi_+(y)$ for $y > 0$, we also have $y_2^- < y_2^+$ and $y_1^- > y_1^+$. Denote by $y_3^- \in (y_1^-, y_2^-)$ a unique positive root of $\partial_y \phi_-(y) = -\lambda_0 - \Psi_1^- y^{-2} + 3\Psi_2^+ y^{-4}$. Notice that $\phi_-(y) > 0$ for $y \in (y_1^-, y_2^-)$ and $\phi_-(y) < 0$ for $y \in (0, \infty) \setminus [y_1^-, y_2^-]$; moreover, $\phi_-(y)$ increases in $(0, y_3^-)$ and decreases in (y_3^-, ∞). The line $z = -\lambda_0 y$ is asymptotic for the graph of function $\phi_-(y)$ when $y \to \infty$, and $\lim_{y \downarrow 0} \phi_-(y) = -\infty$. The function $\partial_y \phi_-(y)$ decreases in $(0, y_4^-)$ and increases in (y_4^-, ∞), where $y_4^- := (6\Psi_2^+/\Psi_1^-)^{1/2} > y_3^-$, and $\lim_{y \to \infty} \partial_y \phi_-(y) = -\lambda_0$, see Figure 5.2. Thus, $y_1^+ < y_1^- < y_3^- < y_2^- < y_2^+$. Hence, the following function is positive for $\sigma \in (0, y_2^- - y_3^-)$:

$$\mu^+(\sigma) := -\sup_{y \ge y_2^- - \sigma} \partial_y \phi_-(y) = \min\{|\partial_y \phi_-(y_2^- - \sigma)|, \lambda_0\}.$$

Define closed in $C(F)$ nonempty sets

$$\mathscr{U}^{\varepsilon,\eta} = \{\tilde{u} \in C(F) : y_2^- - \varepsilon \le \tilde{u}/e_0 \le y_2^+ + \eta\}, \quad \varepsilon \in (0, y_2^- - y_1^-), \ \eta \in (0, \infty).$$

We have $\mathscr{U}_0 \subset \mathscr{U}^{\varepsilon,\eta} \subset \mathscr{U}^{\varepsilon,\infty} \subset \mathscr{U}_1$, where the set $\mathscr{U}_1 = \{\tilde{u} \in C(F) : \tilde{u}/e_0 > y_1^-\}$ is open, and $\mathscr{U}_0 = \{\tilde{u} \in C(F) : y_2^- \le \tilde{u}/e_0 \le y_2^+\}$.

Proposition 5.117 ([RZe3]). *Assume that (5.168) is satisfied. Then*

(i) *for any $u_0 \in \mathscr{U}^{\varepsilon,\eta}$, (5.167) admits a unique global solution, and $\mathscr{U}^{\varepsilon,\eta}$ are invariant sets for associated semigroup $\mathscr{S}_t : u_0 \to u(\cdot,t)$ ($t \ge 0$) in $\mathscr{C}_\infty$;*
(ii) *for any $\sigma \in (0, \varepsilon)$ there exists $t_1 > 0$ such that $\mathscr{S}_t(\mathscr{U}^{\varepsilon,\infty}) \subseteq \mathscr{U}^{\sigma,\infty}$ for $t \ge t_1$.*

Proposition 5.117 supports the following.

Theorem 5.118.

(i) *If (5.168) holds, then (5.166) has in $\mathscr{U}_1 \cap C^\infty(F)$ a unique solution u_*, which obeys $y_1^- \le u_*/e_0 \le y_1^+$; moreover, $u_* = \lim_{t \to \infty} u(\cdot,t)$, where u solves (5.167) with $u_0 \in \mathscr{U}_1$, and for any $\sigma \in (0, y_2^- - y_3^-)$, the set $\mathscr{U}^{\sigma,\infty}$ is attracted by associated semigroup exponentially fast to u_* in C-norm:*

$$\|u(\cdot,t) - u_*\|_{C(F)} \le \delta^{-1}(e_0)\, e^{-\mu^+(\sigma)t} \|u_0 - u_*\|_{C(F)} \quad (t > 0, \ u_0 \in \mathscr{U}^{\sigma,\infty}).$$

(ii) Let β, Ψ_1, Ψ_2 be smooth functions on $F \times \mathbb{R}^n$ with a metric $\langle \cdot, q \rangle$. If (5.168) holds for any $F \times \{q\}$ ($q \in \mathbb{R}^n$), then the solution u_*, see (i), is smooth on $F \times \mathbb{R}^n$.

For $\Psi_2 \equiv 0$ when $\Psi_3 = 0$, $\Psi_1 > 0$, condition (5.168) reduces to $\lambda_0 > 0$, furthermore, each of the functions $\phi_-(y) = -\lambda_0 y + \Psi_1^- y^{-1}$ and $\phi_+(y) = -\lambda_0 y + \Psi_1^+ y^{-1}$ has one positive root $y_2^- = (\Psi_1^- / \lambda_0)^{1/2}$ and $y_2^+ = (\Psi_1^+ / \lambda_0)^{1/2}$; and $\partial_y \phi_-(y) < 0$ for $y > 0$ and $\lambda_0 > 0$. Recall that the quantities $\Psi_i^\pm$ are defined in Section 5.4.5.

References

[A] J. Adams, Vector fields on spheres, Annals of Math. 75 (1962), 603–632

[AG] M.A. Akivis, V.V. Goldberg, Differential geometry of webs. In *Handbook of differential geometry*, Vol. I, pp. 1–152, North-Holland, Amsterdam, 2000

[An] H. Anciaux, *Minimal submanifolds in pseudo-Riemannian geometry*. With a foreword by F. Urbano. World Scientific Publ. Co. Pte. Ltd., Hackensack, NJ, 2011

[AH] B. Andrews and C. Hopper, *The Ricci flow in Riemannian geometry*, Springer, 2011

[ARW] K. Andrzejewski, V. Rovenski and P. Walczak, Integral formulas in foliations theory, 73–82, in [RW1]

[AW1] K. Andrzejewski, P. Walczak, The Newton transformation and new integral formulae for foliated manifolds, Ann. Global Anal. Geom. 37:2 (2010), 103–111

[AW2] —— Extrinsic curvatures of distributions of arbitrary codimension, J. Geom. Phys., 2010, 60 (5), 708–713

[AKN] K. Andrzejewski, W. Kozłowski and K. Niedziałomski, Generalized Newton transformation and its applications to extrinsic geometry, Asian J. Math., 20, No. 2 (2016), 293–322

[Asi] D. Asimov, Average Gaussian curvature of leaves of foliations, Bull. Amer. Math. Soc., 1978, 84(1), 131–133

[Asu] T. Asuke, Transverse projective structures of foliations and infinitesimal derivatives of the Godbillon-Vey class. Int. J. of Math. 26(4), 2015 (29 pp.)

[Au] T. Aubin, *Some nonlinear problems in Riemannian geometry*, Springer, 1998

[BKO] J.L.M. Barbosa, K. Kenmotsu, G. Oshikiri, Foliations by hypersurfaces with constant mean curvature. Mat. Z. 207, (1991) 97–108

© Springer Nature Switzerland AG 2021

V. Rovenski, P. Walczak, *Extrinsic Geometry of Foliations*, Progress in Mathematics 339, https://doi.org/10.1007/978-3-030-70067-6

[BDRS] E. Barletta, S. Dragomir, V. Rovenski, and M. Soret, Mixed gravitational field equations on globally hyperbolic spacetimes, Class. Quantum Grav. 30, (2013), 085015, 26 pp.

[BCS] S. Bácsó, X. Cheng and Z. Shen, Curvature properties of (α, β)-metrics. Adv. Stud. Pure Math., 48, Math. Soc. Japan, Tokyo, (2007), 73–110

[BHV] L. Bedulli, W. He and L. Vezzoni, Second-order geometric flows on foliated manifolds, Geom. Anal. (2018) 28, 697–725

[BEE] J. Beem, P. Ehrlich and K. Easley, *Global Lorentzian geometry*. New York, Dekker, 1996

[BF] A. Bejancu and H. Farran, *Foliations and geometric structures*. Springer-Verlag, 2006

[Ber] M. Berger, *A Panoramic View of Riemannian Geometry*, Springer, 2002

[BS] A. N. Bernal and M. Sánchez, Smoothness of time functions and the metric splitting of globally hyperbolic spacetimes, Commun. Math. Phys. 257 (2005), 43–50

[Bes] A. L. Besse, *Einstein manifolds*, Springer, 1987

[Bl] D. Blair, *Riemannian geometry of contact and symplectic manifolds*, Springer, 2010

[Bo] R. Bott, *Lectures on characteristic classes and foliations*, LNM 279, Springer 1972, 1–94

[Br] S. Brendle: *Ricci Flow and the Sphere Theorem*, Graduate Studies in Math., 111, AMS, 2010

[BW] F. Brito and P. Walczak, On the energy of unit vector fields with isolated singularities, Annales Polonici Mathematici, 73 (3) (2000), 269–274

[BLR] F. Brito, R. Langevin, and H. Rosenberg, Intégrales de courbure sur des variétés feuilletées, J. Diff. Geom., 1981, 16, 19–50

[BN] F. Brito and A. Naveira, Total extrinsic curvature of certain distributions on closed spaces of constant curvature, Ann. Global Anal. Geom., 2000, 18, 371–383

[BP] L. Brunetti and A. M. Pastore, On the classification of Lorentzian Sasaki space forms, Publications De L'Institut Mathématique, Nouvelle série, tome 94 (108), (2013) 163–168

[BM] A. Bucki and A. Miernowski, Almost r-paracontact connections, Acta Math. Hung. 45 (3–4) (1985), 327–336

[CN] C. Camacho and A. Lins Neto, *Geometric Theory of Foliations*, Birkhhäuser, 1985

[CSC] A. Caminha, P. Souza, F. Camargo, Complete foliations of space forms by hypersurfaces, Bull. Braz. Math. Soc., New Series, 41:3 (2010), 339–353

[CC] A. Candel and L. Conlon, *Foliations, I and II*, Grad. Studies in Math. 60, AMS, 2000, 2003

[CC1] J. Cantwell, L. Conlon, Endsets of exceptional leaves; a theorem of G. Duminy, 225–261 in *Foliations, Geometry and Dynamics*, World Sci., Singapore, 2002.

[CM] K. Catino and M. Mastrolia, *A Perspective On Canonical Riemann Metrics*, Progress in Mathematics, 336, Birkhhäuser Basel, 2020

[Ca] J.S. Case, Singularity theorems and the Lorentzian splitting theorem for the Bakry–Emery–Ricci tensor, J. Geom. Phys. 60 (2010), 477–490

[CR] T.E. Cecil, P.J. Ryan, *Geometry of Hypersurfaces*, Springer Monographs in Mathematics. Springer-Verlag, New York, 2015

[Ch] B.-Y. Chen, *Geometry of submanifolds and its applications*, Science Univ. of Tokyo, 1981

[Ch2] —— *Pseudo-Riemannian geometry, δ-invariants and applications*. World Sci., 2011

[Ch3] —— *Differential Geometry of Warped Product Manifolds and Submanifolds*, World Sci. Publ., 2017

[CST] X. Cheng, Z. Shen and Y. Tian, A class of Einstein (α, β)-metrics. Israel J. Math., 192(1) (2012), 221–249

[CK] S. Chern and N. Kuiper, Some theorems on the isometric imbedding of compact Riemannian manifolds in Euclidean Space, Ann. Math., 56 (1952), 422–430

[CG] L. Conlon and S. Goodman, The closed leaf index of foliated manifolds, Trans. Amer. Math. Soc. 233 (1077), 205–221

[CW] M. Czarnecki and P. Walczak, Extrinsic geometry of foliations, 149–167. In Proc. of the Int. Conf., Lodz, Poland, June 12–14, 2005. World Scientific, 2006

[DHS] U. Dierkes, S. Hildebrandt, and F. Sauvigny, *Minimal surfaces*. Fundamental Principles of Math. Sciences, 339. Springer, Heidelberg, 2010

[DV] P.W. Doyle, and P.J. Vassiliou: Separation of variables for the 1-dimensional non-linear diffusion equation, Int. J. Non-Linear Mechanics, 33 (2) (1998), 315–326

[ES] J. Eells and J. Sampson: Harmonic Mappings of Riemannian Manifolds, Amer. J. Math. 86 (1964), 109–160

[ET] Y. Eliashberg and W. Thurston, *Confoliations*, Amer. Math. Soc., 1998.

[FIP] M. Falcitelli, S. Ianus, and A.M. Pastore, *Riemannian submersions and related topics*, World Scientific, Singapore, 2004

[FPM] L.M. Fernández and A. Prieto-Martín, On η-Einstein para-S-manifolds, Bull. Malays. Math. Sci. Soc. 40 (2017), 1623–1637

[Fe] D. Ferus, Totally geodesic foliations, Math. Ann., 188, (1970), 313–316

[FH] P. Foulon and B. Hasselblatt, Godbillon–Vey invariants for maximal isotropic C^2-foliations, Adv. Studies in Pure Mathematics, 72 (2017), 349–366

[F] T. Frankel, Manifolds with positive curvature, Pacific J. Math., 11 (1961), 165–171.

[GP] H. Geiges and J.G. Pérez, Transversely holomorphic flows and contact circles on spherical 3-manifolds. Enseign. Math. 62, no. 3–4 (2016), 527–567

[GLW] E. Ghys, R. Langevin and P. Walczak, Entropie géométrique des feuilletages, Acta Math. 160 (1988), 105–142

[Gl] H. Gluck, Dynamical behavior of geodesic fields, in *Global Theory of Dynamical Systems*, LNM 819, Springer 1980, 190–215

[GZ] H. Gluck and W. Ziller, On the volume of a unit vector field on the three-sphere, Comment. Math. Helvetici, 61 (1986), 177–192

[GV] C. Godbillon and J. Vey, Un invariant des feuilletages de codimension 1, C. R. Acad. Sci. Paris Sér A-B, 273 (1971), A92–A93

[GPS] I. Gordeeva, V.I. Pan'zhenskii and S. E. Stepanov, Riemann–Cartan manifolds, J. of Math. Sci., 169(3), 2010, 342–361

[Gr1] A. Gray, Pseudo-Riemannian almost product manifolds and submersions, J. Math. Mech., 16:7 (1967), 715–737

[Gr2] —— *Tubes*, Birkhäuser, 2004

[GW] D. Gromoll and G. Walschap, *Metric foliations and curvature*, Birkhäuser, 2009

[GKM] D. Gromoll, W. Klingenberg, W. Meyer, *Riemannsche Geometrie im Grossen*, Lect. Notes in Math. 55, Springer Verlag, Berlin–Heidelberg–New York, 1968

[Gu] G. G. Gurevich, *Foundations of the Theory of Algebraic Invariants*, Noordhof, 1964

[Ha] A. Haefliger, Some remarks on foliations with minimal leaves, J. Diff. Geom. 15 (1980), 269–284

[HL] R. Harvey and H. B. Lawson, Calibrated foliations, Aer. J. Math 104 (1980), 607–633

[HH] G. Hector and U. Hirsch, *Introduction to the Geometry of Foliations, Parts A and B*, Vieweg and Sohn, Wiesbaden, 1981 and 1983

[Hu] S. Hurder, Problem Set, 205–213. In Proc. "Foliations 2012", World Sci. Publ., 2013.

[HL] S. Hurder and R. Langevin, Dynamics and the Godbillon-Vey class of C^1-foliations. J. Math. Soc. Jpn. 70, no. 2 (2018), 423–462

[Is] A. Isidori, *Nonlinear Control Systems*, 3-rd ed., Springer Verlag, New York, 1995

[JS] M.A. Javaloyes and M. Sánchez, On the definition and examples of Finsler metrics. Ann. Sc. Norm. Super. Pisa Cl. Sci. 13(3), 813–858 (2014).

[Jo] J. Jost, *Riemannian geometry and geometric analysis*, 7th ed., Universitext, Springer, 2017

[Ka] T. Kashiwada, On a contact 3-structure, Math. Z. 238, no. 4, (2001), 829–832

[KW] J.L. Kazdan and F.W. Warner, Curvature functions for compact 2-manifolds, Ann. of Math. 99 (1974), 14–47

[KN] J. L. Kelley and I. Namioka, *Linear topological spaces*, Springer, New York, 1976

[Kl] W. Klingenberg, *Riemannian Geometry*, Walter de Gruyter, 1995

[Kn] H. Kneser, Reguläre Kurvenscharen auf den Ringflächen, Math. Ann. 91 (1923), 135–154

[LaW] R. Langevin and P. Walczak, Conformal geometry of foliations, Geom. Dedicata 132 (2008), 135–178

[LT] P. Li, and L.-F. Tam, Positive harmonic functions on complete manifolds with nonnegative curvature outside a compact set. Ann. of Math. (2) 125, no. 1, (1987), 171–207

[LMR] M. Lovric, M. Min-OO, and E. Ruh: Deforming transverse Riemannian metrics of foliations, Asian J. Math. 4(2) (2000) 303–314

[LuW] M. Lużyńczyk and P. Walczak, New integral formulae for two complementary orthogonal distributions on Riemannian manifolds, Ann. Glob. Anal. Geom. 48 (2015), 195–209

[Mas] T. Maszczyk, Foliations with rigid Godbillon–Vey class, Math. Z. 230(2) (1999), 329–344

[Mat] M. Matsumoto, Theory of Finsler spaces with (α, β)-metric. Reports on mathematical physics, 31(1) (1992), 43–83

[Miq] V. Miquel, Some examples of Riemannian almost-product manifolds, Pacific J. Math., 111 (1984), 163–178

[Mik] J. Mikeš, et al. *Differential geometry of special mappings*, Palacký Univ., Olomouc, 2015

[MR] A.N. Milgram, P.C. Rosenbloom, Harmonic forms and heat conduction. I. Closed Riemannian manifolds, Proc. Nat. Acad. Sci., 1951, 37, 180–184

[Mol] P. Molino, *Riemannian foliations*, Progress in Math., vol. 73, Birkhäuser, 1988

[Mon] A. Montesinos, On certain classes of almost product structures, Michigan Math. J., 1983, 30(1), 31–36

[Mo2] —— *Geometry of characteristic classes*. Translations of Math. Monographs, 199. Iwanami Series in Modern Math. AMS, Providence, RI, 2001

[Mrv] J. Morvan, Distance of two submanifolds of a manifold with positive curvature, Rend. mat. e appl., 3 (1983), 357–366

[Na] A. Naveira, A classification of Riemannian almost product manifolds, Rend. Math., 3 (1983), 577–592

[N] T. Nora, Seconde forme fondamentale d'une application et d'un feuilletage. Thèse, l'Univ. de Limoges (1983), 115 pp.

[No] S. Novikov, *Topology of foliations*, Trudy Moskov. Mat. Obsc. 14 (1965), 248–278

[Op] B. Opozda, A sectional curvature for statistical structures, Linear Algebra and its Applications, 497, (2016), 134–161

[Os1] G. Oshikiri, Mean curvature functions of codimension-one foliations, Comment. Math. Helv. 65 (1990), 79–84

[Os2] —— Mean curvature functions of codimension-one foliations, II, Comment. Math. Helv. 66 (1991), 512–520

[Os3] —— A characterization of mean curvature functions of codimension-one foliations, Tohoku Math. J. 49, No. 4 (1997), 557–563

[Os4] —— Some properties of mean curvature vectors for codimension-one foliations, Illinois J. Math. 49 (2005), 159–166

[OD] E. Özüsağlam and E. Dikici, Pseudo f-manifolds with complemented frames. Adv. Appl. Clifford Algebr. 26 (2016), no. 1, 305–314

[PM] J. Palis and W. de Melo, *Geometric theory of dynamical systems*, Springer Verlag, 1982

[Pa] G.P. Paternain, *Geodesic Flows*, Birkhäuser, 1999.

[Pe] P. Petersen, *Riemannian geometry*, 3d ed. Springer, 2016.

[PT2] —— On solutions of the Ricci curvature equation and the Einstein equation, Israel J. Math. 171 (2009), 61–76

[PP] R. Pina and M. Pieterzack, Prescribed curvature tensor in locally conformally flat manifolds. J. Geom. Phys. 123 (2018), 438–447

[PR] R. Ponge, and H. Reckziegel: Twisted products in pseudo-Riemannian geometry, Geom. Dedicata 48 (1993), 15–25

[RP] P. Popescu and V. Rovenski, An integral formula for singular distributions, Results in Mathematics, 75, Article number: 18 (2019), DOI: 10.1007/s00025-019-1145-1

[RPS] P. Popescu, V. Rovenski and S. Stepanov, On singular distributions with statistical structure, Mathematics, MDPI, 2020, 20 pp.

[RPS2] P. Popescu, V. Rovenski and S. Stepanov, On the Bochner technique for singular distributions, arXiv:2008.12868, 2020.

[PBM] A. P. Prudnikov, Y. A. Brychkov and O. I. Marichev, *Integrals and Series, Vol. 3*, Gordon and Breach Sci. Publ., New York, 1990

[PRSW] J. I. R. Prieto, M. Saralegi-Aranguren and R. Wolak, Cohomological tautness for Riemannian foliations, Russ. J. Math. Phys. 16, no. 3 (2009), 450–466

[Pu] A. Pulemotov, Maxima of curvature functionals and the prescribed Ricci curvature problem on homogeneous spaces. J. Geom. Anal. 30 : 1 (2020), 987–1010

[Re1] G. Reeb, Sur la courboure moyenne des variétés intégrales d'une équation de Pfaff $\omega = 0$, C. R. Acad. Sci. Paris 231 (1950), 101–102

[Re2] G. Reeb: Sur certaines propriétés topologiques des variétés feuilletées, Actualités, Sci. Ind., no. 1183, Hermann and Cie., Paris, 1952

[Rei] B. Reinhart, *Differential geometry of foliations*, Ergeb. Math., 99, Springer, New York, 1983

[RW] B.L. Reinhart and J.W. Wood, A metric formula for the Godbillon–Vey invariant for foliations, Proc. Amer. Math. Soc., 38, No. 2 (1973), 427–430

[Re] R. Reilly, Applications of the Hessian operator in a Riemannian manifold, Ind. Univ. Math. J. 26, 1977, 459–472

[R1] V. Rovenski, *Foliations on Riemannian Manifolds and Submanifolds*, Birkhäuser, 1998

[R2] —— Foliations, submanifolds and mixed curvature (a survey), J. Math. Sci., New York, 99, No. 6 (2000), 1699–1787

[R3] —— On the role of partial Ricci curvature in geometry of submanifolds and foliations, Ann. Polonici Math., 1998, v. 68 (LXVIII), No. 1, 61–82

[R4] —— Extrinsic geometric flows on codimension-one foliations, J. of Geom. Analysis, 23(3), (2013), 1530–1558

[R5] —— Integral formulae for a Riemannian manifold with two orthogonal distributions, Central European J. Math. 9, No. 3, (2011), 558–577

[R6] —— On solutions to equations with partial Ricci curvature, J. Geom. and Physics, 86, (2014), 370–382

[R7] —— Einstein–Hilbert type action on spacetimes, Publications de l'Institut Mathématique, Issue: (N.S.) 103 (117) (2018), 199–210

[R8] —— The partial Ricci flow for foliations, 125–155, in [RW1]

[R9] —— Integral formulas for a metric-affine manifold with two complementary orthogonal distributions, Global J. Adv. Research Class. and Modern Geom., 6(1), (2017), 7–19

[R10] —— The new Minkowski norm and integral formulae for a manifold with a set of one-forms, Balkan J. of Geometry and Its Applications, 23, No. 1 (2018), 75 – 99

[R11] —— The weighted mixed curvature of a foliated manifold, FILOMAT (Publ. de L'Inst. Math.) 33:4 (2019), 1097–1105

[R12] —— Prescribing the mixed scalar curvature of a foliation. Balkan J. of Geometry and Its Applications, Vol. 24, No. 1, (2019), 73–92

[R13] —— Problems of Extrinsic Geometry of foliations. Results of Science and Technology: Modern mathematics and its applications. Thematic reviews, Vol. 999 (2019), 1–10

[R14] —— Integral formulas for a Riemannian manifold with several orthogonal complementary distributions. Global J. of Advanced Research on Classical and Modern Geometries, Vol. 10, (2021), Issue 1, 32–42

[R15] —— The Einstein–Hilbert type action on almost k-product manifolds. Preprint. arXiv:2009.03212 (2020)

[R16] — A series of integral formulas for a foliated sub-Riemannian manifold, preprint (2021), arXiv:2103.02473

[RS] V. Rovenski and V. Sharafutdinov, The partial Ricci flow on one-dimensional foliations, arXiv:1308.0985, 18 pp. 2013

[RW1] V. Rovenski and P. Walczak (eds.), *Geometry and its Applications*, Springer Proceedings in Mathematics and Statistics, 72, Springer, 2014

[RW2] V. Rovenski and P. Walczak, *Topics in extrinsic geometry of codimension-one foliations*, Springer Briefs in Mathematics, Springer-Verlag, 2011

[RW3] —— Integral fomulae for foliations on Riemannian manifolds, in Proc. of 10th Int. Conference "Diff. Geometry and Its Applications", Olomouc, 203–214, World Sci. Publ., 2008

[RW4] —— Variational formulae for the total mean curvatures of a codimension-one distribution, 83–93. Proc. 8th Int. Colloq., Santiago-de Compostela, Spain, 2008, World Sci. Publ., 2009

[RW5] —— Integral formulae on foliated symmetric spaces, Math. Ann. 352(1), (2012), 223–237

[RW6] —— Deforming convex bodies in Minkowski geometry, arXiv:1910.01854v2 (Accepted: International J. of Math.)

[RW7] —— Integral formulae for codimension-one foliated Randers spaces, Publ. Math. Debrecen, 91/1-2, (2017), 95–110

[RW8] —— A Godbillon-Vey type invariant for a 3-dimensional manifold with a plane field, Differential Geom. and its Applications, 66, (2019), 212–230

[RW9] —— Variations of the Godbillon-Vey invariant of foliated 3-manifolds, Complex Analysis and Operator Theory, 13(6), (2019), 2917–2937

[RW10] —— Variations of the Godbillon-Vey invariant for transversely paralleliz-able foliations, arXiv:1909.13250, 18 pp. 2019

[RWo1] V. Rovenski and R. Wolak, Deforming metrics of foliations, Central European J. Math., 11(6) 2013, 1039–1055

[RWo2] —— The partial Ricci flow on $\mathfrak{g}$-foliations, arXiv:1905.07704, 2019, 16 pp.

[RZ1] V. Rovenski and T. Zawadzki, The Einstein–Hilbert type action on pseudo-Riemannian almost product manifolds, J. of Math. Physics, Analysis, Geometry, 15, No. 1 (2019), 86–121

[RZ2] —— Variations of the total mixed scalar curvature of a distribution, Annals of Global Analysis and Geometry, 54 (2018), 87–122

[RZ3] —— The mixed scalar curvature of almost-product metric-affine manifolds, Results in Mathematics, (2018) 73:23

[RZ4] —— The Einstein–Hilbert type action on metric-affine almost product manifolds, Preprint, arXiv:2007.12406, 2020

[RZe1] V. Rovenski and L. Zelenko, Prescribing the mixed scalar curvature of a foliation, 83–123, in [RW1]

[RZe2] —— The mixed Yamabe problem for harmonic foliations, Europ. J. of Math., 1, (2015), 503–533

[RZe3] —— Prescribing mixed scalar curvature of foliated Riemann–Cartan spaces, J. of Geometry and Physics, 126 (2018) 42–67

[Sc] L. Schwartz, *Théorie des distributions*, Hermann, Paris, 1966

[SW] P. Schweitzer and P. Walczak, Prescribing mean curvature vectors for foliations, Illinois J. Math. 48 (2004), 21–35

[SS] Y.-B. Shen and Z. Shen, *Introduction to modern Finsler geometry*, World Sci. 2016

[SWZ] K. Smoczyk, G. Wang and Y. Zhang, The Sasaki-Ricci flow. Internat. J. Math. 21 (2010), no. 7, 951–969

[St] P. Stefan, Accessible sets, orbits and foliations with singulrities, Proc. London Math. Soc. 29 (1974), 699–713

[SM] S.E. Stepanov and J. Mikeš, Liouvile-type theorems for some classes of Riemannian almost product manifolds and for special mappings of Riemannian manifolds, Differential Geom. and its Appl. 54, Part A (2017), 111–121

[S1] S.E. Stepanov, $O(n) \times O(m-n)$-structures on m-dimensional manifolds, and submersions of Riemannian manifolds. St. Petersburg Math. J. 7 (1996), no. 6, 1005–1016

[S2] —— Liouville-type theorems for twisted and warped products manifolds, arXiv:1608.03590

[Su1] D. Sullivan: Cycles for the dynamical study of foliated manifolds and complex manifolds, Invent. Math. 36 (1976), 225–256

[Su2] —— A homological characterization of foliations consisting of minimal surfaces, Comm. Math. Helv. 54 (1979), 218–223

[Su] H. J. Sussmann, *Orbits of familes of vector fields and integrability of distributions*, Trans. Amer. Math. Soc. 180 (1973), 171–188

[Sv] M. Svensson, Holomorphic foliations, harmonic morphisms and the Walczak formula, J. Lond. Math. Soc., 2003, 68, 781–794

[TY] R. Takagi and S.Yorozu, Minimal foliations on Lie groups, Tohoku Math. J. 36(4) (1984) 541–554

[Tam] I. Tamura, *Topology of Foliations*, Iwananmi Shoten, 1076 (English transl.: AMS, 1992)

[Tar] A. Tarrio, On certain clacces of metric para-ϕ-manifolds with parallelizable kernel, Tebsir, N.S. 57 (1996), 258–267

[Tay] M. E. Taylor: *Partial Differential Equations, III: Nonlinear Equations*, 2nd Edition, Applied Math. Sciences 117, Springer 2011

[Th1] W. Thurston, Noncobordant foliations of S^3, Bull. AMS, 78, No. 4 (1972), 511–514

[Th2] —— Existence of codimension-one foliations, Ann. of. Math., 194 (1976), 347–352

[Th3] —— The theory of foliations in codimension greater than one, Comment. Math. Helv., 49 (1974), 214–231

[To] Ph. Tondeur, *Foliations on Riemannian manifolds*, Springer, 1988

[W1] P. Walczak, *Dynamics of Foliations, Groups and Pseudogroups*, Birkhäuser, 2004

[W2] —— Mean curvature functions for codimension-one foliations with all the leaves compact, Czechoslovak Math. J. 34 (109) (1984), 146–156

[W3] —— Mean curvature functions for foliated bundles, Topics in differential geometry (Debrecen 1984), Colloq. Math. Soc. Janos Bolyai, 46, North Holland, 1988, 1309–1317

[W4] —— An integral formula for a Riemannian manifold with two orthogonal complementary distributions. Colloq. Math., 58 (1990), 243–252

[W5] —— Conformally defined geometry on foliated Riemann manifolds, 431–439. In Proc. of the Int. Conf., Lodz, Poland, June 12–14, 2005. World Scientific, 2006

[W6] —— Integral formulae for foliations with singularities. Coll. Math. 150(1) (2017), 141–148

[W7] —— Tautness and the Godbillon–Vey class of foliations. In Proc. "Foliations 2012", 205–213, World Sci. Publ., 2013.

[Wal] P. Walters, *An Introduction to Ergodic Theory*, Springer Verlag, 1982

[War] F. W. Warner, *Foundations of differentiable manifolds and Lie groups*, Springer, 1983

[WPAH] G.M. Webb, A. Prasad, S.C. Anco and Q. Hu, Godbillon-Vey helicity and magnetic helicity in Magnetohydrodynamics, J. Plasma Phys. (2019), vol. 85, Issue 5, 775850502.

[Wy] W. Wylie, Sectional curvature for Riemannian manifolds with density, Geom. Dedicata, 178 (2015), 151–169

[Yan] K. Yano, On a structure f satisfying $f + f^3 = 0$, Techn. Rep. 12, Univ. of Washington, 1961.

[Yau] S.T. Yau, Some Function-Theoretic Properties of Complete Riemannian Manifolds and their Applications to Geometry. Indiana Univ. Math. J., 25 (1976), 659–670

[YZ] C. Yu, H. Zhu, On a new class of Finsler metrics. Diff. Geom. & Appl. 29 (2011), 244–254

[Za] T. Zawadzki, Existence conditions for conformal submersions with totally umbilical fibers, Differential Geometry and its Applications 35 (2014), 69–85

[Ze] A. Zeghib, Feuilletages géodésiques des variétés localement symétriques, Topology, 36, (1997), 805–828

Index

A

Action
  Einstein–Hilbert, 188
Atlas
  foliated, 3
  nice, 4

B

Boundary
  tangential, 3
  transverse, 3
Bundle
  cotangent, 51
  foliated, 7
  principal, 89
  spatial tangent, 254
  tangent, 50

C

Cartan torsion, 20
Chart
  biregular foliated, 244
  distinguished, 3
Complete transversal, 5
Component
  Novikov, 134
  Reeb, 10, 140
  saturated Novikov, 134
Confoliation, 34, 209

Connection
  Chern, 21, 120
  Levi-Civita, 14
Current
  de Rham, 126
  Dirac, 126
Curvature
  (generalized) mean, 25, 57, 69
  dimension, 41
  extrinsic, 29, 84, 89
  Gaussian, 27, 63, 136, 250
  mixed Ricci, 189, 190
  mixed scalar, 30, 274
  of a curve, 201
  partial Ricci, 27, 36, 154
  Ricci, 18
  scalar, 18
  sectional, 18
  weighted, 90
  weighted scalar, 40

D

De Rham cohomology, 18
Derivative
  covariant, 16
  Lie, 52
Distribution
  curvature-invariant, 156, 258
  geometrically taut, 202

© Springer Nature Switzerland AG 2021
V. Rovenski, P. Walczak, *Extrinsic Geometry of Foliations*, Progress
in Mathematics 339, https://doi.org/10.1007/978-3-030-70067-6

Distribution (*cont.*)
  harmonic, 208
  involutive, 1, 54
  non-degenerate, 24
  singular, 78
  tangent (normal) to $\mathscr{F}$, 79
  topologically taut, 202
  totally geodesic, 208
  totally umbilical, 208
Divergence, 54
  leaf-wise, 85

E
Equation
  Burgers, 277
  Codazzi, 28
  elliptic, 272
  Euler–Lagrange, 164, 167, 171,
      174, 179, 191
  forced Burgers, 278
  heat, 289
  parabolic, 236, 288
  reaction-diffusion, 294
  Riccati, 27
  Ricci, 28
Euler characteristic, 54
Exponential map, 63
Extrinsic geometry, 23

F
Flow, 50
  extrinsic geometric, 236, 238
  geometric, 223
  heat, 223, 225
  mean curvature, 223
  Newton transformation, 243
  partial Ricci, 251
  Ricci, 223
  transverse Ricci, 223
Focal point, 64
Foliation, 3
  k-nullity, 45
  (transversely) orientable, 56
  civilized, 76
  concordant, 210

  conformal, 25
  conformally harmonic, 67
  geometrically taut, 124, 128, 247
  harmonic, 25
  Hirsch, 12
  Hopf, 25, 41
  lift, 81
  purification, 128
  Reeb, 10, 249
  relative nullity, 45
  Riemannian, 25, 215, 236
  singular, 71
  tangentially Lie, 268
  topologically taut, 128
  totally geodesic, 25, 236
  totally umbilical, 25, 236
  transversely harmonic, 25
Form
  closed, 53, 125
  exact, 53
  exterior differential, 53
  harmonic, 18
  index, 205
  second fundamental, 208
  volume, 17, 54
Formula
  integral, 62
  of second variation, 41
  Rummler, 136
  variational, 23, 153
Function
  Dirac delta, 289
  elementary symmetric, 57
  harmonic, 27, 169, 175, 289
  mean curvature, 248
  power sums, 57
  theta, 292

G
Geodesic, 15

H
Hamiltonian, 295
Heat kernel, 289
Hessian, 17

Holonomy
  group, 6
  map, 5
  pseudogroup, 2, 5, 9

I
Immersion, 49
Indicatrix, 20
Integrable in average, 206
Invariant
  Bott, 220
  Godbillon-Vey of (ω, T), 200, 201
  Godbillon-Vey type, 220
Isometry, 13
Isomorphism
  $\flat$ (musical), 13
  $\sharp$ (musical), 13
  of pseudogroups, 5

J
Jacobi identity, 50

L
Lamination, 9
Laplacian, 37
  Hodge, 17
  rough, 17
Leaf (of a foliation), 3
  generic, 11
  proper, 3
Lie bracket, 50

M
Manifold
  almost multi-product, 91
  closed, 7
  Einstein, 18, 69
  Riemann–Cartan, 19
  Riemannian, 13
  statistical, 19, 98, 99
Matrix
  generalized companion, 240
Maximum principle, 293
Metric
  t-dependent, 177
  adapted, 235

bundle-like, 25
compatible, 200
conformally defined, 67
Einstein, 161
Finsler, 20
Lorentz, 24
pseudo-Riemannian, 24
Riemannian, 24
rotational symmetric, 31, 284
Mixed plane, 27
Morphism of pseudogroups, 5

N
Nice covering, 4
Norm, 13
  (α, β)-, 21
  (α, b)-, 104
  exponential, 112
  Kropina, 111
  Minkowski, 20
  quadratic, 112
  Randers, 111
  slope, 111

O
Operator
  Casorati type, 28
  co-nullity (splitting), 24
  divergence, 17
  Jacobi, 27, 82
  Laplace-Beltrami, 17
  Schrödinger, 274, 292
  shape (Weingarten), 63, 208
  star, 17
  weighted Jacobi, 45

P
Plaque, 3
Product
  doubly twisted, 30, 168, 235
  multiply twisted, 96
  multiply warped, 96
  twisted, 31, 169, 214, 235, 264
  warped, 31, 215, 277
Pseudogroup, 2
Pullback, 8

S

Saturation, 9
Semigroup, 294, 300, 301
Set
  exceptional minimal, 9
  invariant, 300, 301
  minimal, 9
  saturated, 9
Space
  Banach, 282, 292
  Hausdorff, 11
  Hilbert, 292
  locally symmetric, 63, 78
  rank one symmetric, 73
  Sobolev, 289
  tangent, 48
Space-time, 188
Structure
  K-contact, 33, 184
  f-, 34
  almost complex, 15
  almost contact, 33
  almost multi-product, 198
  almost-product, 157
  complex, 15
  contact, 56, 184, 267
  Hermitian, 15
  Kähler, 15
  para-f-, 35
  transversely holomorphic,
    219
  weak almost contact, 269
  weak quasi-Sasakian, 268
Subbundle
  horizontal, 82
  vertical, 82
Submanifold, 49
  curvature-invariant, 46
  embedded, 49
  immersed, 49
  minimal (harmonic), 123
  ruled, 45
Submersion, 49
  Riemannian, 15
Suspension, 7

T

Tensor, 51
  adapted biconformal, 173
  contorsion (difference), 19, 120
  curvature, 14
  fundamental, 24
  integrability, 24, 216
  Jacobi, 46
  mixed Einstein, 189
  mixed Ricci, 189
  Nijenhuis torsion, 15
  partial Ricci, 27
  torsion, 255
  truncated, 225, 253, 258
Theorem
  Cayley–Hamilton , 57
  de Rham decomposition, 26
  Divergence, 17, 202
  Elliptic regularity, 293
  Euler, 106
  exchange variable, 64
  Ferus, 43
  Frobenius, 54
  Hahn–Banach, 247
  Hodge, 18
  Hopf-Rinov, 16
  Lebesque, 64
  Sobolev embedding, 282, 290
  Stokes', 74, 202
Torsion, 201
Transformation
  Cole-Hopf, 278
  generalized Newton, 89
  Newton, 57

V

Variation
  $g^{\perp}$-, 158
  $g^{\top}$-, 176
  $g^{\pitchfork}$-, 176, 217
  adapted, 157, 176
Vector field, 50
  characteristic, 82
  complete, 50
  deformation, 28

energy, 122
geodesible, 130
geodesic, 13
Jacobi, 46

Killing, 13
nonsingular, 29
transversely holomorphic, 219

About the Authors

The authors of this book, Prof. Vladimir Rovenski and Prof. Pawel Walczak, are well known scientists—specialists in differential geometry, topology and dynamics of foliations. Their scientific contacts began in May–June 1995 during the International Conference "Foliations: Geometry and Dynamics" in Warsaw, where the second author was one of the organizers. Their common interests in Riemannian geometry of foliations and submanifolds contributed to the beginning of scientific co-operation: the authors formed a common research theme and the idea of a scientific relay race. The scientific relay race was started by P. Walczak who won a Marie Curie grant and conducted his research in the Institute of Mathematics of Burgundy (Dijon, France) in 2003–2005. Then V. Rovenski won a similar Marie Curie grant and conducted his research in co-operation with P. Walczak in the University of Lodz in 2008–2010. These co-operations after the completion of grants are still ongoing, and the scientific relay race is successfully continued by their students. The co-operation and friendship of the authors for over 25 years has led to several scientific works regarding foliations and this is considered in the book.

Many international scientific meetings preceded the writing of this book on differential geometry "in the large", such as workshops "Geometric Structures and Interdisciplinary Applications" in 2008, 2013, 2018 organized by V. Rovenski in Haifa, several conferences "Foliations: Geometry and Dynamics" organized in Poland by P. Walczak (vice-president of Polish Mathematical Society in 2005–2011), and the section "Contemporary Geometry" of Israeli–Polish Mathematical Meeting in 2011, Lodz. Researchers will be able to continue the aforementioned scientific relay race and deepen the understanding of the modern differential geometry.

November 2020   Irina Albinsky, PhD

© Springer Nature Switzerland AG 2021   319
V. Rovenski, P. Walczak, *Extrinsic Geometry of Foliations*, Progress
in Mathematics 339, https://doi.org/10.1007/978-3-030-70067-6

Lightning Source UK Ltd.
Milton Keynes UK
UKHW020719310522
403750UK00002B/20